D1706567

Chuchuan Huang
Jan, 2010 @ Austin

Biology of Inositols and Phosphoinositides

Subcellular Biochemistry
Volume 39

SUBCELLULAR BIOCHEMISTRY

SERIES EDITOR

J. ROBIN HARRIS, University of Mainz, Mainz, Germany

ASSISTANT EDITORS

B.B. BISWAS, University of Calcutta, Calcutta, India

P. QUINN, King's College London, London, U.K

Recent Volumes in this Series

Volume 31 **Intermediate Filaments**
Edited by Harald Herrmann and J. Robin Harris

Volume 32 **alpha-Gal and Anti-Gal: alpha-1,3-Galactosyltransferase, alpha-Gal Epitopes and the Natural Anti-Gal Antibody**
Edited by Uri Galili and Jos-Luis Avila

Volume 33 **Bacterial Invasion into Eukaryotic Cells**
Tobias A. Oelschlaeger and Jorg Hacker

Volume 34 **Fusion of Biological Membranes and Related Problems**
Edited by Herwig Hilderson and Stefan Fuller

Volume 35 **Enzyme-Catalyzed Electron and Radical Transfer**
Andreas Holzenburg and Nigel S. Scrutton

Volume 36 **Phospholipid Metabolism in Apoptosis**
Edited by Peter J. Quinn and Valerian E. Kagan

Volume 37 **Membrane Dynamics and Domains**
Edited by P.J. Quinn

Volume 38 **Alzheimer's Disease: Cellular and Molecular Aspects of Amyloid beta**
Edited by R. Harris and F. Fahrenholz

Biology of Inositols and Phosphoinositides

Subcellular Biochemistry Volume 39

Edited by

A. Lahiri Majumder
Bose Institute
Kolkata
India

and

B. B. Biswas
University College of Science
Kolkata, India and
University of Calcutta
Kolkata
India

Springer

This series is a continuation of the journal Sub-Cellular Biochemistry.
Volume 1 to 4 of which were published quarterly from 1972 to 1975

ISBN-10 0-387-27599-1 (HB)
ISBN-13 978-0-387-27599-4 (HB)
ISBN-10 0-387-27600-1 (e-book)
ISBN-13 978-0-387-27600-7 (e-book)

© 2006 Springer
All rights reserved. This work may not be translated or copied in whole or in part without the written permission of the publisher (Springer Science + Business Media, Inc., 233 Springer street, New York, NY 10013, USA), except for brief excerpts in connection with review or scholarly analysis. Use in connection with any form of information storage and retrieval, electronic adaption, computer software, or by similar or dissimilar methodology now known or hereafter developed is forbidden.
The use in this publication of trade names, trademarks, service marks and similar terms, whether or not they are subject to proprietary rights.

Printed in the Netherlands (BS/DH)

9 8 7 6 5 4 3 2 1

springer.com

INTERNATIONAL ADVISORY EDITORIAL BOARD

R. BITTMAN, Queens College, City University of New York, New York, USA
D. DASGUPTA, Saha Institute of Nuclear Physics, Calcutta, India
H. ENGELHARDT, Max-Planck-Institute for Biochemistry, Munich, Germany
L. FLOHE, MOLISA GmbH, Magdeburg, Germany
H. HERRMANN, German Cancer Research Center, Heidelberg, Germany
A. HOLZENBURG, Texas A&M University, Texas, USA
H-P. NASHEUER, National University of Ireland, Galway, Ireland
S. ROTTEM, The Hebrew University, Jerusalem, Israel
M. WYSS, DSM Nutritional Products Ltd., Basel, Switzerland
P. ZWICKL, Max-Planck-Institute for Biochemistry, Munich, Germany

Preface

From being to becoming important, *myo*-inositol and its derivatives including phosphoinositides and phosphoinositols involved in diversified functions in wide varieties of cells overcoming its insignificant role had to wait more than a century. *Myo*-inositol, infact, is the oldest known inositol and it was isolated from muscle as early as 1850 and phytin (Inositol hexakis phosphate) from plants by Pfeffer in 1872. Since then, interest in inositols and their derivatives varied as the methodology of isolation and purification of the stereoisomers of inositol and their derivatives advanced. Phosphoinositides were first isolated from brain in 1949 by Folch and their structure was established in 1961 by Ballou and his coworkers. After the compilation of scattered publications on cyclitols by Posternak (1965), proceedings of the conference on cyclitols and phosphoinositides under the supervision of Hoffmann-Ostenhof, were published in 1969. Similar proceedings of the second conference on the same subject edited by Wells and Eisenberg Jr was published in 1978. In that meeting at the concluding session Hawthorne remarked "persued deeply enough perhaps even myoinositol could be mirror to the whole universe". This is now infact the scenario on the research on inositol and their phosphoderivatives. Finally a comprehensive information covering the aspects of chemistry, biochemistry and physiology of inositols and their phosphoderivatives in a book entitled Inositol Phosphates written by Cosgrove (1980) was available. Inositol Metabolism in Plants edited by Morre, Boss & Loewus, was published in 1990. In 1996 a special volume of Subcellular Biochemistry (Vol 26) entitled "*Myo*-inositol Phosphates Phosphoinositides and Signal Transduction' edited by Biswas and Biswas was brought out to record the explosion of interest due to discovery of the "phosphoinositide effect". It is thought pertinent to publish another volume of Subcellular Biochemistry taking into an account of the advancement of knowledge in this area during last decade or so and this volume on "Biology of Inositol & Phosphoinositides" is the outcome of the present theme.

Implicitly what is to be mentioned that this volume is not intended to be all inclusive. However, this is aimed at giving a wide coverage starting from the structural aspect of Inositols and their derivatives to functional genomics, genetics of Inositol metabolism and storage; phosphoinositide metabolism in health and disease, in stress signaling and finally evolutionary consideration on the basis of genomics of diversified organisms. Repetitions as appear in some chapters have been retained because of the interest of the similar problems tackled from different angles using different systems following specific pattern of presentation and elucidation of importance of inositols and its derivatives in the diversified cellular functions.

Inositols, in particular *myo*-inositol, plays a central role in cellular metabolism. An array of complicated molecules that incorporate the inositol moiety are found in nature. Structural heterogeneity of inositol derivatives is compounded by the presence of stereo-and regio-isomers of the inositol moiety. Because of the large number of isomeric inositols and their derivatives present in nature, a detailed understanding of the structural, stereochemical and nomenclatures related issues involving inositol and its derivatives is essential to investigate biological aspects. A pertinent discussion of the stereochemical, conformational, prochiral, chiral and nomenclature issues associated with inositols and structural variety of inositol derivatives is presented. Murthy (Chapter 1) has taken painstaking effort in removing the confusion still present in the literature regarding the structure and nomenclature of inositol phosphates, phosphoinositides and glycosyl phosphoinositols. As far as possible along with old nomenclature the new assignment as recommended is also included to remove the confusion about the structural configuaration of inositol phosphates and phosphoinositides.

Loewus (Chapter 2) reviewed extensively *myo*-inositol biosynthesis from Glucose-6-phosphate and its catabolism particularly in plants related to cell wall biogenesis based on experimental data thus far available. Interestingly, *myo*-inositol is synthesized by both eukaryotes and prokaryotes specifically by two enzyme systems, one is M1P synthase and the other is IMP-phosphatase. Once MI is formed it is utilized for many biosynthetic processes in plants including formation of raffinose series of oligosaccharides, biosynthesis of isomeric inositols & their O-methyl ethers and membrane biogenesis. In addition to processes mentioned wherein the inositol structure is conserved, a major catabolic process competes for free MI and it is oxidized to D-Glucuronic acid by the enzyme MIoxygenase (MIOX). D-Glucuronic acid is subsequently converted to D-Glucuronic acid-1-P and UDP-D-Glucuronic acid by GlcUA-1-kinase and GlcUA-1-P uridyl –transferase respectively. UDP-D-GlcUA is the starting point to produce uronosyl and pentosyl component of cell wall polysaccharides. UDP-D-GlcUA was also generated directly from Glc-6-P through Phosphoglucomutase, UTP-Glc-1-P uridyl transferase and UDP-Glc dehydrogenase. Thus an alternative pathway to these cell wall polysaccharides that bypassed UDP-Glc dehydrogenase is recorded. These two pathways are now referred to as the MIOP and SNOP. MI as potential precursor of L-tartarate and oxalate which are linked with L-ascorbate breakdown products has been recently suggested. It is expected that efforts will continue to uncover new molecular and biochemical details of the MIOP and SNOP interrelationship as well as linkage of ascorbic acid biosynthesis and MI metabolism in plants.

The functional genomics of *myo*-inositol metabolism is the new aspect where Tora-Sinajad and Gillaspy (Chapter 3) have detailed the MIPS, IMP and MIOX genes and proteins with an overall focus on determining gene function involved in the inositol anabolic and catabolic pathways in different systems. One central theme what emerges at present is that the genes encoding enzymes of these pathways are present in prokaryotes, unicellular and multicellular eukaryotes. However, the regulations and contribution to specific end products are different. Not only prokaryotes utilize *myo*-inositol in different pathways from yeast, plants or animals but also their genetic diversity for these genes differs. More interestingly, plants appear to exhibit more complexity with respect to the numbers of genes that encode MIPS, IMPase and MIOXase enzymes as well as other regulatory proteins. In prokaryotes, the focus has been onto cell wall and RNA processing.

Genetical studies of *myo*-inositol phosphates and phosphoinositides are at present at an initial stage and the progress recorded has been slow due to difficulties in raising the mutants with respect to metabolism of *myo*-inositol and its phosphoderivatives. A thought-provoking discussion on this aspect has been initiated by Raboy and Bowen (Chapter 4). They started in sequence *myo*-inositol and *myo*-inositol-hexakis-phosphate as focal points for the purpose of metabolism and functions along with evolutionary consideration on the basis of genetics and genomics data available for diversified organisms. Besides participation in signal transduction, the involvement of *myo*-inositol phosphates and phosphoinositides in other functions of basal cellular metabolism and housekeeping, is considered. Differences between divergent species with respect to *myo*-inositol phosphate and phosphoinositide pathways when analyzed through mutations that block specific sites often take alternative metabolic routes to provide that component leading probably to metabolic balancing. Finally, an understanding of how *myo*-inositol phosphates and phosphoinositides are compartmentalized has been elucidated.

Roberts (Chapter 5) has contributed ably about Inositols in Bacteria & Archaea giving in details the identity of varied *myo*-inositol compounds their enzymology and functions including infectivity and virulence. In fact, *myo*-inositol compounds are not ubiquitous in bacteria but restricted to certain classes of these organisms and surprisingly not involved in signal transduction pathways in any of those organisms thus far studied signifying that *myo*-inositol is required for other functions.

A considerable progress has been made in the recent past in genetic regulation of MIP synthase in yeast initiated by isolation of the first *S. cerevisiae* inositol auxotrophs and the subsequent cloning and sequencing of its structural gene INO1. Indepth studies on the regulation of INO1 revealing the complex mechanisms controlling phospholipid metabolism related to cellular signaling pathways have been described succinctly by Nunez and Henry (Chapter 6).

Geiger (Chapter 7) dealt on the structure & mechanism of MIPS and compared the sequence alignment of MIPS from *S.cerevisiae*, *Mycobacterium tuberculosis* and *Archaeoglobus fulgidus*. When the yeast enzyme is aligned with the other two, there are significant differences in domain architecture. Enzymes from other two sources do not contain N-terminal 65 amino acids. However, both of them have C-terminal regions similar to yMIPS that serve to fix the relative orientation between the catalytic domain and Rossmann fold domain. The active site of MIPS is located between the bottom of the Rossmann fold domain and the beta sheet of the catalytic domain. A reasonable hypothesis for the detailed mechanism of the reaction has been discussed taking into consideration of the combination of the inhibitor-bound structure and the modeling approaches. Many significant questions raised remain to fully characterize the mechanism of MIPS in future. The combination of structural, biochemical and genetical studies on enzymes from different sources is leading to the complex mechanism for the conversion of G-6-P to MIP catalyzed by MIP synthase.

The review on phosphoinositide metabolism to understand the subcellular signaling in an organism and the functional coding of phosphoinositide signals deviation specifically in plants is attractive. Inositol phospholipids have multiple effects on cellular metabolism regulating cytoskeletal structure, membrane associated enzymes, ion channels and pumps, vescicle trafficking as well as producing second messenger. Boss, Davis, Im, Galvvo and Perera (Chapter8) discussed several aspects such as lipid-protein interactions, association of subcellular structure with inositol lipids domains, cellular pools of phosphatidylinositol 4,5 bisphosphate in order to understand subcellular signaling network in stress conditions.

In addition, Zonia and Munnik (Chapter 9) dealt with the functional coding of phosphoinositide signals during plant stress taking into consideration the pertinent discoveries on phosphoinositide signaling during cellular homeostasis, difference in phosphoinositide synthesis in different systems or their direct and indirect involvement in eliciting signals in unstressed cells and during both biotic and abiotic stress conditions including plant cell swelling and shrinking process. It is also apparent from discussions therein that the plant signals and cellular responses may differ in variety of ways from those in other organisms. Therefore, more and more new data about phosphoinositide signaling in plants are emerging which has been very ably presented.

A wider perspective on inositols and their metabolites in abiotic and biotic stress responses has been documented by Taji, Takahashi and Shinozaki (Chapter 10). Inositol and its metabolites function as both osmolytes and secondary messengers under biotic and abiotic stresses. The accumulation of different osmolytes during osmotic stress is an ubiquitous biochemical mechanism found in different organisms from bacteria, fungi and algae to plants and animals. Plants accumulate many types of inositol derivatives during abiotic stresses such as drought, low temperature and salinity in contrast most animals accumulate only *myo*-inositol. They have dwelt on molecular basis of osmolyte strategies in animals and plants, *myo*-inositol 1-phosphate, D-ononitol and D-pinitol, galactinol and reffinose as osmolytes in plants, inositol phosphates as signaling molecules with special reference to inositol (1,4,5) trisphosphate levels in response to abiotic stress, involvement of inositol trisphosphate in abscisic acid signaling, enzymes that regulate inositol trisphosphate levels and finally inositol hexakisphosphate as a signaling mediator with special reference to mRNA export, DNA repair, DNA recombination, vasicular trafficking, antioxidants and antitumor compounds. Functional aspects of inositols and its phosphoderivatives thus far mentioned once again support the view that they play a central role in cellular metabolism.

The important role of Inositol phosphate and phosphoinosities in health & disease has been critically tackled by Shi, Azab, Thompson and Greenberg (Chapter 11). They discussed the involvement of Ins(1,4,5)P_3 in neurological disorders such as Bipolar and Alzheimer's diseases. A correlation of abnormal function of IP_3R1 has been found associated with epilepsy and ataxia in mice as well as Huntington's disease in human patients. Moreover alteration in InsP3/Ca^{2+} signaling is one of the suggested mechanisms for malignant hyperthermia in humans. The Ca^{2+} overload due to the increased $InsP_3$ activity was suggested as a major contributor to the severe cardiac arrhythmias seen during the ischemia/reperfusion cycles. $InsP_6$ has been shown to exhibit strong antioxidant properties and is used as a potential anti-neoplastic therapy. In addition, they dwelt on the diseases caused by perturbation of PI metabolism such a P13K/AKT pathway in cancer and in type 2 diabetes as well as P15P in insulin signaling. P1(4,5) P_2 accumulation due to deficiency of OCRL1 gene (Oculo-Cerebro-Rinal syndrome of Lowe) is consistent with loss of function of OCRL1 product [P1(4,5)P2–5 phosphatase] along with other diseases associated with the myotubularin family opening up possibilities for effective drug design.

Mammalian *myo*-inositol 3-phosphate synthase (MIPS) and its role in biosynthesis of brain inositol and its clinical use as psycho-actve agent has been documented by Parthasarathy, Seelan, Tobias, Casanova and Parthasarathy (Chapter 12). They tried to draw attention to inositol homeostasis in mammalian brain by inositol synthase through dietary intake of inositol and continuous hydrolysis of inositol monophosphate by IMPase1. Inositol pathways have been implicated in the pathogenesis of bipolar disorder, with the mood stabilizers, valproate and lithium targeting inositol synthase and IMPase1 respectively. The inhibitory effect of valproate on inositol synthase suggests that this enzyme may be a potential therapeutic target for modulating brain inositol levels. Clinical studies on panic disorder, schizophrenia, obsessive –compulsive disorder, post traumatic stress disorder, attention deficit disorder, autism and alzheimer's disease by supplying inositol to the patient, have been presented implicating the biochemical and genetical regulation of inositol synthase and IMPase in the brain.

Presence of MIPS throughout evolutionary diverse organisms from eubacteria to higher eukaryotes presupposes its early origin. Analysis by multiple sequence alignment, phylogenetic tree generation and comparison of crystal structures thus far available provide new perspective into the origin and evolutionary relationships among the various MIPS proteins. Evolutionary divergence of MIPS and significance of a "core catalytic structure" have been succinctly discussed by GhoshDastidar, Chatterjee, Chatterjee and Lahiri Majumder (Chapter 13). Finally the interesting question arises whether acetolactate synthase homologue in *E.coli* can function as a MIPS present in *Synechocystis* which lacks in similar structure to the known MIPS from other prokaryotic and eukaryotic sources so far compared.

It is thus apparent from foregoing discussion that the importance has been given on the metabolism of inositol and phosphoinositides in different organisms from archaea to mammals aiming at working out metabolomics and to unravel a variety of network of metabolism. Attention has also been focussed on the key enzyme MIPS (*myo*-inositol–3-phosphate synthase), its structure, sequence, mechanism of action and functional genomics, purifying it from variety of organisms such as *Archaeoglobus fulgidus, Mycobacterium tuberculosis, Saccharomyces cerevisiae, Porteresia coarctata* and mammals. The alignment of sequences of this enzyme from different sources helps in working out diversification and evolution of *myo*-inositol–3 phosphate synthase and proposition of different models for its action mechanism. The important role of *myo*-inositol, its phospho-derivatives and phospholipids in health and disease is an emerging aspect which warrant significant attention at this time to work out their implications in pharmacogenomics. The role of *myo*-inositol and its phospho-derivatives under stress conditions opens up a new vista in functional genomics to identify the variety of gene expression under a particular condition in different systems and genetic along with molecular genetic studies have provided a thought provoking discussion implicating inositol phosphates and phosphoinositides functions in signaling other than that in signal transduction in mammals. Finally, the framework of molecular evolution of MIPS from archea to man has been proposed taking into consideration of sequence alignment, comparison of crystal structures and the sequence homology of the core catalytic domains of MIPS. Similar studies on the sequence deviations in IMPase and phytase from variety of organisms may be initiated in future. Our endeavour in organizing this volume of Subcellular Biochemistry will be fruitful if the workers in this emerging field embracing different disciplines such as molecular biology, chemistry, bioinformatics, psychiatry, agriculture, medicine, microbiology and biotechnology are benefitted from the information compiled by the experts in the field of biology of inositols and phosphoinositides.

B.B.Biswas
Arun Lahiri Majumder

Contents

Chapter 1

Structure and Nomenclature of Inositol Phosphates, Phosphoinositides, and Glycosylphosphatidylinositols

Pushpalatha P.N. Murthy
Department of Chemistry, Michigan Technological University, Houghton, MI 49931, U.S.A.

1. INTRODUCTION

Inositol is a cyclohexanehexol, a cyclic carbohydrate with six hydroxyl groups one on each of the ring carbons. *myo*-Inositol (hexahydroxycyclohexane) is the oldest known inositol; it was isolated from muscle extracts by Scherer in 1850 who called it inositol from the Greek word for muscle (Posternak, 1965). Since then, interest in inositols and their derivatives has waxed and waned. However, in the last 20 years we have seen a veritable explosion of interest in this area because of the discovery of new inositol derivatives and widespread recognition of the critical roles that phosphoinositides and inositol phosphates play in cellular signal transduction (Irvine and Schell, 2001; Shears, 2004; Toker, 2002; Toker and Cantley, 1997; Vanhaesebroek *et al.*, 2001). *myo*-Inositol occupies a central position in cellular metabolism – the inositol moiety is utilized by nature to biosynthesize a wide variety of compounds including inositol phosphates, phosphatidylinositides, glycosylphosphatidylinositols, inositol esters, and ethers. In addition, *myo*-inositol is converted into other stereoisomers of inositol and uronose and pentose sugars (Loewus, 1990b; Loewus and Murthy, 2000).

Inositol phosphates, phosphatidylinositides, and glycosylphosphatidylinositols encompass a diverse group of inositol-containing compounds with great structural complexity and heterogeneity. They mediate a myriad of biological processes and ongoing research suggests that they may be actively involved in many more cellular processes (Bernfield *et al.*, 1999; Irvine and Schell, 2001;

A. Lahiri Majumder and B. B. Biswas (eds.), Biology of Inositols and Phosphoinositides, 1–20
© 2006 *Springer. Printed in the Netherlands.*

Low, 2000; Shears, 2001, 2004). Many of these will be detailed in the rest of this book. In this chapter, I will introduce the structural, stereochemical, conformational, and nomenclature aspects of inositol phosphates, phosphatidylinositides, and glycosylphosphatidylinositols.

2. STEREOCHEMISTRY

At first glance inositol appears to be a relatively simple molecule – on closer examination, a host of sophisticated stereochemical issues, chiral, prochiral, and conformational reveal themselves (Parthasarathy and Eisenberg, 1986, 1990; Posternak, 1965). In fact, the complexity of stereochemical issues in cyclitols, namely cyclohexanes with one hydroxyl group on three or more ring atoms, is well recognized (IUPAC, 1976; Posternak, 1965); inositol can serve as a good model for a sophisticated discussion of various forms of isomerism.

2.1 Stereoisomers of inositol

The six secondary hydroxyl groups on the cyclohexane ring can be arranged in one of two orientations, axial or equatorial giving rise to nine stereoisomeric forms (Figure 1). The chair form rather than the Howarth projection

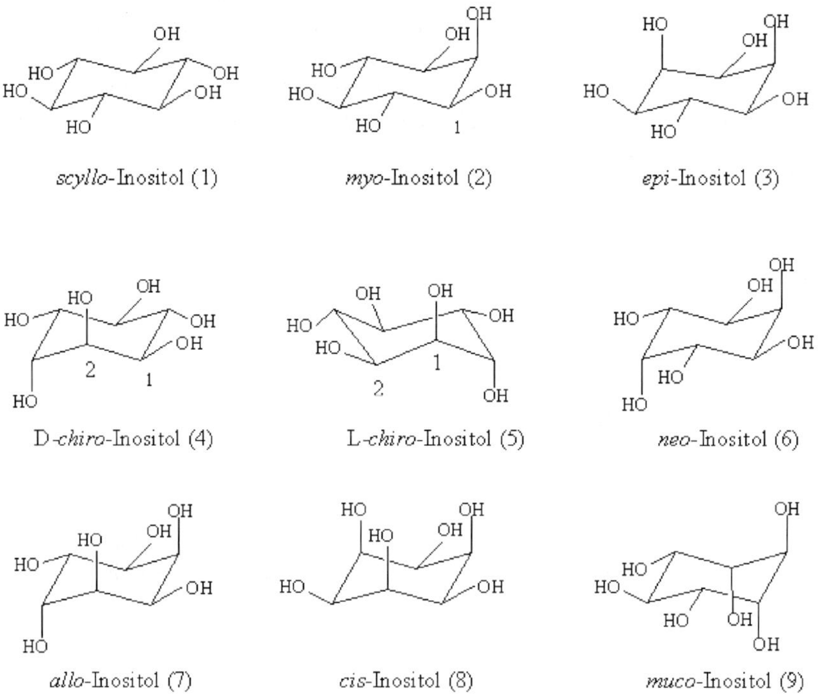

Figure 1. Stereoisomers of inositol.

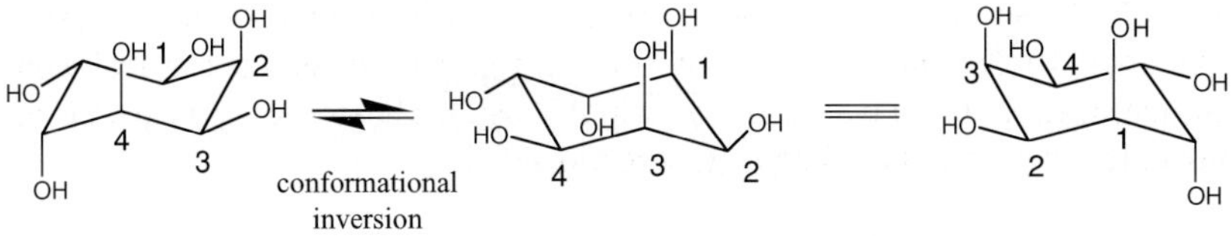

(10) (11) (12)

Figure 2. *allo*-Inositol: Conformational isomers and enantiomers.

will be employed in this review so that axial/equatorial distinctions of substituents are clearly illustrated and the geometrical and stereochemical consequences of the two orientations are obvious; these structural features are lost in the Howarth projection. Additionally, Howarth projections can sometimes be misleading; for example, it incorrectly indicates that *allo*-inositol has a plane of symmetry in the molecule.

Of the nine stereoisomers of inositol, the *scyllo*-isomer has no axial hydroxyl, the *myo*-isomer has one, the *epi*-, *chiro*-, and *neo*-isomers have two, and the *allo*-, *cis*-, and *muco*-isomers have three hydroxyl groups (Figure 1). Of these, six isomers (*scyllo*-, *myo*-, *epi*-, *neo*-, *cis*-, and *muco*-isomers) have one or more planes of symmetry in the molecule (*meso* compounds) and are therefore not chiral. D-*chiro*- and L-*chiro*-isomers do not have a plane of symmetry and are chiral molecules; moreover they are enantiomers of each other. The *allo*-isomer is unique – the conformational isomer of (10) (Figure 2) is (11) which is also its enantiomer! Since interconversion between conformational isomers is rapid, *allo*-inositol exists as a 50/50 mixture of the two enantiomers at room temperature. Therefore, although *allo*-inositol is chiral, the compound is optically inactive at room temperature because it is a racemic mixture; a chiral reagent, such as an enzyme, would be expected to preferentially react with one enantiomer and not the other.

Six isomers of inositol have been found in nature to date; these include *myo*-, *muco*-, *neo*- D-*chiro*-, L-*chiro*-, and *scyllo*-isomers.

2.2 *myo*-Inositol

The *myo*-isomer is the most abundant form in nature. It occupies a unique place in inositol metabolism because this is the only isomer synthesized *de novo* from D-glucose-6-phosphate; all other isomers are derived from *myo*-inositol (Loewus, 1990b; Loewus and Murthy, 2000). In early 1900s, *myo*-inositol was called *meso-inositol* (abbreviated ms-inositol) probably because, like mesotartaric acid, it does not exhibit optical activity and cannot be resolved into optical isomers (Posternak, 1965). However, as discussed above, six other isomers also do not exhibit optical activity so the name seemed inappropriate. In 1954, the name *myo*-inositol was introduced – a name not particularly well coined either since it is a pleonasm – both *myo* and *inositol* mean muscle. Nevertheless, the name persists.

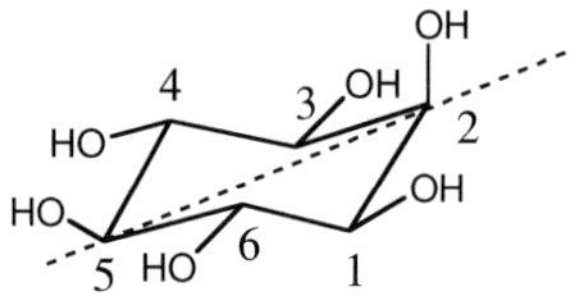

myo-Inositol (13) ----, plane of symmetry

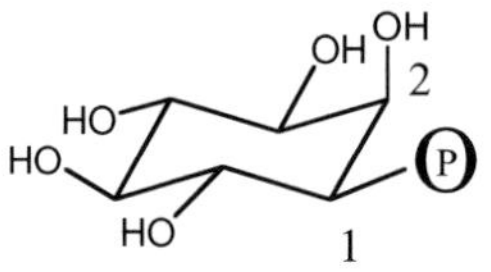

1D-*myo*-inositol-1-monophosphate
(14)

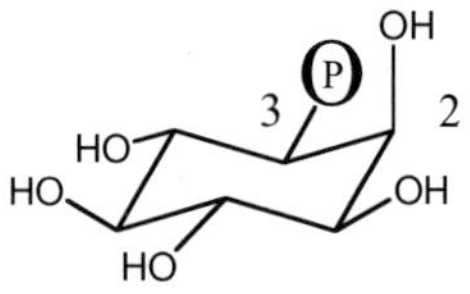

1D-*myo*-inositol-3-monophosphate
(15)

(P) = Phosphate

Figure 3. *myo*-Inositol and phosphorylated derivatives.

myo-Inositol can be divided into two mirror image halves (13) (Figure 3) – a perpendicular plane passes through C-2 and C-5 and splits the molecule into non-superimposable mirror images. Therefore, any modification that will disturb the symmetry of the molecule will render the molecule chiral, thus *myo*-inositol is prochiral. For example, phosphorylation (or any other substitution or reaction) at C-1, C-3, C-4, or C-6 (Figure 3) eliminates the plane of symmetry and leads to a chiral molecule; reaction at C-2 or C-5 preserves the plane of symmetry and the molecule remains achiral. C-1 is enantiopic to C-3 and C-6 is enantiotopic to C-4, thus the product of a substitution at C-1 (14) will be the enantiomer of the same reaction at C-3 (15).

Although the two halves of *myo*-inositol are stereochemically nonequivalent, they are chemically equivalent to an achiral molecule or reagent. In other words, an achiral molecule or reagent does not show preferential selectivity for either side and therefore reacts with both halves at the same rate. However, the biochemical consequence of enantiotopic carbons of most relevance to biochemists is that a chiral molecule, such as an enzyme, can readily distinguish between C-1 and C-3 as well as C-4 and C-6 and preferentially reacts with one enantiotopic carbon or another. For example, *myo*-inositol kinase phosphorylates *myo*-inositol exclusively at the D-3 position and yields the chiral product, 1D-*myo*-inositol-3-monophosphate (15) (Deitz and Albersheim, 1965; Loewus *et al.*, 1982). This is the same isomer produced from glucose-6-phosphate by *myo*-inositolphosphate synthase (reviewed in Loewus, 1990a). The only route by which the enantiomer 1D-*myo*-inositol-1-monophosphate (14) (Figure 3) is biosynthesized is by dephosphorylation of 1D-*myo*-inositol-1,4-bisphosphate by phosphatase. Enantiomers cannot be readily interconverted; conversion of one enantiomer to another involves the cleavage of covalent bonds (about 100 kcal/mol) and reformation with the opposite spatial arrangement. Enzymes react preferentially with one enantiomer or another, so the spatial differences between enantiotopic carbons can have profound biological consequences.

2.3 Nomenclature

The nomenclature of inositol and its derivatives has been a source of confusion for some time. This is partly due to the stereochemical complexity of all cyclitols, including inositols, and partly due to changes in the numbering rules by IUPAC (1976) and IUB (1989). The inositol story is a good illustration of the fact that defining stereochemistry of molecules is a difficult and complex issue and a satisfactory set of rules often evolves after numerous attempts. The underlying rationale behind cyclitol nomenclature and the evolution of the current system of numbering can be found in a number of publications (IUB, 1989; IUPAC, 1976; Murthy, 1996; Parthasarathy and Eisenberg, 1986, 1990). Rules currently followed have been in place since 1986.

myo-Inositol is a *meso* compound with mirror image halves. Therefore, to name the compound two descriptors have to be assigned – the absolute stereochemistry (D or L) and the numbering of the carbon atoms. Consequently, a number of questions arise: which atom should be considered for designating stereochemistry, which atom should be assigned 1, and what is the direction of numbering?

- According to IUPAC recommendations I-4 (IUPAC, 1976), numbers and the direction of numbering in inositols are assigned with reference to the spatial relations and nature of substituents on the ring. In *myo*-inositol, substituents are assigned to two sets, substituents above the ring are assigned to one set and those below to another. Lowest number is assigned to the set with more substituents. In the case of *myo*-inositol [Figure 4, (18)], there are four hydroxyls (C-1, C-2, C-3, and C-5) in one set and two (C-4 and C-6) in the other.
- Carbon-1 could be assigned to either of two enantiotopic carbons (16) and (17) or (18) and (19) (Figure 4). IUPAC recommendation (1976) is as follows: If the molecule is viewed in the vertical (Fischer – Tollens) projection with C-1 at the top with C-2 and C-3 on the front edge of the ring, the configuration is assigned D if the hydroxyl group or other substituent at the lowest-numbered chiral center projects to the right (16) and L if it projects to the left (17). More commonly a horizontal projection is used and the structure is drawn (18) and (19), so that if the substituent on the lowest-numbered asymmetric carbon is above the plane of the ring and the numbering is counterclockwise, the configuration is assigned D (18), and if clockwise,

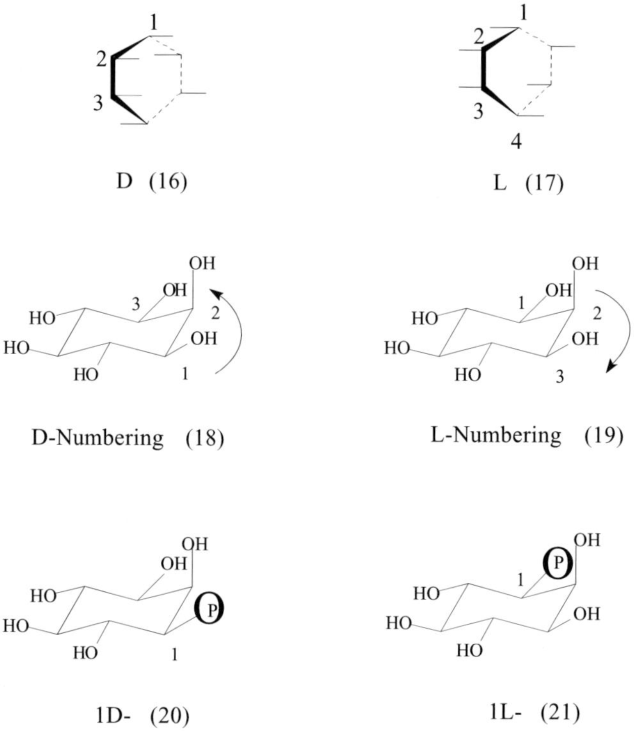

Figure 4. Nomenclature of *myo*-inositol and its derivatives.

the configuration is L (19). Therefore, in substituted inositol phosphates (20) and (21), the starting point could be either of the enantiotopic carbons as shown. The number 1 precedes the D or L to indicate that C-1 is the chiral center considered to define configuration.

- Of the two possibilities, the IUPAC – IUB recommendation of 1976 stipulated that for *meso* compounds, such as *myo*-inositol, the L designation should be applied.

Strict adherence to the lowest-locant rule sometimes obscures straightforward metabolic relationships. For example, consider the dephosphorylation of (22) – (23) (Figure 5). The IUPAC (1976) rules require that compound (22) be labeled 1D-*myo*-inositol-1,3,4,5-tetrakisphosphate rather than 1L-*myo*-inositol-1,3,5,6-tetrakisphosphate, so that substituents are attached to carbons with lower numbers, whereas (23) should be labeled 1L-*myo*-inositol-1,5,6-trisphosphate rather than 1D-*myo*-inositol-3,4,5-trisphosphate. Thus, to describe the dephosphorylation reaction (Figure 5), the necessity to switch between D and L numbering leads to the following statement:

1D-*myo*-inositol-1,3,4,5-tetrakisphoshphate is dephosphorylated to 1L-*myo*-inositol-1,5,6-trisphosphate.

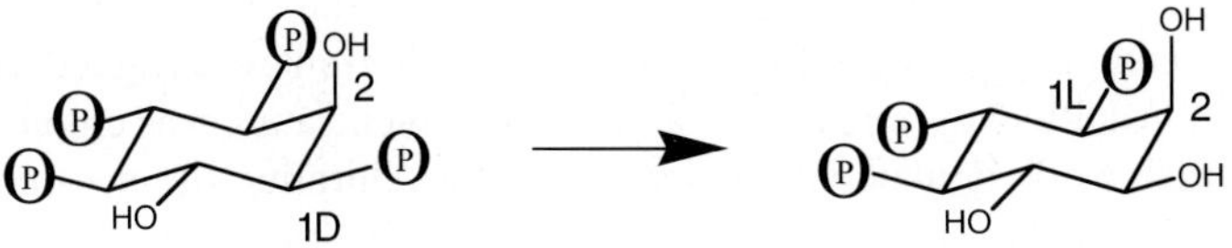

1D-*myo*-inositol-1,3,4,5-tetrakisphosphate

(22)

1L-*myo*-inositol-1,3,5,6-tetrakisphosphate

1L-*myo*-inositol-1,5,6-trisphosphate

(23)

1D-*myo*-inositol-3,4,5-trisphosphate

Figure 5. Hydrolysis of *myo*-inositol phosphates – alternative numbering.

From this statement, it is not immediately obvious which carbon has undergone dephosphorylation. Alternatively, the statement,

> 1D-*myo*-inositol-1,3,4,5-tetrakisphosphate is dephosphorylated to 1D-*myo*-inositol-3,4,5-trisphosphate,

clearly indicates that dephosphorylation has occurred at C-1.

To more easily discern straightforward metabolic relationships, IUB (1989) recommended that the lowest-locant rule be relaxed and either the 1D or the 1L numbering be allowed so long as the prefix 1D or 1L is specified. Thus, the author could use either numbering depending on the relationships that was being stressed.

It was further recommended that the symbol "Ins" be taken to mean *myo*-inositol with numbering proceeding counterclockwise, namely the 1D configuration. These recommendations were in response to the large number of *myo*-inositol phosphates of 1D configuration that were being discovered as hydrolytic products of phosphoinositides in investigations of signal transduction (IUB, 1989).

To remember the numbering of *myo*-inositol phosphates, Agranoff (1978) pointed out the resemblance of the chair conformation to a turtle (Figure 4) and suggested a mnemonic. The head of the turtle represents the C-2 axial hydroxyl and the four limbs and tail represent the five equatorial groups. The right hand limb is designated the D-1 position and the proceeding counterclockwise, the head is D-2, etc. The left front limb is D-3. Alternatively, if the L stereospecific numbering is employed, the left front limb is L-1 and the numbering proceeds clockwise as shown in (19) (Figure 4). In phosphoinositides, the right limb carries the diacylglycerol group.

An additional complicating issue should be kept in mind when reading the seminal pre-1968 literature. *myo*-Inositol derivatives that are currently designated D configuration were assigned L configuration and vice versa in the literature published before 1968. The reasons for the pre-1968 nomenclature are as follows: before 1968, rules of carbohydrate nomenclature dictated that the orientation of highest numbered chiral carbon, C-6 in this case, specify the configuration, D or L. As the hydroxyl groups at C-1 and C-6 are trans to each other, the compound assigned D before circa 1968 are now assigned 1L and vice versa. Additionally, circa 1960, a different nomenclature was employed in the seminal work conducted to determine the chirality of *myo*-inositol-monophosphate, the isomer (15) that resulted from hydrolysis of D-galactinol was designated D-*myo*-inositol-1-monophosphate as it was derived without change of configuration during the reaction (Ballou and Pizer, 1960). The *myo*-inositol monophosphate isomer (14), derived from the hydrolysis of phosphatidylinositides was assigned L as it exhibited optical rotation opposite to that of *myo*-inositol-1-monophosphate (15) assigned D (Ballou and Pizer, 1960). In 1968, as mentioned above, the rule was changed so that the lowest-numbered stereogenic carbon now specifies configuration.

In summary, following are some of the main points regarding inositol nomenclature:

- *myo*-Inositol derivatives that were designed L configuration in the literature before 1968 would be designated D according to current rules.
- Between 1968 and 1986, *myo*-inositol was numbered in L-stereochemistry.
- Currently, both the 1D or 1L designations can be employed.
- The use of the symbol Ins implies *myo*-inositol with numbering counter-clockwise, namely the 1D configuration.

In summary, in the literature before 1968, compound (14) would be labeled L-*myo*-inositol-1-monophosphate in keeping with the then rule that configuration be assigned on the basis of the orientation of substituents on carbon-6. After 1968, it would be designated *myo*-inositol-1-monophosphate (unspecified configuration implies L) or 1D-*myo*-inositol-1-monophosphate. Currently, (14) would be designated Ins(1)P_1.

In this article, the current designation will be followed; the symbol Ins will be used to denote *myo*-inositol derivatives. The symbol I will be used to indicate inositol, the unspecified stereoisomerism.

2.4 Conformation of inositol phosphates

Conformational isomers are structural isomers that are interconverted by rotation about single bonds. Properties of conformational isomers including size, shape, energy and chemical reactivity, and biological properties, including binding interactions with other proteins and chelating ability with metal ions can be significantly different. The importance of the effect of conformational changes on biological activity is well illustrated by the dramatic loss of enzyme activity in going from native conformation to denatured conformation, a process woefully familiar to biochemists; the native and denatured structures of enzymes are conformational isomers (except in cases where cleavage of disulfide bonds are involved). Conformational flexibility of biomolecules has a major impact on binding interactions with enzymes and receptors and thus on biological activity. The energy required for rotation about single (σ) bonds is low, the activation energy for the chair – chair transition of cyclohexane is 45 kJ/mol (10.8 kcal/mol) (Carey and Sundberg, 2000). Thus, in contrast to chiral isomers (Section 2.2), interconversion between conformational isomers can readily occur at room temperature. Because of the facile interconversion of conformational isomers, multiple low-energy conformers exist at room temperature. Therefore, an understanding of the possible low-energy conformations at room temperature is necessary to understand chelating and binding properties as well as ligand – protein interactions at the active site.

Conformational inversion of a chair form leads to an alternative chair form (24) and (25) (Figure 6) through a series of intermediate boat and twisted chair forms (Carey and Sundberg, 2000). During the chair – chair inversion process, the axial and equatorial groups at each of the carbons are interconverted. Generally, of the two alternative chair forms shown below (Figure 6), the conformation in which bulky substituents occupy equatorial positions (24) has less steric hindrance and thus is lower in energy than one in which bulky substituents occupy axial positions (25).

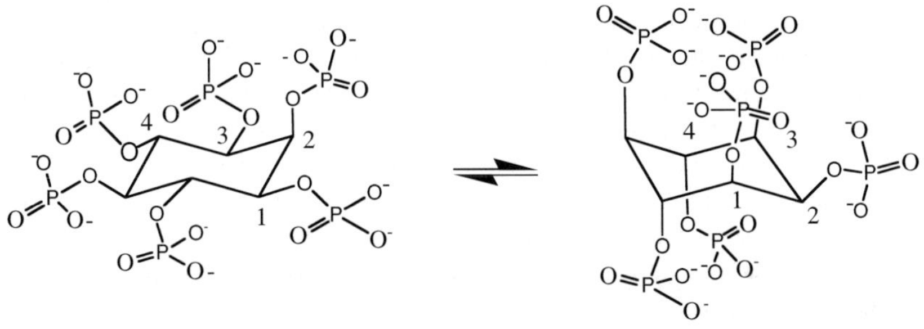

(24) (25)

Figure 6. Conformational interconversion of chair forms of phytate.

The conformation(s) adopted by inositol phosphates, including $InsP_6$, has been the subject of much debate and numerous investigations (Isbrandt and Oertei, 1980; Lasztity and Lasztity, 1990 and references therein; Barrientos and Murthy, 1996; Bauman *et al.*, 1999; Volkmann *et al.*, 2002). Most of the attention has been focused on phytic acid because of its long history and high endogenous concentration. A full understanding has been complicated by the fact that the energy of conformations is influenced by various factors including,

- number, substitution pattern, and stereochemistry of phosphate groups on the inositol backbone,
- physical state (solid or solution) of the inositol phosphate, and
- factors of the solvating media such as pH and counterions.

The shape, size, and charge distribution of the two possible conformations (24) and (25) are significantly different thus influencing chelating and binding properties.

Generally, as stated above, the conformation in which a maximum number of bulky substituents occupy equatorial positions is energetically more favorable. On this basis, at first glance the 1 ax/5 eq (24) conformation in which the five bulky substituents (hydroxyl or phosphate groups compared to hydrogen, the second substituent) are in equatorial orientations and one is in axial position would be expected to be the sterically favored form (24). This is indeed the case for inositol and the lower inositol phosphates (IP_1 to IP_4). However, counter to this generalization, the higher inositol phosphates (IP_6 and IP_5) undergo ring flip to the sterically hindered (5 ax/1 eq) form (25) at high pH (above 9.0) (Barrientos and Murthy, 1996; Isbrandt and Oertei, 1980; Bauman *et al.*, 1999; Volkmann *et al.*, 2002).

2.4.1 Inositol hexakisphosphates

Two isomers of inositol hexakisphosphate, *myo*- and *scyllo*-isomers were investigated at different pH (Barientos and Murthy, 1996; Bauman *et al.*, 1999). The two isomers differ in the orientation of the phosphate at one of six carbons, however, the impact of this on the proclivity toward ring flip to the sterically hindered form differs greatly.

- $InsP_6$ adopts the 1 ax/5 eq (26) (Figure 7) form at pH < 9.5 and the sterically hindered 5 ax/1 eq form exclusively at pH > 9.5 (27) (Barrientos and

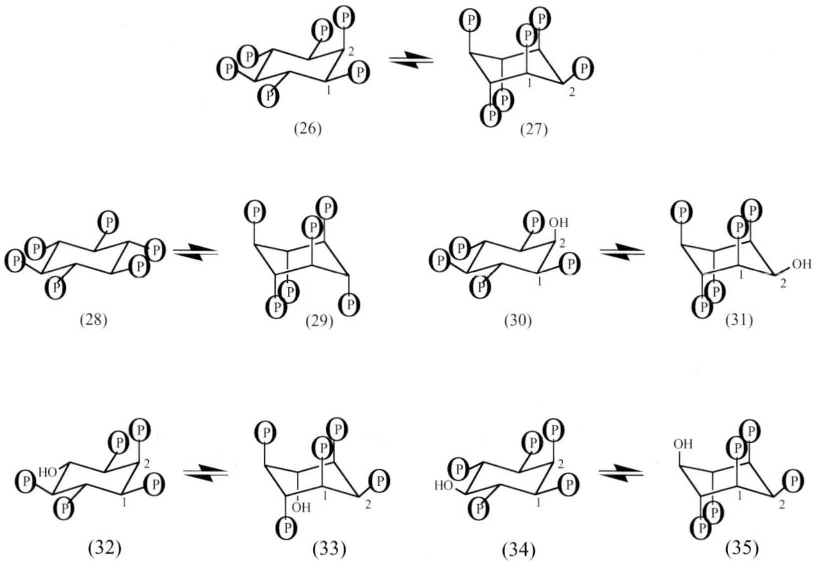

Figure 7. Conformational inversion of inositol phosphates.

Murthy, 1996; Isbrandt and Oertei, 1980). Between pH 9.0 and 9.5 (pK_a range of the three least acidic protons) both conformations are in dynamic equilibrium. Inversion to the sterically hindered form occurs only after the molecule is fully deprotonated to the dodecanionic form, above pH 9.5 (Barrientos and Murthy, 1996; Isbrandt and Oertei, 1980). The driving force for adopting the sterically hindered form (27) may be electrostatic repulsion due to the presence of five contiguous equatorial dianionic phosphates in the 1 ax/5 eq form and reduction of electrostatic repulsion by complexation with counterions in the 5 ax/1 eq form (27). Activation energy for ring flip from the 1 ax/5 eq to 5 ax/1 eq form is 54.8 $\pm$ 0.8 kJ/mol. The size of the counterion plays a critical role in stabilizing the 5 ax/1 eq form. In the presence of large counterions such as hydrated Li^+ (radius = 3.4 Å), tetramethyl ammonium or tetrabutyl ammonium cations (radii >5Å), no evidence for the 5 ax/1 eq form was observed suggesting that the formation of a tight chelation cage is critical to stabilizing the 5 ax/1 eq form. On the other hand, small counterions such as Na^+, K^+, Rb^+, and Cs^+ that facilitate the formation of chelation cages stabilize the 5 ax/1 eq form (Barrientos and Murthy, 1996; Bauman *et al.*, 1999; Isbrandt and Oertei, 1980) at high pH.

- In the solid state, X-ray crystallography (Blank *et al.*, 1971) and Raman spectroscopy (Isbrandt and Oertei, 1980) data suggest that the dodecasodium salt of phytic acid adopts the sterically hindered 5 ax/1 eq conformation. Again, the 5 ax/1 eq form may be adopted to minimize electrostatic repulsion between the five contiguous dianionic phosphates in equatorial positions.
- *scyllo*-Inositol hexakisphosphate [Figure 7, (28) and (29)] undergoes ring inversion from the 6 eq to the 6 ax form more readily than any other inositol phosphate investigated (Volkmann *et al.*, 2002). The $\Delta G^{\#}$ is less than 51.2 $\pm$ 1.0 kJ/mol; for comparison, the $\Delta G^{\#}$ for ring inversion of unsubstituted cyclohexane is 45 kJ/mol. The low activation energy may be due to destabilization caused by six dianionic equatorial phosphates in the 5 eq form (28) and stabilization of the 6 ax form (29) by the formation of chelation cages on both faces of the cyclohexane ring. The presence of syn-1,3,5-triaxial trisphosphates on both faces (29) facilitates the formation of tight chelation cages with counterions.

2.4.2 Inositol pentakisphosphates

All isomers of *myo*-inositol pentakisphosphate investigated, Ins(1,3,4,5,6)P_5, (30); Ins(1,2,3,5,6)P_5, (32); and Ins(1,2,3,4,6)P_5, (34), adopt the 1 ax/5 eq forms at low pH and the 5 ax/1 eq chair forms [(31), (33), and (35), respectively] at high pH. However, the activation energy for ring flip is influenced by the substitution pattern of phosphates.

- Ins(1,3,4,5,6)P_5 exists in the 1 ax/5 eq form (30) up to pH 10.5 and converts to the 5 ax/1 eq form (31) above pH 10.7. Between pH 10.5 and 10.7 both forms are in dynamic equilibrium. Of the different isomers of IP_5, Ins(1,3,4,5,6)P_5

undergoes chair – chair interconversion to the sterically hindered form most readily The activation energy for ring flip is $\Delta G^{\#}$ is 59.6 ± 0.5 kJ/mol. Stabilization of the 5 ax/1 eq form may be due to the five contiguous equatorial phosphates in (30) and the ability of the *syn*-1,3,5-triaxial trisphosphate arrangement (31) to form chelation cages with counterions (Volkmann *et al.*, 2002; Bauman *et al.*, 1999).

- In the case of Ins(1,2,3,5,6)P_5 (32) and Ins(1,2,3,4,6)P_5 (34), the situation is different. Ins(1,2,3,5,6)P_5 exists in the 1 ax/5 eq form (32) up to pH 8.0 and Ins(1,2,3,4,6)P_5 (34) up to pH 9.5. At higher pH, an interconverting mixture of the 1 ax/5 eq and the 5 ax/1 eq forms exist. The exclusive presence of the 5 ax/1 eq form was not observed. NMR spectra indicate that the $\Delta G^{\#}$ for Ins(1,2,3,5,6)P_5 is lower than that of Ins(1,2,3,4,6)P_5 ($\Delta G^{\#} = 73.9 \pm 0.8$ kJ/mol). This may be due to the destabilizing effect of three contiguous equatorial phosphates in the 1 ax/5 eq form of Ins(1,2,3,5,6)P_5 compared to two sets of two in Ins(1,2,3,4,6)P_5, as well as stabilization of the 5 ax/1 eq form by the formation of a tight chelation cage due to the presence of a *syn*-1,3,5-triaxial trisphosphate arrangement on one face of (33) and the lack of such an arrangement in (35) (Barrientos and Murthy, 1996; Volkmann *et al.*, 2002).

The results summarized above were obtained from dynamic nuclear magnetic resonance spectroscopy techniques. Conformational predictions using molecular modeling calculations (Gaussian) agree with experimental results in aqueous solution. Moreover, calculations in gas phase indicated that the sterically hindered forms of charged IP_6s and IP_5s are indeed more stable even in the gas phase (Bauman *et al.*, 1999; Volkmann *et al.*, 2002).

In summary, the electrostatic repulsion due to four or more equatorially oriented dianionic phosphates disfavors the 1 ax/5 eq orientation and induces ring flip to the 5 ax/1 eq form. The proclivity toward ring flip is influenced by the number, position, and orientation of dianionic phosphate substituents on the inositol ring; the facility for ring flip is as follows: (29) > (27) > (31) > (33) > (35).

3. DIPHOSPHORYLATED INOSITOL PHOSPHATES

myo-Inositol phosphates with one or more diphosphate or pyrophosphate groups have been detected in plant and animal cells (Figure 8) (Shears, 1998, 2001, 2004). Compound (36) with diphospho-substituent at the 5-position (5-diphosphoinositol pentakisphosphate) and compound (37) with two diphospho-substituents at 5- and 6-positions (5,6-bisdiphosphoinositol-tetrakisphosphate) have been identified. Two structural features are noteworthy: with the possibility of 12 negative charges on a relatively small cyclohexane ring, IP_6 is probably the most negatively charged small molecule in living cells; heparin also has a high negative charge density but it is a much larger molecule. The fact that

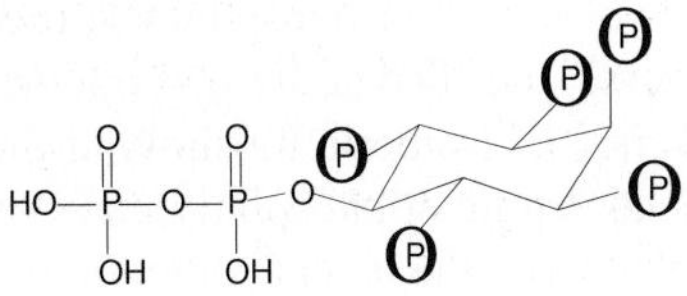

5-diphospho-*myo*-inositol pentakisphosphate

5-[PP]-InsP5 (36)

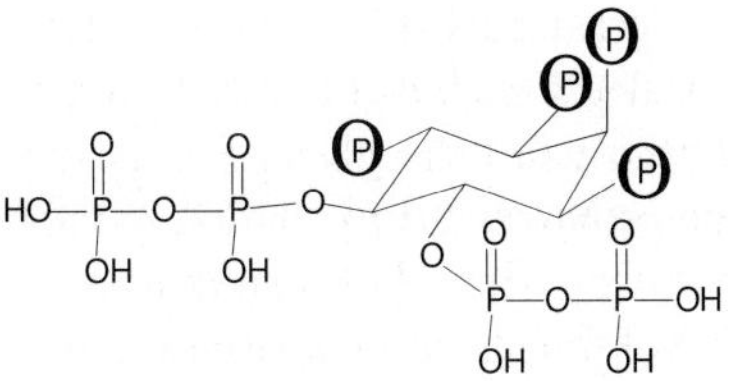

5,6-bisdiphospho-*myo*-inositol tetrakisphosphate

5,6-[PP]2-InsP4 (37)

Figure 8. Diphosphorylated inositol phosphates.

diphosphorylated inositol phosphates can carry higher negative charge density than IP_6 puts it in an unique class. Second, diphosphorylated inositol phosphates contain high-energy phosphoric anhydride bond(s) (ΔG hydrolysis = ~7.5 kcal/mol). Thus, they have the ability to function as phosphate donors and phosphorylate hydroxyl groups in inositol phosphates, proteins, or other molecules. The biological roles of diphosphorylated compounds as well as the enzymes that are involved in their metabolism are being actively investigated (Shears, 1998, 2001, 2004).

4. PHOSPHATIDYLINOSITIDES

Phosphatidylinositides include a group of compounds in which inositol phosphates are esterified to diacylglycerol by a phosphodiester linkage [Figure 9 (38)]. The involvement of phosphoinositides in signal transduction was first observed in the late 1950s (Hokin and Holin, 1953) and the critical roles they play in signal transduction were established in the 1980s (Agranoff *et al.*, 1983, Streb *et al.*, 1983). Since then, phosphoinositides have been intensively investigated and their involvement in a myriad of biological processes has been identified. A number of reviews detail our current understanding of the metabolism and roles of phosphoinositides in biological processes (Irvine and Schell, 2001; Shears, 2004; Stephens *et al.*, 2000; Toker, 2002; Toker and Cantley, 1997; Vanhaesebroek *et al.*, 2001).

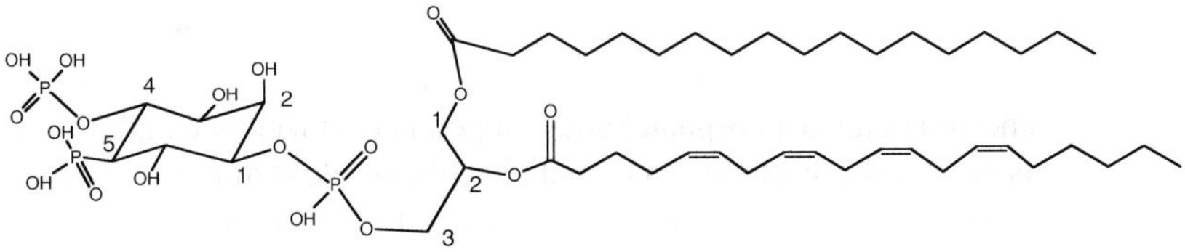

(38)

Figure 9. sn-1-stearoyl-2-arachidonyl-phosphatidylinositol-4,5-bisphosphosphate.

A great deal of structural heterogeneity is observed in phosphoinositides (38). These include (a) variation in the fatty acids esterified at the *sn*-1 and *sn*-2 positions of the glycerol unit, (b) the presence of isomeric inositols in the head group, and (c) variation in the position and degree of phosphorylation on the inositol ring. Phosphatidyl-*myo*-inositol (PtdIns) is the parent compound of most of the phosphoinositides; others are derived from it. The five free hydroxyls on PtdIns can be phosphorylated to yield isomeric phosphatidylinositol phosphates. Three isomers of $PtdInsP_1$ with phosphates at the C-3, C-4, or C-5 positions [$PtdIns(3)P_1$, $PtdIns(4)P_1$, or $PtdIns(5)P_1$] have been identified in a wide variety of living cells, however, interestingly, phosphatidylinositol monophosphate isomers with phosphates at C-2 or C-6 positions have not been discovered in natural sources to date. Three isomers of phosphatidylinositol bisphosphates [$PtdIns(3,4)P_2$, $PtdIns(3,5)P_2$, and $PtdIns(4,5)P_2$] and one isomer of $PtdInsP_3$ [$PtdIns(1,3,4)P_3$] have been characterized in cells. Although the presence of 4- and 5-phosphorylated derivatives has been known for many years (Folch and Woolley, 1942), the discovery of 3-phosphorylated derivatives is relatively recent (Whitman *et al.*, 1988).

myo-Inositol is the predominant isomer in most of the inositol-containing lipids. *scyllo*-Inositol-containing phosphoinositide lipids have been found in plant cells (Narasimhan *et al.*, 1997) and *chiro*-inositol-containing lipids have been characterized in plant and animal cells (Larner *et al.*, 1988; Mato *et al.*, 1987; Pak and Larner, 1992).

PtdIns is the most abundant phosphatidylinositide in cells; it constitutes about 90% of the inositol lipids in cell membranes. The other six mono- and bis-phosphorylated PtdIns make up the remaining 10% with $PtdIns(4)P_1$ and $PtdIns(4,5)P_2$ constituting up about 9% of the lipid pool, each contributing different amounts depending on the tissue. The remaining five lipids make up 1%. $PtdIns(3)P_1$ or $PtdIns(5)P_1$ contribute about 0.4%, and the 3-lipids jointly constitute about 0.1%. These numbers vary slightly from cell to cell (Stephens *et al.*, 2000; Vanhaesebroek *et al.*, 2001), however, it is clear that the quantitatively minor inositol lipids $Ptd(4,5)P_2$ and the 3-phosphorylated lipids play critical roles in signal transduction (Stephens *et al.*, 2000; Toker and Cantley, 1997; Vanhaesebroek *et al.*, 2001).

5. GLYCOSYLPHOSPHATIDYLINOSITOL

Glycosylated forms of PtdIns (GPI) have been found to anchor a variety of cell surface proteins to cell membranes by forming a covalent linkage to the C-terminal end of proteins (Low, 1989, 2000; Low and Saltiel, 1988; Saltiel, 1996). GPI anchors are found in diverse organisms from primitive eukaryotes to mammals and plant cells. GPI molecules are structurally complex, they are made up of variable and non-variable components; the non-variable structural features have been conserved through evolution from protozoa to mammals

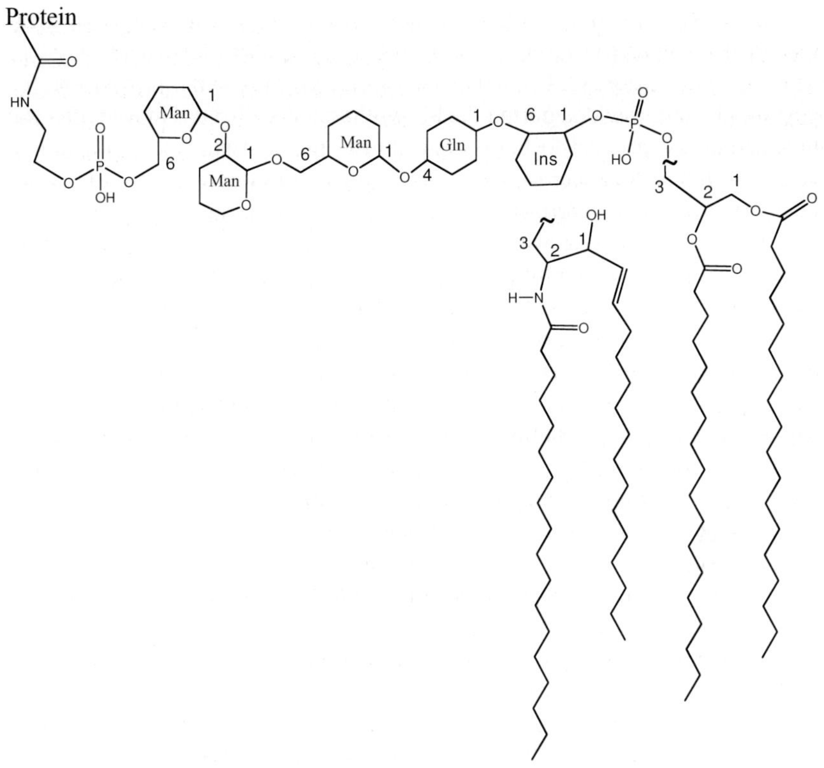

Figure 10. Glycosylphosphatidylinositol. Man, mannose; Gln, glucosamine; Ins, inositol. Sphingosine lipid portion in glycophosphosphingolipid is also shown.

and flowering plants suggesting that it may fulfill a critical role. GPI consists of three distinct units (Figure 10):

- an inositolphospholipid with variable hydrophobic groups,
- a short conserved glycan linking the inositol lipid to the protein, and
- variable substituents on the conserved mannose sugars.

The lipid portion of the molecule has been found to be 1,2-diacylglycerol, 1-alkyl-2-acylglycerol, or a ceramide-based lipid (glycosylinositolphosphorylceramide or glycophosphosphingolipid in which the lipid is a long chain N-acylated amino diol with a single double bond) (Figure 10). In fact, the 1-alkyl, 2-acylglycerol, and ceramide-based unit are more prevalent than the 1,2-diacylglycerol species. Thus, although this group of compounds is referred to as glycosylphosphatidylinositols for convenience, it should be kept in mind that the lipid portion may not be a 1,2-diacylglycerol moiety (Low, 2000).

An unusual structural feature from the inositol perspective is the derivatization of the 2-hydroxyl group of inositol by a long chain fatty acid (palmitic acid). In many mature GPI-proteins, the palmatoyl side chain is not present; the palmatoyl chain is added during the biosynthesis of GPI and removed after the GPI anchor is attached to the protein during posttranslational modification. Although the *myo*-isomer of inositol is the most prevalent form in GPI molecules, the presence of *chiro*-inositol has also been detected.

The conserved core consists of a short glycan chain of four sugar residues glycosidically linked to 6-hydroxyl of the Ins moiety of PtdIns (Figure 10). The four-sugar glycan consists of one glucosamine and three mannose units. The nonreducing end of the mannose moiety is attached to a phosphodiester-linked ethanolamine. The protein is attached to GPI by an amide bond between the amino group of the ethanolamine and the carboxy terminal amino acid of the protein during posttranslational modification in the lumen of endoplasmic reticulum. The lipid part of the molecule is inserted into the lipid bilayer and anchors the protein in the membrane (Low, 2000).

Great variability is observed in the substituents that are attached to the three mannose residues. Substituents include simple units such as ethanolamine, glucose, or mannose or oligoglycans of variable size, structure, and complexity (Low, 2000).

GPI anchors are biosynthesized in the endoplasmic reticulum and added to proteins during posttranslational modification. The biosynthesis of complex GPI molecules is a substantial commitment of resources by the cell and the fact that GPI molecules and the biosynthetic machinery have been conserved through evolution has lead to speculation that these molecules may confer unique advantages to the cell and also fulfill additional functions. GPI anchors provide stable binding of proteins to the lipid bilayer as well as localize proteins in particular regions of membranes. In addition to the role of anchoring cell surface proteins, a number of other functions, including its role in signal transduction, are being pursued by investigators (Low, 2000). The roles of GPI in insulin action and nutrient uptake (nitrate and phosphate uptake) in plants are also being investigated (Kunze *et al.*, 1997; Low, 2000; Saltiel, 1996; Stöhr *et al.*, 1995).

GPI molecules have been found to anchor another important class of proteins, the heparin sulfate proteoglycans (HSPG), at the cell surface (Bernfield *et al.*, 1999). Heparin sulfate, a strongly anionic linear polysaccharide, is covalently attached to proteins; these proteins are called glypicans because of the use of GPI to attach proteins to the cell surface. By way of the heparin sulfate chains, glypicans bind a variety of extracellular signals, form signaling complexes with receptors, and modulate receptor – ligand interactions. Glypicans mediate many cellular processes including cell growth, cell division, cell – cell adhesion, and cell defense (Bernfield *et al.*, 1999). Extracelluar ligands that bind to glypicans include growth factors, cytokines, cell-adhesion molecules, and coagulation factors among others.

6. CONCLUSIONS

Inositol is a deceptively simple molecule. On closer study, a number of sophisticated stereochemical, prochiral, chiral, and conformational issues associated with inositols and their derivatives become evident. Inositols, in particular *myo*-inositol, play a central role in cellular metabolism. An array of complicated molecules that incorporate the inositol moiety are found in nature. Structural heterogeneity of inositol derivatives is compounded by the presence of stereo- and regioisomers of the inositol unit. Because of the large number of isomeric inositols and their derivatives present in nature, a detailed understanding of the structural, stereochemical, and nomenclature issues involving inositol and its derivatives is essential to investigate biological aspects. A discussion of the stereochemical, conformational, prochiral, chiral, and nomenclature issues associated with inositols and the structural variety of insoitol derivatives is presented in this chapter.

REFERENCES

Agranoff, B.W., 1978, Textbook errors: Cyclitol confusion. *Trends Biochem. Sci.* **3:** N283–N285.

Agranoff, B.W., Murthy, P.P.N., and Sequin, E.B., 1983, Thrombin-induced phosphodiesteratic cleavage of phosphatidylinositol bisphosphate in human platelets. *J. Biol. Chem.* **258:** 2076–2078.

Ballou, C.E., and Pizer, L.I., 1960, The absolute configuration of the *myo*-inositol 1-phosphates and a confirmation of the bornesitol configuration. *J. Am. Chem. Soc.* **82:** 3333–3335.

Barrientos, L.G., and Murthy, P.P.N., 1996, Conformational studies of *myo*-inositol phosphates. *Carbohydr. Res.* **296:** 39–54.

Bauman, A.T., Chateauneuf, G.M., Boyd, B.R., Brown, R.E., and Murthy, P.P.N., 1999, Conformational inversion processes in phytic acid: NMR spectroscopic and molecular modeling studies. *Tetrahedron Lett.* **40:** 4489–4492.

Bernfield, M., Gotte, M., Park, P.W., Reizes, O., Fitzgerald, M.L., Lincecumj, J., and Zako, M., 1999, Functions of cell surface heparan sulfate proteoglycans. *Annu. Rev. Biochem.* **68:** 729–777.

Blank, G.E., Pletcher, J., and Sax, M., 1971, The structure of *myo*-inositol hexaphosphate dodecasodium salt octatriacontahydrate: A single crystal X-ray analysis. **44:** 319–325.

Carey, F.A., and Sundberg, R.J., 2000, Advanced Organic Chemistry, Part A, Structure and Mechanism, 4th ed. Kluwer Academic/Plenum Publishers, New York, pp. 123–185.

Deitz, M.P., and Albersheim, P., 1965, The enzymic phosphorylation of *myo*-inositol. *Biochem. Biophys. Res. Commun.* **19:** 598–602.

Folch, J., and Woolley, D.W., 1942, Inositol, a constituent of a brain phosphatide. *J. Biol. Chem.* **142:** 963–964.

Hokin, M.R., and Hokin, L.E., 1953, Enzyme secretion and the incorporation of P32 into phospholipides of pancreatic slices. *J. Biol. Chem.* **203:** 967–977.

Irvine, R.F., and Schell, M.J., 2001, Back in the water: The return of the inositol phosphates. *Nat. Rev. Mol. Cell. Biol.* **2:** 327–338.

Isbrandt, L.R., and Oertei, R.P., 1980, Conformational states of *myo*-inositol hexakis(phosphate)in aqueous solution. A ^{13}C NMR, ^{31}P NMR, and Raman spectroscopic investigation. *J. Am. Chem. Soc.* **102:** 3144–3148.

IUB Nomenclature Committee, 1989, Numbering of atoms in *myo*-inositol. *Biochem. J.* **258:** 1–2.

IUPAC Commission on the Nomenclature of Organic Chemistry and IUPAC-IUB Commission on Biochemical Nomenclature, 1976, Nomenclature of cyclitols. *Biochem. J.* **153:** 23–31.

Kunze, M., Riedel, J., Lange, U., Hurwitz, R., and Tischner, R., 1997, Evidence for the presence of GPI-anchored PM-NR in leaves of *Beta vulgaris* and from PM-NR in barley leaves. *Plant Physiol. Biochem.* **35:** 507–512.

Larner, J., Huang, L.C., Schwartz, C.F.W., Oswald, A.S., Shen, T.-Y., Kinter, M., Tang, G., and Zeller, K., 1988, Rat liver insulin mediator which stimulates pyruvate dehydrogenase phosphatase contains galatosamine and D-chiroinositol. *Biochem. Biophys. Res. Commun.* **151:** 1416–1426.

Lasztity, R., and Lasztity, L., 1990, Phytic acid in cereal technology. *Adv. Cereal Sci. Technol.* **10:** 309–371.

Loewus, F.A., 1990a, Inositol biosynthesis, metabolism: Precursor role and breakdown. In: Morre, D.J., Boss, W.F., and Loewus, F.A. (eds.), Inositol Metabolism in Plants. Wiley-Liss, New York, NY, pp. 13–19.

Loewus, F.A., 1990b, Inositol metabolism: Precursor role and breakdown. In: Morre, D.J., Boss, W.F., and Loewus, F.A. (eds.), Inositol Metabolism in Plants. Wiley-Liss, New York, pp. 21–45.

Loewus, F.A., and Murthy, P.P.N., 2000, *myo*-Inositol metabolism in plants. *Plant Sci.* **150:** 1–19.

Loewus, M.W., Sasaki, K., Laevitt, A.L., Munsell, L., Sherman, W.R., and Loewus, F.A., 1982, The enantiomeric form of *myo*-inositol-1-phosphate produced by *myo*-inositol 1-phosphate synthase and *myo*-inositol kinase in higher plants. *Plant Physiol.* **70:** 1661–1663.

Low, M.G., 1989, The glycosyl-phosphatidylinositol anchor of membrane proteins. *Biochim. Biophys. Acta* **988:** 427–454.

Low, M.G., 2000, Glycosylphosphatidylinositol-anchored proteins and their phospholipases. In: Cockcroft, S. (ed.), Frontiers in Molecular Biology, Vol. 27, Biology of Phosphoinositides. Oxford University Press, New York, pp. 211–238.

Low, M.G., and Saltiel, A.R., 1988, Structural and functional roles of glycosylphosphatidylinositol in membranes. *Science* **239:** 268–275.

Mato, J.M., Kelly, K.L., Abler, A., and Jarett, L. 1987, Identification of a novel insulin-sensitive glycophospholipid from H35 hepatoma cells. *J. Biol. Chem.* **262:** 2131–2137.

Murthy, P.P.N., 1996, Metabolism of inositol phosphates in plants. In: Biswas, B.B. and Biswas, S. (eds.), Inositol Phosphates, Phosphoinositides, and Signal Transduction, Subcellular Biochemistry Series, Vol. 26, Plenum Press, New York, pp. 227–255.

Narasimhan, B., Pliska-Matyshak, G., Kinnard, R., Carstensen, S., Ritter, M.A., von Weymarn L., and Murthy, P.P.N., 1997, Novel phosphoinositides in barley aleurone cells, additional evidence for the presence of phosphatidyl-*scyllo*-inositol. *Plant Physiol.* **113:** 1385–1393.

Pak, Y., and Larner, J., 1992, Identification and characterization of chiroinositol-containing phospholipids form bovine liver. *Biochem. Biophys. Res. Commun.* **184:** 1042–1047.

Parthasarathy, R., and Eisenberg, F., Jr., 1986, The inositol phospholipids: A stereochemical view of biological activity. *Biochem. J.* **235:** 313–322.

Parthasarathy, R., and Eisenberg, F., Jr., 1990, Biochemistry, stereochemistry, and nomenclature of the inositol phosphates, In: Reitz, A.B. (ed.), Inositol Phosphates and Derivatives: Synthesis, Biochemistry, and Therapeutic Potential. ACS Symposium Series 463. American Chemical Society, Washington, DC, pp. 1–19.

Posternak, T., 1965, The Cyclitols. Holden-Day, Inc., Publishers, San Francisco, CA, pp. 7–48.

Saltiel, A. R., 1996, Structural and functional roles of glycosylphosphoinositides. In: Biswas, B.B. and Biswas, S. (eds.), Inositol Phosphates, Phosphoinositides, and Signal Transduction, Subcellular Biochemistry Series, Vol. 26. Plenum Press, New York, pp. 165–185.

Shears, S.B., 1998, The versatility of inositol phosphates as cellular signals. *Biochim. Biophys. Acta* **1436:** 49–67.

Shears, S.B., 2000, Inositol pentakis- and hexakisphosphate metabolism adds versatility to the actions of inositol polyphosphates novel effects on ion channels and protein traffic. In: Biswas, B.B. and Biswas, S. (eds.), Inositol Phosphates, Phosphoinositides, and Signal Transduction, Subcellular Biochemistry Series, Vol. 26. Plenum Press, New York, pp. 187–226.

Shears, S.B., 2001, Assessing the omnipotence of inositol hexakisphosphate. *Cell Signal.* **13:** 151–158.

Shears, S.B., 2004, How versatile are inositol phosphate kinases? *Biochem. J.* **377:** 265–280.

Stephens, L., McGregor, A., and Hawkins, P., 2000, Phosphoinositide 3-kinases: Regulation by cell-surface receptors and function of 3-phosphorylated lipids. In: S. Cockcroft (ed.), Frontiers in Molecular Biology, Vol. 27, Biology of phosphoinositides. Oxford University Press, New York, pp. 32–108.

Stöhr, C., Schuler, F., and Tischner, R., 1995, Glycosyl-phosphatidylinositol-anchored proteins exist in the plasma membranes of *Chlorella Saccharophila* (Kruger) Nadson: Plasma-membrane-bound nitrate reductase as an example. *Planta* **196:** 284–287.

Streb, H., Irvine, R.F., Berridge, M.J., and Schulz, I., 1983, Release of Ca^{2+} from a nonmitochondrial intracellular store in pancreatic acinar cells by inositol-1,4,5-trisphosphate. *Nature* **306:** 67–69.

Toker, A., 2002, Phosphoinositides and signal transduction. *Cell. Mol. Life Sci.* **59:** 761–779.

Toker, A., and Cantley, L.C., 1997, Signalling through the lipid products of phosphoinositide-3-OH kinase. *Nature.* **387:** 673–676.

Vanhaesebroek, B., Leevers, S., Ahmadi, K., Timms, J., Katso, R., Driscoll, P.C., Woscholski, R., Parker, P.J., and Waterfield, M.D., 2001, Synthesis and function of 3-phosphorylated inositol lipids. *Annu. Rev. Biochem.* **70:** 535–602.

Volkmann, C.J., Chateauneuf, G.M., Pradhan, J., Bauman, A.T., Brown, R.E., and Murthy, P.P.N., 2002, Conformational flexibility of inositol phosphates: Influence of structural characteristics. *Tetrahedron Lett.* **43:** 4853–4856.

Whitman, M., Downes, C.P., Keeler, M., Keeler, T., and Cantley, L.C., 1988, Type-1 phosphatidylinostiol kinase makes a novel inositol phospholipid, phosphatidylinostiol-3-phosphate. *Nature* **332:** 644–646.

Chapter 2

Inositol and Plant Cell Wall Polysaccharide Biogenesis

Frank A. Loewus
Institute of Biological Chemistry, Washington State University, Pullman, WA 99164-6340, USA

1. INTRODUCTION

To the best of our knowledge, cyclization of D-glucose-6-phosphate (Glc-6-P) to 1L-*myo*-inositol-1-phosphate (MI-1-P) by *myo*-inositol-1-P synthase (MIPS, EC 5.5.1.4), followed by dephosphorylation of the latter by a specific MI monophosphatase (MIPase, EC 3.1.1.25), constitutes the sole *de novo* route to free MI in plants. Other sources of free MI involve salvage mechanisms on metabolic products bearing an intact MI structure. Examples include phosphoinositide biochemistry (Stevenson *et al.*, 2000) and hydrolysis of MI-containing polyphosphates or glycosides such as phytic acid and galactinol (Ercetin and Gillaspy, 2002; Hitz *et al.*, 2002; Loewus, 1973a,b, 2002; Loewus and Loewus, 1980; Loewus and Murthy, 2000; Morré *et al.*, 1990; Styer *et al.*, 2004).

Free MI undergoes phosphorylation by a Mg^{2+}-activated, ATP-dependent MI kinase (MIK, EC 2.7.1.64) to yield MI-1-P of the same stereoisomeric form produced by MIPS (English *et al.*, 1966; Loewus, M.W. *et al.*, 1982). For an introduction to current recommendations on rules for numbering atoms in *myo*-inositol see http://www.chem.qmul.ac.uk/iupac/cyclitol/myo.html.

Whether separate pools of MI-1-P are generated, one from *de novo* biosynthesis, another from the action of MIK on recycled free MI, needs to be explored. Evidence is accumulating to suggest the presence of several independent sites of MI-1-P formation within cells and tissues (Benaroya *et al.*, 2004; Hegeman *et al.*, 2001; Lackey *et al.*, 2002, 2003; Yoshida *et al.*, 1999). Multiple forms of MIPase are also found (Gillaspy *et al.*, 1995; Styer *et al.*, 2004). Quite possibly,

A. Lahiri Majumder and B. B. Biswas (eds.), Biology of Inositols and Phosphoinositides, 21–46
© 2006 *Springer. Printed in the Netherlands.*

free MI from both sources intermingle once produced. Alternatively, dedicated metabolic processes limit such mixing.

Free MI is required for many biosynthetic processes in plants including formation of the raffinose series of oligosaccharides (Obendorf, 1997), biosynthesis of isomeric inositols and their *O*-methyl ethers (Miyazaki *et al.*, 2004), and membrane biogenesis (Collin *et al.*, 1999). Many of the inter-relationships that govern production and utilization of free MI have yet to be sorted out Karner *et al.*, 2004. In addition to processes noted above wherein the inositol structure is conserved, at least one major catabolic process competes for free MI. MI oxygenase (MIOase, EC 1.13.99.1) was first discovered in kidney tissue (Charalampous and Lyras, 1957; Howard and Anderson, 1967). Here, the carbocyclic ring of MI is oxidized between carbon 1 and carbon 6 with incorporation of a single atom of O_2 exclusively into CO_2H of the product, D-glucuronic acid (GlcUA). In animals, GlcUA is successively converted in subsequent steps to L-gulonate, 3-oxo-L-gulonate, L-xylulose, xylitol, D-xylulose, and D-xylulose-5-P, which then enters the pentose phosphate cycle.

Evidence for MIOase in plants has relied largely on experiments involving the comparative use of radiolabeled MI and D-glucose (Glc) as markers to follow their relative roles as precursors of UDP-GlcUA and its metabolic products (Loewus and Loewus, 1980). The recent report of a MIOase gene in chromosome 4 (*miox4*) of *Arabidopsis* and confirmation of its enzymatic activity as bacterially expressed recombinant protein provides important new evidence in this regard (Lorence *et al.*, 2004).

This review will examine experimental evidence for participation of MI in plant cell wall biogenesis in an attempt to consolidate a long-standing viewpoint (Loewus, 1973a) that oxidation of MI provides an alternate starting point to a pathway furnishing uronosyl and pentosyl residues for cell wall biogenesis.

2. HISTORICAL PERSPECTIVE

Serendipitous discovery of a MI oxidation pathway (MIOP) to uronosyl and pentosyl units of plant cell wall polysaccharides emerged from experiments on the biosynthesis of L-ascorbic acid (AsA) which sought to generate GlcUA from labeled MI *in situ* in plant tissues. Isherwood *et al.* (1954) had proposed a biosynthetic pathway (D-galactose → D-galacturonic acid → L-galactonic acid → L-galactono-1,4-lactone → L-ascorbic acid) in plants similar to one (D-glucose → D-glucuronic acid → L-gulonic acid → L-gulono-1,4-lactone → L-ascorbic acid) proposed earlier for ascorbic aciD-synthesizing animals (summarized by Burns, 1967). Both schemes predicted an inversion of the six-carbon chain between the sugar precursor and the product, AsA. While the animal pathway did exhibit such an inversion (Loewus *et al.*, 1960), no inversion was observed when [1-^{14}C]Glc or D-[1-^{14}C]galactose ([1-^{14}C]Gal) was supplied to detached strawberry fruit (Loewus, 1961). Subsequent studies on this tissue with D-[1-^{14}C]- or D-[6-^{14}C]glucuronolactone resulted in AsA labeled predominately in carbon 6 or carbon 1, respectively, clearly an inversion of the carbon chain. In a similar experiment, D-[1-^{14}C]galacturonic acid-labeled strawberry fruit also produced AsA labeled predominately in carbon 6 (Loewus and Kelly, 1961). D-[1-^{14}C]Glucuronolactone-labeled berries also produced ^{14}C-labeled uronosyl and pentosyl residues in cell wall polysaccharides (Finkle *et al.*, 1960). These novel studies provided first evidence of at least two separate pathways of AsA formation in higher plants, findings now established by identification and characterization of key enzymes involved in these pathways (Agius *et al.*, 2003; Lorence *et al.*, 2004; Smirnoff *et al.*, 2001, 2004).

To examine relative roles of hexose and uronic acid as AsA precursors, the possibility of generating labeled GlcUA from [2-^{3}H]- or [2-^{14}C]MI *in situ* in strawberry fruit was tested (Loewus, 1965; Loewus *et al.*, 1962). In [2-^{3}H] MI-labeled strawberry fruit, 40% of the ^{3}H was recovered in free D-xylose (Xyl) and uronosyl and pentosyl residues of pectin. In [2-^{14}C] MI-labeled strawberry fruit, 33% of the ^{14}C was recovered in these products. Only trace amounts of ^{3}H or ^{14}C appeared in AsA (Loewus *et al.*, 1962). D-Galacturonosyl, D-xylosyl, and L-arabinosyl residues of cell wall polysaccharides as well as free Xyl were degraded to establish the location of the radiolabel. In each instance, it was at carbon 5 (Loewus and Kelly, 1963).

To explore the potential of MI as a precursor of uronosyl and pentosyl residues in hemicellulose, germinating barley, a tissue rich in this cell wall polymer was used (Loewus, 1965). Two-day-old seedlings were labeled with [2-^{14}C] MI by placing the radioactive solution directly on root hairs. After 32 h, tissues were repeatedly extracted with 70% ethanol to remove soluble ^{14}C, and then treated successively with pectinase, rumen bacterial extract, and dilute sulfuric acid. Washed residue was extracted with 3.5 N NaOH for 16 h at 25°. Insoluble residues were removed and the clear supernatant neutralized and freed of salts. A slight precipitate of hemicellulose A was removed and three volumes of ethanol added to the neutral solution to recover hemicellulose B. This fraction contained four times as much ^{14}C per unit weight as the preceding hydrolytic steps. Hydrolysis of a portion of this hemicellulose B with 3 N HCl followed by chromatographic separation showed ^{14}C confined to uronosyl, arabinosyl, and xylosyl residues. Virtually all ^{14}C present in each of these three products was in carbon 5.

The similarity between products of MI metabolism and those of GlcUA metabolism supported the view that the first step in MI catabolism in plants was an oxidative cleavage to form GlcUA. It also provided an experimental means of demonstrating that Glc cyclized to MI with no conformation changes in the carbon chain (Fischer, 1945; Loewus, 1974). Additional experiments were undertaken to confirm this finding. Stem-fed detached parsley leaves were labeled with [1-^{14}C]Glc (Loewus, 1965; Loewus and Kelly, 1962). After a period of metabolism, labeled MI, sucrose, pectin, and AsA were recovered from leaf extracts. The glucose moiety of sucrose, galacturonosyl residues of pectin, and AsA each had about 80% of their ^{14}C in carbon 1, the original position of the supplied radiolabeled glucose. Partial redistribution of ^{14}C, primarily between carbon 1 and carbon 6 of labeled constituents, is a process characteristic of hexose/triose phosphate metabolism (Krook *et al.*, 1998, 2000; Shibko and Edelman, 1957). The labeled MI was injected into ripening strawberry fruits where it was utilized for pectin biosynthesis (Loewus and Kelly, 1963). This labeled pectin was hydrolyzed to recover its labeled galacturonosyl and pentosyl residues, which were degraded to determine the location of ^{14}C in the carbon chain. Again, about 80% of the ^{14}C in the galacturonate was in carbon 1 and corresponded to carbon 6 of the injected MI. In other words, MI recovered from [1-^{14}C]Glc-labeled parsley leaves had 80% of its ^{14}C in the same position as the labeled Glc used for tagging the leaves.

Prior to the discovery that MI catabolism in plants produced uronosyl and pentosyl residues of cell wall polysaccharides such as pectin and hemicellulose, it was largely assumed that the sole pathway to these products arose from oxidation of UDP-Glc to UDP-GlcUA (Davies and Dickinson, 1972) and its subsequent metabolism (Feingold, 1982; Loewus and Dickinson, 1982). Once evidence for cyclization of Glc to MI in plants and animals was obtained (Eisenberg *et al.*, 1964; Loewus and Kelly, 1962) and the enzyme MIPS that catalyzed this process was isolated and characterized (Loewus and Loewus, 1971, 1973a,b; 1974; Sherman *et al.*, 1981), it was possible to construct an alternative pathway to these cell wall polysaccharides that bypassed UDP-Glc dehydrogenase (UDP-GlcDHase, EC 1.1.1.22). These two pathways, conveniently referred to as the MIOP and the sugar nucleotide oxidation pathway (SNOP), are shown in Figure 1. Absent from this figure is a proposed scheme linking oxidation of MI to AsA biosynthesis (Lorence *et al.*, 2004). Section 4.3 of this review provides an overview of this new development and its possible involvement in products of AsA catabolism, notably, oxalic acid, and tartaric acid (Bánhegyi and Loewus, 2004; DeBolt *et al.*, 2004).

From a practical viewpoint, radiolabeled MI is a useful tool for marking uronosyl and pentosyl residues of pectin, hemicellulose, and related plant cell wall polysaccharides. A wide variety of plant tissues have been studied with this procedure including cell and algal cultures, aquatic plants, germinating seeds, root tips, vascular and leaf tissue, floral parts and exudates, germinating pollen, ripening fruit, and seed development (Albersheim, 1962; Asamizu and Nishi, 1979; Harran and Dickinson, 1978; Imai and Terashima, 1991, 1992; Imai *et al.*, 1997, 1998, 1999; Knee, 1978; Kroh and Loewus, 1968; Kroh *et al.*, 1970a,b, 1971; Labarca and Loewus, 1970, 1972, 1973; Labarca *et al.*, 1973; Loewus, 1965; Loewus and Kelly, 1963; Loewus and Labarca, 1973; Loewus *et al.*, 1962, 1973; Maiti and Loewus, 1978a,b; Manthey and Dickinson, 1978; Mattoo and Lieberman, 1977; Roberts and Loewus, 1966, 1968, 1973; Roberts *et al.*, 1967a,b, 1968; Sasaki and Loewus, 1980, 1982; Sasaki and Taylor, 1984, 1986; Seitz *et al.*, 2000; Verma and Dougall, 1979; Wakabayashi *et al.*, 1989). Furthermore, synthesis of perdeuterated MI has provided a marker for stoichiometric evidence of the MIOP (Sasaki and Nagahashi, 1990; Sasaki *et al.*, 1989).

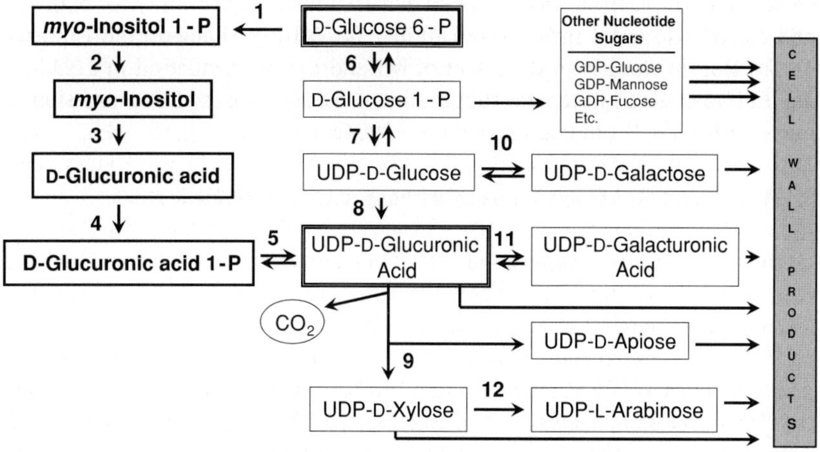

Figure 1. Alternative pathways from Glc-6-P to UDP-GlcUA and its products in plants are identified as: the MIOP (**bold font**) and the SNOP (normal font). Enzymes are numbered: (1) MI-1-P synthase, EC 5.5.1.4; (2) MI-1-P phosphatase, EC 3.1.3.25; (3) MI oxygenase, EC 1.13.99.1; (4) GlcUA-1-kinase, EC 2.7.1.43; (5) GlcUA-1-P uridylyltransferase, EC 2.7.7.44; (6) phosphoglucosemutase, EC 5.4.2.2; (7) UTP-Glc-1-P uridylyltransferase, EC 2.7.7.9; (8) UDP-Glc dehydrogenase, EC 1.1.1.22; (9) UDP-Glc decarboxylase, EC 4.1.1.35; (10) UDP-Glc 4-epimerase, EC 5.1.3.2; (11) UDP-GlcUA 4-epimerase, EC 5.1.3.6; and (12) UDP-L-arabinose 4-epimerase, EC, 5.1.3.5.

3. ALTERNATIVE PATHWAYS FOR UDP-GLcUA BIOSYNTHESIS IN PLANTS

3.1 MI oxidation pathway (MIOP, Steps 1–5 in Figure 1)

3.1.1 Step 1: 1L-*myo*-inositol-1-P synthase (MIPS, EC 5.5.1.4)

This enzyme is a highly conserved protein by which Glc-6-P is cyclized to 1L-*myo*-inositol-1-P (MI-1-P) in three partial reactions involving two enzyme-bound intermediates (Gumber *et al.*, 1984; Loewus, 1990; Loewus, M.W. *et al.*, 1980, 1984; Majumder *et al.*, 1997, 2003; Stieglitz *et al.*, 2005). It is the sole pathway of MI biosynthesis in bacteria, algae, fungi, plants, and animals and the first committed step via the MIOP to MI and GlcUA. MIPS is expressed in cytosol- and membrane-bound organelles, possibly by unique sorting signals within the primary structures (Lackey *et al.*, 2002, 2003).

3.1.2 Step 2: MI monophosphatase (MIPase, EC 3.1.1.25)

This enzyme is a relatively substrate-specific, Mg^{2+}-dependent, alkaline phosphatase (Parthasarathy *et al.*, 1994), which commonly accompanies MIPS during early stages of purification from plant extracts. It has high affinity for both 1D- and 1L-MI-1-P and a much lower affinity for MI-2-P (Loewus and Loewus, 1982). Recent studies on the effect of MI and Li^+ on regulation of *LeIMP-1* and *LeIMP-2* genes in tomato suggest that these substances alter expression as measured by GUS staining (Styer *et al.*, 2004).

3.1.3 Step 3: MI oxygenase (MIOase, EC 1.13.99.1)

Of the five enzymatic steps in the MIOP in plants, four have been isolated and their properties reported. Plant MIOase remains a challenge although its counterpart in animal tissues has been actively investigated for over 37 years (Arner *et al.*, 2001; Howard and Anderson, 1967; Reddy *et al.*, 1981). Arner isolated and sequenced a cDNA clone encoding MIOase from pig kidney and expressed the rMIO protein in bacteria. Their enzyme, a 32.7-kDa protein, lacked significant sequence to other known proteins. Native pig MIOase appears to complex with GlcUA reductase to produce L-gulonate, the second intermediate leading to AsA in AsA-synthesizing animals.

The pathogenic yeast, *Cryptococcus neoformans*, synthesizes MI and catabolizes this cyclitol to GlcUA. These pathways regulate in opposite modes, repressing conditions for one are inducing conditions for the other (Molina *et al.*, 1999). More recently, a non-pathogenic species, *C. lactativorus*, which grows on MI as its sole energy source, provided the advantage that MIOase can be induced by MI. MIOase is absent in Glc-grown cells but is present in MI-grown cells. This organism was used to purify, characterize, and clone MIOase (Kanter *et al.*, 2003).

Recently, Kanter *et al.* (2005) discovered a gene family of MIOases in *Arabidopsis* that contribute to the pool of nucleotide sugars for synthesis of plant cell wall polysaccharides.

Molecular evidence that a functionally unassigned open reading frame in *Arabidopsis* does, in fact, encode a putative MIOase has just been reported (Lorence *et al.*, 2004). In this study, MIOase provides a possible entry point into AsA biosynthesis. Those concerned with MIOase and its role in the MIOP will now have the opportunity to extend these findings to breakdown products of AsA catabolism (Bánhegyi and Loewus, 2004) as well as to cell wall biogenesis.

2-*O,C*-Methylene-MI (MMO) is one of several MI antagonists that produce morphological modifications in *Schizosaccharomyces pombe*, a fission yeast with an absolute requirement for MI (Schopfer *et al.*, 1969). When this epoxide is injected into rats, MIOase is inactivated and necrotic lesions develop in the kidneys. Simultaneous administration of MI prevents enzyme inactivation and reduces cytotoxicity (Weinhold and Anderson, 1967). The epoxide also represses pollen germination, inhibits pollen tube elongation, and delays seed germination (Chen *et al.*, 1977; Maiti and Loewus, 1978a,b).

3.1.4 Step 4: GlcUA 1-kinase (GlcUAKase, EC 2.7.1.43)

This enzyme was first isolated from mung bean seedlings (Neufeld *et al.*, 1959). Ungerminated lily pollen (Dickinson, 1982) is an abundant source and the activity of GlcUAKase does not increase during germination (Dickinson *et al.*, 1973). The enzyme is competitively inhibited by its product, α-D-glucuronate-1-P, and by UDP-GlcUA (Gillard and Dickinson, 1978; Leibowitz *et al.*, 1977). Since both inhibitors are intermediates in the MIOP, their effects may have regulatory significance in cell wall formation.

3.1.5 Step 5: GlcUA-1-P uridylyltransferase (GlcUAUase, EC 2.7.7.44)

UDP-GlcUA is a product common to two pathways (Figure 1, Steps 1–5 and Steps 6–8). In the MIOP, it is produced from GlcUA-1-P by GlcUAUase while in the SNOP it is produced from UDP-Glc by UDP-Glc dehydrogenase. It is not known if the products from these two pathways commix or if the pathways involved represent separate intracellular compartments. The plant enzyme was partially purified from barley seedlings and characterized by Roberts (1971). Subsequently, a modification in the assay for enzymatic activity provided enhanced detection (Dickinson *et al.*, 1977). A survey of GlcUAUase levels in a number of plant tissues suggests that this enzyme is present in amounts well in excess of that required to maintain requirements for cell wall biosynthesis and in this respect points to Steps 3 and 4 as potential rate-limiting steps (Dickinson *et al.*, 1977; Roberts and Cetorelli, 1973).

3.2 Sugar nucleotide oxidation pathway (SNOP, Steps 6–8)

3.2.1 Steps 6 and 7: phosphoglucomutase (PGMase, EC 5.4.2.2) and UTP:Glc-1-P uridylyltransferase (EC 2.7.7.9)

Activation of Glc-6-P to form UDP-D-Glc (Steps 6 and 7) provides the glucosyl donor requirement for UDP-D-glucuronosyl units and related nucleotides in polysaccharide biosynthesis via the SNOP (Feingold, 1982; Feingold and Avigad, 1980). Both cytosolic and plastidic isoforms of PGMase occur in plants (Davies *et al.*, 2003; Periappuram *et al.*, 2000). A recent histochemical analysis of PGMase in *Arabidopsis* found several organ-specific quantitative trait loci (Sergeeva *et al.*, 2003).

UDP-Glc is the principal sugar nucleotide and UTP:Glc 1-P uridylyltransferase is the major sugar nucleotide transferase in plant tissues (Feingold, 1982). Apart from its role in the SNOP, the transferase functions as glycosyl donor to sucrose, starch, callose, and cellulose (Hopper and Dickinson, 1972; Schlüpmann *et al.*, 1994; Tenhaken and Thulke, 1996). UDP-Glc inhibits this enzyme, contributing to control of its own production and thus regulating flow of hexosyl units toward formation of major cellular glycosides as well as toward cell wall polysaccharides.

3.2.2 Step 8: UDP-D-glucose dehydrogenase (EC 1.1.1.22)

UDP-D-Glucuronate (UDP-GlcUA), product of Step 8, is also the final step (Step 5) of the MIOP. As such, the potential impact of this biosynthetic juncture on cell wall growth and development cannot be ignored. Historically, efforts to probe relative contributions of the MIOP and SNOP to UDP-GlcUA weighed heavily on the side of the SNOP (Davies and Dickinson, 1972; Feingold, 1982; Gibeaut *et al.*, 2001; Hinterberg *et al.*, 2002; Johansson *et al.*, 2002; Loewus and Dickinson, 1982; Robertson *et al.*, 1995; Stewart and Copeland, 1998; Tenhaken and Thulke, 1996) but this is likely to change as more chemical and molecular data *a propos* the MIOP emerge (Kanter *et al.*, 2005; Kärkönen, 2005; Kärkönen/et al/., 2005; Loewus and Loewus, 1980; Morré *et al.*, 1990; Seitz *et al.*, 2000).

It is worth noting here that ". . . UDP-Glc dehydrogenase is a good candidate for a control point in the metabolic pathway of cell wall synthesis not only because it is in low concentration relative to other enzymes in the pathway, operates far from equilibrium, and because so much of the cell wall carbohydrate is acted on by this enzyme . . ." (Gibeaut *et al.*, 2001) but also because it is strongly inhibited by its product, as well as a subsequent product, UDP-Xyl, leading to speculation that tissue-specificity determines functional contributions of the MIOP and SNOP (Davies and Dickinson, 1972; Harper and Bar-Peled, 2002; Hinterberg *et al.*, 2002).

3.3 Relative contributions of the MIOP and SNOP to plant cell wall biogenesis

Ambivalence regarding a functional role for the MIOP in plant cell wall polysaccharide biosynthesis still lingers in current literature (Doblin *et al.*, 2003; Gibeaut, 2000; Mellerowicz *et al.*, 2001; Reiter, 2002; Reiter and Vanzin, 2001; Ridley *et al.*, 2001), in large part due to scarce attention given to molecular aspects of this alternative pathway (Kanter *et al.*, 2005; Seitz *et al.*, 2000). This despite a seemingly ubiquitous occurrence of the MIOP in plants as emphasized by a growing list of plant tissues that absorb labeled MI and metabolize it to cell wall uronides and pentoses (see Section 2). A survey of earlier literature is summarized in following sections.

3.3.1 Pectin synthesis in oat seedlings and ripening apples

Experiments with oat seedlings concluded that MI and Glc pass through a common precursor of pectin and that the rate of pectin synthesis was limited by a reaction subsequent to that intermediate. Moreover, the rate of incorporation of MI into GalUA residues of pectin was equivalent to that of Glc (Albersheim, 1962). Studies involving pectin synthesis in cortical slices from ripening apples incubated in sucrose media containing [2-^{3}H]MI or [^{14}C]methyl-L-methionine also gave comparable rates (Knee, 1978).

3.3.2 Influence of MI on redistribution of ^{14}C from [1-^{14}C]Glc in parsley leaf pectin and starch

When detached parsley leaves were labeled with [1-^{14}C]Glc, then transferred either to water or to 1% MI and allowed to metabolize for 42 h, both MI and pectin-derived galacturonosyl residues from such leaves retained about 80% of the ^{14}C in the original position while remaining label appeared in the other terminal position. In contrast to this, the distribution pattern of ^{14}C in sucrose-derived Glc from these same leaves was greatly influenced by the presence of excess MI which caused redistribution of 40% of the ^{14}C from carbon 1 into carbon 6 (Loewus, 1965). These results may be interpreted as evidence favoring operation of the MIOP where, in the presence of excess MI, more labeled Glc is available to equilibrate with triose phosphate (Krook *et al.*, 1998, 2000). The fact that this excessive redistribution of ^{14}C in the 1% MI enriched experiment, occurred only in sucrose-derived Glc, a product of UDP-Glc, but not in galacturonosyl residues of pectin, a product of UDP-GlcUA, is significant (Roberts *et al.*, 1968). In a comparable study involving [6-^{14}C]Glc-labeled root tips from three-day-old *Zea mays* seedlings, raising the internal concentration of MI did not greatly influence the pattern of Glc uptake and CO_2 release yet greatly reduced the flow of label into galacturonosyl units of pectin as well as glucuronosyl units of hemicellulose. As expected, pentosyl units of pectin and hemicellulose which arose from decarboxylation of UDP-GlcUA at Step 9 (Figure 1) were essentially unlabeled (Roberts and Loewus, 1973).

3.3.3 Hydrogen isotope effect in MIPS biosynthesis

In another study designed to probe the functionality of the MIOP, use was made of a hydrogen isotope effect at carbon 5 of Glc-6-P by MIPS (Loewus, M.W 1977). MIPS prepared from sycamore maple cell- (Loewus and Loewus, 1971) or rice cell-cultures (Funkhouser and Loewus, 1975; Loewus *et al.*, 1978) converted [5-^{3}H]Glc-6-P to [2-^{3}H]MI at rates ranging from 0.2 to 0.5 that of unlabeled substrate, an isotope effect indicating involvement of carbon 5 of Glc-6-P in MI biosynthesis (Loewus *et al.*, 1978). Comparison of ^{3}H/^{14}C ratios in glucosyl and galacturonosyl residues of starch and pectin, respectively, from germinating lily pollen digests, provided meaningful evidence of a functional role for the MIOP (Loewus, M.W. and Loewus, 1980).

3.3.4 Comparative study of MMO inhibition of MIOase in germinating lily pollen and wheat

As mentioned earlier (Section 3.1.3), MMO, an inhibitor of MIOase, repressed lily pollen germination and tube elongation. When excess MI was included in the germination medium, MMO effects were partially blocked or, if MI was supplied subsequent to MMO inhibition, reversed (Chen and Loewus, 1977; Chen *et al.*, 1977). MMO did not inhibit MIPS or UDP-Glc dehydrogenase. When [2-^{3}H]MI was included in the growth medium of lily pollen, ^{3}H rapidly incorporated into uronosyl and pentosyl units of tube wall polysaccharides, primarily pectic components. MMO blocked this process. Uptake of Glc by germinating lily pollen was not altered in the presence of MMO. Pollen grains germinated in pentaerythritol-balanced, sucrose-free media (Dickinson, 1978) containing [1-^{14}C]Glc produced labeled pollen tubes with ^{14}C-labeled glucosyl, galactosyl, uronosyl, and pentosyl units in their tube wall polysaccharides. When MMO was present in the media, incorporation of ^{14}C into glucosyl and galactosyl units was unaffected but incorporation of ^{14}C into uronosyl and pentosyl units was greatly repressed (Loewus *et al.*, 1973). These results provided further evidence in support of a functional role for the MIOP in pollen tube wall biogenesis.

3.3.5 Comparative labeling of germinating lily pollen with [2-^{3}H]MI or [1-^{14}C]Glc

Further evidence for the MIOP was obtained by germinating lily pollen in pentaerythritol media, a non-metabolized poly-hydroxylated osmoticum (Dickinson, 1978). Pollen tubes, grown for 3 h to deplete endogenous levels of MI and Glc, were resuspended in growth media containing 5.6–28 mM Glc with a trace of [2-^{3}H]MI or in media containing 5.6–28 mM MI with a trace of [1-^{14}C]Glc. After 3 h in labeled media, tubes were ground in 70% ethanol and tube walls recovered. Walls were treated successively with amyloglucosidase and pectinase to recover starch-derived Glc and pectin-derived uronosyl and pentosyl units by chromatography. In the presence of high levels of Glc, the [2-^{3}H]MI-labeled tubes continued to label L-arabinose (Ara) and Xyl units of pectin as well as Ara units from amyloglucosidase-treated polysaccharides. When the labeled marker was [1-^{14}C]Glc, only the control (no added MI) contained [^{14}C[Ara. With increasing levels of Glc, only hexosyl units (Glc, Gal, Rha) were labeled (Maiti and Loewus, 1978b). A functional MIOP in lily pollen tubes was the simplest interpretation of these results.

3.3.6 Role of phytate-derived MI in pectin and hemicellulose biosynthesis

Another approach to the question of MIOP functionality was taken by examining utilization of [2-^{3}H]MI by wheat kernels either by imbibition during germination or by injection into partially digested endosperm of 72 h seedlings. Results indicated that MI reserves (phytate) within the caryopsis provided a significant portion of the carbon requirements for pectin and hemicellulose biosynthesis during germination (Maiti and Loewus, 1978a,b). Imbibition of an aqueous solution of MMO at 40 μg per caryopsis delayed germination by 50 h. At 400 μg of MMO per caryopsis, only 14% of the kernels germinated in 94 h. If MI was included in the germination medium, it partially reversed the inhibition. MMO failed to alter the relative distribution of ^{3}H from MI into galacturonsyl, arabinosyl, or xylosyl units of pectin and hemicellulose although the total amount of ^{3}H incorporated diminished.

3.3.7 Comparative study of ^{3}H-labeled *myo-* and *scyllo*-inositol metabolism in maturing wheat

In this study, [2-^{3}H]MI or *scyllo*-[randomly positioned-^{3}H]inositol ([R-^{3}H]SI) was injected into hollow peduncles of post-anthesis developing wheat spikes. There was rapid translocation and accumulation of ^{3}H in developing kernels. In the case of [2-^{3}H]MI, 50–60% of the ^{3}H from MI was found in cell wall polysaccharides that were recovered from the stem-region of the injection. That portion translocated to kernels was recovered in cell wall polysaccharides, phytate, galactinol, and MI. In the case of [R-^{3}H]SI, most of the ^{3}H was translocated to kernels where it accumulated as [R-^{3}H]SI and *O*-α-galactosyl-SI. No ^{3}H from [R-^{3}H]SI was found in cell wall polysaccharides or phytate (Sasaki and Loewus, 1980).

When kernels from labeled plants were germinated, most of the tritiated galactinol or galactosyl-SI hydrolyzed within 1 day. Phytate from [2-^{3}H]MI-labeled kernels released free [2-^{3}H]MI over several days. [2-^{3}H]MI-labeled cell wall polysaccharides were also degraded during germination and their sugar residues were reutilized for new cell wall formation in developing shoot and roots. Most of the newly released [^{3}H]MI from phytate and galactinol recycled over the MIOP into pentosyl residues of cell wall polysaccharides (Sasaki and Loewus, 1982).

3.3.8 The MIOP and glucogenesis

Cyclization of Glc-6-P to MI-1-P (Step 1, Figure 1) is essentially irreversible (Loewus and Loewus, 1974). Subsequent MIOP steps leading to UDP-GluUA and its metabolic conversion to uronosyl and pentosyl components of cell wall polysaccharides offer no direct metabolic routes to glycosyl products such as starch. Nevertheless, prolonged tube growth of lily pollen in media containing [2-^{3}H]MI led to accumulation of a significant fraction of ^{3}H in ^3H_2O and starch-derived [^{3}H]Glc. Pollen germinated in media containing [2-^{14}C]MI stored much less ^{14}C in starch (Rosenfield and Loewus, 1975). Both [2-^{3}H]- and [2-^{14}C]MI were converted to C5-labeled pentosyl products without

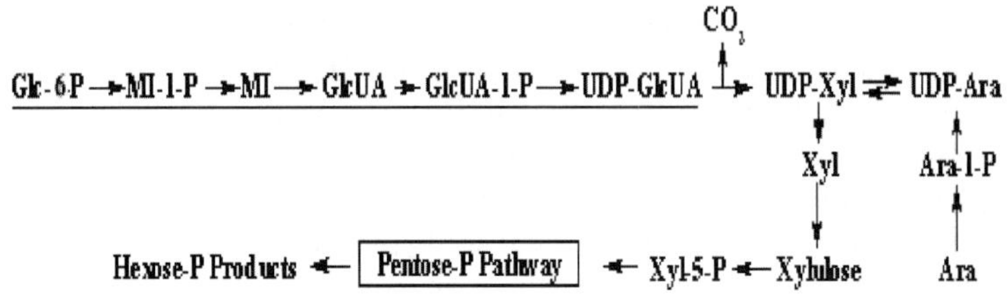

Figure 2. Metabolic route from MI to starch-derived Glc via the MIOP (*reaction sequence underlined*).

redistribution or loss of label during passage over the MIOP. It was during subsequent interconversion of pentose to hexose that exchange of ^{3}H with water and redistribution of ^{14}C within starch-derived Glc occurred. This could be demonstrated by growing lily pollen tubes in media containing [1-^{14}C]Ara, [5-^{14}C]Ara, or [5-^{3}H]Xyl. Metabolic pathways (Figure 2) incorporating these pentoses into cell wall polysaccharides or into starch via pentose-P/hexose-P interconversion occurred beyond the MIOP (Rosenfield and Loewus, 1978a,b; Rosenfield *et al.*, 1978).

3.3.9 MI and xylan biosynthesis

Non-cellulosic polysaccharides are intimately associated with cellulose in plants and often account for over half of the polysaccharide content of the primary cell wall, much of it as pectin and xyloglucan (O'Neill and York, 2003). In non-graminaceous plants, xyloglucan, and xylan are the most abundant hemicellulosic polysaccharides of mature wood, as much as 20–30% of dry wall-mass (Mellerowicz *et al.*, 2001). UDP-GluUA, sole source of UDP-Xyl for xylan biosynthesis, is the ultimate product common to both the MIOP and SNOP. In the latter pathway, UDP-GlcUA is produced by UDP-Glc dehydrogenase, an enzyme strongly inhibited by UDP-Xyl, whereas in the MIOP this product is formed by UDP-GluUA-1-P urylyltransferase which is not inhibited by UDP-Xyl. In the event that free MI or a metabolically addressable source of MI is available, the MIOP might provide sufficient UDP-Xyl to limit oxidation of UDP-Glc via the SNOP, in essence, redirecting utilization of UDP-Glc toward biosynthesis of UDP-glycans while the MIOP continues to produce UDP-GluUA for xylan biosynthesis. Studies described by Imai and associates support such a scheme (Imai and Terashima, 1991, 1992; Imai *et al.*, 1997, 1998, 1999). They found that immature, differentiating, xylem tissue of magnolia produced xylan-containing [^{3}H]Xyl-residues within 24 h after administration of [2-^{3}H]MI to growing stems and this product was selectively retained as xylan in mature cell walls 5 months later (Imai *et al.*, 1999).

4. UDP-GlcUA: PRECURSOR OF URONOSYL AND PENTOSYL COMPONENTS IN CELL WALL POLYSACCHARIDES

In view of a potential regulatory role for the MIOP associated with MI biosynthesis or intermediates leading to UDP-GluUA via the MIOP, renewed attention needs to be given to these processes (Kärkönen, 2005). They include use of MIPS mutants that alter free MI availability, hormonal effects on growth or development, modifications of enzymes involved in the MIOP, use of specific inhibitors such as MMO, and application of a host of newly developed tools of molecular biology. A model of such approaches is found in evidence for coexistence and use of both MIOP and SNOP during UDP-GluUA formation in young *Arabidopsis* seedlings (Kanter *et al.*, 2005; Seitz *et al.*, 2000). Moreover, their data suggest that each pathway exhibits temporal and spatial regulatory responses reflecting unique qualities of that pathway.

4.1 Effects of plant growth regulators on MI metabolism

Fifteen-day-old suspension cultures of *Acer pseudoplatanus* grown in Murashige and Skoog medium with 3% Glc, 4.4 μM 6-benzylaminopurine (BA) and 0.45 μM 2,4-dichlorophenoxyacetic acid (2,4-D) readily took up 100 mg/l of [2-^{3}H]MI over 24 h and utilized up to 20% for pectin biosynthesis. Virtually all of this ^{3}H was recovered in galacturonosyl and pentosyl residues. Increasing the BA level 10-fold drastically blocked [2-^{3}H]MI uptake and little was available for pectin biosynthesis. Increasing the 2,4-D level 10-fold had little or no effect on [2-^{3}H]MI uptake but did diminish the amount of ^{3}H appearing in pectin (Verma *et al.*, 1976).

Exogenously applied abscisic acid (ABA) suppressed elongation of squash hypocotyl segments and inhibited incorporation of [1-^{14}C]Glc and [2-^{3}H]MI into cell wall fractions (Wakabayashi *et al.*, 1989). In the absence of ABA, the extent of incorporation of [1-^{14}C]Glc was much greater than that of [2-^{3}H]MI but label from the former appeared uniformily in hexosyl, pentosyl, and uronosyl residues while the latter was limited to just pentosyl and uronosyl residues of pectin and hemicellulose B suggesting that the MIOP was involved.

In studies on ABA-promoted responses in the duckweed, *Spirodela polyrrhiza*, it was found that such treatment activates a developmental pathway culminating in formation of turions, modified resting-state fronds distinguishable from vegetative fronds due to thicker walls, absence of aerenchyma, and accumulation of anthocyanin, MI, starch, and phytic acid (Flores and Smart, 2000; Smart and Fleming, 1993; Smart and Flores, 1997; Smart and Trewavas, 1983). The last-mentioned citation found, in addition to phytic acid and its expected precursors, three novel plant inositol phosphates, two of which were diphosphoinositol pentakis and hexakis phosphates ($InsP_7$ and $InsP_8$), putative functional high-energy intermediates (Safrany *et al.*, 1999; Saiardi *et al.*,

2002). Continued exposure to ABA led to less ^{3}H in turion cell walls relative to that in inositol polyphosphates. It is tempting to speculate that diverse regulatory functions involving ABA are involved. These might include one or more of the following: altering contributions from SNOP or MIOP to UDP-GluUA, regulating phosphorylation of [2-^{3}H]MI to [2-^{3}H]MI-1-P (initial step in production of polyphosphrylated MIs) or altering demands on the free MI pool that must supply a host of other MI requirements involved in turion development. Only further study of the broad effects of ABA on turion development will resolve the current impetus.

4.2 UDP-GlcUA decarboxylase (EC 4.1.1.35)

UDP-GlcUA decarboxylase (UDP-GlcUADase) which catalyzes the conversion of UDP-GlcUA to UDP-xylose (UDP-Xyl) has been purified and cloned from a fungus, *C. neoformans* (Bar-Peled *et al.*, 2001), pea seedlings, *Pisum sativum* L. (Kobayashi *et al.*, 2002), *Arabidopsis* (Harper and Bar-Peled, 2002), and immature rice seeds, *Oryza sativa* cv. Nipponbare (Suzuki *et al.*, 2003). It has also been purified and characterized from differentiating tobacco cells (*Nicotiana tabacum* L.) where its 87-kDa isoform is localized to cytoplasm (Wheatley *et al.*, 2002). As Reiter and Vanzin (2001) point out, UDP-GluUADase represents a major branch point in the biosynthesis of UDP-sugars, involving epimerization, decarboxylation, and rearrangement reactions; the source of uronosyl and pentosyl residues found in pectin and hemicellulose as well as products such as free Xyl (Loewus *et al.*, 1962; Rosenfield and Loewus, 1978a), apiosyl residues of cell wall polysaccharide from *Lemna* and apiin (Roberts *et al.*, 1967b). While these reactions are beyond discussion of processes dealt with in this chapter, it is well to consider their competitive roles apropos their common source. As noted above (Section 3.3.9), feedback inhibition of UDP-Glc dehydrogenase by UDP-Xyl is one such instance.

4.3 Putative role of MI as a source of D-galacturonate for AsA biosynthesis via the MIOP

Interest in AsA biosynthesis was greatly motivated by comparative studies on plants and AsA-synthesizing animals (Isherwood *et al.*, 1954). Subsequent findings stemming from radioisotopic experiments suggested that two pathways to AsA occurred in plants, one referred to as the "non-inversion" or "direct" route, since it conserved the original carbon chain sequence of Glc, appeared to be the biosynthetic pathway while the other, the "inversion" route, seemingly a salvage pathway (see Section 2). Subsequently, enzymatic and molecular evidence were obtained for the "direct" route, now referred to as the "D-mannose/L-galactose" pathway (Laing *et al.*, 2004; Running *et al.*, 2003; Smirnoff *et al.*, 2004; Wheeler *et al.*, 1998). The putative "inversion" route (Loewus, 1963) gained fresh interest when Agius *et al.* (2000) isolated and characterized GalUR, a gene from strawberry encoding a NADPH-dependent D-galacturonate reductase. They examined strawberry fruit over the full range of development and found expression of GalUR to be positively correlated to AsA content with highest levels of expression in the fully ripe berry where hydrolyzed pectic products rich in D-galacturonic acid (GalUA) occur.

In another significant development, evidence was obtained that GDP-Man 3′,5′-epimerase produces novel intermediates, GDP-L-gulose, (Wolucka and Van Montagu, 2003) as well as GDP-L-Gal (Smirnoff *et al.*, 2004) and both of these nucleotides function as intermediates in AsA biosynthesis in plants. Over 40 years ago, Calvin's group reported the isolation of a diphosphate ester of 2-keto-L-gulonic acid from *Chlorella pyrenoidosa* metabolizing $^{14}CO_2$ in the light and speculated on its possible role in AsA biosynthesis (Moses *et al.*, 1962). Detached bean apices and ripening strawberry fruits stem-fed aqueous solutions of L-gulono- or L-galactono-1,4-lactone readily convert these compounds to AsA (Baig *et al.*, 1970). At the time, three possibilities were offered for these results: (1) utilization of either substrate by a single oxidizing enzyme, (2) existence of separate oxidizing enzymes for each substrate, or (3) a two-step process wherein epimerization at carbon 3 precedes oxidation. Wolucka and Van Montagu's findings now suggest that both L-gulono- and L-galactono-1,4-lactone may be present as intermediates and that oxidation to AsA is a site-specific process. Jain and Nessler (2000) found that constitutive expression of L-gulonolactone oxidase cDNA in *vtc1-1* mutant *A. thaliana* plants increased AsA content twofold (Conklin, 2001). In an extension of this finding, L-gulonolactone oxidase constructs were also expressed in other *A. thaliana* lines defective in AsA production but unrelated to the defect unique to the *vtc1-1* mutant. All five *vtc* mutant lines that were tested rescued AsA content, equal or higher than that of wild-type plants, suggesting existence of alternative AsA pathways (Radzio *et al.*, 2003).

As noted earlier, experiments leading to a bacterially expressed recombinant protein from chromosome 4 of *Arabidopsis* with the properties of MIOase have provided opportunity to test the full potential for AsA biosynthesis through the "inversion" pathway (Lorence *et al.*, 2004). It remains to be determined whether this route is ancillary to the D-mannose/L-galactose (alternatively, L-gulose) pathway or is a major source of AsA via MIOase in plants.

5. CONCERNING A METABOLIC ROLE FOR *MYO*-INOSITOL IN PLANT CELL WALL BIOGENESIS

This synoptic review of studies involving MI metabolism as applied to formation of uronosyl and pentosyl residues of cell wall polysaccharides is intended as a reminder to those involved in research on cell wall structure and function of need to give closer scrutiny to the MIOP as an alternative pathway. A recent book (Rose, 2003), prefaced as "...written at professional and reference level...," dismisses the MIOP with a single reference (Doblin *et al.*, 2003: page 205) to a review (Feingold and Avigad, 1980) that was prepared over 24 years ago. Gibeaut (2000) cites Schlüpmann *et al.* (1994) as an example of a sucrose requirement for normal *in vitro* growth of *Nicotiana* pollen but fails to mention the findings of Dickinson (1965, 1967, 1978) as regards tolerance toward pentaerythritol as non-metabolizable osmoticum for normal *in vitro* growth of *Lilium* pollen in the absence of sucrose. Under such growth conditions, *Lilium* pollen utilized $[^{14}C]$MI for tube wall pectic substance biosynthesis for periods up to 8 h (Kroh and Loewus, 1968). Gibeaut cited a study on the uptake and metabolism of $[1\text{-}^{14}C]$Glc versus $[2\text{-}^{3}H]$MI by etiolated squash hypocotyl segments (Wakabayashi *et al.*, 1989) as evidence for selective cell wall labeling by Glc, ignoring the fact that they were comparing labeling processes in intact pollen tube germination/elongation to that of elongation of hypocotyl segments from squash.

Reiter and Vanzin (2001) and Mellerowicz *et al.* (2001) offer refreshing excursions into an emerging viewpoint that includes both the SNOP and MIOP contributions to UDP-GluUA biosynthesis and subsequent steps of epimerization, decarboxylation and rearrangements to furnish UDP-Gal, UDP-Xyl, UDP-Ara, and UDP-apiose (Burget *et al.*, 2003; Mølhø *et al.*, 2003). Hopefully, the pioneering efforts of Tenhaken and his colleagues (Kanter *et al.*, 2005; Seitz *et al.*, 2000; Tenhaken and Thulke, 1996) will continue to uncover new molecular and biochemical details on the inter-relationships of the MIOP and SNOP and Nessler's group (Jain and Nessler, 2000; Lorence *et al.*, 2004; Radzio *et al.*, 2003) will extend their studies on the biosynthesis of AsA and its relationship to MI metabolism.

6. *MYO*-INOSITOL AS POTENTIAL PRECURSOR OF L-TARTRATE AND OXALATE (BREAKDOWN PRODUCTS OF L-ASCORBATE)

Experiments have yet to be performed which tie production of oxalate and L-threonate/L-tartrate in plants to a specific pathway of AsA biosynthesis (Bánhegyi and Loewus, 2004) but the ease with which AsA is synthesized in phloem and transported via vascular processes from source to sink (Fransceshi and Tarlyn, 2002; Hancock *et al.*, 2003; Tedone *et al.*, 2004) suggests that MI-linked AsA biosynthesis (Lorence *et al.* 2004) may play a significant role in this process. It is of interest to note that an invertebrate, the marine demosponge, *Chondriosia reniformis*, which sheds substantial amounts of crystalline calcium oxalate, contains the same level of AsA as is found in plants (Cerrano *et al.*, 1999). The biosynthetic pathway of AsA in this organism and its putative role as a precursor of oxalate have yet to be determined.

REFERENCES

Agius, F., Gonzáles-Lamonthe, R., Caballero, J.L., Muñoz-Blanco, J., Botella, M.A., and Valpuesta, V., 2003, Engineering increased vitamin C levels in plants by overexpression of a D-galacturonic acid reductase. *Nat. Biotechnol.* **21:** 177–181.

Albersheim, P., 1962, Hormonal control of *myo*-inositol incorporation into pectin. *J. Biol. Chem.* **238:** 1608–1610.

Arner, R.J., Prabhu, S., Thompson, J.T., Hildenbrandt, G.R., Liken, A.D., and Reddy, C.C., 2001, *myo*-Inositol oxygenase: Molecular cloning and expression of a unique enzyme that oxidizes *myo*-inositol and D-*chiro*-inositol. *Biochem. J.* **360:** 313–320.

Asamizu, T., and Nishi, A., 1979, Biosynthesis of cell-wall polysaccharides in cultured carrot cells. *Planta* **146:** 49–54.

Baig, M.M., Kelly, S., and Loewus, F., 1970, L-Ascorbic acid biosynthesis in higher plants from L-gulono- or L-galactono-1,4-lactone. *Plant Physiol.* **46:** 277–280.

Bánhegyi, G., and Loewus, F.A., 2004, Ascorbic acid catabolism: Breakdown pathways in animals and plants. In: Asard, H., May, J.M., and Smirnoff, N. (eds.), Vitamin C. Functions and Biochemistry in Animals and Plants. BIOS Sci. Publ., London and New York, pp. 31–48.

Bar-Peled, M., Griffith, C.L., and Doering, T.L., 2001, Functional cloning and characterization of a UDP-glucuronic acid decarboxylase: The pathogenic fungus *Cryptococcus neoformans* elucidates UDP-xylose synthesis. *Proc. Natl. Acad. Sci. U.S.A.* **98:** 12003–12008.

Benaroya, R.O., Zamski, E., and Tel-Or, E., 2004, L-*myo*-inositol 1-phosphate synthase in the aquatic fern *Azolla filiculoides*. *Plant Physiol. Biochem.* **42:** 97–102.

Burget, E.G., Verma, R., Mølhø, M., and Reiter, W.-D., 2003, The biosynthesis of L-arabinose in plants: Molecular cloning and characterization of a Golgi-localized UDP-D-xylose 4-epimerase encoded by the *MUR4* gene of *Arabidopsis*. *Plant Cell* **15:** 523–531.

Burns, J.J., 1967, Ascorbic acid. In: Greenberg, D.M. (ed.), Metabolic Pathways, Vol. 1, 3rd ed. Academic Press, New York, pp. 394–411.

Cerrano, C., Bravestrello, G., Arillo, A., Benatti, U., Bonpadre, S., Cattaneo-Vietti, R., Gaggero, L., Giovine, M., Leone, L., Lucchetti, G., and Sarà, M., 1999, Calcium oxalate production in the marine sponge *Chondrosia reniformis*. *Mar. Ecol. Prog. Ser.* **179:** 297–300.

Charalampous, F.C., and Lyras, C., 1957, Biochemical studies on inositol. IV. Conversion of inositol to glucuronic acid by rat kidney extracts. *J. Biol. Chem.* **228:** 1–13.

Chen, M., and Loewus, F.A., 1977, *myo*-Inositol metabolism in *Lilium longiflorum* pollen. Uptake and incorporation of *myo*-inositol-2-^{3}H. *Plant Physiol.* **59:** 653–657.

Chen, M., Loewus, M.W., and Loewus, F.A., 1977, The effect of *myo*-inositol antagonist 2-*O*, *C*-methylene-*myo*-inositol on the metabolism of *myo*-inositol-2-^{3}H and D-glucose-1-^{14}C in *Lilium longiflorum* pollen. *Plant Physiol.* **59:** 658–663.

Collin, S., Justin, A.-M., Cantrel, C., Arondel, V., and Kader, J.-C., 1999, Identification of AtPIS, a phosphatidylinositol synthase from *Arabidopsis*. *Eur. J. Biochem.* **262:** 652–658.

Conklin, P.L., 2001, Recent advances in the role and biosynthesis of ascorbic acid in plants. *Plant Cell Environ.* **24:** 383–394.

Davies, M.D., and Dickinson, D.B., 1972, Properties of uridine diphosphoglucose dehydrogenase from pollen of *Lilium longiflorum*. *Arch. Biochem. Biophys.* **152:** 53–61.

Davies, E.J., Tetlow, I.J., Bowsher, C.G., and Emes, M.J., 2003, Molecular and biochemical characterization of cytosolic phosphoglucomutase in wheat endosperm (*Triticum aestivum* L. cv. Axona). *J. Exp. Bot.* **54:** 1351–1360.

DeBolt, S., Hardie, J., Tyerman, S., and Ford, C.M., 2004, Comparison and synthesis of raphide crystals and druse crystals in berries of *Vitis vinifera* L. cv. Cabernet Sauvignon: Ascorbic acid as precursor for both oxalic and tartaric acids as revealed by radiolabelling studies. *Aust. J. Grape Wine Res.* **10:** 134–142.

Dickinson, D.B., 1965, Germination of lily pollen: Respiration and tube growth. *Science* **150:** 1818–1819.

Dickinson, D.B., 1967, Permeability and respiratory properties of germinating pollen. *Physiol. Plant.* **20:** 118–127.

Dickinson, D.B., 1978, Influence of borate and pentaerythritol concentrations on germination and tube growth of *Lilium longiflorum* pollen. *J. Am. Soc. Hortic. Sci.* **103:** 413–416.

Dickinson, D.B., 1982, Occurrence of glucuronokinase in various plant tissues and comparison of enzyme activity of seedlings and green plants. *Phytochemistry* **21:** 843–844.

Dickinson, D.B., Hopper, J.E., and Davies, M.D., 1973, A study of pollen enzymes involved in sugar nucleotide formation. In: Loewus, F. (ed.), Biogenesis of Plant Cell Wall *Polysaccharides*. Academic Press, NY, pp. 29–48.

Dickinson, D.B., Hyman, D., and Gonzales, J.W., 1977, Isolation of uridine 5(-pyrophosphate glucuronic acid pyrophosphorylase and its assay using ^{32}P-pyrophosphate. *Plant Physiol.* **59:** 1082–1084.

Doblin, M.S., Vergara, C.E., Read, S., Newbigin, E., and Bacic, A., 2003, Plant cell wall biosynthesis: Making the bricks. In: Rose, J.K.C. (ed.), The Plant Cell Wall. CRC Press, Boca Raton, FL, pp. 183–222.

Ercetin, M.E., and Gillaspy, G.E., 2002, Molecular characterization of an *Arabidopsis* gene encoding a phospholipids-specific inositol polyphosphate 5-phosphatase. *Plant Physiol.* **135:** 938–946.

Eisenberg, F., Jr., Bolden, A.H., and Loewus, F.A., 1964, Inositol formation by cyclization of glucose chain in rat testis. *Biochem. Biophys. Res. Commun.* **14:** 419–424.

English, P.D., Deitz, M., and Albersheim, P., 1966, *myo*-Inositol kinase: Partial purification and identification of product. *Science* **151:** 198–199.

Feingold, D.S. 1982, Aldo (and keto) hexoses and uronic acids. In: Loewus, F.A. and Tanner, W. (eds.), Plant Carbohydrates I. Intracellular Carbohydrates, Vol. 13A, New Series, Encyclopedia of Plant Physiology. Springer-Verlag, Berlin, pp. 3–76.

Feingold, D.S., and Avigad, D. 1980, Sugar nucleotide transformations in plants. In: Preiss, J. (ed.), The Biochemistry of Plants, Vol. 3, Carbohydrates: Structure and Function. Academic Press, New York, pp. 101–170.

Finkle, B.J., Kelly S., and Loewus, F.A., 1960, Metabolism of D-[1-^{14}C]- and D-[6-^{14}C]glucuronolactone by the ripening strawberry. *Biochim. Biophys. Acta* **38:** 332–339.

Fischer, H.O.L., 1945, Chemical and biochemical relationships between hexoses and inositol. *Harvey Lect.* **40:** 156–178.

Flores, S., and Smart, C.C., 2000, Abscisis acid-induced changes in inositol metabolism in *Spirodela polyrrhiza*. *Planta* **211:** 823–832.

Fransceshi, V.R., and Tarlyn, N.M., 2002, L-Ascorbic acid is accumulated in source leaf phloem and transported to sink tissues in plants. *Plant Physiol.* **130:** 649–656.

Funkhouser, E.A., and Loewus, F.A., 1975, Purification of *myo*-inositol 1-phosphate synthase from rice cell culture by affinity chromatography. *Plant Physiol.* **56:** 786–790.

Gibeaut, D.M., 2000, Nucleotide sugars and glycosyltransferases for synthesis of cell wall matrix polysaccharides. *Plant Physiol. Biochem.* **38:** 69–80.

Gibeaut, D.M., Cramer, G.R., and Seeman, J.R., 2001, Growth, cell walls, and UDP-glc dehydrogenase activity of *Arabidopsis thaliana* grown in elevated carbon dioxide. *J. Plant Physiol.* **158:** 569–576.

Gillard, D.F., and Dickinson, D.B., 1978, Inhibition of glucuronokinase by substrate analogs. *Plant Physiol.* **62:** 706–709.

Gillaspy, G.E., Keddie, J.S., Oda, K., and Guissem, W., 1995, Plant inositol monophosphatase is a lithium-sensitive enzyme encoded by a multigene family. *Plant Cell* **7:** 2175–2185.

Gumber, S.C., Loewus, M.W., and Loewus, F.A., 1984, *myo*-Inositol-1-phosphate synthase from pine pollen: Sulfhydryl involvement at the active site. *Arch. Biochem. Biophys.* **231:** 372–377.

Hancock, R.D., McRae, D., Haupt, S., and Viola, R., 2003, Synthesis of L-ascorbic acid in the phloem. *BMC Plant Biol.* **3:** 7.

Harper, A.D., and Bar-Peled, M., 2002, Biosynthesis of UDP-xylose, cloning and characterization of a novel *Arabidopsis* gene family, *UXS*, encoding soluble and putative membrane-bound UDP-glucuronic acid decarboxylase isoforms. *Plant Physiol.* **130:** 2188–2198.

Harran, S., and Dickinson, D.B., 1978, Metabolism of *myo*-inositol and growth in various sugars of suspension-cultured tobacco cells. *Planta* **141:** 77–82.

Hegeman, C.E., Good, L.L., and Grabau, E.A., 2001, Expression of D-*myo*-inositol-3-phosphate synthase in soybean. Implications for phytic acid biosynthesis. *Plant Physiol.* **125:** 1941–1948.

Hinterberg, B., Klos, C., and Tenhaken, R., 2002, Recombinant UDP-glucose dehydrogenase from soybean. *Plant Physiol. Biochem.* **40:** 1011–1017.

Hitz, W.D., Carlson, T.J., Kerr, P.S., and Sebastian, S.A., 2002, Biochemical and molecular characterization of a mutation that confers a decreased raffinose saccharide and phytic acid phenotype on soybean seeds. *Plant Physiol.* **128:** 650–660.

Hopper, J.E., and Dickinson, D.B., 1972, Partial purification and sugar nucleotide inhibition of UDP-glucose pyrophosphorylase from *Lilium longiflorum* pollen. *Arch. Biochem. Biophys.* **148:** 523–535.

Howard, C.F., and Anderson, L., 1967, Metabolism of *myo*-inositol in animals, complete catabolism of *myo*-inositol-^{14}C by rat kidney slices. *Arch. Biochem. Biophys.* **118:** 332–339.

Imai, T., Guto, H., Matsumura, H., and Yasuda, S., 1998, Determination of the distribution and reaction of polysaccharides in wood cell-walls by the isotope tracer techniques. VII. Double radiolabeling of xylan and pectin in magnolia (*Magnolia kobus* DC) and comparison of their behaviors during kraft pulping by radiotracer technique. *J. Wood Sci.* **44:** 106–110.

Imai, T., Guto, H., Matsumura, H., and Yasuda, S., 1999, Determination of the distribution and reaction of polysaccharides in wood cell-walls by the isotope tracer techniques. VIII. Selective radiolabeling of xylan in mature cell walls of magnolia (*Magnolia kobus* DC) and visualization of its distribution by microautoradiography. *J. Wood Sci.* **45:** 164–169.

Imai, T., and Terashima, N., 1991, Determination of the distribution and reaction of polysaccharides in wood cell-walls by the isotope tracer techniques. II. Selective radio-labeling of pectic substances in mitsumata (*Edgeworthia papyrifera*). *Mokuzai Gakkaishi* **37:** 733–740 (in English).

Imai, T., and Terashima, N., 1992, Determination of the distribution and reaction of polysaccharides in wood cell-walls by the isotope tracer techniques. IV. Selective radio-labeling of xylan in magnolia (*Magnolia kobus*) and visualization of its distribution in differentiating xylem by microautoradiography. *Mokuzai Gakkaishi* **38:** 693–699 (in English).

Imai, T., Yasuda, S., and Terashima, N., 1997, Determination of the distribution and reaction of polysaccharides in wood cell-walls by the isotope tracer techniques. V. Behavior of xylan during kraft pulping studied by the radiotracer technique. *Mokuzai Gakkaishi* **43:** 2241–246 (in English).

Isherwood, F.A., Chen, Y.T., and Mapson, L.W., 1954, Synthesis of L-ascorbic acid in plants and animals. *Biochem. J.* **56:** 1–15.

Jain, A.K., and Nessler, C.L., 2000. Metabolic engineering of an alternative pathway for ascorbic acid biosynthesis in plants. *Mol. Breeding* **6:** 73–78.

Johansson, H., Sterky, F., Amini, B., Lundberg, J., and Kleczkowski, L.A., 2002, Molecular cloning and characterization of a cDNA encoding poplar UDP-glucose dehydrogenase, a key gene of hemicellulose-/pectin formation. *Biochim. Biophys. Acta* **1576:** 53–58.

Kanter, U., Becker, M., Friauf, E., and Tenhaken, R., 2003, Purification, characterization and functional cloning of inositol oxygenase from *Cryptococcus*. *Yeast* **20:** 1317–1329.

Kanter, U., Usadel, B., Guerineau, F., Li, Y., Pauly, M., and Tenhaken, R., 2005, The inositol oxygenase gene family of *Arabidopsis* is involved in the biosynthesis of nucleotide sugar precursors for cell-wall matrix polysaccharides. *Planta* **221:** 243–254.

Kärkönen, A., 2005. Biosynthesis of UDP-AlcA: Via UDPGDH or the/myo/-inositol oxidation pathway? Plant Biosystems, 139:46–49.

Kärkönen, A., Murigneux, A., Martinant, J-P., Pepey, E., Tatout, C., Dudley, B.J. and Fry, S.C., 2005. UDP-glucose dehydrogenases of maize: a role in cell wall pentose biosynthesis. Biochem. J., 391:409–415.

Karner, U., Peterbauer, T., Raboy, V., Jones, D.A., and Hedley, C.L., 2004, *myo*-Inositol and sucrose concentrations affect the accumulation of raffinose family oligosaccharides in seeds. *J. Exp. Bot.* **55:** 1982–1987.

Knee, M., 1978, Metabolism of polymethylgalacturonate in apple fruit cortical tissue during ripening. *Phytochemistry* **17:** 1262–1264.

Kobayashi, M., Nakagawa, H., Suda, I., Miyagawa, I., and Matoh, T., 2002, Purification and cDNA cloning of UDP-D-glucuronate carboxy-lyase (UDP-D-xylose synthase) from pea seedlings. *Plant Cell Physiol.* **43:** 1259–1265.

Kroh, M., Labarca, C., and Loewus, F., 1971, Utilization of pistil exudates for pollen tube wall biosynthesis in *Lilium longiflorum*. In: Heslop-Harrison, J. (ed.), *Pollen: Development and Physiology*. Butterworths, London, pp. 273–278.

Kroh, M., and Loewus, F., 1968, Biosynthesis of pectic substance in germinating pollen: Labeling with *myo*-inositol-2-^{14}C. *Science* **160:** 1352–1354.

Kroh, M., Miki-Hirosige, H., Rosen, W., and Loewus, F., 1970a, Inositol metabolism in plants. VII. Distribution and utilization of label from *myo*-inositol-U-^{14}C and -2-^{3}H by detached flowers and pistils of *Lilium longiflorum*. *Plant Physiol.* **45:** 86–91.

Kroh, M., Miki-Hirosige, H., Rosen, W., and Loewus, F., 1970b, Incorporation of label into pollen tube walls from *myo*-inositol labeled *Lilium longiflorum* pistils. *Plant Physiol.* **45:** 92–94.

Krook, J., Vreugdenhil, D., Dijkema, C., and van der Plas, L.H.W., 1998, Sucrose and starch metabolism in carrot (*Daucus carota* L.) cell suspensions analyzed by ^{13}C-labeling: indications for a cytosol and a plastid-localized oxidative pentose phosphate pathway. *J. Exp. Bot.*, **49:** 1917–1924.

Krook, J., Vreugdenhil, D., Dijkema, C., and van der Plas, L.H.W., 2000, Uptake of ^{13}C-glucose by cell suspensions of carrot (*Daucus carota*) measured by in vivo NMR: Cycling of triose-, pentose- and hexose-phosphates. *Physiol. Plant.* **108:** 124–133.

Labarca, C., Kroh, M., and Loewus, F., 1970, The composition of stigmatic exudate from *Lilium longiflorum*. Labeling studies with *myo*-inositol, D-glucose and L-proline. *Plant Physiol.* **46:** 150–156.

Labarca, C., and Loewus, F., 1972, The nutritional role of pistil exudates in pollen tube wall formation in *Lilium longiflorum*, I. Utilization of injected stigmatic exudates. *Plant Physiol.* **50:** 7–14.

Labarca, C., and Loewus, F., 1970, The nutritional role of pistil exudate in pollen tube wall formation in *Lilium longiflorum*. II. Production and utilization of exudate from stigma and stylar canal. *Plant Physiol.* **52:** 87–92.

Lackey, K.H., Pope, P.M., and Johnson, M.D., 2002, Biosynthesis of inositol phosphate in organelles of *Arabidopsis thaliana*. *SAAS Bull. Biochem. Biotech.* **15:** 8–15.

Lackey, K.H., Pope, P.M., and Johnson, M.D., 2003, Expression of 1L-myoinositol-1phosphate synthase (EC 5.5.1.4) in organelles of *Phaseolus vulgaris*. *Plant Physiol.* **132:** 2240–2247.

Laing, W.A., Fearson, N., Bulley, S., and MacRae, E., (2004), Kiwifruit L-galactose dehydrogenase: Molecular, biochemical and physiological aspects of the enzyme. *Funct. Plant Biol.* **31:** 1015–1025.

Leibowitz, M.D., Dickinson, D.B., Loewus, F.A., and Loewus, M.W., 1977, Partial purification and study of pollen glucuronokinase. *Arch. Biochem. Bioiphys.* **179:** 559–564.

Loewus, F.A., 1961, Aspects of ascorbic acid biosynthesis in plants. *Ann. N. Y. Acad. Sci.* **92:** 57–78.

Loewus, F.A., 1963, Tracer studies of ascorbic acid formation in plants. *Phytochemistry* **2:** 109–128.

Loewus, F.A., 1965, Inositol metabolism and cell wall formation in plants. *Fed. Proc. Fed. Am. Soc. Exp. Biol.* **24:** 655–862.

Loewus, F.A. (ed.), 1973a, Biogenesis of Plant Cell Wall Polysaccharides. Academic Press, New York.

Loewus, F.A., 1973b, Metabolism of inositol in higher plants. *Ann. N. Y. Acad. Sci.* **165:** 577–598.

Loewus, F.A., 1974, The biochemistry of *myo*-inositol in plants, a review. *Recent Adv. Phytochem.* **8:** 179–207.

Loewus, F.A., 1990, Inositol biosynthesis. In: Marré, D.J., Boss, W.F., and Loewus, F.A. (eds.), Inositol Metabolism in Plants. Wiley-Liss, New York, pp. 13–19.

Loewus, F.A., 2002, Biosynthesis of phytate in food grains and seeds. In: Reddy, N.R. and Sathe, S.K. (eds.), Food Phytates. CRC Press, Boca Raton, FL, pp. 53–61.

Loewus, F.A., and Dickinson, D.B., 1982, Cyclitols. In: Loewus, F.A. and Tanner, W. (eds.), Plant Carbohydrates I. Intracellular Carbohydrates, Vol. 13A, New Series, Encyclopedia of Plant Physiology. Springer-Verlag, Berlin, pp. 192–216.

Loewus, F.A., and Kelly, S., 1961. The metabolism of D-galacturonic acid and its methyl ester in the detached ripening strawberry. *Arch. Biochem. Biophys.* **95:** 483–493.

Loewus, F.A., and Kelly, S., 1962, Conversion of glucose to inositol in parsley leaves. *Biochem. Biophys. Res. Commun.* **7:** 204–208.

Loewus, F.A., and Kelly, S., 1963, Inositol metabolism in plants, I. Labeling patterns in cell wall polysaccharides from detached plants given *myo*-inositol-2-t or -2-^{14}C. *Arch. Biochem. Biophys.* **102:** 96–105.

Loewus, F., Chen, M.-S., and Loewus, M.W., 1973, The *myo*-inositol oxidation pathway to cell wall polysaccharides. In: Loewus, F. (ed.), Biogenesis of Plant Cell Wall Polysaccharides. Academic press, New York, pp. 1–27.

Loewus, F.A., Kelly, S., and Hiatt, H.H., 1960. Ascorbic acid synthesis from D-glucose-2-^{14}C in the liver of the intact rat. *J. Biol. Chem.* **235:** 937–939.

Loewus, F., and Labarca, C., 1973, Pistil secretion product and pollen tube wall formation. In: Loewus, F. (ed.), Biogenesis of Plant Cell Wall Polysaccharides. Academic Press, New York, pp. 175–193.

Loewus, F.A., Kelly, S., and Neufeld, E.F., 1962, Metabolism of *myo*-inositol in plants: Conversion to pectin, hemicellulose, D-xylose and sugar acids. *Proc. Natl. Acad. Sci. U.S.A.* **48:** 421–425.

Loewus, F.A., and Loewus, M.W., 1980, *myo*-Inositol: biosynthesis and metabolism. In: Preiss, J. (ed.), The Biochemistry of Plants, Vol. 3, Carbohydrates: Structure and Function. Academic Press, New York, pp. 43–76.

Loewus, F.A., and Murthy, P.P.N., 2000, *myo*-Inositol metabolism in plants, (a review). *Plant Sci.* **150:** 1–19.

Loewus, F.A., Loewus, M.W., Maiti, I.B., and Rosenfield, C.-L., 1978, Aspects of *myo*-inositol metabolism and biosynthesis in higher plants. In: Wells, W.W. and Eisenberg, F. (eds.), Cyclitols and Phosphoinositides. Academic Press, New York, pp. 249–267.

Loewus, M.W., 1977, Hydrogen isotope effects in the cyclization of D-glucose-6-phosphate by *myo*-inositol-1-phosphate synthase. *J. Biol. Chem.* **252:** 7221–7223.

Loewus, M.W., and Loewus, F.A., 1971, The isolation and characterization of D-glucose 6-phosphate cycloaldolase (NAD-dependent) from *Acer pseudoplatanus* L. cell cultures. Is occurrence in plants. *Plant Physiol.* **48:** 255–260.

Loewus, M.W., and Loewus, F.A., 1973a, D-Glucose 6-phosphate cycloaldolase: Inhibition studies and aldolase function. *Plant Physiol.* **51:** 263–266.

Loewus, M.W., and Loewus, F.A., 1973b, Bound NAD$^+$ in glucose 6-phosphate cycloaldolase of *Acer pseudoplatanus. Plant Sci. Lett.* **1:** 65–69.

Loewus, M.W., and Loewus, F.A., 1974, *myo*-Inositol 1-phosphate synthase inhibition and control of uridine diphosphate-D-glucuronic acid biosynthesis. *Plant Physiol.* **54:** 367–371.

Loewus, M.W., and Loewus, F.A., 1980, The C-5 hydrogen isotope effect in *myo*-inositol 1-phosphate synthase as evidence for the *myo*-inositol oxidation pathway. *Carbohydr. Res.* **83:** 333–342.

Loewus, M.W., and Loewus, F.A., 1982, *myo*-Insoitol-1-phosphatase from the pollen of *Lilium longiflorum* thunb. *Plant Physiol.* **70:** 765–770.

Loewus, M.W., Bedgar, D.L., and Loewus, F.A., 1984, 1L-*myo*-inositol 1-phosphate synthase. An ordered sequential mechanism. *J. Biol. Chem.* **259:** 7644–7647.

Loewus, M.W., Loewus, F.A., Brillinger, G.-U., Otsuka, H., and Floss, H.G., 1980, Stereochemistry of the *myo*-inositol-1-phosphate synthase reaction. *J. Biol. Chem.* **255:** 11710–11712.

Loewus, M.W., Sasaki, K., Leavitt, A.L., Munsell, L., Sherman, W.R., and Loewus, F.A., 1982, The enantiomeric form of *myo*-inositol 1-phosphate produced by *myo*-inositol 1-phosphate synthase and *myo*-inositol kinase in higher plants. *Plant Physiol.* **70:** 1661–1663.

Lorence, A., Chevone, B.I., Mendes, P., and Nessler, C.L., 2004, *myo*-Inositol oxygenase offers a possible entry point into plant ascorbate biosynthesis. *Plant Physiol.* **134:** 1200–1205.

Maiti, I.B., and Loewus, F.A., 1978a, *myo*-Inositol metabolism in germinating wheat. *Planta* **142:** 55–60.

Maiti, I.B., and Loewus, F.A., 1978b, Evidence for a functional *myo*-inositol oxidation pathway in *Lilium longiflorum* pollen. *Plant Physiol.* **62:** 280–283.

Majumder, A.L., Chatterjee, A., Dastidar, K.G., and Majee, M., 2003, Diversification and evolution of L-myo-inositol 1-phosphate synthase. *FEBS Lett.* **553:** 3–10.

Majumder, A.L., Johnson, M.D., and Henry, S.A., 1997, 1L-*myo*-Inositol-1-phosphate synthase. *Biochim. Biophys. Acta* **1348:** 245–256.

Manthey, A.E., and Dickinson, D.B., 1978, Metabolism of *myo*-inositol by germinating *Lilium longiflorum* pollen. *Plant Physiol.* **61:** 904–908.

Mattoo, A.K., and Lieberman, M., 1977, Localization of the ethylene-synthesizing system in apple tissue. *Plant Physiol.* **60:** 794–799.

Mellerowicz, E.J., Baucher, M., Sundberg, B., and Woerjan, W., 2001, Unravelling cell wall formation in the woody dicot stem. *Plant Mol. Biol.* **47:** 239–274.

Miyazaki, S., Rice, M., Quigley, F., and Bohnert H.J., 2004, Expression of plant inositol transporters in yeast. *Plant Sci.* **166:** 245–252.

Mølhø, M., Verma, R., and Reiter, W.-D., 2003, The biosynthesis of the branched-chain sugar D-apiose in plants: Functional cloning and characterization of a UDP-D-apiose/UDP-D-xylose synthase from *Arabidopsis*. *The Plant J.* **35:** 693–703.

Molina, Y., Ramos, S.E., Douglass, T., and Klig, L.S., 1999, Inositol synthesis and catabolism in *Cryptococcus neoformans*. *Yeast* **15:** 1657–1667.

Morré, D., Boss, W.F., and Loewus, F.A. (eds.), 1990, Inositol Metabolism in Plants. Wiley-Liss, New York.

Moses, V., Ferrier, R.J., and Calvin, M., 1962, Characterization of the photosynthetically synthesized "γ-keto" phosphate ester of 2-keto-L-gulonic acid. *Proc. Natl. Acad. Sci. U.S.A.* **48:** 1644–1652.

Neufeld, E.F., Feingold, D.S., and Hassid, W.Z., 1959, Enzymic phosphorylation of D-glucuronic acid by extracts from seedlings of *Phaseolus aureus* seedlings. *Arch. Biochem. Biophys.* **83:** 96–100.

Obendorf, R.L., 1997, Oligosaccharides and galactosyl cyclitols in seed desiccation tolerance. *Seed Sci. Res.* **7:** 63–74.

O'Neill, M.A., and York, W.S., 2003, The composition and structure of primary cell walls. In: Rose, J.K.C. (ed.), The Plant Cell Wall. CRC Press, Boca Raton, FL, pp. 1–54.

Parthasarathy, L., Vadnal, R.E., Parthasarathy, R., and Shamala Devi, C.S., 1994, Biochemical and molecular properties of lithium-sensitive monophosphatase. *Life Sci.* **54:** 1127–1142.

Periappuram, C., Steinhauer, L., Barton, D.L., and Zon, J., 2000, The plastidic phosphoglucomutase from *Arabidopsis*. A reversible enzyme reaction with an important role in metabolic control. *Plant Physiol.* **122:** 1193–1200.

Radzio, J.A., Lorence, A., Chevone, B.I., and Nessler, C.L., 2003, L-Gulono-1,4-lactone oxidase expression rescues vitamin C-deficient *Arabidopsis* (*vtc*) mutants. *Plant Mol. Biol.* **53:** 837–844.

Reddy, C.C., Swan, J.S., and Hamilton, G.A., 1981, *myo*-Inositol oxygenase from hog kidney. I. Purification and characterization of the oxygenase and of an enzyme complex containing the oxygenase and D-glucuronate reductase. *J. Biol. Chem.* **256:** 8510–8518.

Reiter, W.-D., 2002, Biosynthesis and properties of the plant cell wall. *Curr. Opin. Plant Biol.* **5:** 536–542.

Reiter, W.-D., and Vanzin, G.F., 2001, Molecular genetics of nucleotide sugar interconversion pathways in plants. *Plant Mol. Biol.* **47:** 95–113.

Ridley, B.L., O'Neill, M.A., and Mohnen, D., 2001, Pectins: Structure, biosynthesis, and oligogalacturonide-related signaling. *Phytochemistry* **57:** 929–967.

Roberts, R.M., 1971, The formation of uridine diphosphate-glucuronic acid in plants. *J. Biol. Chem.* **246:** 4995–5002.

Roberts, R.M., and Cetorelli, J.J., 1973, UDP-D-glucuronic acid pyrophosphorylase and the formation of UDP-D-glucuronic acid in plants. In: Loewus, F. (ed.), Biogenesis of Plant Cell Wall Polysaccharides. Academic Press, New York, pp. 49–68.

Roberts, R.M., Deshusses, J., and Loewus, F., 1968, Inositol metabolism in plants. V. Conversion of *myo*-inositol to uronic acid and pentose units of acidic polysaccharides in root tips of *Zea mays*. *Plant Physiol.* **43:** 979–989.

Roberts, R.M., and Loewus, F., 1966, Inositol metabolism in plants. III. Conversion of *myo*-inositol-2-^{3}H to cell wall polysaccharides in sycamore (*Acer pseudoplatanus* L.) cell culture. *Plant Physiol.* **41:** 1489–1498.

Roberts, R.M., and Loewus, F., 1968, Inositol metabolism in plants. VI. Conversion of *myo*-inositol to phytic acid in *Wolffiella floridana*. *Plant Physiol.* **43:** 1710–1716.

Roberts, R.M., and Loewus, F., 1973, The conversion of D-glucose-6-^{14}C to cell wall polysaccharide material in *Zea mays* in the presence of high endogenous levels of *myo*-inositol. *Plant Physiol.* **52:** 646–650.

Roberts, R.M., Shah, R., and Loewus, F., 1967a, Conversion of *myo*-inositol-2-^{14}C to labeled 4-*O*-methyl-glucuronic acid in the cell wall of maize root tips. *Arch. Biochem. Biophys.* **119:** 590–593.

Roberts, R.M., Shah, R., and Loewus, F., 1967b, Inositol metabolism in plants. IV. Biosynthesis of apiose in *Lemna* and *Petroselinum*. *Plant Physiol.* **42:** 659–666.

Robertson, D., McCormack, B.A., and Bolwell, G.P., 1995, Cell wall polysaccharide and related metabolism in elicitor-stressed cells of French bean (*Phaseolus vulgaris* L.). *Biochem. J.* **306:** 745–750.

Rose, J.K.C. (ed.), 2003, The Plant Cell Wall. CRC Press, Boca Raton, FL.

Rosenfield, C.-L., and Loewus, F.A., 1975, Carbohydrate interconversions in pollen–pistil interactions of the lily. In: Mulcahy, D.L. (ed.) Gamete Competition in Plants and Animals. North Holland Publ. Co., Amsterdam, The Netherlands, pp. 151–160.

Rosenfield, C.-L., and Loewus, F.A., 1978a, Metabolic studies on intermediates in the *myo*-inositol oxidation pathway in *Lilium longiflorum* pollen. II. Evidence for the participation of uridine diphosphoxylose and free xylose as intermediates. *Plant Physiol.* **61:** 96–100.

Rosenfield, C.-L., and Loewus, F.A., 1978b, Metabolic studies on intermediates in the *myo*-inositol oxidation pathway in *Lilium longiflorum* pollen. III. Polysaccharidic origin of labeled glucose. *Plant Physiol.* **61:** 101–103.

Rosenfield, C.-L., Fann, C., and Loewus, F.A., 1978, Metabolic studies on intermediates in the *myo*-inositol oxidation pathway in *Lilium longiflorum* pollen. I. Conversion to hexoses. *Plant Physiol.* **61:** 89–95.

Running J.A., Burlingame, R.P., and Berry, A., 2003, The pathway of L-ascorbic acid biosynthesis in the colourless microalga *Prototheca moriformis*. *J. Exp. Bot.* **54:** 1841–1849.

Safrany, S.T., Caffrey, J.J., Yang, X., and Shears, S.B., 1999, Diphosphoinositol polyphosphates: The final frontier for inositide research? *Biol. Chem.* **380:** 945–951.

Saiardi, A., Sciambi, C., McCaffery, M., Wendland, B., and Snyder, S.H., 2002, Inositol pyrophosphates regulate endocytic trafficking. *Proc. Natl. Acad. Sci. U.S.A.* **99:** 14206–14211.

Sasaki, K., and Loewus, F.A., 1980, Metabolism of *myo*-[2-^{3}H]-inositol and *scyllo*-[R-^{3}H]-inositol in ripening wheat kernels. *Plant Physiol.* **66:** 740–746.

Sasaki, K., and Loewus, F.A., 1982, Redistribution of tritium during germination of grain harvested from *myo*-[2-^{3}H]-inositol- and *scyllo*-[R-^{3}H]-inositol-labeled wheat. *Plant Physiol.* **69:** 220–225.

Sasaki, K., and Nagahashi, G., 1990, A deuterium-labeling technique to study *myo*-inositol metabolism. In: Marré, D.J., Boss, W.F., and Loewus, F.A. (eds.), Inositol Metabolism in Plants. Wiley-Liss, New York, pp. 47–54.

Sasaki, K., Nagahashi, G., Gretz, M.R., and Taylor, I.E.P., 1989, Use of per C-deuterated *myo*-inositol for the study of cell wall synthesis in germinating beans. *Plant Physiol.* **90:** 686–689.

Sasaki, K., and Taylor, I.E.P., 1984, Specific labeling of cell wall polysaccharides with *myo*-[2-^{3}H]inositol during germination and growth of *Phaseolus vulgaris* L. *Plant Cell Physiol.* **25:** 989–997.

Sasaki, K., and Taylor, I.E.P., 1986, *myo*-Inositol synthesis from [1-^{3}H]glucose in *Phaseolus vulgaris* L. during early stages of germination. *Plant Physiol.* **81:** 493–496.

Seitz, B., Klos, C., Wurm, M., and Tenhaken, R., 2000, Matrix polysaccharide precursors in *Arabidopsis* cell walls are synthesized by alternate pathways with organ-specific expression patterns. *Plant J.* **21:** 537–546.

Schlüpmann, H., Bacic, A., and Read, S.M., 1994, Uridine diphosphate glucose metabolism and callose synthesis in cultured pollen tubes of *Nicotiana alata* Link et Otto. *Plant Physiol.* **105:** 650–670.

Schopfer, W.H., Deshusses, J., Wustenfeld, D., and Posternak, T., 1969, Growth inhibitors of *Schizosaccharomyces pombe*. *Arch. Sci.* **22:** 651–665.

Sergeeva, L.I., Vonk, J., Keurentjes, J.J.B., van der Plas, L.H.W., Koornneef, M., and Vreugdenhil, D., 2004, Histochemical analysis reveals organ-specific quantitative trait loci for enzyme activities in *Arabidopsis*. *Plant Physiol.* **134:** 237–245.

Sherman, W.R., Loewus, M.W., Piña, M.Z., and Wong, Y.-H.H., 1981, Studies on *myo*-inositol-1-phosphate synthase from *Lilium longiflorum* pollen, *Neurospora crassa* and bovine testis. Further evidence that a classical aldolase is not utilized. *Biochim. Biophys. Acta* **660:** 299–305.

Shibko, S., and Edelman, J., 1957, Randomization of the carbon atoms in glucose and fructose during their metabolism in barley seedlings. *Biochim. Biophys. Acta* **25:** 642–644.

Smart, C.C., and Fleming, A.J., 1993, A plant gene with homology to D-*myo*-inositol-3-phosphate synthase in rapidly up-regulated during an abscisic-acid-induced morphogenic response in *Spirodela polyrrhiza*. *Plant J.* **4:** 279–293.

Smart, C.C., and Flores, S., 1997, Overexpression of D-*myo*-inositol-3-phosphate synthase leads to elevated levels of inositol in *Arabidopsis*. *Plant Mol. Biol.* **33:** 811–820.

Smart, C.C., and Trewavas, A.J., 1983, Abscisis-acid-induced turion formation in *Spirodela polyrrhiza* L. II. Ultrastructure of the turion; a stereological analysis. *Plant Cell Environ.* **6:** 515–120.

Smirnoff, N., Conklin, P.I., and Loewus, F.A., 2001, Biosynthesis of ascorbic acid in plants: A renaissance. *Ann. Rev. Plant Physiol. Plant Mol. Biol.* **53:** 437–467.

Smirnoff, N., Running, J.A., and Gatzek, S, 2004, Ascorbic acid biosynthesis: A diversity of pathways. In: Asard, H., May, J.M. and Smirnoff, N. (eds.), Vitamin C. Functions and Biochemistry in Animals and Plants. BIOS Sci. Publ., London and New York, pp. 7–29.

Stevenson, J.M., Perera, I.Y., Hellmann, I., Persson, S., and Boss, W.F., 2000, Inositol signaling and plant growth. *Trends Plant Sci.* **5:** 252–258; Erratum, *ibid.* 2000, **5:** 357.

Stewart, D.C., and Copeland, L., 1998, Uridine 5(-diphosphate-glucose dehydrogenase from soybean nodules. *Plant Physiol.* **116:** 149–355.

Stieglitz, K.A., Yang, H., Roberts, M.F., and Stec, B., 2005, Reaching for mechanistic consensus across life kingdoms: Structure and insight into catalysis of the *myo*-inositol-1-phosphate synthase (mIPS) from *Archaeoglubus fulgidus*. *Biochemistry* **44:** 213–224.

Styer, J.C., Keddie, J., Spence, J., and Gillaspy, G.E., 2004, Gneomic organization and regulation of the *LeIMP-1* and *LeIMP-2* genes encloning *myo*-inositol monophosphatase in tomato. *Gene* **326:** 35–41.

Suzuki, K., Suzuki, Y., and Kitamura, S., 2000, Cloning and expression of a UDP-glucuronic acid decarboxylase gene in rice. *J. Exp. Bot.* **54:** 1997–1999.

Tedone, L., Hancock, R.D., Alberino, S., Haupt, S., and Viola, R., 2004, Long-distance transport of L-ascorbic acid in potato. *BMC Plant Biol.* **4:** 16.

Tenhaken, R., and Thulke, O., 1996, Cloning of an enzyme that synthesizes a key nucleotide–sugar precursor of hemicellulose biosynthesis from soybean: UDP-glucose dehydrogenase. *Plant Physiol.* **112:** 1127–1134.

Verma, D.C., and Dougall, D.K., 1979, Biosynthesis of *myo*-inositol and its role as a precursor of cell-wall polysaccharides in suspension culture of wild-carrot cells. *Planta* **146:** 55–62.

Verma, D.C., Tarvares, J., and Loewus, F.A., 1976, Effect of benzyladenine, 2,4-dichlorophenoxyacetic acid, and D-glucose in *myo*-inositol metabolism in *Acer pseudoplatanus* L. cells grown in suspension culture. *Plant Physiol.* **37:** 241–244.

Wakabayashi, K., Sakurai, N., and Kuraishi, S., 1989, Effects of ABA on synthesis of cell-wall polysaccharides in segments of etiolated squash hypocotyls. I. Changes in incorporation of glucose and *myo*-inositol into cell-wall components. *Plant Cell Physiol.* **30:** 99–105.

Weinhold, P.A., and Anderson, L., 1967, Metabolism of *myo*-inositol in animals. III. Action of the antagonist 2-*O*,*C*-methylene-*myo*-inositol. *Arch. Biochem. Biophys.* **122:** 529–36.

Wheatley, E.R., Davies, D.R., and Bolwell, G.P., 2002, Charaterization and immunolocation of an 87 kDa polypeptide associated with UDP-glucuronic acid decarboxylase activity from differentiating tobacco cells (*Nicotiana tabacum* L.). *Phytochemistry* **61:** 771–780.

Wheeler, G.L., Jones, M.A., and Smirnoff, N., 1998, The biosynthetic pathway of vitamin C in higher plants. *Nature* **303:** 365–369.

Wolucka, B.A., and Van Montagu, M., 2003, GDP-Mannose 3′,5′-epimerase forms GDP-L-gulose, a putative intermediate for the *de novo* biosynthesis of vitamin C in plants. *J. Biol. Chem.* **278:** 47483–47490.

Yoshida, K.T., Wada, T., Koyama, H., Mizobuchi-Fukuoka, R., and Naito, S., 1999, Temporal and spatial patterns of accumulation of the transcript of *myo*-inositol 1-phosphate synthase and phytin-containing particles during seed development in rice. *Plant Physiol.* **119:** 65–72.

Chapter 3

Functional Genomics of Inositol Metabolism

Javad Torabinejad and Glenda E. Gillaspy
Department of Biochemistry, Virginia Tech, 306 Fralin Biotechnology Center, Blacksburg, VA 24061, USA

1. INTRODUCTION

myo-Inositol (inositol) is synthesized by both eukaryotes and prokaryotes by what is unofficially known as the Loewus pathway. This single documented biosynthetic route to produce inositol (Figure 1) resulted from pioneering work in the 1960s from several biochemists including the substantial efforts of Frank Loewus (Loewus and Murthy, 2000). The delineated pathway of synthesis of inositol from glucose-6-phosphate involves the action of the 1L-*myo*-inositol 1-phosphate synthase (MIPS; EC 5.5.1.4) (Majumder *et al.*, 1997) and *myo*-inositol monophosphatase (IMPase; EC 3.1.3.25) (Parthasarathy *et al.*, 1994). At first, the reaction catalyzed by these two enzymes seems to be more complex than required for synthesis of a polyol: Glucose-6-phosphate must first be converted to inositol 1-phosphate by an internal oxidation, enolization, followed by an aldol condensation and reduction by NADH, all of which are catalyzed by MIPS (Jin *et al.*, 2004). Then the IMPase enzyme dephosphorylates inositol 1-phosphate resulting in what is called *free* inositol. Inositol is required for many cellular processes, especially in eukaryotes. In yeast and animal cells, inositol is primarily incorporated into phosphatidylinositol (PI), phosphatidylinositol phosphates (PIP)s, and inositol phosphates that function in signal transduction pathways (Odorizzi *et al.*, 2000; Payrastre *et al.*, 2001; Tolias and Cantley, 1999). Synthesis of other important cellular components such as glycerophosphoinositide anchors and sphingolipids also utilize the carbon skeleton of inositol (Dunn *et al.*, 2004; Loertscher and Lavery, 2002; Sims *et al.*, 2004). In addition to the use of phosphoinositides for their signaling processes (Meijer and Munnik, 2003; Stevenson *et al.*, 2000), plants in particular utilize free

A. Lahiri Majumder and B. B. Biswas (eds.), Biology of Inositols and Phosphoinositides, 47–70
© 2006 *Springer. Printed in the Netherlands.*

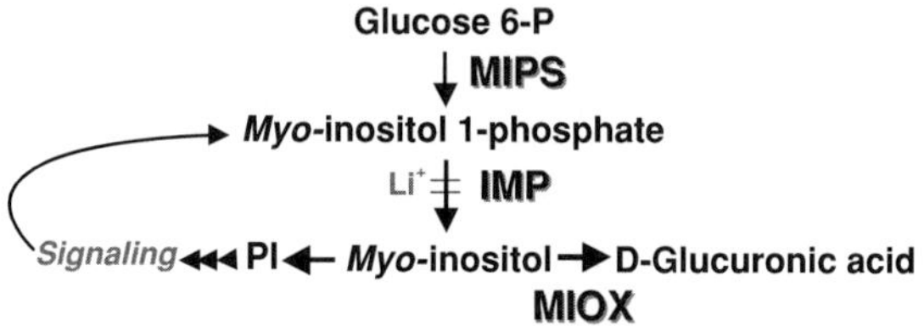

Figure 1. Inositol synthesis and catabolism. Inositol is synthesized from glucose-6-phosphate (Glucose 6-P) by the action of L-*myo*-inositol 1-phosphate synthase (MIPS) and *myo*-inositol monophosphatase (IMP). IMPase is also required for the last step of inositol (1,4,5)P_3 second messenger breakdown. The first step in inositol catabolism utilizes the *myo*-inositol oxygenase (MIOX) enzyme, which cleaves the inositol ring to form D-glucuronic acid.

inositol to synthesize several other crucial cellular compounds including those involved in hormone regulation (indole acetic acid–inositol conjugates), stress tolerance (ononitol, pinitol), and oligosaccharide synthesis (galactinol) (Loewus and Murthy, 2000). Moreover, plants also utilize large amounts of the phosphorus-storing inositol hexakisphosphate, $InsP_6$, which is also synthesized in other organisms. $InsP_6$ has been recently shown to modulate the activities of several chromatin-remodeling complexes *in vitro* which can alter gene expression (Shen *et al.*, 2003). With respect to $InsP_6$, it is worth noting that a third enzyme of importance, named *myo*-inositol kinase (MI Kinase), can phosphorylate inositol, and this is regarded as a possible beginning step of $InsP_6$ synthesis (English *et al.*, 1966). This phosphorylation by MI Kinase, which seems to counteract the dephosophorylation by IMPase has resulted in some to label this part of the pathway a futile cycle. However, it is the belief of many who study this pathway that complex regulatory mechanisms in eukaryotic cells control how free inositol flows into downstream pathways.

An additional factor that is important in the maintenance of the free inositol pool is the enzyme that oxidizes inositol to D-glucuronic acid. This enzyme, *myo*-inositol oxygenase (MIOX; EC 1.13.99.1), has been purified and its gene cloned recently (Arner *et al.*, 2001; Lorence *et al.*, 2004; Reddy *et al.*, 1981). The identification of the MIOX enzyme from various organisms thus creates the opportunity to understand both the synthesis and the catabolism of inositol in organisms and to unravel the mechanisms involved. Because the genes encoding MIPS, IMPase, and MIOX enzymes are conserved and somewhat easily identified in eukaryotes, the complex regulation of inositol synthesis and metabolism can now be dissected at the molecular level. In particular, the tools of functional genomics can be applied to this pathway, so that we can begin to understand gene and enzyme function. This chapter will highlight the groundwork and progress in using functional genomic tools to dissect inositol synthesis and catabolism. Of special importance are the questions of MIPS, IMP, and MIOX gene number and diversity, whether these genes encode biochemically distinct enzyme isoforms, how these genes are regulated, and the impact of individual MIPS, IMP, and MIOX genes on inositol synthesis and catabolism.

2. MIPS

2.1 MIPS gene diversity

MIPS is generally regarded as the rate-limiting enzyme in inositol synthesis. As such, it is an important enzyme for eukaryotes, and as it turns out also for prokaryotes. Since an excellent phylogenetic analysis has been recently published (Majumder *et al.*, 2003) and Chapter 13 of this volume deals with the known and putative MIPS genes, only a brief discussion is presented here. MIPS genes from many species of prokaryotes and eukaryotes have been isolated and studied (Hirsch and Henry, 1986; Johnson and Sussex, 1994; Wong *et al.*, 1982). Unique amongst all of these data is the fact that only plants appear to contain a gene family encoding different MIPS isoforms. In crop species, such as corn and soybean (which contain larger genomes), there are at least seven and four gene members, respectively (Hegeman *et al.*, 2001). In contrast, our analysis of the *Arabidopsis thaliana* (Arabidopsis) genome indicates the presence of three MIPS genes. Therefore, the multiplicity of MIPS genes may indicate the presence of a complex temporal and/or spatial regulation of the first step of inositol synthesis in plants.

2.2 MIPS gene expression

Although a compelling picture of regulation of inositol synthesis has not been forthcoming from any multicellular eukaryote, important studies on this pathway in yeast (*Saccharomyces cerevisiae*) have been of interest (Majumder *et al.*, 1997). Inositol auxotrophic (*ino*) mutants exist which do not synthesize inositol and thus are defective in membrane PIs required for viability. The *ino1* mutants contain a dysfunctional copy of the single yeast MIPS gene (Dean-Johnson and Henry, 1989), while *ino2* and *ino4* encode transcriptional activators of *ino1* (Ambroziak and Henry, 1994). A negative regulatory protein, Opi1 represses *ino1* transcription when inositol levels are high (White *et al.*, 1991). Another regulator, *reg1*, identified in a suppressor screen, functions in the glucose availability pathway, linking glucose metabolism and inositol synthesis (Ouyang *et al.*, 1999). Our genome analyses of the available plant species indicate that there are no strong candidates for *ino2*, *ino4*, Opi1, or *reg1* homologs in the plant kingdom. Thus, although this does not rule out the possibility that transcriptional activators and repressors are involved in regulation of inositol synthesis in plants, currently there is no evidence to support this. Further, many of the inositol metabolites that are produced in plant cells are not present in yeast, thus plants may utilize a more complex regulatory network to balance inositol synthesis with the varying inositol needs of the cell.

Expression studies of plant MIPS genes have revealed the possibility of specialized roles for individual MIPS isoforms (Dean-Johnson and Wang, 1996; Hegeman *et al.*, 2001; Smart and Fleming, 1993; Yoshida *et al.*, 1999; Yoshida *et al.*, 2002). Transcription of individual MIPS genes is regulated by development (Dean-Johnson and Wang, 1996), some sugars (Yoshida *et al.*, 2002), hormones (Smart and Fleming, 1993), photoperiod (Hait *et al.*, 2004; Hayama *et al.*, 2002), and salt stress (Hait *et al.*, 2004; Ishitani *et al.*, 1996). An *in silico* expression analysis of the three Arabidopsis genes can easily be performed using the MPSS database (http://mpss.udel.edu/). Our analysis reveals that for the most part, AtMIPS1 is expressed in siliques, leaves, and inflorescences, and AtMIPS2 shows a similar expression pattern except it has much lower expression in leaves. In contrast, the AtMIPS3 gene is mainly expressed in roots and to some degree in inflorescences including immature buds. This analysis suggests that the three Arabidopsis MIPS isoforms are differentially regulated.

In plants, regulation of inositol synthesis has been linked to stress in three separate plant species. The MIPS gene (*tur1*) from *Spirodela polyrrhiza* (duckweed) was first identified in a screen for rapidly upregulated genes during the ABA-induced morphogenic process of turion development (Smart and Fleming, 1993). Turions are dormant structures and expression of *tur-1* was localized to the stolon tissue that connects the developing turion to the node of the mother frond. In this tissue, increased MIPS expression resulted in a threefold increase in inositol, which may be involved in the synthesis of inositol phosphate signaling molecules (Flores and Smart, 2000).

Bohnert's research group has done pioneering work on inositol metabolites in plants that are thought to act as osmolytes. These compounds are derived from inositol and include ononitol (4-*O*-methyl inositol) and pinitol (1D-4-*O*-methyl-*chiro* inositol) (Nelson *et al.*, 1998). Both molecules are synthesized from inositol after action of an inositol methyltransferase (IMT) (Rammesmayer *et al.*, 1995) and an inositol epimerase, respectively. Production of these molecules is tightly linked to salt stress in the halophyte, *Mesembryanthemum crystallium* (ice plant); after salt stress, pinitol can constitute over two-thirds of the soluble carbohydrate fraction (Paul and Cockburn, 1989). Bohnert's group found that salt stress increased transcription of the two ice plant MIPS genes and the single IMT gene by fivefold (Ishitani *et al.*, 1996). In contrast, in Arabidopsis, which is a glycophyte, upregulation of inositol synthesis via increased MIPS expression in response to salt stress has not been documented. Thus it is possible that halophytic plants have a unique regulation of the inositol synthesis pathway allowing them to increase inositol levels in response to stress (Ishitani *et al.*, 1996). Recently, the *PINO1* gene was isolated for a salt tolerant MIPS from *Porteresia coarctata*, a halophytic rice species (Majee *et al.*, 2004). Transgenic tobacco plants containing the *PINO1* gene grew at high salt concentrations and accumulated inositol compared to unstressed plants.

Lastly, in rice, both ABA and sugar synergistically upregulate a MIPS gene in cultured cells from scutellum (Yoshida *et al.*, 2002). Together, the data from aforementioned studies indicate that stress or ABA increases expression of MIPS genes in different plants. Since the MIPS enzyme catalyzes the first committed step in inositol synthesis, upregulation at this point in the pathway can act to increase inositol production. Furthermore, many of the inositol metabolites that are produced in plant cells are not present in yeast, thus plants may utilize a complex regulatory network to balance inositol synthesis with the varying inositol needs of the cell.

Compared to plants and yeast, little is known about the regulation of MIPS expression in mammals. MIPS has been isolated from various mammalian tissues (Maeda and Eisenberg, 1980). Earlier work had demonstrated the regulation of MIPS activity was influenced by a number of drugs and conditions in the reproductive organs of rats (Eisenberg, 1967; Hasegawa and Eisenberg, 1981; Rivera-Gonzalez *et al.*, 1998; Whiting *et al.*, 1979). For example, Eisenberg (1967) measured the MIPS activity in various rat organs and found that the activity was equal for the samples obtained from brain, liver, lung, and spleen. He also demonstrated that the activity in these organs was only 13% of the total activity in the testis. Another group investigated the effect of Li^+ on MIPS and IMP gene expression in tissue obtained from mouse frontal cortex and hippocampus. Both MIPS and IMPaseA1 mRNA were upregulated, indicating regulation of both gene products in the hippocampus (Shamir *et al.*, 2003). Guan *et al.* (2003) cloned a full-length cDNA of the human MIPS gene and investigated enzyme activity and mRNA expression in various tissues exposed to a number of chemical compound and hormones. A higher level of MIPS mRNA was found in testis, ovary, heart, placenta, and pancreas compared to other tissues such as thymus, skeletal muscle, colon, and blood leukocytes. These authors suggested that a G-protein might be involved in signal transduction in the regulation of the human MIPS gene. Li^+ exposure resulted in a 50% suppression of the MIPS mRNA expression in cultured HepG2 cells, suggesting a possible feedback regulation from a higher concentration of inositol 1-phosphate due to the inhibition of IMP (Guan *et al.*, 2003). Blasting this human MIPS cDNA sequence against Arabidopsis suggests a closer identity to AtMIPS2 in plants than AtMIPS1 or AtMIPS3. It is worth noting that AtMIPS2, likewise, shows a much higher expression in the reproductive tissues.

2.3 Altering MIPS expression

There continues to be sustained interest in controlling MIPS expression to increase or decrease inositol synthesis in plants. Since inositol may function as an osmolyte in plants, increasing inositol levels in transgenic plants could lead to better stress protection in crop species (Bohnert *et al.*, 1995). Further, decreases in inositol synthesis may allow for decreased production of environmental contaminants such as $InsP_6$ (Hegeman and Grabau, 2001). In support of this, a *Glycine max* (soybean) mutant has been described that alters lysine 396 (lys396asn) of a soybean MIPS isoform (Hitz *et al.*, 2002). This amino acid alteration was shown to reduce MIPS activity by 90%, and lower seed inositol levels to 20% of wildtype levels. $InsP_6$ levels in the mutant were reduced by one-half and inorganic phosphate was concomitantly increased. These results indicate that changes in inositol synthesis can impact the synthesis of inositol metabolites. Thus strategies aimed at manipulating inositol synthesis via expression of transgenes are likely to be successful in altering seed $InsP_6$ levels.

The first reported transgenic plant altered in inositol metabolism was an Arabidopsis plant transformed with the duckweed *tur-1* gene (Flores and Smart, 2000). Constitutive expression of the *tur-1* gene utilizing the 35S cauliflower mosaic virus promoter resulted in a fourfold increase in inositol levels. These MIPS overexpressing plants had no obvious phenotype, and were not tolerant to salt stress as might be expected. These results indicate that engineering salt tolerance via upregulation of inositol synthesis may require more than just alterations in MIPS expression. However, they also indicate that MIPS expression can be altered to increase the inositol pool size *in planta*.

A transgenic reduction in MIPS activity has also been reported. Antisense suppression of MIPS in transgenic potato resulted in a sevenfold reduction in inositol, and resulted in reduced apical dominance, altered leaf morphology, precocious leaf senescence and a decrease in overall tuber yield. The altered leaf morphology was found to be due to cell enlargement in the leaves. The authors of this study found that in addition to inositol, galactinol, and raffinose levels were reduced. In contrast, increases were observed for hexose phosphates, sucrose, and starch. The authors further concluded that many different inositol metabolites may have contributed to the development of the phenotypes observed (Keller *et al.*, 1998).

3. IMPase AND IMPase-RELATED PROTEINS

3.1 Regulation by Li^+

The IMPase gene was first cloned from bovine brain tissue, and this work was facilitated by many excellent studies on the IMPase enzyme from Lily pollen and other tissues (Chen and Charalampous, 1966; Eisenberg, 1967; Eisenberg *et al.*, 1964; Hallcher and Sherman, 1980; Loewus *et al.*, 1984; Majerus *et al.*, 1999). The impetus for IMPase gene identification from brain tissue centers around the unique role of IMPase in both *de novo* inositol synthesis and PI signaling. Figure 1 shows that PI signaling requires IMPase to dephosphorylate the last product of inositol $(1,4,5)P_3$ ($InsP_3$) second messenger catabolism. This unique position in both pathways may allow for coordination of inositol synthesis and $InsP_3$ recycling. The best evidence for this comes from the use of LiCl in both biochemical and physiological studies. In animals, yeast, and plants, IMPase activity can be inhibited by Li^+. In certain animal systems and in yeast, application of LiCl not only alters development, but also reduces the free pool of inositol (Moore *et al.*, 1999; Maslanski *et al.*, 1992; Shaldubina *et al.*, 2002; Vaden *et al.*, 2001). For example, when *Xenopus laevis* embryos are treated with Li^+ a process referred to as dorsalization occurs (Kao *et al.*, 1986) which is accompanied by a decrease in both free inositol and $InsP_3$ levels (Maslanski *et al.*, 1992). Dorsalization is thought to be the result of inhibition of PI signaling on the ventral side of the embryo, thus allowing the dorsal information to be recapitulated throughout the embryo. The dorsalization effects of Li^+ can be prevented or "rescued" by co-introduction of inositol with Li^+ (Busa and Gimlich, 1989). This implicates the reaction catalyzed by IMPase as the target of Li^+ action in frogs.

In humans, Li^+ therapy is used in the clinical treatment of certain bipolar disorders (*e.g.*, manic depression), where its action is presumed to be inhibition of abnormal PI signaling, due to a decrease in inositol substrate availability (Lachman and Papolos, 1989). More recent studies have indicated the presence of other Li^+-sensitive enzymes in animals that may also function as *in vivo* targets for Li^+ action in animal development. These include the inositol polyphosphate-1 phosphatase (IPPase) (York *et al.*, 1995) and glycogen synthase kinase-3β (Klein and Melton, 1996). However, why inositol would rescue the phenotypes putatively caused by the inhibition of such enzymes remains unclear; thus, this area is still ripe for experimentation that could reconcile these issues.

3.2 The inositol P domain: Phylogenetic analysis

Interestingly, the amino acid sequence or domain that imparts Li^+-sensitivity is known (Neuwald *et al.*, 1991; York *et al.*, 1995). The Inositol P domain (as defined by the Pfam protein domain web-based database: www.pfm. wustl.edu) is a 155 amino acid residue domain found in many prokaryotic and eukaryotic proteins. We used amino acid sequences from *Lycopersicon esculentum* (tomato) (LeIMP1), *Escherichia coli* (EcSuhB), and cow (BtIPPase) IMPases in BLASTp searches of both fully and partially sequenced genomes. We identified over 73 proteins from various organisms that contain the Inositol P domain. We then utilized ClustalX and PAUP 4.0b software to create an amino acid alignment and phylogenetic tree, respectively, of these proteins (Figure 2). Our results indicate that there are four different types of proteins that contain the Inositol P domain: fructose-1,6-bisphosphate 1-phosphatases (FBPases), IMPase enzymes, the IMP-related enzymes, and the IPPases. For clarity, we removed the FBPases from our future analyses as these proteins are more distantly related compared to the other three types of proteins. However, one protein annotated as an FBPase (NpFbpase1) was included as it is more related to

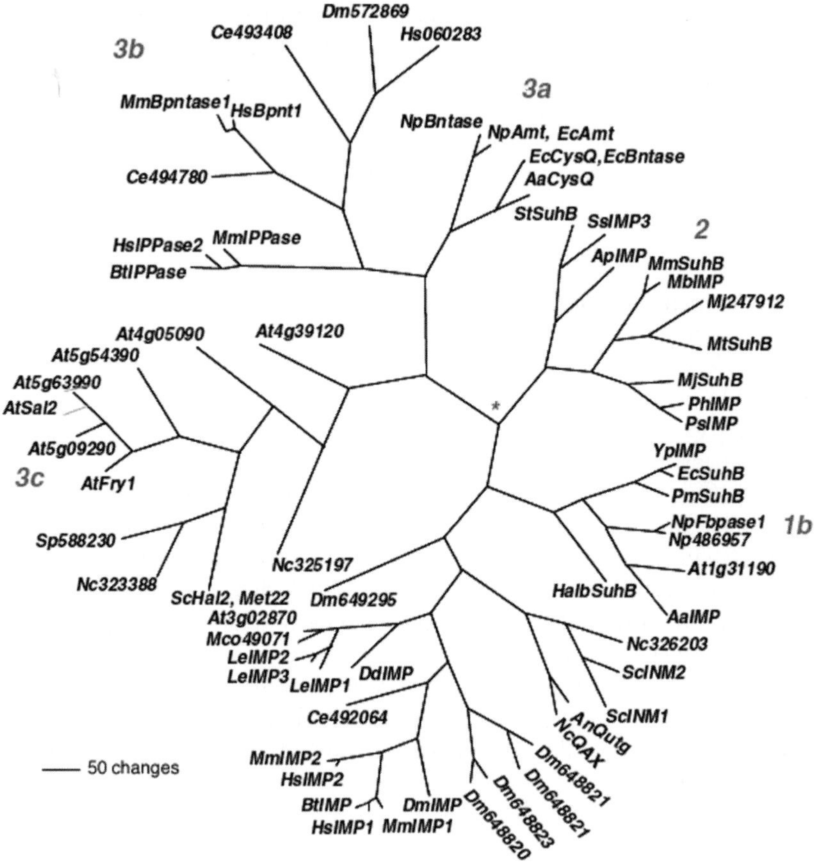

Figure 2. Phylogenetic analysis of proteins containing the Inositol P Domain. ClustalX and PAUP 4.0b were used to create amino acid alignments and an unrooted bootstrapped phylogenetic tree. Assignment of branch numbers are indicated by numbers. The asterisk denotes a possible placement for a common ancestor protein. The first two initials of each protein name represent the genus and species. Numbers that follow are from Genbank protein accession numbers or in the case of Arabidopsis, the gene number. A complete list of accession numbers can be found at (http://www.biochem.vt.edu/gillaspy/Research.html). *Abbreviations*: Ap, *Aeropyrum pernix*; Aa, *Aquifex aeolicus*; An, *Aspergillis nidulans*; Bt, *Bos taurus*; Ce, *C. elegans*; Dm, *D. melanogaster*; Dd, *Dictyostelium discoideum*; Ec, *E. coli*, Hal, *Halobacterium* sp. NRC-1; Hs, *Homo sapiens*; Le, *Lycopersicon esculentum*; Mc, *Mesembryanthemum crystallium*; Mj, *Methanococcus jannaschii*; Mb, *Methanosarcina barkeri*; Mm, *Methanosarcina mazei*; Mt; *Methanothermobacter thermautotrophicus*; Mm, *Mus musculus*; Nc, *Neurospora crassa*; Np, *Nostoc punctiforme*; Ns, *Nostoc* sp. PCC 7120; Pm, *Pasteurella multocida*; Pa, *Pyrococcus abyssi*; Pf, *Pyrococcus furiosus*; Ph, *Pyrococcus horikoshii*; Ps, *Pirellula* sp.; Rn, *Rattus norvegicus*; Sc, *S. cerevisiae*; Ss, *Sulfolobus solfataricus*; St, *Sulfolobus tokodaii*; Sp, *Schizosaccharomyces pombe*; Yp, *Yersinia pestis*.

the IMPase proteins than to other characterized FBPases. It should be noted that we found many incorrect annotations in the NCBI gene database, so we urge caution when using gene annotations without supplementary analyses. For example, any new protein identified that contains the Inositol P domain is annotated as "inositol monophosphatase family protein," yet the protein may in fact be more similar to the IPPases. Our final tree supports division of IMPases, IMP-related enzymes and IPPases into three main categories as seen by the presence of three main branches (Figure 2).

3.2.1 IMP gene and protein diversity

Previously characterized IMPases are found on branch 1a and 1b (Figure 2). This includes the bovine gene product (BtIMPase), and the gene products encoding the two human and mouse IMPases (HsIMPase1 and HsIMPase2, MmIMPase1 and MmIMPase2) (McAllister *et al.*, 1992; Shamir *et al.*, 2001), and the three tomato IMPases (LeIMPase1, 2 and 3) (Gillaspy *et al.*, 1995). The animal IMPase proteins cluster into one clade that includes one putative IMPase each from *Drosophila melanogaster* (Drosophila) (DmIMPase), and *Caenorhabditis elegans* (Ce492064). Since the sequences of the Drosophila and *C. elegans* genomes are complete, we can conclude that worms contain a single gene that encodes IMPase, whereas Drosophila contains five other IMP gene products that are located on other clades of this tree. Whether these genes are expressed, and whether they encode active IMPase enzymes is unknown, but there exists the possibility that IMPase isoform diversity plays a unique role in Drosophila.

The multigene family of tomato IMP is found in a separate clade along with the single *Dictyostelium discoideum* IMPase (Van Dijken *et al.*, 1996). We have confirmed IMPase activity for the single Arabidopsis IMPase (At3g02870; G. Gillaspy, unpublished data), while the ice plant IMPase (Mco49071) has not yet been examined. The tomato IMPs have been characterized (Gillaspy *et al.*, 1995) and the differences in gene number between tomato and Arabidopsis may be due to overall genome size. A third clade of IMPase enzymes includes the two *S. cerevisiae* IMPases (ScINM1 and ScINM2) which have been shown to encode active IMPase (Lopez *et al.*, 1999). The closest relatives to these yeast proteins are the putative IMPases from other fungi (*Aspergillis nidulans* and *Neurospora crassa*).

Prokaryotic proteins related to IMPase have been characterized and a group of these are found on branch 1b. The *E. coli* SuhB gene was isolated as a suppressor of many diverse temperature-sensitive mutations (Chang *et al.*, 1991; Matsuhisa *et al.*, 1995). Surprisingly, suppression of these unrelated phenotypes does not require active SuhB protein (Chen and Roberts, 2000). These authors characterized the SuhB protein and found that it too catalyzes the hydrolysis of inositol 1-phosphate with similar kinetics to the animal IMPases. However, EcSuhB can exist as a monomer unlike the dimers of other IMPases, and is more hydrophobic than animal and plant IMPases. Together, these data suggest that the suppressive ability of SuhB may rely on these physical differences, rather than catalytic differences (Chen and Roberts, 2000). Of special interest to us is the presence of a single Arabidopsis gene within this clade (At1g31190). No other multicellular eukaryote contains both an IMPase and SuhB homolog, which suggests that plants differ in their IMPase activity needs as compared to animals.

3.2.2 IMPase-related gene and protein diversity

The second main branch of our tree contains a group of archael IMPase-related proteins (branch 2). Annotation issues are apparent in this group as proteins are annotated either as IMPase or SuhB homologs. This group appears to be phylogenetically distinct from the prokaryotic IMPases and SuhBs (branch 1b). However, there are currently no biochemical data that would explain why these IMPase-related proteins are distinct from their counterparts on branches 1a and 1b. It has been shown that the *Methanococcus jannaschii* IMPase (MjIMP) has a broader substrate specificity than other IMPases (Chen and Roberts, 1998), but this quality is probably shared by other IMPases from thermophiles (Chen and Roberts, 1999), some of which are present on branch 1b (*e.g.*, AaIMP). Of special interest on this branch is the Mj247912 protein. This predicted protein has a unique protein domain arrangement. In addition to the Inositol P domain, this protein also contains an NAD/ATP kinase domain. This juxtaposition of domains appears to be unique amongst all genomes sequenced to date, including several other sequenced archael genomes.

3.2.3 IPPase gene and protein diversity

IPPase enzymes catalyze the removal of 1-phosphate from inositol (1,4)P_2 substrates, which is a required step to recycle inositol in the cell (York *et al.*, 1995). The third branch of our tree contains the characterized IPPases from humans (HsIPPase) (York *et al.*, 1993), cows (BtIPPase) (York and Majerus, 1990), and mice (MmIPPase). This small clade is located on branch 3b along with additional animal proteins such as the HsBnt1 protein. The HsBpnt1 protein is a dual function enzyme that has both IPPase and 3′-phosphoadenosine-5′-phosphate (PAP) phosphatase (PAPase) activities (Spiegelberg *et al.*, 1999). PAPases are thought to be crucial for the function of enzymes sensitive to inhibition by PAP, such as sulfotransferases and RNA processing enzymes (Dichtl *et al.*, 1997). These enzymes sometimes are annotated or named as bisphosphonucleotidases (Bpnt or Bpntase). Our tree indicates a separation between the characterized IPPases and the dual function IPPases/PAPases.

Branch 3a contains related enzymes from algae and prokaryotes and this group most likely has only PAPase activity as deduced from studies of the

E. coli CysQ protein. The EcCysQ gene was first identified as a required gene in cysteine biosynthesis (Neuwald *et al.*, 1992). Since PAP is an intermediate in assimilation of sulfate during cysteine biosynthesis, it has been speculated that the CysQ protein breaks down PAP to prevent toxic amounts from accumulating in the cell (Neuwald *et al.*, 1992).

The last part of branch 3 contains a group of plant and fungal proteins that we predict are also dual function PAPases/IPPases. The yeast Hal2 protein (ScHal2, Met22) was isolated on the basis of imparting improvement of growth under NaCl and LiCl stresses (Murguia *et al.*, 1996). PAPase activity is required for achieving salt tolerance, although ScHal2 also contains an IPPase activity. AtFry1 (or AtSal1) is a closely related Arabidopsis gene whose protein product exhibits PAPase activity and a less efficient IPPase activity (Quintero *et al.*, 1996). Furthermore, the *Km* value for inositol-1,4-bisphosphate is much higher (90 μM) than the *Km* value for PAP (2–10 μM). The AtSal2 protein has also been shown to possess both PAPase and IPPase activities (Gil-Mascarell *et al.*, 1999). Interestingly, AtFry1 was subsequently identified genetically as a mutation that disrupts $InsP_3$ signaling in plants (Xiong *et al.*, 2001). Whether the PAPase activity of AtFry1 plays a role in the mutant phenotype is still unclear. However, one fact stands out from the phylogenetic tree: plants contain many more of these PAPases/IPPases than do other multicellular eukaryotes. This may point to a unique role of the encoded enzymes or a complex regulation of individual genes in plants.

3.3 IMP gene expression

As described earlier, in the yeast *S. cerevisiae*, Li^+ is known to inhibit IMPase and decrease mass inositol levels (Vaden *et al.*, 2001). Yeast contains two IMP genes (INM1 and INM2). INM1 gene transcription has been shown to be upregulated by *myo*-inositol, but not by *epi*-inositol, and downregulated by Li^+ (Murray and Greenberg, 1997; Murray and Greenberg, 2000; Shaldubina *et al.*, 2002). It is speculated that INM1 expression is tied to the need for recycling inositol from the breakdown products of PI signaling molecules. Thus, when there is a surplus of inositol and when PIP synthesis is high, INM1 expression may increase to provide INM1 protein to facilitate inositol recycling. Accordingly, Li^+ treatment may decrease INM1 expression through a feedback inhibition mechanism involving a decreased need for inositol recycling (Murray and Greenberg, 2000).

In plants, the three different tomato IMP genes are differentially and developmentally regulated (Gillaspy *et al.*, 1995). Expression of the LeIMP2 gene is spatially restricted to a discreet domain within the plant and is found only within epidermal and cortex cells of specific stem/leaf junctions in an abaxial-specific pattern and in the shoot apical meristem. Furthermore, unlike yeast, inositol and Li^+ repress expression of the LeIMP2 promoter (Styer *et al.*, 2004).

Very little is known about IMP or IMP-related gene expression in prokaryotes. Only two published reports exist. First, in *Rhizobium leguminosarum*, the IMP genes have been shown to be regulated in response to various environmental factors (Janczarek and Skorupska, 2004). Second, studies show that the *E. coli* SuhB gene is autoregulated and that the SuhB transcript in the SuhB mutant has a much greater half-life than that of the wildtype strain. Thus, the SuhB protein is implicated in the control of gene expression by modulating mRNA turnover (Inada and Nakamura, 1996.)

In mammals, the mouse IMPA1 gene is upregulated in the hippocampus. When mice are exposed to Li^+, both the MIPS and the IMPA1 gene become upregulated (Shamir *et al.*, 2003). This may reflect a compensatory response of both genes to inositol depletion. In humans, two pieces of evidence link IMPA2 gene expression with bipolar disorder, at least in males. IMPA2 mRNA levels of B lymphoblast cell lines from male patients with bipolar disorder and high Ca^{2+} were lower than those from healthy males or females with bipolar disorder. However, IMPA2 mRNA levels from postmortem temporal cortex of males with bipolar disorder were significantly higher as compared with male controls (Yoon *et al.*, 2001). Collectively, these observations suggest a potential complex role of IMP gene expression in animal brain tissue and in the pathophysiology of bipolar disorder.

3.4 Altering IMP expression

Since the inositol depletion hypothesis (Berridge *et al.*, 1989) first suggested that a reduction in IMPase activity might deplete normal free inositol pools and limit production of key substrates for PI signaling, investigators have sought to decrease IMP expression. In yeast a double gene disruption in ScINM1 and ScINM2 had no apparent effect and mutants were not inositol auxotrophs (Lopez *et al.*, 1999). Overexpression of yeast IMPs increased Li^+ and sodium tolerance, elevated intracellular Ca^{2+}, and reduced the intracellular accumulation of Li^+. This phenotype could be blocked by a mutation in a Ca^{2+}-activated-extrusion ATPase, but it was not affected by inositol supplementation (Lopez *et al.*, 1999). This suggests that the yeast IMPs are limiting for the optimal operation of the inositol cycle of Ca^{2+} signaling, and that the yeast IMPs can indirectly modulate Ca^{2+}-activated pumps.

In plants, the IMP gene products have been found to be Li^+-sensitive (Gillaspy *et al.*, 1995), although the effects of Li^+ on development and free inositol levels may be more complex than in animal embryos or yeast. LiCl application disrupts the development of tomato seedlings by inducing hypertrophic growth on stems and suppressing postembryonic shoot development (Gillaspy and Gruissem, 2001). These phenotypes cannot be reversed by co-introduction of inositol with LiCl. It has been speculated that inositol cannot reverse the observed effects of Li^+ application in whole seedlings because (1) either multiple enzymes and physiological targets are affected and some of these do not involve inositol metabolism; or (2) the spatial distribution of inositol and complexity of inositol metabolic events in the seedling limit the ability for added inositol to reach areas affected by Li^+ application. A clear way to resolve these issues would be to limit IMP expression in the plant either through antisense suppression or isolation of an IMP "knock-out" loss-of-function mutant, and measure the effect on both inositol synthesis and PI signaling.

In prokaryotes, the focus of IMP alterations has not been on signaling, but has lead investigators to the cell wall and to RNA processing. A mutation in the pssB gene of *R. leguminosarum* caused exopolysaccharide overproduction, suggesting that the pssB gene is normally involved in regulating exopolysaccharide synthesis (Janczarek *et al.*, 1999). In *Mycobacterium smegmatis*, an IMP mutation slowed growth, increased clumping in liquid culture, and increased resistance to several antibiotics. This mutant contained a 50% reduction in phosphatidylinositol dimannoside, a precursor of cell wall material (Nigou and Besra, 2002). Thus, IMP function in prokaryotes may be related to cell wall production/regulation. Another interesting story surrounds the *E. coli* SuhB gene. The SuhB mutant by itself is cold-sensitive, and results in defects in protein synthesis (Chang *et al.*, 1991). Cold-resistant suppressors were isolated and mapped to the gene encoding RNase III, which is a double-strand RNA processing enzyme. Two known RNase III mutations, both defective in RNA cleavage activity, restored growth of SuhB mutants. These mutations did not alter the level of SuhB expression, which suggests that RNase III and SuhB proteins bind to one another and are antagonistic (Inada and Nakamura, 1995). Inada and Nakamura have suggested that the SuhB protein has an anti-RNase III activity that affects mRNA stability (Inada and Nakamura, 1996). It is not known if IMP is involved in similar activities in eukaryotes, however.

There have been no reports of altering IMP gene expression through the use of transgenes in animals. Of special interest would be an IMP "knock-out" mouse, which could be used to test Berridge's inositol depletion hypothesis (Berridge *et al.*, 1989). Data concerning the downregulation of IMPase activity in animals via Li^+ action and its relationship to bipolar disorder has already been discussed in this chapter. However, there is an interesting account which links IMP function to another human disorder, galactosemia. The human IMP gene was isolated as a genetic suppressor of galactose toxicity in a gal7 yeast strain that lacks galactose 1-phosphate uridyl transferase (Mehta *et al.*, 1999). Interestingly, the suppression by IMP was sensitive to Li^+. In this mutant and in humans with galactosemia, high galactose 1-phosphate impedes normal physiology. The ability of HsIMP to suppress galactose toxicity has led to speculation that high galactose 1-phosphate levels during galactosemia impair IMPase activity by acting as an effective substrate competitor. If so, this would make galactosemia a disorder in IMP function.

4. MIOX: OXIDATION OF INOSITOL

Since the oxidation of inositol can also affect the *in vivo* levels of inositol, it is important to mention the oxidative cleavage of the inositol ring which produces D-glucuronic acid (Figure 1). Loewus was the first to address the question of what happens to inositol once synthesized by the cell (Roberts *et al.*, 1967). Labeling maize root tips with inositol-2-(^{14}C) showed that between 50% and 71% of the incorporated label was oxidized to D-glucuronic acid and acted on by the inositol oxidation pathway to produce pectic cell wall polysaccharides. These experiments have been repeated on numerous types of plant samples with similar results (Loewus and Murthy, 2000). The MIOX enzyme was first measured in rat kidney (Charalampous and Lyras, 1957) and later shown to be present in plants as well (Loewus *et al.*, 1962). This enzyme directly and irreversibly oxidizes inositol in the presence of molecular oxygen without the requirement for other cofactors.

There is a continuing interest in inositol catabolism as it relates to research on the diabetic state. Rats given streptozotocin develop insulin resistance and this model system is often used to study the diabetic state in mammals. Changes in both *myo*- (Zhu and Eichberg, 1990) and *chiro*-inositol (Larner, 2002) have been noted in diabetic animals. Depletion of *myo*-inositol is associated with nephropathy, retinopathy, neuropathy, and cataracts. There appears to be differences in how various tissues control inositol oxidation with respect to diabetes. Levels of inositol in diabetic animals are lower in the central nervous system (Whiting *et al.*, 1979), whereas renal clearance of inositol is higher (Whiting *et al.*, 1979). At the same time, kidney MIOX activity levels have been shown to decrease (Whiting *et al.*, 1979). Thus, the question of inositol catabolism and diabetes is a complex one, to which functional genomics approaches will be valuable.

4.1 MIOX enzyme diversity

The gene encoding MIOX was first isolated from pigs via purification of MIOX activity (Arner *et al.*, 2001). Similar work was performed in *Cryptococcus neoformans*, the human pathogenic yeast, which requires a voluminous polysaccharide capsule that contains D-glucuronic acid for pathogenicity (Doering, 2000). The MIOX gene sequences from pigs and *C. neoformans* were used to identify a plant (Arabidopsis) MIOX gene product which has been shown to be catalytically active *in vitro* (Lorence *et al.*, 2004). Using these sequences, we performed a comprehensive search for MIOX genes from several sequenced and other genomes and have phylogenetically analyzed a group of 21 proteins that we identify as MIOX isoforms (Figure 3). All identified, putative MIOX genes encode proteins that range from 217 to 354 amino acids in length. Each of these proteins contains a conserved domain referred to as the DUF706 domain in the Pfam database. Since all proteins containing the

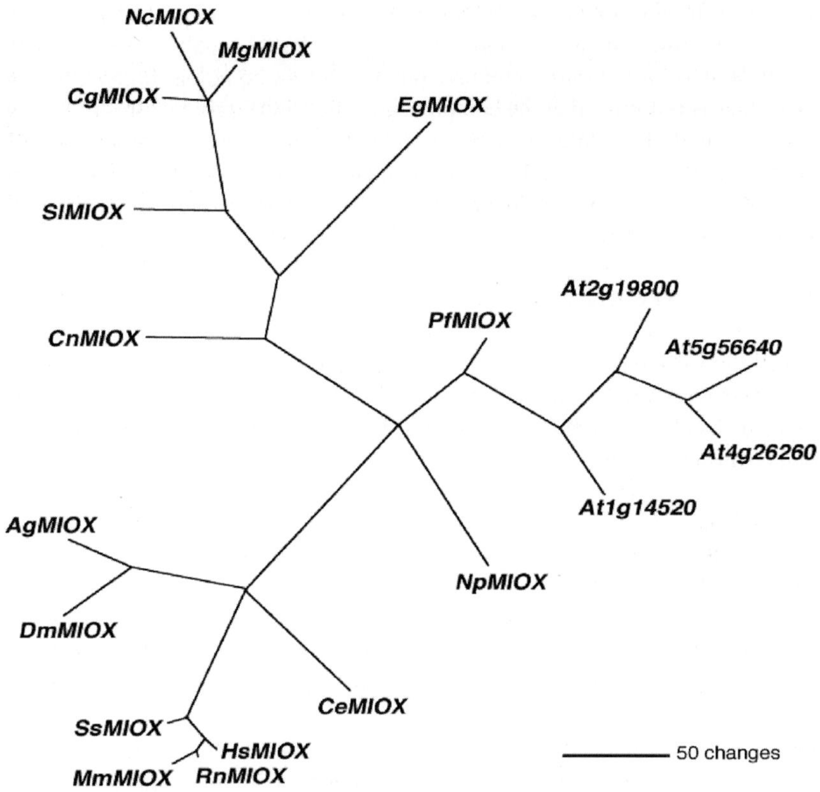

Figure 3. Phylogenetic analysis of MIOX proteins. ClustalX and PAUP 4.0b were used to create amino acid alignments and an unrooted, bootstrapped phylogenetic tree. Assignment of branch numbers are indicated by numbers. The asterisk denotes a possible placement for a common ancestor. The first two initials of each protein name represent the genus and species. Numbers that follow are from Genbank protein accession numbers or in the case of Arabidopsis, the gene number. A complete list of accession numbers can be found at (http://www.biochem.vt.edu/gillaspy/Research.html). *Abbreviations:* Ag, *Anopheles gambiae*; At: *Arabidopsis thaliana*; Ce, *C. elegans*; Cg, *Colletotrichum gloeosporioides*; Cn, *Cryptococcus neoformans*; Dm, *D. melanogaster*; Eg, *Eremothecium gossypii*; Hs, *Homo sapiens*; Mm, *Mus musculus*; Mg, *Magnaporthe grisea*; Nc, *Neurospora crassa*; Np, *Nostoc punctiforme*; Pf, *Monterey pine*; Rn, *Rattus norvegicus*; Ss, *Sus scrofa*; Sl, *Sporopachydermia lactativora*.

DUF706 domain also contain significant homology to known MIOX proteins, we speculate that the DUF706 domain is the equivalent of a MIOX activity domain.

Our phylogenetic analysis revealed three main branches of MIOX proteins corresponding to animal (1), plant (2) and fungal (3) groups. Interestingly, all organisms except for plants contain only a single MIOX gene. Indeed, Arabidopsis contains four MIOX genes which are most related to the pine MIOX (Figure 3). An additional interesting fact is that yeasts such as *S. cerevisae* and

S. pombe do not contain a conserved MIOX gene, but several other mostly pathogenic fungi contain a single MIOX gene. Further, only one blue green algae, *Nostoc punctiforme*, contains a putative MIOX homolog. Taken together, this analysis indicates that the complexity of the MIOX protein family is more similar to that of the MIPS protein family in that most species contain a single MIPS and MIOX gene product. The abundance of plant MIOX isoforms is at present unclear, but may indicate a complex regulation of inositol catabolism in the plant kingdom.

4.2 MIOX gene expression

Very little work has been done of the expression of MIOX genes as they have only been recently identified. Expression of the At4g26260 gene from Arabidopsis has been examined. This gene is expressed in flowers, leaves, seedpods, and stems. Levels of this mRNA were highest in flowers (Lorence *et al.*, 2004).

4.3 Altering MIOX expression

Loewus was the first to suggest and then investigate the *myo*-inositol oxidation pathway (MIOP) (Loewus, 1965; Loewus, 1969; Loewus *et al.*, 1962; Roberts *et al.*, 1967). This pathway bears directly on two important aspects of plant physiology. While pectic cell wall polysaccharides are of vital importance to the plant cell rendering structural support (Loewus, 1965), production of ascorbic acid (AsA) from this pathway function as an antioxidant. Cell wall polysaccharides can also be produced from the sugar nucleotide oxidation pathway (SNOP) via glucose. In the SNOP, production of UDP-glucose requires the UDP-glucose dehydrogenase enzyme which is strongly inhibited by both UDP-glucuronate and UDP-xylose (Gainey and Phelps, 1975; Ordman and Kirkwood, 1977). This means that under certain conditions (*i.e.*, when inositol levels are high or glucose levels are low), the MIOP may play a major role in cell wall polysaccharide formation. This role is supported by evidence from labeling studies in pollen (Loewus and Loewus, 1980), and recent studies on Arabidopsis showing that expression of the single UDP-glucose dehydrogenase gene is virtually absent from young seedlings, thereby implicating the MIOP at this developmental stage (Seitz *et al.*, 2000). The exact contribution of each pathway (MIOP vs. SNOP) during plant development has been unexplored to date, but is an important aspect of plant physiology that deserves attention.

The second possible product of major importance from inositol catabolism is AsA. In animals (except humans), D-glucuronic acid (resulting from either the SNOP or MIOP) is converted to L-gulonic acid and then to gulono 1,4 lactone (Banhegyi *et al.*, 1997). The conversion of L-gulonic acid to AsA is catalyzed by the L-gulonolactone oxidase enzyme, which is missing in humans

(Nishikimi *et al.*, 1994). Humans, therefore, require plant-derived AsA to provide this important water-soluble antioxidant. This explains the fact that sailors crossing the Atlantic Ocean were dependent on intake of citrus containing AsA, but stowaway rats, with a functional L-gulonolactone oxidase enzyme, were not. The route to AsA synthesis in plants has been debated for many years (Smirnoff *et al.*, 2001), and remains one of the few unsolved primary metabolic pathways in plants. The animal pathway just described requires inversion of the glucose carbon skeleton. Yet from radiolabeled tracer studies in plants, it was shown that AsA produced in plants results from a non-inversion pathway (Loewus, 1963), ruling out D-glucuronic acid as a major precursor. A significant advance in this pathway was made by Wheeler *et al.* by showing that L-galactono-1,4-lactone, another potential precursor for AsA production, could be synthesized from glucose via GDP-D-mannose without the carbon skeleton conversion (Wheeler *et al.*, 1998). Further, in this elegant work, the authors essentially reconstituted the pathway *in vitro*, showing that all proposed steps were possible in Arabidopsis. The Smirnoff/Wheeler pathway was nicely confirmed at the genetic level by the identification of Arabidopsis mutants deficient in AsA (~fivefold reduction) that are impaired in the GDP-mannose pyrophosphorylase enzyme (Conklin *et al.*, 2000). Since this enzyme is required for GDP-D-Man synthesis these two studies provide ample evidence for AsA synthesis from this precursor, although they do not rule out the existence of alternate pathways. For example, Jain and Nessler expressed the rat cDNA encoding L-gulonolactone oxidase in transgenic tobacco and lettuce plants. The transgenic plants accumulated up to seven times more AsA than untransformed plants, indicating that endogenous L-gulonic acid substrate can contribute to AsA synthesis (Jian and Nessler, 2000).

A new intriguing set of facts is of interest to this story. The Arabidopsis MIOX gene has been overexpressed in transgenic Arabidopsis plants. The resulting transgenic plants contained two to threefold more AsA than control plants (Lorence *et al.*, 2004). This suggests that increased catabolism of inositol leads to a greater amount of D-glucuronic acid being shunted into AsA synthesis. This would seem to offer evidence of a direct route from inositol to AsA in plants. An alternative explanation of these data is that altering MIOX led to a decrease in free inositol that affected AsA synthesis indirectly. Levels of inositol were not measured in the MIOX overexpressing transgenic plants, so a detailed understanding of the inositol metabolism pathway in these plants awaits further study.

5. CONCLUSIONS

We have detailed the MIPS, IMP, and MIOX genes and proteins with an overall focus on determining gene function within the inositol synthesis and catabolism pathway. Each of these enzymes contributes to the overall cellular need for inositol and its breakdown products. One central theme to emerge from our analysis is that while examples exist of prokaryotes, and unicellular and multicellular eukaryotes that contain genes encoding these enzymes, the regulation and contribution to specific end products is different. That is to say that prokaryotes utilize inositol in different pathways from yeast, plants, or humans, and that their genetic diversity for these genes also differs. Another interesting finding is that plants appear to contain more complexity with respect to the numbers of genes that encode MIPS, IMPase, and MIOX enzymes. This probably reflects a more complex regulation of these genes and/or a more complex repertoire of inositol-containing end products.

There will certainly be new genes identified in the future that impact inositol synthesis and catabolism. The yeast model system, in particular, is a rich system to identify new genetic targets that regulate inositol synthesis (Chapter 6). Other enzymes yet to be purified that impact the pathway have been or are being studied currently. The MI Kinase, which phosphorylates free inositol producing inositol 1L-phosphate (English *et al.*, 1966), is important to understanding the flux of inositol into inositol phosphate production. The product of this reaction may be the substrate for subsequent phosphorylation events resulting in the production of the phosphorus-storing compound, $InsP_6$, which is abundant in seeds and in other non-plant organisms as well. MI kinase and phospho inositol kinases have been partially purified from plant, bacterial, and animal sources (English *et al.*, 1966; Majumder *et al.*, 1972). In plants, the MI kinase protein is most likely a 95 kDa soluble, ATP/Mg^{++} dependent kinase (Loewus and Loewus, 1983). Another enzyme that is important in both plants and animals is the *myo*-inositol epimerase. This enzyme has never been purified, but is speculated to convert *myo*-inositol into other epimers of inositol. Foremost is *chiro*-inositol, which has been linked to Type 2 diabetes in humans (Larner, 2002) and to stress protection in plants (Nelson *et al.*, 1998). Thus, efforts to explore existing and new avenues of inositol synthesis and catabolism hold significant promise in the coming years.

REFERENCES

Ambroziak, J., and Henry, S.A., 1994, *Ino2* and *ino4* gene products, positive regulators of phospholipid biosynthesis in *Saccharomyces cerevisiae*, form a complex that binds to the inol promoter. *J. Biol. Chem.* **269:** 15344–15349.

Arner, R.J., Prabhu, K.S., Thompson, J.T., Hildenbrandt, G.R., Liken, A.D., and Reddy, C.C., 2001, *myo*-Inositol oxygenase: Molecular cloning and expression of a unique enzyme that oxidizes *myo*-inositol and D-*chiro*-inositol. *Biochem. J.* **360:** 313–320.

Banhegyi, G., Braun, L., Csala, M., Puskas, F., and Mandl, J., 1997, Ascorbate metabolism and its regulation in animals. *Free Radic. Biol. Med.* **23:** 793–803.

Berridge, M.J., Downes, C.P., and Hanley, M.R., 1989, Neural and developmental actions of lithium: A unifying hypothesis. *Cell* **59:** 411–419.

Bohnert, H.J., Nelson, D.E., and Jensen, R.G., 1995, Adaptations to environmental stresses. *Plant Cell* **7:** 1099–1111.

Busa, W., and Gimlich, R., 1989, Lithium-induced teratogenesis in frog embryos prevented by a polyphosphoinositide cycle intermediate or a diacylglycerol analog. *Dev. Biol.* **132:** 315–324.

Chang, S.F., Ng, D., Baird, L., and Georgopoulos, C., 1991, Analysis of an *Escherichia coli* DNAb temperature-sensitive insertion mutation and its cold-sensitive extragenic suppressor. *J. Biol. Chem.* **266:** 3654–3660.

Charalampous, F.C., and Lyras, C., 1957, Biochemical studies on inositol. IV. Conversion of inositol to glucuronic acid by rat kidney extracts. *J. Biol. Chem.* **228:** 1–13.

Chen, I.W., and Charalampous, C.F., 1966, Biochemical studies on D-inositol 1-phosphate as an intermediate in the biosynthesis of inositol from glucose-6-phosphate, and characteristics of two reactions in this biosynthesis. *J. Biol. Chem.* **241:** 2194–2199.

Chen, L., and Roberts, M.F., 1998, Cloning and expression of the inositol monophosphatase gene from *Methanococcus jannaschii* and characterization of the enzyme. *Appl. Environ. Microbiol.* **64:** 2609–2615.

Chen, L., and Roberts, M.F., 1999, Characterization of a tetrameric inositol monophosphatase from the hyperthermophilic bacterium *Thermotoga maritima*. *Appl. Environ. Microbiol.* **65:** 4559–4567.

Chen, L., and Roberts, M.F., 2000, Overexpression, purification, and analysis of complementation behavior of *E. coli* Suhb protein: Comparison with bacterial and archaeal inositol monophosphatases. *Biochemistry* **39:** 4145–4153.

Conklin, P.L., Saracco, S.A., Norris, S.R., and Last, R.L., 2000, Identification of ascorbic acid-deficient *Arabidopsis thaliana* mutants. *Genetics* **154:** 847–856.

Dean-Johnson, M., and Henry, S.A., 1989, Biosynthesis of inositol in yeast. Primary structure of *myo*-inositol-1-phosphate synthase (EC 5.5.1.4) and functional analysis of its structural gene, the *ino1* locus. *J. Biol. Chem.* **264:** 1274–1283.

Dean-Johnson, M., and Wang, X., 1996, Differentially expressed forms of 1L-*myo*-inositol-1-phosphate synthase in *Phaseolus vulgaris*. *J. Biol. Chem.* **271:** 17215–17218.

Dichtl, B., Stevens, A., and Tollervey, D., 1997, Lithium toxicity in yeast is due to the inhibition of RNA processing enzymes. *EMBO J.* **16:** 7184–7195.

Doering, T., 2000, How does *Cryptococcus* get its coat? Trends Microbiol. **8:** 545–551.

Dunn, T.M., Lynch, D.V., Michaelson, L.V., and Napier, J.A., 2004, A post-genomic approach to understanding sphingolipid metabolism in *Arabidopsis thaliana*. *Ann. Bot. (Lond.)* **93:** 483–497.

Eisenberg, F.J., 1967, D-*myo* inositol 1-phosphate as product of cyclization of glucose 6-phosphate and substrate for a specific phosphatase in rat testis. *J. Biol. Chem.* **242:** 1375–1382.

Eisenberg, F., Bolden, A.H., and Loewus, F.A., 1964, Inositol formation by cyclization of glucose chain in rat testis. *Biochem. Biophys. Res. Commun.* **14:** 419–424.

English, P.D., Deitz, M., and Albersheim, P., 1966, Myoinositol kinase: Partial purification and identification of product. *Science* **151:** 198–199.

Flores, S., and Smart, C.C., 2000, Abscisic acid-induced changes in inositol metabolism in *Spirodela polyrrhiza*. *Planta* **211:** 823–832.

Gainey, P.A., and Phelps, C.F., 1975, Interactions of uridine diphosphate glucose dehydrogenase with the inhibitor uridine diphosphate xylose. *Biochem. J.* **145:** 129–134.

Gillaspy, G., and Gruissem, W., 2001, Li^+ induces hypertrophic growth and downregulation of IMP activity in tomato. *J. Plant Growth Regul.* **20:** 78–86.

Gillaspy, G.E., Keddie, J.S., Oda, K., and Gruissem, W., 1995, Plant inositol monophosphatase is a lithium-sensitive enzyme encoded by a multigene family. *Plant Cell* **7:** 2175–2185.

Gil-Mascarell, R., Lopez-Coronado, J.M., Belles, J.M., Serrano, R., Rodriguez, P.L., Murguia, J.R., Quintero, F.J., Garciadeblas, B., and Rodriguez-Navarro, A., 1999, The Arabidopsis Hal2-like gene family includes a novel sodium-sensitive phosphatase. *Plant J.* **17:** 373–383.

Guan, G., Dai, P., and Shechter, I., 2003, cDNA cloning and gene expression analysis of human *myo*-inositol 1-phosphate synthase. *Arch. Biochem. Biophys.* **417:** 251–259.

Hallcher, L.M., and Sherman, W.R., 1980, The effects of lithium ion and other agents on the activity of *myo*-inositol 1-phosphatase from bovine brain. *J. Biol. Chem.* **255:** 10896–10901.

Hasegawa, R., and Eisenberg, F., Jr., 1981, Selective hormonal control of *myo*-inositol biosynthesis in reproductive organs and liver of the male rat. *Proc. Natl. Acad. Sci. U.S.A.* **78:** 4863–4866.

Hayama, R., Izawa, T., and Shimamoto, K., 2002, Isolation of rice genes possibly involved in the photoperiodic control of flowering by a fluorescent differential display method. *Plant Cell Physiol.* **43:** 494–504.

Hegeman, C.E., Good, L.L., and Grabau, E.A., 2001, Expression of D-*myo*-inositol-3-phosphate synthase in soybean. Implications for phytic acid biosynthesis. *Plant Physiol.* **125:** 1941–1948.

Hegeman, C.E., and Grabau, E.A., 2001, A novel phytase with sequence similarity to purple acid phosphatases is expressed in cotyledons of germinating soybean seedlings. *Plant Physiol.* **126:** 1598–1608.

Hirsch, J.P., and Henry, S.A., 1986, Expression of the *Saccharomyces cerevisiae* inositol-1-phosphate synthase (*ino1*) gene is regulated by factors that affect phospholipid synthesis. *Mol. Cell Biol.* **6:** 3320–3328.

Hitz, W.D., Carlson, T.J., Kerr, P.S., and Sebastian, S.A., 2002, Biochemical and molecular characterization of a mutation that confers a decreased raffinosaccharide and phytic acid phenotype on soybean seeds. *Plant Physiol.* **128:** 650–660.

Inada, T., and Nakamura, Y., 1995, Lethal double-stranded RNA processing activity of ribonuclease III in the absence of Suhb protein of *Escherichia coli. Biochimie* **77:** 294–302.

Inada, T., and Nakamura, Y., 1996, Autogenous control of the Suhb gene expression of *Escherichia coli. Biochimie* **78:** 209–212.

Ishitani, M., Majumder, A.L., Bornhouser, A., Michalowski, C.B., Jensen, R., and Bohnert, H., 1996, Coordinate transcription induction of *myo*-inositol metabolism during environmental stress. *Plant J.* **9:** 537–548.

Janczarek, M., Krol, J., and Skorupska, A., 1999, The pssb gene product of *Rhizobium leguminosarum* bv. Trifolii is homologous to a family of inositol monophosphatases. *FEMS Microbiol. Lett.* **173:** 319–325.

Janczarek, M., and Skorupska, A., 2004, Regulation of pssa and pssb gene expression in *Rhizobium leguminosarum* bv. Trifolii in response to environmental factors. *Antonie Van Leeuwenhoek* **85:** 217–227.

Jian, A., and Nessler, C., 2000, Metabolic engineering of an alternative pathway for ascorbic acid biosynthesis in plants. *Mol. Breeding* **6:** 73–78.

Jin, X., Foley, K.M., and Geiger, J.H., 2004, The structure of the 1L-*myo*-inositol-1-phosphate synthase-NAD2+-deoxy-D-glucitol 6-(e)-vinylhomophosphonate complex demands a revision of the enzyme mechanism. *J. Biol. Chem.* **279:** 13889–13895.

Johnson, M.D., and Sussex, I.M., 1994, Il-*myo*-inositol 1-phosphate synthase from *Arabidopsis thaliana. Plant Physiol.* **107:** 613–619.

Kao, K.R., Masiu, R.P., and Elinson, R., 1986, Respecification of pattern in *Xenopus laevis* embryos – a novel effect of lithium. *Nature* **322:** 371–373.

Keller, R., Brearley, C., Trethewey, R., and Muller-Rober, B., 1998, Reduced inositol content and altered morphology in transgenic potato plants inhibited for 1D-*myo*-inositol 3-phosphate synthase. *Plant J.* **16:** 403–410.

Klein, P.S., and Melton, D.A., 1996, A molecular mechanism for the effect of lithium on development. *Proc. Natl. Acad. Sci. U.S.A.* **93:** 8455–8459.

Lachman, H.M., and Papolos, D.F., 1989, Abnormal signal transduction: A hypothetical model for bipolar affective disorder. *Life Sci.* **45:** 1413–1426.

Larner, J., 2002, D-*chiro*-inositol – its functional role in insulin action and its deficit in insulin resistance. *Int. J. Exp. Diabetes Res.* **3:** 47–60.

Loertscher, R., and Lavery, P., 2002, The role of glycosyl phosphatidyl inositol (gpi)-anchored cell surface proteins in T-cell activation. *Transpl. Immunol.* **9:** 93–96.

Loewus, F., 1963, Tracer studies of ascorbic acid formation in plants. *Phytochemistry* **2:** 109–128.

Loewus, F., 1965, Inositol metabolism and cell wall formation in plants. *Fed. Proc.* **24:** 855–862.

Loewus, F., 1969, Metabolism of inositol in higher plants. *Ann. N. Y. Acad. Sci.* **165:** 577–598.

Loewus, M.W., Bedgar, D.L., and Loewus, F.A., 1984, 1L-*myo*-inositol 1-phosphate synthase from pollen of *Lilium longiflorum*. An ordered sequential mechanism. *J. Biol. Chem.* **259:** 7644–7647.

Loewus, F., Kelly, S., and Neufeld, E., 1962, Metabolism of *myo*-inositol in plants: Conversion to pectin, hemicellulose, D-xylose, and sugar acids. *Proc. Natl. Acad. Sci. U.S.A.* **48:** 421–425.

Loewus, M.W., and Loewus, F.A., 1980, The C-5 hydrogen isotope-effect in *myo*-inositol 1-phosphate synthase as evidence for the *myo*-inositol oxidation-pathway. *Carbohydr. Res.* **82:** 333–342.

Loewus, F.A., and Loewus, M.W., 1983, *myo*-Inositol: Its biosynthesis and metabolism. *Annu. Rev. Plant Physiol.* **34:** 137–161.

Loewus, F.A., and Murthy, P.P.N., 2000, *myo*-Inositol metabolism in plants. *Plant Sci.* **150:** 1–19.

Lopez, F., Leube, M., Gil-Mascarell, R., Navarro-Avino, J.P., and Serrano, R., 1999, The yeast inositol monophosphatase is a lithium- and sodium-sensitive enzyme encoded by a non-essential gene pair. *Mol. Microbiol.* **31:** 1255–1264.

Lorence, A., Chevone, B.I., Mendes, P., and Nessler, C.L., 2004, *myo*-Inositol oxygenase offers a possible entry point into plant ascorbate biosynthesis. *Plant Physiol.* **134:** 1200–1205.

Maeda, T., and Eisenberg, F., Jr., 1980, Purification, structure, and catalytic properties of l-*myo*-inositol-1-phosphate synthase from rat testis. *J. Biol. Chem.* **255:** 8458–8464.

Majee, M., Maitra, S., Ghosh Dastidar, K., Pattnaik, S., Chatterjee, A., Hait, N.C., Das, K.P., and Majumder, A.L., 2004, A novel salt-tolerant L-*myo*-inositol 1-phosphate synthase from *Porteresia coarctata (Roxb.)Tateoka*, a halophytic wild rice. Molecular cloning, bacterial overexpression, characterization and functional introgression into tobacco conferring salt-tolerance phenotype. *J. Biol. Chem.* **279:** 28539–28552.

Majerus, P.W., Kisseleva, M.V., and Norris, F.A., 1999, The role of phosphatases in inositol signaling reactions. *J. Biol. Chem.* **274:** 10669–10672.

Majumder, A.L., Chatterjee, A., Ghosh Dastidar, K., and Majee, M., 2003, Diversification and evolution of l-*myo*-inositol 1-phosphate synthase. *FEBS Lett.* **553:** 3–10.

Majumder, A.L., Johnson, M.D., and Henry, S.A., 1997, 1L-*myo*-inositol-1-phosphate synthase. *Biochim. Biophys. Acta* **1348:** 245–256.

Majumder, A.L., Mandal N.C., and Biswas, B.B., 1972, Phosphoinositol kinase from germinating mung bean seeds. *Phytochemistry* **11:** 503–508.

Maslanski, J.A., Leshko, L., and Busa, W.B., 1992, Lithium-sensitive production of inositol phosphates during amphibian embryonic mesoderm induction. *Science* **256:** 243–245.

Matsuhisa, A., Suzuki, N., Noda, T., and Shiba, K., 1995, Inositol monophosphatase activity from the *Escherichia coli* Suhb gene product. *J. Bacteriol.* **177:** 200–205.

McAllister, G., Whiting, P., Hammond, E.A., Knowles, M.R., Atack, J.R., Bailey, F.J., Maigetter, R., and Ragan, C. I., 1992, cDNA cloning of human and rat brain *myo*-inositol monophosphatase: Expression and characterization of the human recombinant enzyme. *Biochem. J.* **284:** 749–754.

Mehta, D.V., Kabir, A., and Bhat, P.J., 1999, Expression of human inositol monophosphatase suppresses galactose toxicity in *Saccharomyces cerevisiae*: Possible implications in galactosemia. *Biochim. Biophys. Acta* **1454:** 217–226.

Meijer, H.J., and Munnik, T., 2003, Phospholipid-based signaling in plants. *Annu. Rev. Plant Physiol. Plant Mol. Biol.* **54:** 265–306.

Moore, G.J., Bebchuk, J.M., Parrish, J.K., Faulk, M.W., Arfken, C.L., Strahl-Bevacqua, J., and Manji, H.K., 1999, Temporal dissociation between lithium-induced changes in frontal lobe *myo*-inositol and clinical response in manic-depressive illness. *Am. J. Psychiatry* **156:** 1902–1908.

Murguia, J.R., Belles, J.M., and Serrano, R., 1996, The yeast Hal2 nucleotidase is an *in vivo* target of salt toxicity. *J. Biol. Chem.* **271:** 29029–29033.

Murray, M., and Greenberg, M.L., 1997, Regulation of inositol monophosphatase in *Saccharomyces cerevisiae*. *Mol. Microbiol.* **25:** 541–546.

Murray, M., and Greenberg, M.L., 2000, Expression of yeast INM1 encoding inositol monophosphatase is regulated by inositol, carbon source and growth stage and is decreased by lithium and valproate. *Mol. Microbiol.* **36:** 651–661.

Nelson, D.E., Rammesmayer, G., and Bohnert, H.J., 1998, Regulation of cell-specific inositol metabolism and transport in plant salinity tolerance. *Plant Cell* **10:** 753–764.

Neuwald, A.F., Krishnan, B.R., Brikun, I., Kulakauskas, S., Suziedelis, K., Tomcsanyi, T., Leyh, T.S., and Berg, D.E., 1992, CysQ, a gene needed for cysteine synthesis in *Escherichia coli* K-12 only during aerobic growth. *J. Bacteriol.* **174:** 415–425.

Neuwald, A.F., York, J.D., and Majerus, P.W., 1991, Diverse proteins homologous to inositol monophosphatase. *FEBS Lett.* **294:** 16–18.

Nigou, J., and Besra, G.S., 2002, Characterization and regulation of inositol monophosphatase activity in *Mycobacterium smegmatis*. *Biochem. J.* **361:** 385–390.

Nishikimi, M., Fukuyama, R., Minoshima, S., Shimizu, N., and Yagi, K., 1994, Cloning and chromosomal mapping of the human nonfunctional gene for L-gulono-gamma-lactone oxidase, the enzyme for L-ascorbic acid biosynthesis missing in man. *J. Biol. Chem.* **269:** 13685–13688.

Odorizzi, G., Babst, M., and Emr, S.D., 2000, Phosphoinositide signaling and the regulation of membrane trafficking in yeast. *Trends Biochem. Sci.* **25:** 229–235.

Ordman, A.B., and Kirkwood, S., 1977, UDP glucose dehydrogenase. Kinetics and their mechanistic implications. *Biochim. Biophys. Acta* **481:** 25–32.

Ouyang, Q., Ruiz-Noriega, M., and Henry, S.A., 1999, The *reg1* gene product is required for repression of *ino1* and other inositol-sensitive upstream activating sequence-containing genes of yeast. *Genetics* 152: 89–100.

Parthasarathy, L., Vadnal, R.E., Parthasarathy, R., Shyamala, and Devi, C.S., 1994, Biochemical and molecular properties of lithium-sensitive *myo*-inositol monophosphatase. *Life Sci.* **54:** 1127–1142.

Paul, M., and Cockburn, W., 1989, Pinitol, a compatible solute in *Mesembryanthemum crystallinum* L. *J. Exp. Bot.* **40:** 1093–1098.

Payrastre, B., Missy, K., Giuriato, S., Bodin, S., Plantavid, M., and Gratacap, M., 2001, Phosphoinositides: Key players in cell signalling, in time and space. *Cell. Signal.* **13:** 377–387.

Quintero, F.J., Garciadeblas, B., and Rodriguez-Navarro, A., 1996, The Sal1 gene of *Arabidopsis*, encoding an enzyme with 3′(2′),5′(-bisphosphate nucleotidase and inositol polyphosphatase 1-phosphatase activities, increases salt tolerance in yeast. *Plant Cell* **8:** 529–537.

Rammesmayer, G., Pichorner, H., Adams, P., Jensen, R., and Bohnert, H.J., 1995, Characterization of IMT1, *myo*-inositol *O*-methyltransferase, from *Mesembryanthemum crystallinum*. *Arch. Biochem. Biophys.* **322:** 183–188.

Reddy, C.C., Swan, J.S., and Hamilton, G.A., 1981, *myo*-Inositol oxygenase from hog kidney. I. Purification and characterization of the oxygenase and of an enzyme complex containing the oxygenase and D-glucuronate reductase. *J. Biol. Chem.* **256:** 8510–8518.

Rivera-Gonzalez, R., Petersen, D.N., Tkalcevic, G., Thompson, D.D., and Brown, T.A., 1998, Estrogen-induced genes in the uterus of ovariectomized rats and their regulation by droloxifene and tamoxifen. *J. Steroid Biochem. Mol. Biol.* **64:** 13–24.

Roberts, R.M., Shah, R., and Loewus, F., 1967. Conversion of *myo*-inositol-2-^{14}C to labeled 4-*O*-methyl-glucuronic acid in the cell wall of maize root tips. *Arch. Biochem. Biophys.* **119:** 590–593.

Seitz, B., Klos, C., Wurm, M., and Tenhaken, R., 2000, Matrix polysaccharide precursors in arabidopsis cell walls are synthesized by alternate pathways with organ-specific expression patterns. *Plant J.* **21:** 537–546.

Shaldubina, A., Ju, S., Vaden, D.L., Ding, D., Belmaker, R.H., and Greenberg, M.L., 2002, *epi*-Inositol regulates expression of the yeast *ino1* gene encoding inositol-1-P synthase. *Mol. Psychiatry* **7:** 174–180.

Shamir, A., Shaltiel, G., Greenberg, M.L., Belmaker, R.H., and Agam, G., 2003, The effect of lithium on expression of genes for inositol biosynthetic enzymes in mouse hippocampus; a comparison with the yeast model. *Brain Res. Mol. Brain Res.* **115:** 104–110.

Shamir, A., Sjoholt, G., Ebstein, R.P., Agam, G., and Steen, V.M., 2001, Characterization of two genes, impa1 and impa2 encoding mouse *myo*-inositol monophosphatases. *Gene* **271:** 285–291.

Shen, X., Xiao, H., Ranallo, R., Wu, W.H., and Wu, C., 2003, Modulation of ATP-dependent chromatin-remodeling complexes by inositol polyphosphates. *Science* **299:** 112–114.

Sims, K.J., Spassieva, S.D., Voit, E.O., and Obeid, L.M., 2004, Yeast sphingolipid metabolism: Clues and connections. *Biochem. Cell Biol.* **82:** 45–61.

Smart, C., and Fleming, A., 1993, A plant gene with homology to D-*myo*-inositol-3-phosphate synthase is rapidly and spatially up-regulated during ABA-induced morphogenic response in *Spirodela polrrhiza*. *Plant J.* **4:** 279–293.

Smirnoff, N., Conklin, P.L., and Loewus, F.A., 2001, Biosynthesis of ascorbic acid in plants: A renaissance. *Annu. Rev. Plant Physiol. Plant Mol. Biol.* **52:** 437–467.

Spiegelberg, B.D., Xiong, J.P., Smith, J.J., Gu, R.F., and York, J.D., 1999, Cloning and characterization of a mammalian lithium-sensitive bisphosphate 3′-nucleotidase inhibited by inositol 1,4-bisphosphate. *J. Biol. Chem.* **274:** 13619–13628.

Stevenson, J.M., Perera, I.Y., Heilmann, I.I., Persson, S., and Boss, W.F., 2000, Inositol signaling and plant growth. *Trends Plant Sci.* **5:** 357.

Styer, J.C., Keddie, J., Spence, J., and Gillaspy, G.E., 2004, Genomic organization and regulation of the LeIMP1 and LeIMP2 genes encoding *myo*-inositol monophosphatase in tomato. *Gene* **326:** 35–41.

Tolias, K.F., and Cantley, L.C., 1999, Pathways for phosphoinositide synthesis. *Chem. Phys. Lipids* **98:** 69–77.

Vaden, D.L., Ding, D., Peterson, B., and Greenberg, M.L., 2001, Lithium and valproate decrease inositol mass and increase expression of the yeast *ino1* and *ino2* genes for inositol biosynthesis. *J. Biol. Chem.* **276:** 15466–15471.

Van Dijken, P., Bergsma, J.C., Hiemstra, H.S., De Vries, B., Van Der Kaay, J., and Van Haastert, P.J., 1996, *Dictyostelium discoideum* contains three inositol monophosphatase activities with different substrate specificities and sensitivities to lithium. *Biochem. J.* **314:** 491–495.

Wheeler, G.L., Jones, M.A., and Smirnoff, N., 1998, The biosynthetic pathway of Vitamin C in higher plants. *Nature* **393:** 365–369.

White, M.J., Hirsch, J.P., and Henry, S.A., 1991, The *opi1* gene of *Saccharomyces cerevisiae*, a negative regulator of phospholipid biosynthesis, encodes a protein containing polyglutamine tracts and a leucine zipper. *J. Biol. Chem.* **266:** 863–872.

Whiting, P.H., Palmano, K.P., and Hawthorne, J.N., 1979, Enzymes of *myo*-inositol and inositol lipid metabolism in rats with streptozotocin-induced diabetes. *Biochem. J.* **179:** 549–553.

Wong, Y.H., Mauck, L.A., and Sherman, W.R., 1982, L-*myo*-inositol-1-phosphate synthase from bovine testis. *Methods Enzymol.* **90**(Pt E): 309–314.

Xiong, L., Lee, B., Ishitani, M., Lee, H., Zhang, C., and Zhu, J.K., 2001, *Fiery1* encoding an inositol polyphosphate 1-phosphatase is a negative regulator of abscisic acid and stress signaling in Arabidopsis. *Genes Dev.* **15:** 1971–1984.

Yoon, I.S., Li, P.P., Siu, K.P., Kennedy, J.L., Cooke, R.G., Parikh, S.V., and Warsh, J.J., 2001, Altered IMPA2 gene expression and calcium homeostasis in bipolar disorder. *Mol. Psychiatry* **6:** 678–683.

York, J.D., and Majerus, P.W., 1990, Isolation and heterologous expression of a cDNA encoding bovine inositol polyphosphate 1-phosphatase. *Proc. Natl. Acad. Sci.* **87:** 9548–9552.

York, J.D., Ponder, J.W., and Majerus, P.W., 1995, Definition of a metal-dependent/Li(+)-inhibited phosphomonoesterase protein family based upon a conserved three-dimensional core structure. *Proc. Natl. Acad. Sci. U.S.A.* **92:** 5149–5153.

York, J.D., Veile, R.A., Donis-Keller, H., and Majerus, P.W., 1993, Cloning, heterologous expression, and chromosomal localization of human inositol polyphosphate 1-phosphatase. *Proc. Natl. Acad. Sci. U.S.A.* **90:** 5833–5837.

Yoshida, K.T., Fujiwara, T., and Naito, S., 2002, The synergistic effects of sugar and abscisic acid on *myo*-inositol-1-phosphate synthase expression. *Physiol. Plant.* **114:** 581–587.

Yoshida, K.T., Wada, T., Koyama, H., Mizobuchi-Fukuoka, R., and Naito, S., 1999, Temporal and spatial patterns of accumulation of the transcript of *myo*-inositol-1-phosphate synthase and phytin-containing particles during seed development in rice. *Plant Physiol.* **119:** 65–72.

Zhu, X., and Eichberg, J., 1990, A *myo*-inositol pool utilized for phosphatidylinositol synthesis is depleted in sciatic nerve from rats with streptozotocin-induced diabetes. *Proc. Natl. Acad. Sci. U.S.A.* **87:** 9818–9822.

Chapter 4

Genetics of Inositol Polyphosphates

Victor Raboy and David Bowen
USDA-ARS and University of Idaho, Aberdeen, Idaho 83210, U.S.A.

1. INTRODUCTION

myo-Inositol (Ins, Figure 1 top center) is the cyclic alcohol that, in terms of the distribution of hydroxyl groups around the 6-carbon ring, retains the basic stereochemistry of glucose, the "D-gluco" configuration (Loewus and Murthy, 2000). The Ins ring is basically one synthetic step away from glucose, in that glucose 6-P is converted to Ins (3) P_1 in a reaction catalyzed by one enzyme (see below). Therefore it would seem to make biological common sense that this compound would itself in turn represent a major metabolite, the parent of all other cyclic alcohols and of many other compounds, and that its use would be important to a number of biological processes and functions. While this is true for most organisms, plant cells in particular do a lot with Ins (Loewus and Murthy, 2000; Figure 1). This review will focus on Ins and one branch of derivatives of Ins, the soluble Ins phosphates and their related lipid phosphatidylinositol (PtdIns) phosphates. The canonical pathway for transient production of Ins phosphates and PtdIns phosphates, central as second messengers in cellular signaling, proceeds via the production of PtdIns(4,5)P_2 and two of its derivatives, Ins(1,4,5)P_3 and PtdIns(3,4,5)P_3 (Berridge and Irvine, 1989). In particular, this review will focus on the genetics of what is now known as an important additional pool in cellular Ins/PtdIns phosphate metabolism, Ins-1,2,3,4,5,6-P_6 (Ins P_6, Figure 1 bottom right; Sasakawa *et al.*, 1995), the hexamonoester of Ins. Ins P_6 is the most abundant Ins phosphate in nature.

In the fields of agriculture and nutrition Ins P_6 is commonly referred to as "phytic acid" (Raboy, 2001). This compound was first identified as a major P-containing component of seeds, and thus called phytic acid (Raboy, 1997), but subsequently its occurrence was found to be widespread in eukaryotes, and its metabolism to have many functions (Shears, 2001, 2004). It represents a major

A. Lahiri Majumder and B. B. Biswas (eds.), Biology of Inositols and Phosphoinositides, 71–102
© 2006 *U.S. Government. Printed in the Netherlands.*

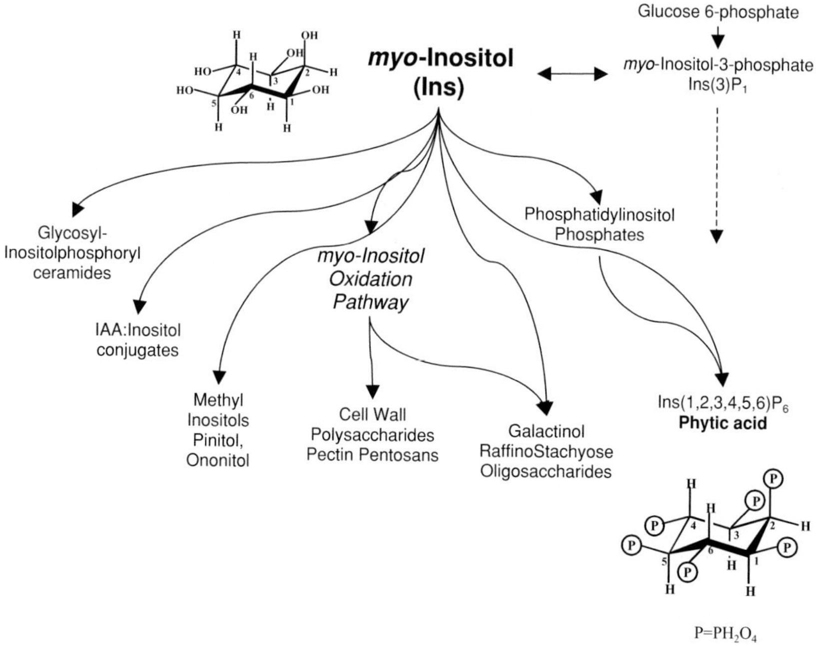

Figure 1. Pathways that utilize *myo*-inositol (Ins). The six carbons of the Ins ring are numbered according to the "D-numbering convention."

metabolic pool in the Ins phosphate/PtdIns phosphate pathways, of importance to development, cellular structure, membrane cycling, signaling, DNA repair, and RNA export (Hanakahi *et al.*, 2000, 2002; Odom *et al.*, 2000; Saiardi *et al.*, 2000a, 2002; York *et al.*, 1999). Ins P_6 might function itself as an effecter molecule, a second messenger, in signal transduction. Studies in plant systems (Lemtiri-Chlieh *et al.*, 2000, 2003) have shown that Ins P_6 might function as a signal regulating K^+-inward rectifying conductance in guard cells, and in mobilizing calcium flux in guard cells. These processes are critical to the function of guard cells in the regulation of transpiration and the maintenance of turgor in plant tissues. Of particular relevance to this review is that Ins P_6 is the most abundant phosphate in seeds, representing from 65% to 80% of seed total P (Raboy, 1997). To put this simple fact in perspective, consider the following: most of the phosphate taken up by crop plants is ultimately translocated to seeds during their development; most of phosphate translocated to seeds is ultimately synthesized into Ins P_6. As a result, the amount of Ins P_6 that accumulates in seeds represents a sum equivalent to more than 50% of all fertilizer phosphate applied annually worldwide (Lott *et al.*, 2000). Thus Ins P_6 represents a bottle-neck in the flux of P through the world-wide agricultural ecosystem.

Applied interest in seed phytic acid concerns its roles, both negative and positive, in human and animal nutrition, and health (Raboy, 2001). The fact that most seed P is found as phytic acid P can have a negative impact on P utilization by livestock. Excretion of dietary phytic acid can contribute to mineral deficiencies in humans. On the other hand, dietary phytic acid might play a positive role as an anti-oxidant and anti-cancer agent. This review will not directly address in detail questions relating to dietary phytic acid, other than to point out that studying and possibly dealing with these issues represents the applied rational for understanding the genetics and biology of seed Ins P_6 metabolism, and for the isolation of the so-called *low phytic acid* (*lpa*) mutations in a number of major seed crop species. The second rational for isolating *lpa* mutations is to advance basic knowledge concerning Ins P_6 metabolism. Initially *lpa* mutations were thought of as falling into two classes, reflecting the fact that the structural metabolic pathways to Ins P_6 can be thought of as consisting of two parts; the first part being the synthesis and supply of Ins substrate, the second being the conversion of Ins to Ins P_6. Thus plant *lpa* mutations were initially hypothesized to be mutations that perturb Ins supply, or those that impact the conversion of Ins to Ins P_6 (Raboy *et al.*, 2000).

At the beginning of the 1990s there was essentially no Mendelian genetics of the Ins phosphate pathways (Raboy and Gerbasi, 1996). There were few if any reports at that time demonstrating that a particular mutant phenotype was due to altered Ins phosphate pathway metabolism, or that altered levels of Ins phosphates were due to a particular mutation. The molecular genetics of these pathways was still only partially developed. Sequences encoding Ins phosphate kinases and phosphatases were just beginning to be identified. Furthermore, at least one important component of the Ins phosphate pathways, the pyrophosphate containing Ins phosphates, had not yet been discovered (Europe-Finner, 1989; Safrany *et al.*, 1999). One decade later a great deal of progress has been made in the biochemistry and molecular genetics of these pathways. There have also been numerous studies describing mutants and mutations in most components of these pathways, such as the plant *lpa* mutations mentioned above and others in many organisms ranging from *Dictyostelium* to the yeast *Saccharomyces*, *Drosophila*, and mammals. Studies that include in-part classical genetics of simply inherited mutant phenotypes continue to contribute to our understanding of these pathways, often yielding unexpected insights. With the acquisition of the complete genome of a number of species, along with large amounts of genomic sequence data for a number of additional species, and large amounts of information concerning expressed sequences, genomics has now also contributed greatly to this field. A selection of recent genetics and genomics studies that provide new insights into the field of Ins and Ins phosphate biology will be reviewed here.

2. THEMES

This review will use in sequence Ins and the conversion of Ins to $InsP_6$ as focal points for a discussion of questions into the metabolism of Ins phosphates, their biological function, and its evolution. Throughout this discussion there will be a few recurrent themes. First, rapidly advancing genomics has shown that with a few important and informative exceptions, most eukaryotes inherit a similar, ancestrally related set of sequences, most of which are expressed, encoding a relatively similar set of enzymes involved in Ins and Ins phosphate metabolism. This raises the question; is the biology of these pathways across evolutionarily divergent microbes, plants, and mammals, largely similar, or greatly different? Is there a conscrved similarity, a conserved set of multiple functions, overlooked in some important ways?

An increasing number of studies indicate that, across these divergent organisms, Ins phosphates, and metabolites involved in their lipid precursor metabolism can serve in signal transduction and the regulation of development, according to the well-known signal-transduction paradigm (Berridge and Irvine, 1989; Irvine and Schell, 2001; Shears, 2004). But they can also serve in other functions that don not perfectly fit the signal-transduction paradigm. Thus Ins phosphates or related lipids also have functions in basal cellular metabolism and housekeeping. A given compound can be an "essential ubiquitous and abundant intermediate," in the same cells, tissues, or organs that it also serves via transient production in second messenger metabolism (Loewen *et al.*, 2004). Recent developments will be discussed that reveal the linkage of cellular Ins, Ins lipid and phosphate metabolism, demonstrating that they together represent a coordinated component of basic cellular nutrition and structure. Does the ability of cells to sense changes in internal concentrations of phosphate, phosphatidic acid, or its breakdown product glycerophosphoinositol, even if those changes are an outcome of external conditions, represent "signal transduction," or simply an example of the molecular mechanisms for feed-back regulation and maintenance of cellular nutritional status? In other words, phosphate, Ins phosphate, Ins lipids and their metabolites probably *first* are important in the own right, and *second*, sometimes important as second messengers.

A third theme concerns the biological significance, or occasional lack thereof, of observed differences in Ins phosphates or their lipid intermediates, or of observed differences in the copy number of genes encoding enzymes in these pathways. These differences might exist between cells within a species, or across species, families, or kingdoms. There may be differences in the apparent presence or absence of a given protein or differences in gene expression. Does this always serve some selective advantage in a Darwinian sense? Does it always have some important biological purpose? Some of these differences, however real, might be neutral in terms of fitness of an organism. Perhaps there is a propensity for researchers to "over interpret" the differences in genes, gene products, and metabolites that they observe. However, in some cases differences between divergent species are highly informative. One aspect of this third theme is the observation that when researchers probe the Ins/PtdIns phosphate pathways via analysis of mutations that block specific components, they often find that alternative metabolic routes still allow the cell to produce important metabolites (Acharya *et al.*, 1998). The Ins/PtdIns pathways sometimes appear to perform a metabolic balancing act.

3. *MYO*-INOSITOL

A simple, two-step biosynthetic pathway to Ins has been established for all organisms. The sole synthetic source of the Ins ring is via the conversion of glucose 6-P to D-Ins(3)P_1 (Figure 2 top), catalyzed by the enzyme D-*myo*-inositol(3)P_1 synthase ("MIPS," EC 5.5.1.4). This enzyme and its activity's product were originally referred to using the "L" numbering convention, thus as an L-Ins(1)P_1 synthase (for a review, see Loewus and Murthy, 2000). However, it has become the accepted convention to refer to Ins phosphates using the "D"-numbering convention, where D-1 and D-3 refer to the same positions on the Ins ring as L-3 and L-1, respectively. The D-numbering convention will be used uniformly here. The ubiquity and importance of Ins has resulted in studies of MIPS from a large number of prokaryotes and eukaryotes (Majumder *et al.*, 2003). In addition to cytosolic forms, chloroplastic isoforms of MIPS have also been reported (Lackey *et al.*, 2003; reviewed in Majumder *et al.*, 2003). The second step in the pathway to Ins is to hydrolysis of Ins(3)P_1 to Ins and P_i (inorganic P), catalyzed by Ins monophosphatase (IMP, EC 3.1.3.25). A summary of the variety of pathways in plant cells that utilize Ins is illustrated in Figure 1.

For the purposes of this review, it is informative to compare the history of our understanding of the regulation of Ins supply in the budding yeast (*Saccharomyces cerevisiae*), versus higher plants. In yeast studies a major focus of research has addressed the regulation of Ins supply as central in the regulation of general phospholipid metabolism (Carman and Henry, 1989). The yeast genome contains one gene, *INO1*, encoding MIPS (Donahue and Henry, 1981). *INO1* as well as a large fraction of all yeast genes involved in phospholipid synthesis, are regulated at the transcriptional level via a feedback-loop system sensitive to cellular Ins and to a lesser extant choline levels (Loewen *et al.*, 2004). The machinery of this regulatory system consists of the *cis*-acting Upstream Activating Sequence (UAS^{INO}) found in the 5′-untranslated sequence of *INO1* and lipid pathway genes, and three transcription factors that interact with UAS^{INO} (Greenberg and Lopes, 1996). *INO1* and the lipid pathway genes are activated via binding to UAS^{INO} of two transcription activators, Ino2p and Ino4p. These in turn are repressed via binding by the transcription factor Opi1p, encoded by the *Overproduction of Inositol1* (*OPI1*) gene. How does this system sense Ins levels? Opi1p is itself repressed via binding by phosphatidic acid, and localization in the ER. At low cellular Ins levels, phosphatidic acid binds to Opi1p, inactivating it in

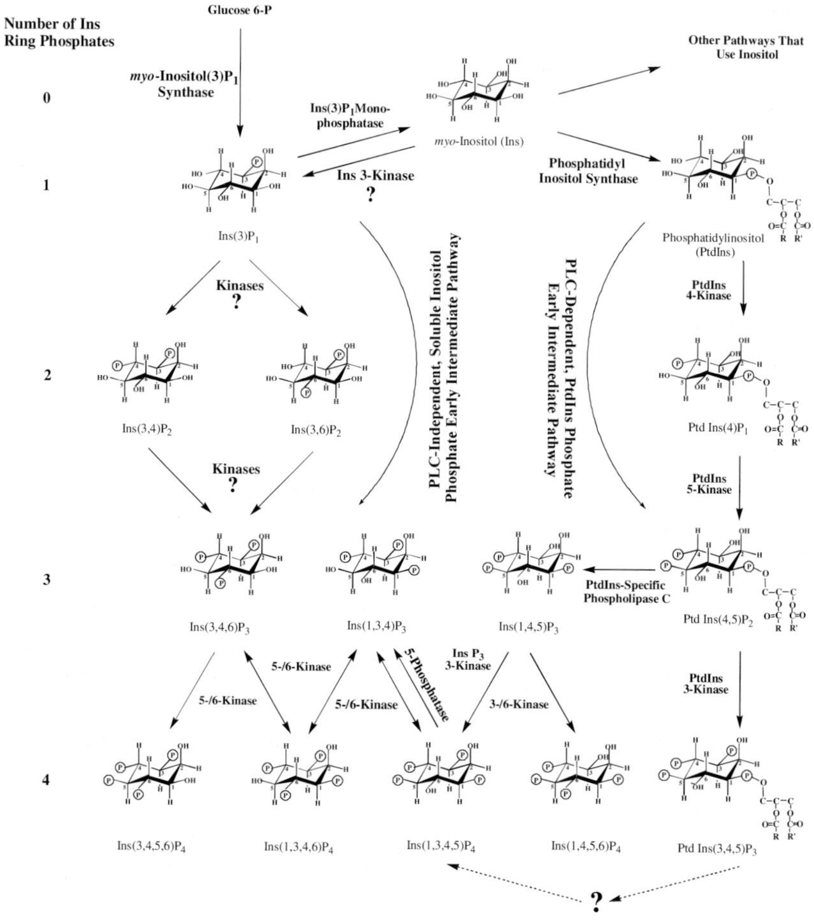

Figure 2. Biosynthetic pathways from glucose 6-P to *myo*-inositol tetra*kis*phosphates. The six carbons of the Ins ring are numbered according to the "D-numbering convention." The pathways to the center-left proceed entirely via soluble *myo*-inositol (Ins) phosphates. Pathways to the center-right proceed via phosphatidylinositol (PtdIns) phosphates. Questionmarks indicate synthetic steps that have not been confirmed in chemical, molecular or genetic analyzes. In addition, the hydrolysis of PtdIns(3,4,5)P_3 to yield Ins(1,3,4,5)P_4 has not been observed in any experiments, and is entirely speculative. P = PH_2O_4.

a localization in the ER, thus in turn allowing the activation of *Ino1* transcription via Ino2p and Ino4p, the translation of MIPS and production of Ins. Ins production in the presence of PtdIns synthase activity (EC 2.7.8.11), in turn consumes phosphatidic acid. Consumption of phosphatidic acid releases Opi1P which is transported to the nucleus, where it represses Ino2p and Ino4p, repressing Ins synthesis. This story is relevant to how compounds involved in signaling pathways, and Ins phosphates in particular, are viewed. Phosphatidic acid "appears to be both an essential and ubiquitous metabolic intermediate and a signaling lipid" (Loewen *et al.*, 2004). Perhaps it will become apparent that, in terms of Ins phosphates, the eukaryotic cell has not evolved separate systems for basal, housekeeping metabolism and transient signaling metabolism, but rather manages to use components of one system for both.

The feedback-loop system links cellular Ins, phosphatidic acid, and lipid levels, but is it the only system in place in yeast that regulates *INO1* expression and cellular Ins levels? In addition to transcriptional regulation via binding of the Ino2p/Inso4p/Opi1p complex, *INO1* is also transcriptionally regulated via the "chromatin-remodeling" complexes *SNF2*, and *INO80*. Shen *et al.* (2003) and Steger *et al.* (2003) together demonstrated that in yeast, induction of genes in response to phosphate starvation, such as *PHO5*, requires the same chromatin-remodeling complexes as does *INO1*. *PHO5* represents a gene encoding one of the structural components of the *PHO* system, in this case a secreted phosphatase, and the regulation of its expression is particularly relevant to this discussion. The *SNF/INO80* remodeling in response to phosphate starvation requires the synthesis of Ins P_4 and Ins P_5 produced by the activity of an Ins(1,4,5)P_3 3-/6-kinase, alternatively called Ins phosphate multikinase (IPMK), in yeast encoded by the *Arg82/IPK2* gene (Shears, 2004; York *et al.*, 1999). This observation was confirmed and extended by El Alami *et al.* (2003), who showed that *PHO* gene repression by phosphate (i.e., repression of gene expression when cells are grown in "high" nutrient phosphate) requires both the Ins phosphate kinase catalytic activity of *Arg82*, a kinase that ultimately produces an Ins P_5, and Ins P_6 kinase activity encoded by the *KCS1* gene, that converts Ins P_6 to pyrophosphate-containing Ins P_7. These Ins phosphate kinases and the gene(s) encoding them will be discussed in more detail below. The relevant observation here is that Ins, phosphate, and Ins phosphate/PtdIns phosphate pathways clearly are interlinked in yeast. However, the authors of these studies uniformly interpret roles and results in terms of signal transduction, even when it appears clear that the sensing mechanisms have to do with cellular nutrition and metabolism.

The compartmentalization/transport mechanism for reversible inactivation of the transcription factors Ino2p and Ino4p has a parallel in the regulation *PHO5*, in this case via compartmentalization of the transcription factor Pho4p in the cytoplasm. Studies of *PHO5* contribute to a developing understanding that in yeast, it is not just Ins and global lipid metabolism that are intimately linked, but there is coordination of phosphate and lipid metabolism via the common component Ins. The *PHO5* gene is regulated in part by the transcription factors encoded by the *PHO2* and *PHO4* genes, Pho2p and Pho4p, respectively. Pho4p is maintained in a phoshorylated state when yeast is grown under high nutrient P, and is localized in the cytoplasm. When yeast is grown under low nutrient P, Pho4p is not phosphorylated and is translocated to the nucleus, where it binds cooperatively with the second transcription factor Pho2p, and thus activates transcription of *PHO5* (O'Neill *et al.*, 1996).

The linkage in coordination of gene expression and metabolism of the phosphate, Ins phosphate, and PtdIns Phosphate pathways in yeast, in response to either phosphate or Ins nutritional status change, is supported and extended by two additional recent studies of the metabolite glycerophosphoinositol and the *Glycerophosphoinositol transporter1* (*GTI1*) gene (Almaguer *et al.*, 2003, 2004). Glycerophosphoinositol is a breakdown product of PtdIns, and is excreted from the yeast cell. When yeast is grown either under P or Ins limitation, *GTI1* is expressed, and glycerophosphoinositol is transported into the cell. This can provide yeast cells with sufficient Ins or P. The transcription factors Pho2p and Pho4p are required for the induction of *GTI1* in response to P starvation. In addition to Pho2p and Pho4p, the transcription factors Ino2p and Ino4p are required for transcription induction in response to Ins starvation.

Thus in studies in yeast, initial interest in the regulation of Ins synthesis concerned its role in phospholipid metabolism, but recent studies have demonstrated clear links between phosphate and lipid metabolism via the common component Ins, and with a role for Ins phosphates. How does this compare with studies in plant systems? While Ins is clearly important in plants as a head-group for the Ins lipids, in plant systems the regulation of expression of MIPS-encoding genes has not been viewed as central to the regulation of phospholipid metabolism. Instead, interest in the regulation of Ins production has ranged from roles of Ins in pathways unique to the plant cell, such as in cell wall polysaccharide synthesis, raffinose/stachyose synthesis, and IAA conjugation (reviewed in Loewus and Murthy, 2000), to roles in hormone and stress response (Ishitani *et al.*, 1996; Smart and Fleming, 1993), and also to the localization of Ins P_6 synthesis and accumulation (Yoshida *et al.*, 1999). Thus, the coordinated expression of the pathway from glucose 6-P, through Ins, to its cyclitol derivatives D-ononitol and D-pinitol, involving MIPS, IMP, and also methylases and epimerases, is central to how the ice plant (*Mesembryanthemum crystallinum*) achieves salt/drought/cold tolerance (Ishitani *et al.*, 1996). The ice plant is used as a model system for studies of stress tolerance, but this role of Ins metabolism is presumed to be widespread in plant stress response. One component of the signal transduction pathway involved in plant stress response, plant guard cell response, and plant development, is the plant hormone abscisic acid, or ABA. Ins synthesis via rapid MIPS induction in response to ABA has been observed in studies of *Spirodela polyrrhiza* (Smart and Fleming, 1993).

Recently the *PINO1* gene of the salt-tolerant wild rice *Porteresia coarctata* (Roxb.) was found to encode a salt-tolerant MIPS enzyme, as compared with the MIPS encoded by the salt-sensitive cultivated rice (*Oryza sativa* L.) (Majee *et al.*, 2004). Comparison of MIPS sequences in these species revealed distinct differences in a 37 amino acid stretch, which upon deletion from the salt-tolerant MIPS rendered it salt-sensitive. Introgression of the *PINO1* gene via transformation rendered tobacco salt-tolerant. The genome of cultivated rice contains one gene encoding MIPS, termed *RINO1*. In addition to constitutive and induced expression that must be essential for the variety of cellular, tissue, and organismal functions discussed above, *RINO1* expression is also induced at the site of $InsP_6$ accumulation in the developing seed (Yoshida *et al.*, 1999). Thus localized *de novo* synthesis of Ins contributes substrate, at least in part, for Ins P_6 accumulation in developing seeds. This was confirmed by the isolation of a mutation, "LR33," in a soybean MIPS gene that confers both a "low phytic acid" and "low raffinose/stachyose" seed phenotype (Hitz *et al.*, 2002). The soybean genome contains at least four MIPS-encoding sequences, one of which appears to have seed-specific expression (Hegeman et al., 2001). One interesting and at present unexplained observation concerning the LR33 mutation is that both germination rate and field emergence in a temperate environment like Iowa USA is greatly reduced when the seed itself was produced in a sub-tropical versus temperate environment (Meis *et al.*, 2003). Thus, subsequent field emergence of non-mutant control lines was 77% for seed produced in temperate environments and 83% for seed produced in sub-tropical environments, whereas field emergence was 63% for LR33 seed produced in temperate environments and only 8% for LR33 seed produced in sub-tropical environments. A similar effect of seed production environment, sub-tropical versus temperate, on subsequent germination rates was also observed. These studies reveal that seed-specific Ins production is critical to both germination and subsequent seedling emergence, but the exact effect of the block in Ins production on these processes is not yet known.

The maize genome contains up to seven sequences with homology to the canonical MIPS (Larson and Raboy, 1999). One of these, the maize chromosome 1S-MIPS, is linked to the maize *low phytic acid* 1 (*lpa*1) locus, also found on chromosome 1S (Raboy *et al.*, 2000). Homozygosity for recessive alleles of the maize *lpa*1 locus result in reductions in seed Ins P_6 ranging from 50% to >90% (Raboy *et al.*, 2001), but have little or no effect on seed total P, nor result in the accumulation of Ins phosphates with five or fewer phosphate esters, compounds that might be precursors to Ins P_6. Instead, the reductions in seed Ins P_6 are accompanied by increases in inorganic P, so that the sum of Ins P_6 and inorganic P in normal and *lpa*1 seed is similar. In contrast, maize *low phytic acid* 2 (*lpa*2) mutations cause reductions in seed Ins P_6 accompanied by increases in both seed inorganic P and in Ins phosphates with five or fewer esters, such as specific Ins *tris*-, tetra*kis* and penta*kis*phosphates (Raboy *et al.*, 2000). Based on these biochemical phenotypes, it was hypothesized that the maize *lpa*1 locus plays some role in substrate Ins supply in the developing seed, whereas maize *lpa*2 plays some role in the conversion of Ins to Ins P_6. This hypothesis has been confirmed by subsequent studies. In maize *lpa*1-1 seed (seed homozygous for the first recessive allele of the *lpa*1 locus), MIPS expression is reduced in proportion to the reduction in seed Ins P_6 (Shukla *et al.*, 2004). However, no change in the maize chromosome 1S MIPS-encoding sequence has been found. Thus maize *lpa*1 is a gene that is closely linked to the 1S MIPS and that impacts 1S MIPS expression. It might be a regulatory locus that plays some role in the expression of many genes in the Ins and Ins phosphate pathways, or a sequence that specifically regulates MIPS expression. Maize *lpa*2 and the enzyme it encodes, an Ins(1,3,4)P_3 5-/6-kinase (Shi *et al.*, 2003), will be discussed in more detail below. Of relevance here is that maize *lpa*2 mutations represent blocks in the conversion of Ins to Ins P_6. As a result *lpa*2 mutant seed have elevated levels of free Ins, indicating that Ins P_6 synthesis in seed does represent a sink for seed Ins.

In terms of copy number of MIPS-encoding sequences, the barley genome is similar to rice, but differs from maize and soybean. The barley genome contains only one sequence with homology to the canonical MIPS gene, which maps to barley chromosome 4H (Larson and Raboy, 1999). Four barley *lpa* mutations have been studied and genetically mapped in some detail to date (Dorsch *et al.*, 2003; Larson *et al.*, 1998; Ockenden *et al.*, 2004): barley *lpa*1-1 whose expression has been shown to be aleurone-specific and which maps to barley chromosome 2H; barley *lpa*2-1 which maps to barley chromosome 7H and which is phenotypically similar to maize *lpa*2-1, in that reductions in seed Ins P_6 are accompanied by increases in Ins tetra*kis*- and penta*kis*phosphate; barley *lpa*3-1 which maps to barley chromosome 1H; and barley M955 (a gene symbol has not been assigned to this mutation yet), which also maps to chromosome 1H and which may be allelic to barley *lpa*3-1. Barley M955 is of particular interest since seeds homozygous for M955 have Ins P_6 levels reduced by >90%, as compared with non-mutant seeds, but retain near-wild type viability.

These chromosomal mapping results clearly demonstrate that none of these barley *lpa* mutations are MIPS mutations. Barley *lpa*1-1 seed have Ins levels similar to non-mutant seed, whereas in barley *lpa*2-1, *lpa*3-1, and M955 seed, Ins levels are twice that observed in non-mutant seed (Karner *et al.*, 2004). This confirms the results with maize *lpa*2 mutations (Shi *et al.*, 2003) that indicate that seed Ins P_6 synthesis does represent a significant sink for seed Ins.

In the barley *lpa* mutants *lpa*2-1 and M955, the elevation in seed Ins is accompanied by elevations in galactinol, raffinose and to a lesser extent sucrose (Karner *et al.*, 2004). This result is relevant to understanding the regulation of the synthesis of raffinose family oligosaccharides in seeds. This class of compounds is important in its own right in seed biology, protecting seed cellular structure during desiccation and serving as carbon reserves during germination. It is also important to the quality of seeds for their major end-use in foods. The synthesis of raffinosaccharides was thought to be regulated via the activity of a key synthetic enzyme, galactinol synthase (EC 2.4.1.123). However, the results with both the soybean MIPS mutation, where reduced Ins production results in reduced raffinosaccharides (LR33; Hitz *et al.*, 2002) and the barley mutations, where reduced conversion of Ins to Ins P_6 increases free Ins and increases raffinose, clearly indicate that substrate supply is at least as importance as enzymatic regulation.

The second enzyme in the Ins synthesis pathway, IMP, is important both to Ins synthesis and to recycling of Ins from Ins phosphates generated during signal-transduction. IMP is inhibited by lithium, and inhibition studies have shown that suppression of IMP greatly impacts signal-transduction (Berridge and Irvine, 1989). In terms of Ins synthesis, studies such as those discussed above have focused on MIPS as the key regulatory site. In this context, IMP is often viewed as a constitutively expressed enzyme. Of relevance to this discussion, molecular genetics studies have shown that in all plant and non-plant organisms studied to date, IMP is encoded by a multigene family (Gillaspy *et al.*, 1995; Styer *et al.*, 2004). In plant systems detailed studies of IMP have used tomato (*Lycopersicon esculentum*) as the model system (Gillaspy *et al.*, 1995), and these studies have shown that at least two tomato IMP genes are differentially expressed (Styer *et al.*, 2004). The barley genome contains only copy of an IMP-encoding sequence (J. Fu, M. Guttieri, E. Souza, V. Raboy, unpublished results). Therefore it is incorrect to suggest that the multigene copy number of IMP is fundamentally important to differential expression and function in all higher eukaryote.

The molecular genetics studies of both IMP and MIPS-encoding sequences in various plant species provides for this review a first case study for questions pertaining to the biological role of gene copy number. First we must briefly review the general question of gene duplication. Genetics and genomics studies have shown that gene duplication is perhaps the rule rather than the exception, ranging from small duplications of single genes or parts of genes to large duplications of chromosomal sequence, to genomic doubling events and polyploidy. Both genomic doubling events and polyploidy are common in higher plants, whereas only genomic doubling is common in animals. In terms of sequences themselves, and not taking into account epigenetics, following duplication there are perhaps four alternative fates for each of a given ancestral gene's multiple copies (Force *et al.*, 1999; Lynch and Katju, 2004). First, in "nonfunctionalization," a copy might simply provide redundant function, and following accumulation of deleterious "loss-of-function" function mutations, become a non-functional "pseudo-gene." Second, in "neo-functionalization," a duplication may provide the genetic material necessary for the evolution of entirely new functions. Third, in "subfunctionalization," following the "duplication-degeneration-complementation" model, each of the two or more copies of an ancestral gene might assume subsets of that gene's original multiple functions. This last fate would include heritable events that lead to differences in tissue-specificity of gene expression. A fourth possibility might be relevant in plant biology, where polyploidy is so common. Perhaps a copy may provide redundant function, but that redundant function itself might add to an organism's robustness, and therefore be selected for. This fourth possibility has not been given a name following the "nonfunctionalization/neofunctionalization/subfunctionalization" model. Perhaps "superfunctionalization" would be appropriate.

In the case of MIPS, the multiple copy number observed in maize and soybean led to the hypothesis that a seed-specific MIPS is required in large-seeded cultivated species in order to provide sufficient substrate for additional nutrient accumulation (Hegeman *et al.*, 2001). Smaller-seeded species like barley and rice do have only one copy of MIPS. However, it is not readily clear that the relatively small-seeded barley and rice are less fit in general terms than is the relatively closely related maize. Consideration of IMP gene copy number provides a second example of this argument. Tomato and Arabidopsis have three copies of IMP and barley has only one, but barley functions perfectly well in terms of fitness, the ability to make seeds and to do signal transduction. An alternative view is that the benefit of gene duplication in terms of fitness is genome-wide and not gene-specific. In this view, the evolutionary benefits of genome-duplication are in neo-functionalization and super-functionalization, and both non-functionalization and sub-functionalization are essentially neutral in terms of fitness. Barley represents an excellent model system for studies of Ins synthesis since in both the case of MIPS and IMP the complex regulation of the expression of an original, ancestral single-copy gene provides all the functions necessary for the organism's fitness. If there are two closely related species where in one species one copy of gene satisfies all needs, whereas in the sister species multiple copies exist and some have subsets of expression, or "tissue-specific" expression, and if both species clearly are robust, then the evolution of multiple copies and tissue-specific expression is clearly seen as neutral in terms of selection value.

4. INOSITOL TO INS P_6

A brief review of the pathways from Ins to Ins P_6, necessarily limited in details concerning any given enzyme or gene, will provide a framework for discussion of biology and function. In a pathway proposed by Biswas *et al.* (1978a), obtained from studies of Ins P_6 synthesis is mung bean seeds, the product of MIPS, Ins(3)P_1, is directly converted to Ins(1,3,4,5,6)P_5 via sequential phosphorylation catalyzed by a "phosphoinositol kinase" of which two electrophoretic forms were identified. The conversion of Ins(1,3,4,5,6)P_5 to Ins P_6 is then catalyzed by a phytic acid-ADP phosphotransferase (Biswas *et al.*, 1978b). This enzyme is now commonly referred to as Ins(1,3,4,5,6)P_5 2-kinase (Phillippy *et al.*, 1994). Phillippy *et al.* (1994) showed that in the presence of high Ins P_6 and high ADP, a soybean Ins(1,3,4,5,6)P_5 2-kinase could regenerate ATP. This provides biochemical evidence in support of the original hypothesis of Morton and Raison (1963) that Ins P_6 serves as a phosphate donor of sufficient transfer potential for the regeneration of ATP in seeds. This is a question of relevance to seed biology. During the initial stage of germination sufficient membrane integrity has not been established for ATP regeneration that requires generation of the proton-motive force. However, mature seed tissues do contain high concentrations of Ins P_6. The hypothesis that Ins P_6 might function as a high-energy phosphate bond substrate for regeneration of ATP was initially dismissed. More recent studies of pyrophosphate containing Ins phosphates (see below) indicate that this early hypothesis was possibly more accurate than first thought.

A pathway to Ins P_6 that begins with Ins as initial substrate and proceeds through site-specific sequential phosphorylation steps of defined, soluble Ins phosphates (Figure 2 middle left), was described in studies of the cellular slime mold *Dictyostelium discoideum* (Stephens and Irvine, 1990), and of the monocot *Spirodela polyrhiza* (Brearley and Hanke, 1996a). The first step in these pathways is the phosphorylation of Ins at the "3" position via an Ins 3-kinase activity (EC 2.7.1.6.4; English *et al.*, 1966; Loewus *et al.*, 1982). Ultimately of course the source of Ins has to be via MIPS and IMP activity. Next, Ins(3)P_1 is converted to Ins(1,3,4,5,6)P_5 via sequential phosphorylation. The *Dictyostelium* and *Spirodela* pathways are similar in that they share the common intermediate Ins(3,4,6)P_3, but differ in their Ins *bis*phosphate intermediates (Ins(3,6)P_2 versus Ins(3,4)P_2) and Ins tetra*kis*phosphate intermediates (Ins(1,2,4,6)P_4 versus Ins(3,4,5,6)P_4). Finally, Ins(1,3,4,5,6)P_5 is converted to Ins P_6 via an Ins(1,3,4,5,6)P_6 2-kinase (Figure 3). There is little genetic or molecular evidence of an Ins 3-kinase, that converts Ins to Ins(3)P_1. There is one study (Drayer *et al.*, 1994; discussed below) of Ins phosphates in a phospholipase C-null *Dictyostelium* line that provides indirect genetic proof that a pathway proceeding entirely via soluble Ins phosphates and the intermediate Ins(3,4,6)P_3 actually exists.

The above pathway to Ins P_6 that proceeds entirely via soluble Ins phosphates has been described as the phospholipase C (PLC)-independent pathway, to distinguish it from the pathway to Ins P_6 in yeast described below (Stevenson-Paulik *et al.*, 2002). This PLC-independent pathway can be summarized as follows: Ins $\rightarrow$ Ins(3)P_1, catalyzed by Ins 3-kinase *or* glucose 6-P to Ins(3)P_1 catalyzed by MIPS; Ins(3)P_1 $\rightarrow$ Ins(3,4)P_2 or Ins(3,6)P_2 $\rightarrow$ Ins(3,4,6)P_3 $\rightarrow$ Ins(3,4,5,6)P_4 or Ins(1,3,4,6)P_4 $\rightarrow$ Ins(1,3,4,5,6)P_5, catalyzed by Biswas' "phosphoinositol kinases" or via a series of as yet not well described Ins monophosphate, *bis*phosphate and polyphosphate kinases; Ins(1,3,4,5,6)P_5 $\rightarrow$ Ins P_6 via a 2-kinase. The alternative routes between Ins(3)P_1 and InsP_5 in the slime mold versus plant pathways provide a first example of the importance or lack thereof of differences in Ins phosphate positional esters; i.e., Ins(3,4)P_2 versus Ins(3,6)P_2, or Ins(3,4,5,6)P_4 versus Ins(1,3,4,6)P_4. Generally in this field, every observed difference in positional esters is considered of biological significance. Is there any real biological significance to these particular differences?

There is growing genetic evidence that in many organisms Ins P_6 synthesis proceeds via pathways that involve the PtdIns phosphate intermediates PtdIns(4)P_1 and PtdIns(4,5)P_2, followed by PtdIns-specific phospholipase C activity and release of Ins(1,4,5)P_3 (Figure 2 middle right). Up to this point this represents the canonical signal-transduction pathway to Ins(1,4,5)P_3 (Berridge and Irvine, 1989). For this compound to function as a specific and transient signal, it must be rapidly metabolized. This can be accomplished by further phosphorylation, or by dephosphorylation, the later catalyzed by 1-phosphatase and

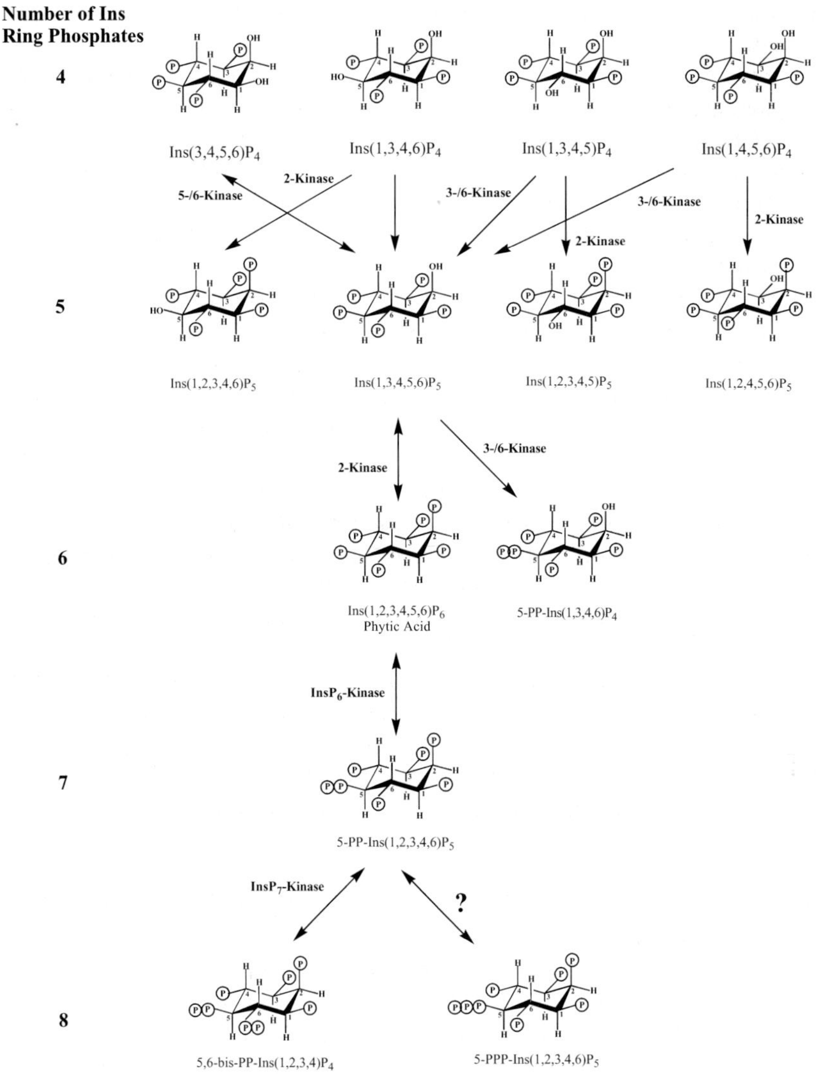

Figure 3. Biosynthetic pathways from *myo*-inositol (Ins) tetra*kis*phosphates to Ins(1,2,3,4,5,6)P_6 ($InsP_6$ or "phytic acid") and the pyrophosphate-containing Ins phosphates. The six carbons of the Ins ring are numbered according to the "D-numbering convention." Questionmarks indicate synthetic steps that have not been confirmed in chemical, molecular or genetic analyzes. In addition, the tri-phosphate-containing 5-PPP-Ins(1,2,3,4,6)P_5 is purely speculative. P = PH_2O_4.

5-phosphatases. There are a growing number of studies that provide genetic proofs that this second messenger pathway functions in plants in ways very similar to its function in other higher eukaryotes. A screen for *Arabidopsis* genes induced by the plant hormone ABA or by stress identified the *FIERY1* (*FRY1*) gene (Xiong *et al.*, 2001). This gene encodes an Ins polyphosphate

1-phosphatase. Mutation of this gene results in the accumulation of Ins(1,4,5)P_3 and in super-induction of ABA- and stress response. Mutations in the *Arabidopsis cotyledon vascular pattern2* (*cvp2*) gene perturb plant development, altering venation patterns (Carland and Nelson, 2004). The *cvp2* gene encodes an Ins polyphosphate 5-phosphatase. The *Arabidopsis* genome contains 15 putative Ins polyphosphate 5-phosphatases (Ercetin and Gillaspy, 2002). Thus there is ample substrate for the evolution of differential substrate specificity, and tissue/developmental/functional specificity of gene expression in signal termination in this plant genome.

In addition to its clear role in signal transduction in plants, Ins(1,4,5)P_3 also may be an important intermediate in pathways to Ins P_6. Depending on the tissue, cellular site or process, this Ins P_6 might function simply as a pool in Ins/PtdIns phosphate pathways, as an effector molecule itself, or as a storage metabolite. In a pathway to Ins P_6 expressed in the nucleus of yeast (York *et al.*, 1999), PtdIns(4,5)P_2 is hydrolyzed to yield Ins(1,4,5)P_3, which is then phosphorylated directly to Ins(1,3,4,5,6)P_5 by an Ins(1,4,5)P_3 3-/6-kinase. This enzyme is also referred in the literature as an IPMK and, in the yeast is encoded by the *IPK2* gene. To avoid confusion we will refer to this enzyme solely as Ins(1,4,5)P_3 3-/6-kinase. Stevenson-Paulik *et al.* (2002) isolated two *Arabidopsis thaliana* genes related to the yeast Ins(1,4,5)P_3 3-/6-kinase, termed AtIpk2α and AtIpk2β An independent study demonstrated that the second of these enzymes is primarily found in the plant nucleus, but is also detected at lower levels in the cytoplasm (Xia *et al.*, 2003). In yeast there is also genetic evidence (York *et al.*, 1999) that Ins(1,3,4,5,6)P_5 is the penultimate Ins phosphate in the pathway to Ins P_6 (Figure 3), and that its conversion is catalyzed by a 2-kinase similar to that described by Biswas *et al.* (1978a,b). The yeast Ins(1,3,4,5,6) 2-kinase is encoded by the *IPK1* gene (York *et al.*, 1999).

The essentials of the major pathway to Ins P_6 in yeast nuclei can be summarized as follows: Ins → PtdIns, catalyzed by PtdIns synthase; PtdIns →→PtdIns(4,5)P_2, catalyzed sequentially by two specific kinases; PtdIns(4,5)P_2→Ins(1,4,5)P_3, catalyzed by a PtdIns-specific phospholipase C; Ins(1,4,5)P_3→→ Ins(1,3,4,5,6)P_5, catalyzed by an Ins(1,4,5)P_3 3-/6-kinase; Ins(1,3,4,5,6)P_5 → Ins P_6, catalyzed by an Ins(1,3,4,5,6)P_5 2-kinase. This is now viewed as the canonical PLC-dependent pathway. Recent genetics studies have shown that this pathway is also critical to Ins P_6 synthesis in *Drosophila* and rat cells (Fujii and York, 2004; Seeds *et al.*, 2004). In contrast, the sole genetic evidence in any organism for a PLC-independent pathway like that described above in *Dictyostelium* and *Spirodela*, one that proceeds solely via soluble Ins phosphates, consists of one elegant study using *Dictyostelium* (Drayer *et al.*, 1994). In this study the presence and levels of the whole series of Ins phosphates typical of a wild-type *Dictyostelium*, including Ins(1,4,5)P_3 and Ins P_6, where shown to be essentially identical in a PLC-null line. Thus some pathway to Ins(1,4,5)P_3 and Ins P_6 *independent of* PLC must exist in this organism.

A PLC-dependent pathway representing an alternative to the yeast PLC-dependent pathway has been described in studies of human cells (Chang *et al.*,

2002; Verbsky *et al.*, 2004; Wilson and Majerus, 1996). These studies described a pathway where Ins(1,4,5)P_3 is first converted to Ins(1,3,4,5)P_4 via an Ins(1,4,5)P_3 3-kinase. This enzyme's activity is very specific for the substrate Ins(1,4,5)P_3. Ins(1,3,4,5)P_4 is then converted to Ins(1,3,4)P_3 via a 5-phosphatase. The 3-kinases and 5-phosphatases are critical to signal transduction pathways in that their activity represents signal termination events, necessary for the transience of the second messenger Ins(1,4,5)P_3. Ins(1,3,4)P_3 is then phosphorylated to Ins(1,3,4,5,6)P_5 via what was first defined as an Ins(1,3,4)P_3 5-/6-kinase (Wilson and Majerus, 1996, 1997). At present there are no genetic proofs that an Ins(1,4,5)P_3 3-/6-kinase of the type identified in Arabidopsis by Stevenson-Paulik *et al.* (2002) is in fact important for the bulk of seed Ins P_6 synthesis or accumulation. That is, mutations in genes encoding an Ins(1,4,5)P_3 3-/6-kinase have not yet been shown to impact seed InsP_6 accumulation, as they have been shown to do in yeast. However, there is genetic evidence that an Ins(1,3,4)P_3 5-/6-kinase has some role in seed Ins P_6 accumulation. The maize *lpa*2 gene encodes an Ins(1,3,4)P_3 5-/6-kinase (Shi *et al.*, 2003). Homozygosity for recessive alleles of maize *lpa*2 reduce seed Ins P_6 by 30% (Raboy *et al.*, 2000). One class of barley "low-phytate" mutants, the "A-Type," accumulate Ins(1,3,4,5)P_4, and it has been proposed that these are mutations in an Ins(1,3,4)P_3 5-6-kinase (Hatzack *et al.*, 2001).

While plant genomes contain sequences encoding both Ins(1,4,5)P_3 3-/6-kinases and Ins(1,3,4)P_3 5-/6-kinases, the yeast genome apparently contains only an Ins(1,4,5)P_3 3-/6-kinase. However, careful analyzes of Ins phosphates in a yeast line null for Ins(1,4,5)P_3 3-/6-kinase indicate that they still accumulate Ins(1,3,4,5)P_4 and low levels of Ins P_6 (Saiardi *et al.*, 2000b). Therefore there must be a second, minor (in terms of yeast) route to Ins P_6 in this organism. The observation that a block in one particular branch of the Ins phosphate pathways reveals an alternative branch, is reminiscent of the findings of Acharya *et al.* (1998). In this study, a *Drosophila* null for Inositol polyphosphate 1-phosphatase (IPP), was shown to be incapable of metabolizing Ins(1,4)P_2, a step critically important to signal transduction via termination of the Ins(1,4,5)P_3 transient signal. However, *Drosophila* IPP nulls "demonstrate compensatory upregulation of an alternative branch in the inositol-phosphate metabolic tree, thus providing a means of ensuring continued availability of inositol" (Acharya *et al.*, 1998). In this alternative branch Ins(1,4,5)P_3 is first converted to Ins(1,3,4)P_3, which is then broken down via activity of a 4-phosphatase, a 3-phosphates, and ultimately the IMP described above.

Perhaps the alternative pathway to Ins(1,3,4,5)P_4 in the yeast Ins(1,4,5)P_3 3-/6-kinase null results from low levels of activity of other enzymes towards various Ins or PtdIns phosphate substrates found in yeast, in a parallel to the results of Acharya *et al.* (1998) concerning the *Drosophila* IPP null. For example, perhaps the accumulation of Ins(1,3,4,5)P_4 results from phospholipase C action on PtdIns(3,4,5)P_3. While no such phospholipase C activity has been documented in any organism, this hypothesis provides a way to discuss a very interesting recent study of PtdIns(3,4,5)P_3 in the fission yeast, *Schizosaccharomyces pombe* (Mitra *et al.*, 2004).

PtdIns(3,4,5)P_3 is a lipid critically important to signal transduction in mammalian cells, and is synthesized from PtdIns(4,5)P_2 primarily via the activity of "class I" phosphoinositide 3-kinases (Cantley, 2002). Thus the pathway to PtdIns(3,4,5)P_3 would proceed as follows: PtdIns → PtdIns(4)P_1 → PtdIns(4,5)P_2 → PtdIns(3,4,5)P_3, the last step catalyzed by the "class I" phosphoinositide 3-kinase. Until recently the lipid PtdIns(3,4,5)P_3 was not observed in yeast cells, and the yeast genome lacks a "class I" phosphoinositide 3-kinase gene. Just as cells need a mechanism to both generate and efficiently breakdown Ins(1,4,5)P_3 if it is to serve as a transient second messenger, mammalian cells express a phosphatase, encoded by the PTEN ("phosphatase and tensin homologue deleted on chromosome 10"), that dephosphorylates PtdIns(3,4,5)P_3, and that is critical to its role in the regulation of cellular development. The fission yeast genome was found to contain one sequence with homology to PTEN, and its deletion resulted in accumulation of PtdIns(3,4,5)P_3. This prompted a re-examination of the pathway to PtdIns(3,4,5)P_3 in the fission yeast. Wild-type fission yeast cells make PtdIns(3,4,5)P_3 via an alternative pathway to that observed in mammalian cells, one that does not utilize "class 1" phosphoinositide 3-kinase. Instead, fission yeast first convert PtdIns to PtdIns(3)P_1, via the activity of a "class III" PtdIns 3-kinase. PtdIns(3)P_1 is then sequentially phosphorylated to yield PtdIns(3,4,5)P_3, whose metabolism via PTEN activity is so rapid that there is no detectable steady state levels. Since the pathway from PtdIns(3)P_1 to PtdIns(3,4,5)P_3 existed in yeast prior to the evolution of "class I" PtdIns 3-kinases in higher eukaryotes, in terms of evolution it has a more ancient function.

The study of Mitra *et al.* (2004) illustrates that the lack of observation of a given compound does not represent proof that that compound has no role in a given organism or tissue. Prior lack of detection of PtdIns(3,4,5)P_3 in yeast simply reflected the fact that this compound is rapidly metabolized, and therefore doesn't accumulate to any detectable steady-state levels. The "class III" PtdIns 3-kinase was first identified as the protein encoded by yeast *vps34* gene, a mutant of which was first isolated in a screen for "vacuolar protein sorting" mutants. The *A. thaliana* genome contains a homolog of this gene, *AtVps34* (Welters *et al.*, 1994). Reduced expression of the *A. thaliana* homolog, achieved via transformation with and expression of an "anti-sense" construct, resulted in severely inhibited growth and development. Clearly, the PtdIns 3-kinase function this gene encodes is important to plant growth and development.

It is interesting that yeast accumulates the linear polyphosphate form of storage phosphate, also found in prokaryotes, but plant cells do not. This suggests one possible explanation for the lack of Ins(1,3,4)P_3 5-/6-kinases in yeast. Perhaps the pathway to Ins P_6 via the Ins(1,4,5)P_3 3-/6-kinase route is primarily nuclear and found in all eukaryotes, whereas the pathway to InsP_6 via the

Ins(1,3,4)P_3 5-/6-kinase route is cytoplasmic, found in those eukaryotes that synthesize Ins P_6 as a non-nuclear storage or inert cellular deposit, and only lost in yeast since this organism uses an alternative pathway to store or sequester excess cellular P.

A recent structural study of the Ins(1,4,5)P_3 3-kinase illustrated that a unique lobe contains four helixes that embrace each of the phosphates in Ins(1,4,5)P_3, explaining both why this enzyme is so substrate specific, and why it cannot phosphorylate PtdIns(4,5)P_2, the "membrane-resident analog of Ins(1,4,5)P_3" (González *et al.*, 2004). Genes encoding this enzyme, along with those encoding the Ins(1,4,5)P_3 3-/6-kinases and the Ins P_6 kinases discussed below, all belong to one gene family descendent from an ancestral Ins phosphate kinase-encoding sequence (Schell *et al.*, 1999). The study of Schell *et al.* (1999) is relevant to this discussion for two reasons. A rabbit cDNA was found to stimulate inorganic P uptake upon injection into a *Xenopus* oocyte, and was termed *Pi Uptake Stimulator* (*PiUS*). This cDNA was found to encode a sequence with homology to Ins(1,4,5)P_3 3-kinase, but in fact the enzyme it encoded had InsP_6 kinase activity. This indicates a link between cellular phosphate nutrition and Ins phosphate metabolism. This link has subsequently been further emphasized by the studies of Almaguer *et al.* (2003, 2004) concerning glycerophosphoinositols, and Steger *et al.* (2003) and El Alami *et al.* (2003), that link Ins phosphates and *Pho* gene expression, as discussed above.

Second, Schell *et al.* (1999) conclude that the Ins(1,4,5) 3-kinase is in evolutionary terms the youngest member of this family of Ins phosphate kinases, its evolution representing a late addition that followed the evolution of Ins(1,4,5)P_3 as a second messenger. An alternative interpretation might be that that the evolution of the function of Ins(1,4,5)P_3 as a second messenger followed the earlier evolution of functions having to do with basic cellular structure and nutrition. This parallels the discussion of Mitra *et al.* (2004) concerning the evolution of the pathway to PtdIns(1,4,5)P_3 in yeast. In that case too what was thought of as the classic signal transduction pathway evolved subsequent to other, more ancient pathways involving these molecules.

The Ins(1,3,4)P_3 5-/6-kinase-encoding genes belong to a second separate and distinct family of Ins phosphate kinases (Cheek *et al.*, 2002). While plant genomes contain Ins(1,4,5)P_3 3-/6-kinases and Ins P_6 kinases, they do not contain members of the gene family encoding the Ins(1,4,5)P_3 3-kinase. How can this be reconciled with the fact that there is solid genetic evidence that Ins P_6 accumulation in seeds requires at least in part a pathway involving Ins(1,3,4)P_3 5-/6-kinase, a pathway that in human cells is thought to require Ins(1,4,5)P_3 3-kinase in order to produce substrate for the Ins(1,3,4)P_3 5-/6-kinase?

There is at least one plausible explanation for the fact that plant genomes don't contain Ins(1,4,5)P_3 3-kinase but require Ins(1,3,4)P_3 5-6-kinase for Ins P_6 production, at least in part. While the Ins(1,4,5)P_3 3-kinase only recognizes Ins(1,4,5)P_3 as substrate, in contrast, both the Ins(1,3,4)P_3 5-/6-kinase and the

Ins(1,4,5)P_3 3-/6-kinase can phosphorylate multiple Ins phosphates. The Ins(1,4,5)P_3 3-/6-kinase has been referred to as a "multikinase," but the Ins(1,3,4)P_3 5-/6-kinase also has "multikinase" and phosphatase activities. For example, the Ins(1,4,5)P_3 3-/6-kinase also can have 5-kinase activity (Chang *et al.*, 2002; Stevenson-Paulik *et al.*, 2002), and can also catalyze the conversion of Ins(1,3,4,5,6)P_5 to a pyrophosphate-containing "5-diphosphoinositol Ins(1,3,4,5)P_4," a "non-phytic acid" InsP_6 (Figure 3; Saiardi *et al.*, 2001a). Similarly, the Ins(1,3,4)P_3 3-/6-kinase also has 1-kinase activity (Yang and Shears, 2000), and can function as a reversible kinase/phosphatase. These two types of Ins polyphosphate mutlikinases can first be distinguished on the basis that one class recognizes Ins(1,4,5)P_3 as substrate (the Ins(1,4,5)P_3 3-/6-kinases) and the second recognizes Ins(1,3,4)P_3 (the Ins(1,3,4)P_3 5-/6-kinases). However, each kind of kinase actually recognizes a unique set of Ins phosphates that ultimately have structural similarities to these two different Ins *tris*phosphates (reviewed in Shears, 2004; Saiardi *et al.*, 2000b, 2001a; Yang and Shears, 2000). It has been hypothesized that the individual Ins phosphates comprising each set contain functionally similar "recognition motifs" consisting of a specific distribution of phosphate and hydroxyl moieties around the Ins ring. One example of this that is relevant to this discussion is that if one rotates and flips Ins(1,3,4)P_3, the distribution of phosphates and one hydroxyl group appear similar to that of Ins(3,4,6)P_3 (Fig. 4). Thus Ins(1,3,4)P_3 5-/6-kinase can convert Ins(1,3,4)P_3 to Ins(1,3,4,6)P_4, but might also be an Ins(3,4,6)P_3 1-kinase, converting it to Ins(1,3,4,6)P_4. Taking the above into consideration, it is possible that one, or perhaps both of these kinases working together, might be able to convert Ins(3,4,6)P_3 to Ins(1,3,4,5,6)P_5. Thus a pathway to Ins P_6 that proceeds entirely via soluble Ins phosphates might in fact utilize enzymes thought of as being part of the PtdIns-intermediate pathway to Ins P_6 (Raboy, 2003; Stevenson-Paulik *et al.*, 2002). These enzymes might in fact represent the two isoforms of "phosphoinositol kinase" first described by Biswas *et al.* (1978a,b). However, it is important to emphasize that to date no single enzyme has been identified that is analogous to Biswas' phosphoinositol kinase, in that it can convert Ins(3)P_1 to Ins(1,3,4,5,6)P_5.

Except for the lack of detailed studies in plant systems, Ins phosphates more highly phosphorylated than Ins P_6, containing pyrophosphate moieties, such as 5-diphosphoinostol(1,2,3,4,6)P_5 ("PP-InsP_5," an Ins P_7) and bis-diphosphoinositol(1,2,3,4)P_4 (bis-PP-InsP_4, an Ins P_8; Figure 3 bottom), have been documented to occur relatively widely in eukaryotic cells (Laussmann *et al.*, 2000; Safrany *et al.*, 1999; Stephens *et al.*, 1993). Enzymes that can generate Ins pyrophosphates include Ins P_6 kinases that produce PP-InsP_5 (Saiardi *et al.*, 1999, 2000a, 2001b), and PP-InsP_5 kinases that produce bis-PP-InsP_4 (Huang *et al.*, 1998; Figure 3 bottom). Ins P_6 kinases appear to be members of the Ins(1,4,5)P_3 3-/6-kinase family that also includes Ins(1,4,5)P_3 3-kinase. There has to date been very little progress in the study of these pyrophosphate-containing compounds in plant systems, with few reports and little detailed

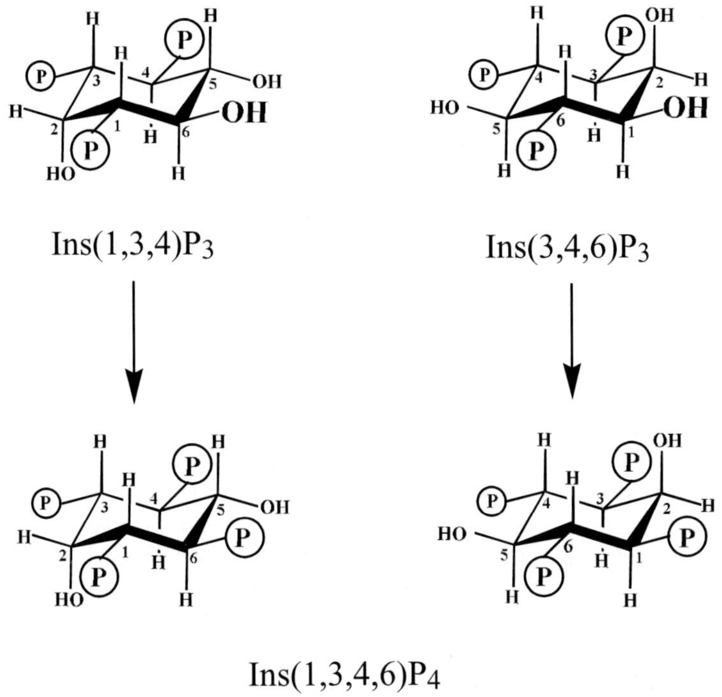

Figure 4. Two proposed modes of action for the *myo*-inositol(1,3,4)P_3 "5-/6-kinase" (Shears, 2004). This enzyme was originally defined as an Ins(1,3,4)P_3 5-/6-kinase, and the 6-kinase step is shown to the left. If one rotates and flips Ins(3,4,6)P_3, as shown to the right, the distribution of phosphates and hydroxyl groups, representing recognition sites for the enzyme, appears similar to that of Ins(1,3,4)P_3. This enzymes putative 1-kinase activity using Ins(3,4,6)P_3 as substrate is shown to the right. P = PH_2O_4.

evidence documenting their presence (Brearley and Hanke, 1996b; Dorsch *et al.*, 2003; Flores and Smart, 2000). There is at present no genetics or molecular genetics of pyrophosphate-containing Ins phosphates in plant systems of any kind.

The PP-InsP_5 kinase of Huang *et al.* (1998) can also act in the reverse direction, using bis-PP-InsP_4 as a phosphate donor in the regeneration of ATP from ADP, "an indication of the high phosphoryl group transfer potential of bis-PP-InsP_4." This is very reminiscent of the original hypothesis of Morton and Raison (1963) that InsP_6 could serve as a donor for the regeneration of ATP, and the subsequent work of Biswas *et al.* (1978a,b) and Phillippy *et al.* (1994) demonstrating that what became known as the Ins(1,3,4,5,6)P_5 2-kinase could, under the right substrate concentrations (high Ins P_6, high ADP), catalyze the reverse reaction and generate ATP from ADP. A second possible function for the pyrophosphate-containing Ins phosphates is that they might bind as effectors that bind to various proteins, acting as a molecular switch in signal transduction (reviewed in Shears, 2004). One example of this is the role Ins phosphates may play as effectors in the *PHO* pathway in yeast. El Alami *et al.* (2003) demonstrated that activity of the yeast $InsP_6$ kinase encoded by the *KCS1* gene is required for repression of *PHO* genes. A third possibility is that pyrophosphate-containing Ins phosphates might indeed serve in protein activation but not directly as effectors that bind to protein. Rather, it has recently been shown that they can non-enzymatically phosphorylate, and thus activate, eukaryotic proteins (Saiardi *et al.*, 2004). A fourth possibility pertains to the classical function of pyrophosphate breakdown. The tri-phosphate-containing 5-triphosphoinositol(1,2,3,4,6)P_5 (another potential Ins P_8 illustrated in Figure 1 bottom right; Shears, 2004) is purely speculative, and has not been documented to exist in any species. It is shown here because its hydrolysis might directly yield pyrophosphate and Ins P_6, providing a model for Ins P_6 transport across a membrane, where pyrophosphate breakdown is used as a driving force (discussed below).

Mammalian "diphosphoinositol polyphosphate phosphohydrolases" ("DIPPs," Figure 1 bottom left) have been identified that cleave Ins pyrophosphates back to Ins P_6 (Caffrey *et al.*, 2000; Hidaka *et al.*, 2002; Hua *et al.*, 2003; Safrany *et al.*, 1998). However Ins P_6 kinases can act in the reverse direction, utilizing Ins pyrophosphates as phosphate donor to regenerate ATP from ADP (Huang *et al.*, 1998), whereas DIPPS act only in Ins pyrophosphate breakdown.

5. LOCALIZATION AND DEPOSITION OF INS P_6 IN THE SEED

The bulk of Ins P_6 that accumulates in seeds is deposited as a mixed salt in discreet inclusions referred to as globoids (Lott, 1984; Raboy, 1997). In the cereal grains Ins P_6 is deposited primarily as a mixed a salt of K and Mg. Globoids are found in one class of storage microvacuoles referred to as Protein Storage Vacuoles (PSVs). PSVs might primarily contain storage proteins, or mixed salts of Ins P_6, or both. In many plant species Ins P_6 deposits occur in specific tissues of the seed. Of particular relevance here is that in the cereal grains, most Ins P_6 deposition occurs within the aleurone layer, the outer layer of the endosperm, and within the germ, consisting of the embryo and scutellum. In normal cereal grains, the central starchy endosperm contains little or no phosphate or Ins P_6 at maturity. Interestingly, even though the cotyledonary tissues of the legume yellow lupine seed are not as differentiated as are the germ and aleurone layer of cereal grains, Ins P_6 deposits still are concentrated in several outer layers of cells (Sobolev *et al.*, 1976). Even though these outer parenchyma cells are otherwise indistinguishable from the inner parenchyma of the cotyledon, they do form a tissue analogous to the cereal aleurone layer, in that they accumulate Ins P_6.

Ogawa *et al.* (1979) demonstrated that early in the development of the rice grain, phosphate, K and Mg are evenly distributed throughout the tissues of the seed. As development progress phosphate becomes more concentrated in the germ and aleurone layer, first accompanied by Mg and later by K. It was hypothesized that is differential timing of deposition of Mg and K simply reflects the changing physiological needs of the developing seed. Thus in seeds, Ins P_6 deposition is intimately linked with mineral storage and deposition. The mechanism by which Ins P_6 synthesis is localized within the developing seed may be central to how mineral deposits are localized. Many other mineral cations can be found as salts of Ins P_6, including Fe, Ca, Mn, and Zn (Lott, 1984). In a study of mineral deposition in developing *A. thaliana* seeds, Otegui *et al.* (2002) showed that while PSVs accumulate K/Mg Ins P_6 salts that are retained till maturity, transient deposits of Mn and Zn salts of Ins P_6 occur during different stages of seed development and are subsequently remobilized. Thus in seeds Ins P_6 appears to function as a counter-ion to minerals and heavy metals, functioning both in transient sequestration and long-term deposition, and the regulation of its synthesis may determine patterns of localization.

Many questions remain as to how Ins P_6 accumulation is localized in seed tissues, how its synthesis is compartmentalized within a given cell, and how deposits are compartmentalized in globoids. One study indicated that Ins P_6 deposits are first observed in the cytoplasm, possibly in association with the endoplasmic reticulum, but that these deposits are remobilized for ultimate deposition with in globoids within PSVs (Greenwood and Bewley, 1984). This study would indicate that Ins P_6 is synthesized in the cytoplasm, and transported for deposition in the globoid. While the focus in this discussion is the plant seed, Ins P_6 deposition is not restricted to the seed in plants, nor to higher plants. Other plant tissues that accumulate nutrient stores, or enter a resting stage and subsequently germinate, such as roots, tubers, turions, and pollen, accumulate Ins P_6 (reviewed in Raboy, 1997). While there is no evidence for deposition of Ins P_6 salts as discrete inclusions in human or yeast cells, recent studies have shown that the parasitic cestode *Echinococcus granulosus* synthesizes Ins P_6, then secretes it into its hydatid cyst wall, depositing it as discreet inclusions consisting of a Ca salt, similar to Ins P_6 mixed salt deposits in plant cells as globoids (Irigoin *et al.*, 2002, 2004).

One mechanism for the localization of Ins P_6 synthesis in a given tissue is via the localized coordinated expression of genes encoding biosynthetic enzymes. Thus Yoshida *et al.* (1999) demonstrated that MIPS expression is localized to the site of Ins P_6 within the rice seed. Shi *et al.* (2003) demonstrated that expression of the maize *lpa*2 gene which encodes an Ins(1,3,4)P_3 5-/6-kinase appears specific to the developing germ tissue. Also, the barley *lpa*1-1 mutation impacts Ins P_6 accumulation in the barley aleurone layer, but has not effect on germ tissue, thus expression of the gene it encodes is tissue specific. While Ins P_6 synthesis probably occurs in most cells of the plant, its accumulation and deposition also represents one component of embryogenesis, and as such is regulated by coordinated gene expression. The *A. thaliana Pickle* (*PKL*) gene encodes a chromatin remodeling factor that normally functions in vegetative tissues to suppress the expression of genes involved in embryogenesis (Ogas *et al.*, 1999). In *pkl* mutants derepression of embryogenesis-related genes results in the expression of embryonic traits in roots, producing "pickle roots." Rider *et al.* (2004) recently demonstrated that this includes increased accumulation of Ins P_6, as compared with normal roots.

If tissue-specific expression of genes contributes to the localization of Ins P_6 synthesis in plant seeds tissues, how is this localized within the cell, and how does deposition of mixed Ins P_6 salts within globoids occur? Figure 5 provides a model for the synthesis of Ins P_6, and its transport with minerals into the PSVs. There is some argument as to whether or not globoids are membrane-bound. Most recently Jiang *et al.* (2001) have provided evidence that globoids are contained within a membrane-bound compartment that itself is contained within the PSV, making the PSV a "compound organelle." For the purposes of this discussion, only a single membrane-bound compartment will be considered. First, water must be transported into the PSV or globoid-containing compartment. This is accomplished via transport through "aquaporins," one of the several isoforms of "Tonoplast Intrinsic Proteins" (TIPs). TIPs are encoded by a large gene family. Different TIPs define functionally distinct vacuoles, and one or more TIPs are probably specific to the globoid containing PSV (Jauh *et al.*, 1999; Takahashi *et al.*, 2004). Second, the membrane delimiting the PSV

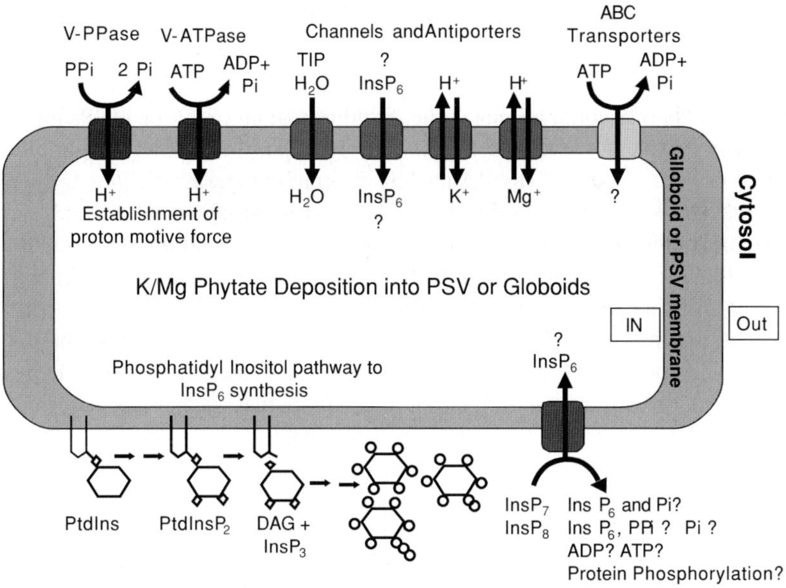

Figure 5. Components and processes of a membrane-bound organelle, representing either a Protein Storage Vacuole (PSV) or a membrane-bound globoid found within a compound PSV. V-PPase = vacuolar inorganic pyrophosphatase. V-ATPase = vacuolar ATPase. TIP = Tonoplast Intrinsic Protein. PtdIns = phosphatidylinositol. DAG = diacylglycerol. Questionmarks indicate purely speculative aspects of the diagram.

(or globoid) is known to contain the vacuolar pyrophosphatase (V-PPase; Jiang *et al.* 2001), which breaks down inorganic pyrophosphatase and in so doing pumps protons into the internal vacuolar space (Maeshima, 2000). Similarly, the vacuolar ATPase breaks down ATP and thereby pumps protons into the vacuolar internal space. Thus a concentration gradient of protons is established that can, via channels and antiporters, drive transport of various solutes from the cytoplasm into the PSV. These solutes must include the counterions K and Mg, and possibly other minerals. Additionally, ABC transporters, which directly use the energy provided by ATP breakdown to transport a variety of solutes, might play some role (Jasinski *et al.*, 2003; Martinoia *et al.*, 2000).

A genetics study of *Arabidopsis* lines that differ quantitatively in the levels of phosphate and Ins P_6 in vegetative and seed tissues identified a Quantitative Trait Locus (QTL) that accounts for a significant amount of the variation observed (Bentsink *et al.*, 2003). Contained within the 99-kb chromosomal segment represented by this QTL were 13 ORFs, one of which encoded a putative V-ATPase. Bentsink *et al.* (2003) hypothesized that the variation in phosphate and Ins P_6 levels observed among the *Arabidopsis* lines in their study was in large part due to variation in phosphate transport caused by heritable differences the V-ATPase, providing experimental evidence for this aspect of the model described in Figure 5.

A pathway to Ins P_6 that proceeds via PtdIns phosphates suggests an mechanism for localization to a specific membrane (Figure 5). However, the question remains as to how Ins P_6 is transported into the PSV or the globoid? Ins polyphosphates containing pyrophosphate moieties may play a key role in this transport. This role could be metabolic. Synthesis of an Ins P_7 or Ins P_8 might first occur in the cytoplasm. Cleavage of the pyrophosphate moiety might then directly drive transport of Ins P_6 into the internal compartment, in a fashion analogous to the function of V-PPases. However, Ins pyrophosphates could also function directly via protein phosphorylation (Saiardi *et al.*, 2004). In this alternative, the Ins pyrophosphate directly phosphorylates the channel protein, resulting in its activation/opening and transport of Ins P_6 to the internal compartment, in a fashion analogous in some ways to Ins (1,4,5)P_3 binding proteins.

If the localization of InsP_6 synthesis is critical to patterns of mineral deposition in the developing seed, then one might hypothesize that the pattern of mineral deposition would be altered in *lpa* mutants. For example, in cereal grains the bulk of Ins P_6 and most minerals is found in the germ and aleurone. The central starchy endosperm normally has very low levels of these seed components. A simple hypothesis would then be that in *lpa* mutants the block in Ins P_6 synthesis would also block the localization of mineral deposition in the germ and aleurone layer, increasing the levels of these minerals in the central endosperm. In a first analysis (Liu *et al.*, 2004) of mineral deposition in the rice *lpa*1-1 mutant (Larson *et al.*, 2000) seeds were dissected into two fractions: a germ fraction including the embryo and scutellum, versus a "rest-of-grain" fraction including the endosperm and aleurone layer. The concerntrations of phosphorus and minerals in these two fractions were similar in normal versus *lpa*1-1 seed. However, this experimental approach cannot detect differences between the central endosperm and aleurone layer, since both were contained in the "rest-of-grain" fraction. A second approach could provide such a distinction. In this second approach, normal and *lpa* grain are milled to varying degrees. Milling produces milled products either enriched in central endosperm, such as white rice, or enriched in aleurone and germ, such as the bran fraction. Analysis of milling fractions in normal versus *lpa*1-1 rice did indicate that the concentrations of K and Mg, but not other minerals such as Fe and Zn, were in fact higher in *lpa*1-1 milled rice (white rice), as compared with white rice produced from the non-mutant control (Bryant *et al.*, 2005). This provides evidence that the localization of mineral deposits in seeds, particularly with reference to K and Mg, is in some part dependent on Ins P_6 synthesis. Since milled rice and wheat products, like white rice and refined wheat flour, are so critical to human nutrition world-wide, enhancing the mineral concentration in milled products could provide an important tool for the improvement of nutritional quality. The increase in K and Mg in white rice produced from the *lpa*1-1 mutant, as compared with that produced from normal rice, where moderate. A similar study of the first wheat *lpa*1-1 mutant (Guittieri *et al.*, 2004) found no substantial changes in mineral distribution in wheat milled products. However, the reduction in seed Ins P_6 in both the first rice and wheat *lpa* mutants are moderate (about 45%). Therefore, additional analyzes of mineral distribution in wheat or rice *lpa* mutants that have more substantial reductions in seed Ins P_6, such as those like the barley M955 mutation, along with analyzes of the set of barley mutants,(Dorsch *et al.*, 2003) would provide a test for the role of Ins P_6 synthesis and mineral deposition in seeds.

6. CONCLUSION

The closer one looks, the biology of Ins, phosphate, PtdIns phosphates, and Ins phosphates, do appear more similar than different across widely divergent organisms. It is possible that distinct pathways exist to Ins P_6 in nuclei versus cytoplasm, but there really isn't strong evidence that divergent eukaryotes differ greatly in their use of these pathways. The differences observed might represents exceptions, not rules. It is possible that the perception of differences results from studies that proceed using paradigms for certain functions and pathways. For example in yeast, the regulation of Ins synthesis was first viewed as critical to lipid metabolism, but now it is becoming clear that the regulation of the yeast *INO1*gene is linked to phosphate nutrition as well. Perhaps in plant systems it will turn out that the regulation of Ins synthesis has importance to lipid metabolism, just as it does in yeast.

The field of Ins phosphate/PtdIns phosphate metabolism and biology has benefited greatly, in fact is largely defined, by the ability to distinguish different positional isomers of these various phosphate esters. However, this paradigm might obscure at times the irrelevance of some differences. Similarly, the signal-transduction paradigm might obscure at times the perhaps more central roles of these metabolites in basal cellular biology.

Finally, new studies often illuminate the importance of pathways and metabolites that simply were not initially observed in first-generation studies. A great example of this is the study of Martin *et al.* (2000) which showed that the Ins phosphate pathways in the amoeba *Entamoeba histolytic* in fact use *neo*-inositol and not *myo*-inositol as the backbone for Ins phosphate synthesis. This suggests that a reanalysis of basic assumptions, leading to a new synthesis of the nature and role of these pathways, might be achieved in the coming years.

REFERENCES

Acharya, J.K., Labarca, P., Delgado, R., Jalink, K., and Zuker, C.S., 1998, Synaptic defects and compensatory regulation of inositol metabolism in inositol polyphosphate 1-phosphatase mutants. *Neuron* **20:** 1219–1229.

Almaguer, C., Cheng, W., Nolder, C., and Patton-Vogt, J., 2004, Glycerophosphoinositol, a novel phosphate source whose transport is regulated by multiple factors in *Saccharomyces cerevisiae*. *J. Biol. Chem.* **279:** 31937–31942.

Almaguer, C., Mantella, D., Perez, E., and Patton-Vogt, J., 2003, Inositol and phosphate regulate *GIT1* trancscription and glycerophosphoinositol incorporation in *Saccharomyces cerevisiae*. *Eukaryot. Cell* **2:** 729–736.

Bentsink, L., Yuan, K., and Koornneef, V., 2003, The genetics of phytate and phosphate accumulation in seeds and leaves of *Arabidopsis thaliana*, using natural variation. *Theor. Appl. Genet.* **106:** 1234–1243.

Berridge, M.J., and Irvine, R.F., 1989, Inositol phosphates and cell signaling. *Nature* **341:** 388–389.

Biswas, B.B., Biswas, S., Chakrabarti, S., and De, B.P., 1978a, A novel metabolic cycle involving *myo*-inositol phosphates during formation and germination of seeds. In Wells, W.W. and Eisenberg, F., Jr. (eds.), Cyclitols and Phosphoinositides. Academic Press, New York, pp. 57–68.

Biswas, S., Maity, I.B., Chakrabarti, S., and Biswas, B.B., 1978b, Purification and characterization of myo-inositol hexaphosphate-adenosine diphosphate phosphotransferase from *Phaseolus aureus*. *Arch. Biochem. Biophys.* **185:** 557–566.

Brearley, C.A., and Hanke, D.E., 1996a, Metabolic evidence for the order of addition of individual phosphate esters to the *myo*-inositol moiety of inositol hexakisphosphate in the duckweed *Spirodela polyrhiza* L. *Biochem. J.* **314:** 227–233

Brearley, C.A., and Hanke, D.E., 1996b, Inositol phosphates in barley (Hordeum vulgare L.) aleurone tissue are sterochemically similar to the products of breakdown of Ins P6 in vitro by wheat bran phytase. *Biochem. J.* **318:** 279–286.

Bryant, R.J., Dorsch, J.A., Rutger, J.N., and Raboy. V., 2005, Amount and distribution of phosphorus and minerals in *low phytic acid* 1 rice seed fractions. *Cereal Chem.* **82:** 517–522.

Caffrey, J.J., Safrany, S.T., Yang, X., and Shears, S.B., 2000, Discovery of molecular and catalytic diversity among human disphosphoinositol polyphosphate phosphohydrolases: An expanding NUDT family. *J. Biol. Chem.* **275:** 12730–12736.

Cantley, L.C., 2002, The phosphoinositide 3-kinase pathway. *Science* **296:** 1655–1657.

Carland, F.M., and Nelson, T., 2004, *COTYLDEDON VASCULAR PATTERN2*-mediated Inositol (1,4,5) trisphosphate signal transduction is essential for closed venation patterns of Arabidopsis foliar organs. *Plant Cell* **16:** 1263–1275.

Carman, G.M., and Henry, S.A., 1989, Phospholipid biosynthesis in yeast. *Annu. Rev. Biochem.* **58:** 635–669.

Chang, S.-C., Miller, A.L., Feng, Y., Wente, S.R., and Majerus, P.W., 2002, The human homolog of the rat inositol phosphate multikinase is an inositol 1,3,4,6-tetrakisphosphate 5-kinase. *J. Biol. Chem.* **277:** 43836–43843.

Cheek, S., Zhang, H., and Grishin, N.V., 2002, Sequence and structure classification of kinases. *J. Mol. Biol.* **320:** 855–881.

Donahue, T.F., and Henry, S.A., 1981, *myo*-Inositol-1-phosphate synthase: Characteristics of the enzyme and identification of its structural gene in yeast. *J. Biol. Chem.* **256:** 7077–7085.

Dorsch, J.A., Cook, A., Young, K.A., Anderson, J.M., Bauman, A.T., Volkmann, C.J., Murthy, P.P.N., and Raboy, V., 2003, Seed phosphorus and inositol phosphate phenotype of barley *low phytic acid* genotypes. *Phytochem.* **62:** 691–706.

Drayer, A.L., Van der Kaay, J., Mayr, G.W., and Van Haastert, P.J.M., 1994, Role of phospholipase C in *Dictyostelium*: Formation of inositol 1,4,5-trisphosphate and normal development in cells lacking phospholipase C activity. *EMBO J.* **13:** 1601–1609.

El Alami, M., Messenguy, F., Scherens, B., and Dubois, E., 2003, Arg82p is a bifunctional protein whose inositol polyphosphate kinase activity is essential for nitrogen and *PHO* gene expression but not for Mcm1p chaperoning in yeast. *Mol. Microbiol.* **49:** 457–468.

English, P.D., Dietz, M., and Albersheim, P., 1966, Myoinositol kinase: Partial purification and identification of product. *Science* **151:** 198–199.

Ercetin, E.E., and Gillaspy, G.E., 2002, Molecular characterization of an Arabidopsis gene encoding a phospholipid-specific inositol polyphosphate 5-phosphatase. *Plant Physiol.* **135:** 938–946.

Europe-Finner, G.N., Gammon, B., Wood, C.A., and Newell, P.C., 1989, Inositol tris- and polyphosphate forming during chemotaxis of *Dictyostelium*. *J. Cell. Sci.* **93**, 585–592.

Flores, S., and Smart, C.C., 2000, Abscisic acid-induced changes in inositol metabolism in Spirodela polyrrhiza. *Planta* **211:** 823–832.

Force, A., Lynch, M., Pickett, F.B., Amores, A., Yan, Y., and Postlethwait., 1999, Preservation of duplicate genes by complementary, degenerative mutations. *Genetics* **151:** 1531–1545.

Fujii, M., and York, J.D., 2004, A role for rat inositol polyphosphate kinases, rIpk2 and rIpk1, in inositol pentakisphosphate and inositol hexakisphosphate production in Rat-1 cells. *J. Biol. Chem.*: In Press.

Gillaspy, G.E., Keddie, J.S., Oda, K., and Gruissem, W., 1995, Plant inositol monophosphatase is a lithium-sensitive enzyme encoded by a multigene family. *Plant Cell* **7:** 2175–2185.

González, B., Schell, M.J., Letcher, A.J., Veprintsev, D.B., Irvine, R.F., and Williams, R.L., 2004, Structure of a human inositol 1,4,5-trisphosphate 3-kinase: Substrate binding reveals why it is not a phosphoinositide 3-kinase. *Mol. Cell* **15:** 689–701.

Greenberg, M.L., and Lopes, J.M., 1996, Genetic regulation of phospholipids biosynthesis in *Saccharomyces cerevisiae*. *Microbiol. Rev.* **60:** 1–20.

Greenwood, J.S., and Bewley, J.D., 1984, Subcellular distribution of phytin in the endosperm of developing castor bean: A possibility for its synthesis in the cytoplasm prior to deposition within protein bodies. *Planta* **160:** 113–120.

Guttieri, M., Bowen, D., Dorsch, J.A., Souza, E. and Raboy, V., 2004, Identification and characterization of a low phytic acid wheat. *Crop Sci.* **44**.

Hanakahi, L.A., Bartlet-Jones, M., Chappell, C., and West, S.C., 2000, Binding of inositol phosphate to DNA-PK and stimulation of double-strand break repair. *Cell* **102:** 721–729

Hanakahi, L.A., and West, S.C., 2002, Specific interaction of IP6 with human Ku70/80, the DNA-binding subunit of DNA-PK. *EMBO J.* **21**: 2038–2044.

Hatzack, F., Hubel, F., Zhang, W., Hansen, P.E., and Rasmussen, S.K., 2001, Inositol phosphates from barley low-phytate grain mutants analyzed by metal-dye detection HPLC and NMR. *Biochem. J.* **354:** 473–480.

Hegeman, C.E., Good, L.L., and Grabau, E.A., 2001, Expression of D-*myo*-Inositol-3-phosphate synthase in soybean. Implications for phytic acid biosynthesis. *Plant Physiol.* **125:** 1941–1948.

Hidaka, K., Caffrey, J.J., Hua, L., Zhang, T., Falck, J.R., Nickel, G.C., Carrel, L., Barnes, L.D., and Shears, S.B., 2002, An adjacent pair of human NUDT genes on chromosome X are preferentially expressed in testis and encode tow new isoforms of diphosphoinositol polyphosphate phosphohydrolase. *J. Biol. Chem.* **277:** 32730–32738.

Hitz, W.D., Carlson, T.J., Kerr, P.S, and Sebastian, S.A., 2002, Biochemical and molecular characterization of a mutation that confers a decreased raffinosaccharide and phytic acid phenotype on soybean seeds. *Plant Physiol.* **128:** 650–660.

Hua, L.V., Hidaka, K., Pesesse, X., Barnes, L.D., and Shears, S.B., 2003, Paralogous murine *Nudt10* and *Nudt11* genes have differential expression patterns but encode identical proteins that are physiologically competent disphosphoinositol polyphosphate phosphohydrolases. *Biochem. J.* **373:** 81–89.

Huang, C.-F., Voglmaier, S.M., Bembenek, M.E., Saiardi, A., and Snyder, S.H., 1998, Identification and purification of diphosphoinositol pentakisphosphate kinase, which synthesizes the inositol pyrophosphate bis(diphospho)inositol tetrakisphosphate. *Biochem.* **37:** 14998–15004.

Ishitani, M., Majumder, A.L., Bornhouser, A., Michalowski, C.B., Jensen, R.G., and Bohnert, H.J., 1996, Coordinate transcriptional induction of *myo*-inositol metabolism during environmental stress. *Plant J.* **9:** 537–548.

Irigoin, F., Casaravilla, C., Iborra, F., Sim, R.B., Ferreira, F., and Diaz, A., 2004, Unique precipitation and exocytosis of a calcium salt of *myo*-inositol hexaphosphate in larval *Echinococcus granulosus*. *J. Cellular Biochem.* **93:** 1272–1281.

Irigoin, F., Ferreira, F., Fernandez, C., Sim, R.B., and Diaz, A., 2002, *myo*-inositol hexakisphosphate is a major component of an extracellular structure in the parasitic cestode *Echinococcus granulosus*. *Biochem J.* **362:** 297–304.

Irvine, R.F., and Schell, M.J., 2001, Back in the water: The return of the inositol phosphates. *Nature Rev. Mol. Cell Biol.* **2:** 327–338.

Jasinski, M., Ducos, E., Martinoia, E., and Boutry, M., 2003, The ATP-binding cassette transporters: Structure, function, and gene family comparison between rice and Arabidopsis. *Plant Physiol.* **131:** 1169–1177.

Jauh, G-Y., Phillips, T.E., and Rogers, J.C., 1999, Tonoplast intrinsic protein isoforms as markers for vacuolar functions. *Plant Cell* **11:** 1867–1882.

Jiang, L., Phillips, T.E., Hamm, C.A., Drozdowicz, Y.M., Rea, P.A., Maeshima, M., Rogers, S.W., and Rogers, J.C., 2001, The protein storage vacuole: A unique compound organelle. *J. Cell Biol.* **155:** 991–1002.

Karner, U., Peterbauer, T., Raboy, V., Jones, D.A., Hedley, C.L., and Richter. A., 2004, *myo*-Inositol and sucrose concentrations affect the accumulation of raffinose family oligosaccharides in seeds. *J. Exp. Bot.* **55:** 1981–1987.

Lackey, K.H., Pope, P.M., and Dean Johnson, M., 2003, Expression of 1L-myoinositol-1-phosphate synthase in organelles. *Plant Physiol.* 2240–2247.

Larson, S.R., and Raboy, V., 1999, Linkage mapping of maize and barley *myo*-inositol 1-phosphate synthase DNA sequences: Correspondence with a *low phytic acid* mutation. *Theor. Appl. Genet.* **99:** 27–36.

Larson, S.R., Rutger, J.N., Young, K.A., and Raboy, V., 2000, Isolation and genetic mapping of a non-lethal rice (*Oryza sativa* L.) *low phytic acid* mutation. *Crop Sci.* **40:** 1397–1405.

Larson, S.R., Young, K.A., Cook, A., Blake, T.K., and Raboy, V., 1998, Linkage mapping two mutations that reduce phytic acid content of barley grain. *Theor. Appl. Genet.* **97:** 141–146.

Laussmann, T., Pikzack, C., Thiel, U., Mayr, G.W., and Vogel, G., 2000, Diphospho-*myo*-inositol phosphates during the life cycle of *Dictyostelium* and *Polysphondylium*. *Eur. J. Biochem.* **267:** 2447–2451.

Lemtiri-Chlieh, F., MacRobbie, E.A.C., and Brearley, C.A., 2000, Inositol hexakisphosphate is a physiological signal regulating the K^+-inward rectifying conductance in guard cells. *Proc. Natl. Acad. Sci. U.S.A.* **97:** 8687–8692.

Lemtiri-Chlieh, F., MacRobbie, E.A.C., Webb, A.A.R., Mansion, N.F., Brownlee, C., Skepper, J.N., Chen, J., Prestwich, G.D., and Brearley, C.A., 2003, Inositol hexaphosphate mobilizes an endomembrane store of calcium in guard cells. *Proc. Natl. Acad. Sci. U.S.A* **100:** 10091–10095.

Liu, J.C., Ockenden, I., Truax, M., and Lott, J.N.A., 2004, Phytic acid-phosphorus and other nutritionally important mineral nutrient elements in grains of wild-type and *low phytic acid* (*lpa*1-1) rice. *Seed Sci. Res.* **14:** 109–116.

Loewen, C.J.R., Gaspar, M.L., Jesch, S.A., Delon, C., Ktistakis, N.T., Henry, S.A., and Levine, T.P., 2004, Phospholipid metabolism regulated by a transcription factor sensing phosphatidic acid. *Science* **304:** 1644–1647.

Loewus, F.A., and Murthy, P.P.N., 2000, *myo*-Inositol metabolism in plants. *Plant Sci.* **150:** 1–19.

Loewus, M.W., Sasaki, K., Leavitt, A.L., Munsell, L., Sherman, W.R., and Loewus, F.A., 1982, The enantiomeric form of *myo*-inositol-1-phosphate produced by *myo*-inositol 1-phosphate synthase and *myo*-inositol kinase in higher plants. *Plant Physiol.* **70:** 1661–1663.

Lott, J.N.A., 1984, Accumulation of seed reserves of phosphorus and other minerals. In: Murray, D.R. (ed.), Seed Physiology. Academic Press, New York, pp. 139–166.

Lott, J.N.A., Ockenden, I., Raboy, V., and Batten, G.D., 2000, Phytic acid and phosphorus in crop seeds and fruits: A global estimate. *Seed Sci. Res.* **10:** 11–33.

Lynch, M., and Katju, V., 2004, The altered evolutionary trajectories of gene duplicates. *Trends Genet.* **20:** 544–549.

Maeshima, M., 2000, Vacuolar H^+-pyrophosphatase. *Biochimica. Biophysica. Acta.* **1465:** 37–51.

Majee, M., Maitra, S., Dastidar, K.G., Pattnaik, S., Chatterjee, A., Hait, N.C., Das, K.P., and Majumder, A.L., 2004, A novel salt-tolerant L-*myo*-inositol-1-phosphate synthase from *Porteresia coarctata* (Roxb.) Tateoka, a halophytic wild rice: Molecular cloning, bacterial overexpression, characterization, and functional introgression into tobacco-conferring salt tolerance phenotype. *J. Biol. Chem.* **279:** 28539–28552.

Majumder, A.L., Chatterjee, A., Dastidar, K.G., and Majee, M., 2003, Diversification and evolution of L-*myo*-inositol 1-phosphate synthase. *FEBS Lett.* **553:** 3–10.

Martin, J.-B., Laussmann, T., Bakker-Grunwald, T., Vogel, G., and Klein, G., 2000, *neo*-Inositol polyphosphates in the amoeba *Entamoeba histolytica*. *J. Biol. Chem.* **275:** 10134–10410.

Martinoia, E., Massonneau, A., and Frangne, N., 2000, Transport processes of solutes across the vauolar membrane of higher plants. *Plant Cell Physiol.* **41:** 1175–1186.

Meis, S.J., Fehr, W.R., and Schnebly, S.R., 2003, Seed source effect on field emergence of soybean lines with reduced phytate and raffinose saccharides. *Crop Sci.* **43:** 1336–1339.

Mitra, P., Zhang, Y., Rameh, L.E., Ivshina, M.P., McCollum, D., Nunnari, J.J., Hendricks, G.M., Kerr, M.L., Field, S.J., Cantley, L.C., and Ross, A.H., 2004, A novel phosphatidylinositol(3,4,5)P_3 pathway in fission yeast. *J. Cell Biol.* **166:** 205–211.

Morton, R.K., and Raison, J.K., 1963, A complete intracellular unit for incorporation of amino-acid into storage protein utilizing adenosine triphosphate generated from phytate. *Nature* **200:** 429–433.

Ockenden, I., Dorsch, J.A., Reid, M.M., Lin, L., Grant, L.K., Raboy, V., and Lott, J.N.A., 2004, Characterization of the storage of phosphorus, inositol phosphate and cations in grain tissues of four barley (*Hordeum vulgare* L.) *low phytic acid* genotypes. *Plant Sci.* **167:** 1131–1142.

Odom, A.R., Stahlberg, A., Wente, S.R., and York, J.D., 2000, A role for nuclear inositol 1,4,5-trisphosphate kinase in transcriptional control. *Science* **287:** 2026–2029.

Ogas, J., Kaufmann, S., Henderson, J., and Somerville, S., 1999, Pickle is a CHD3 chromatin-remodeling factor that regulates the transition from embryonic to vegetative development in *Arabidopsis*. *Proc. Natl. Acad. Sci. U.S.A.* **96:** 13839–13844.

Ogawa, M., Tanaka, K., and Kasai, Z., 1979, Accumulation of phosphorus, magnesium, and potassium in developing rice grains: Followed by electron microprobe X-ray analysis focusing on the aleurone layer. *Plant Cell Physiol.* **20:** 19–27.

O'Neill, E.M., Kaffman, A., Jolly, E.R., O'Shea, E.K. 1996. Regulation of PHO4 nuclear localization by the PHO80-PHO85 cyclin-CDK complex. *Science* **271:** 209–212.

Otegui, M.S., Capp, R., and Staehelin, L.A., 2002, Developing seeds of Arabidopsis store different minerals in two types of vacuoles and in the endosplasmic reticulum. *Plant Cell* **14:** 1311–1327.

Phillippy, B.Q., Ullah, A.H.J., and Ehrlich, K.C., 1994, Purification and some properties of inositol 1,3,4,5,6-penta*kis*phosphate 2-kinase from immature soybean seeds. *J. Biol. Chem.* **269:** 28393–28399.

Raboy, V., 1997, Accumulation and storage of phosphate and minerals. In: Larkins, B.A., Vasil, I.K. (eds.), Cellular and Molecular Biology of Plant Seed Development. Kluwer Academic Publishers, Dordrecht Netherlands, pp. 441–477.

Raboy, V., 2001, Seeds for a better future: "Low phytate" grains help to overcome malnutrition and reduce pollution. *Trends in Plant Sci.* **6:** 458–462.

Raboy,V., and Gerbasi, P., 1996, Genetics of *myo*-inositol phosphate synthesis and accumulation. In: Biswas, B.B., Biswas, S. (eds.), *myo*-Inositol Phosphates, Phosphoinositides, and Signal Transduction. Plenum Press, New York, pp. 257–285.

Raboy, V., Gerbasi, P.F., Young, K.A., Stoneberg, S.D., Pickett, S.G., Bauman, A.T., Murthy, P.P.N., Sheridan, W.F., and Ertl, D.S., 2000, Origin and seed phenotype of maize *low phytic acid 1-1* and *low phytic acid 2-1*. *Plant Physiol.* **124:** 355–368.

Raboy, V., Young, K.A., Dorsch, J.A., and Cook, A., 2001, Genetics and breeding of seed phosphorus and phytic acid. *J. Plant Physiol.* **158:** 489–497.

Rider, S.D., Jr., Hemm, M.R., Hostetler, H.A., Li, H.-C., Chapple, C., and Ogas. J., 2004, Metabolic profiling of the *Arabidopsis pkl* mutant reveals selective derepression of embryonic traits. *Planta* **219:** 489–499.

Safrany, S.T., Caffrey, J.J., Yang, X., Bembenek, M.E., Moyer, M.B., Burkhart, W.A., and Shears, S.B., 1998, A novel context for the 'MutT' module, a guardian of cell integrity, in a diphosphoinositol polyphosphate phosphohydrolase. *EMBO J.* **17:** 6599–6607.

Safrany, S.T., Caffrey, J.J., Yang, X., and Shears, S.B., 1999, Diphosphoinositol polyphosphates: The final frontier for inositide research? *Biol. Chem.* **380:** 945–951.

Saiardi, A., Bhandari, R., Resnick, A.C., Snowman, A.M., and Snyder, S.H., 2004, Phosphorylation of proteins by inositol pyrophosphates. *Science* **306:** 2101–2105.

Saiardi, A., Caffrey, J.J., Snyder, S.H., and Shears, S.B., 2000a, The inositol hexakisphosphate kinase family: Catalytic flexibility and function in yeast vacuole biogenesis. *J. Biol. Chem.* **275:** 24686–24692.

Saiardi, A., Caffrey, J.J., Snyder, S.H., and Shears, S.B., 2000b, Inositol polyphosphate multikinase (ArgRIII) determines nuclear mRNA export in *Saccharomyces cerevisiae*. *FEBS Lett.* **468:** 28–32.

Saiardi, A., Erdjument-Bromage, H., Snowman, A.M., Tempst, P., and Snyder, S.H., 1999, Synthesis of diphosphoinositol pentakisphosphate by a newly identified family of higher inositol polyphosphate kinases. *Curr. Biol.* **9:** 1323–1326.

Saiardi, A., Nagata, E., Luo, H.R., Sawa, A., Luo, X., Snowman, A.M., and Snyder, S.H., 2001a, Mammalian inositol polyphosphate multikinase synthesizes inositol 1,4,5-trisphosphate and an inositol pyrophosphate. *Proc. Natl. Acad. Sci. U.S.A* **98:** 2306–3211.

Saiardi, A., Nagata, E., Luo, H.R., Snowman, A.M., and Snyder, S.H., 2001b, Identification and characterization of a novel inositol hexakisphosphate kinase. *J. Biol. Chem.* **276:** 39179–39185.

Saiardi, A., Sciambi, C., McCaffery, J.M., Wendland, B., and Snyder, S.H., 2002, Inositol polyphosphates regulate endocytic trafficking. *Proc. Natl. Acad. Sci. U.S.A.* **99:** 14206–14211.

Sasakawa, N., Sharif, M., and Hanley, M.R., 1995, Metabolism and biological activities of inositol pentakisphosphate and inositol hexakisphosphate. *Biochem. Pharmacol.* **50:** 137–146.

Schell, M.J., Letcher, A.J., Brearley, C.A., Biber, J., Murer, H., and Irvine, R.F., 1999, PiUS (Pi uptake stimulator) is an inositol hexaphosphate kinase. *FEBS Lett.* **461:** 169–172.

Seeds, A.M., Sandquist, J.C., Spana, E.P., and York, J.D., 2004, A molecular basis for inositol polyphosphate synthesis in *Drosophila melanogaster*. *J. Biol. Chem.* **279:** 47222–47232.

Shears, S.B., 2001, Assessing the omnipotence of inositol hexakisphosphate. *Cell. Signal.* **13:** 151–158.

Shears, S.B., 2004, How versatile are inositol phosphate kinases? *Biochem. J.* **377:** 265–280.

Shen, X., Xiao, H., Ranallo, R., Wu, W.-H., and Wu, C., 2003, Modulation of ATP-dependent chromatin-remodeling complexes by inositol phosphates. *Science* **299:** 112–114.

Shi, J., Wang, H., Wu, Y., Hazebroek, J., Meeley, R.B., and Ertl, D.S., 2003, The maize low phytic acid mutant *lpa2* is caused by mutation in an inositol phosphate kinase gene. *Plant Physiol.* **131:** In Press.

Shukla, S., VanToai, T.T., and Pratt, R.C., 2004, Expression and nucleotide sequence of an INS(3)P_1 synthase gene associated with low-phytate kernels in maize (*Zea mays* L.). *J. Agric. Food Chem.* **52:** 4565–4570.

Smart, C.C., and Fleming, A.J., 1993, A plant gene with homology to D-*myo*-inositol-3-phosphate synthase is rapidly and spatially up-regulated during an abscisic-acid-induced morphogenic response in *Spirodela polyrrhiza*. *Plant J.* **4:** 279–293.

Sobolev, A.M., Buzulukova, N.P., Dmitrieva, M.I., and Barbashova, A.K., 1976, Structural-biochemical organization of aleurone grains in yellow lupin. *Soviet Plant Physiol.* **23:** 739–746.

Steger, E.J., Haswell, E.S., Miller, A.L., Wente, S.R., and O'Shea, E.K, 2003, Regulation of chromatin remodeling by inositol polyphosphates. *Science* 299**:** 114–116.

Stephens, L.R., and Irvine, R.F., 1990, Stepwise phosphorylation of *myo*-inositol leading to *myo*-inositol hexakisphosphate in *Dictyostelium*. *Nature* **346:** 580–583.

Stephens, L., Radenberg, T., Thiel, U., Vogel, G., Khoo, K.-H., Dell, A., Jackson, T.R., Hawkins, P.T., and Mayr, G.W., 1993, The detection, purification, structural characterization, and metabolism of diphosphoinositol pentakisphosphate(s) and bisdiphosphoinositol tetrakisphosphate(s). *J. Biol. Chem.* **268:** 4009–4015.

Stevenson-Paulik, J., Odom, A.R., and York, J.D., 2002, Molecular and biochemical characterization of two plant inositol polyphosphate 6-/3-/5-kinases. *J. Biol. Chem.* **277:** 42711–42718.

Styer, J.C., Keddie, J., Spence, J., and Gillaspy, G.E., 2004, Genomic organization and regulation of the *LeIMP-1* and *LeIMP-2* genes encoding *myo*-inositol monophosphatase in tomato. *Gene* **326:** 35–41.

Takahashi, H., Rai, M., Kitagawa, T., Morita, S., Masumura, T., and Tanaka, K., 2004, Differential localization of tonoplast intrinsic proteins on the membrane of protein body type II and aleurone grain in rice seeds. *Biosci. Biotchnol. Biochem.* **68:** 1728–1736.

Verbsky, J.W., Chang, S.-C., Wilson, M.P., Mochizuki, Y., and Majerus, P.W., 2004, The pathway for the production of inositol hexakisphosphate (InsP_6) in human cells. *J. Biol. Chem.*: In Press.

Welters, P., Takegawa, K., Emr, S.D., and Chrispeels, M.J., 1994, ATVPS34, a phosphatidylinositol 3-kinase of *Arabidopsis thaliana*, is an essential protein with homology to a calcium-dependent lipid binding protein. *Proc. Natl. Acad. Sci. U.S.A.* **91:** 11398–11402.

Wilson, M.P., and Majerus, P.W., 1996, Isolation of inositol 1,3,4-trisphosphate 5/6-kinase, cDNA cloning and expression of the recombinant enzyme. *J. Biol. Chem.* **271:** 11904–11910.

Wilson, M.P., and Majerus, P.W., 1997, Characterization of a cDNA encoding *Arabidopsis thaliana* Insoitol 1,3,4-trisphosphate 5/6-kinase. *Biochem. Biophys. Res. Comm.* **232:** 678–681.

Xia, H.-J., Brearley, C., Elge, S., Kaplan, B., Fromm, H., and Mueller-Roeber, B., 2003, Arabidopsis inositol polyphosphate 6-/3-kinase is a nuclear protein that complements a yeast mutant lacking a functional Arg-Mcm1 transcription complex. *Plant Cell* **15:** 449–463.

Xiong, L., Lee, B., Ishitani, M., Lee, H., Zhang, C., and Zhu, J.-K., 2001, *Fiery1* encoding an inositol polyphosphate 1-phosphatase is a negative regulator of abscisic acid and stress signaling in *Arabidopsis*. *Genes Devel.* **15:** 1971–1984.

Yang, X., and Shears, S.B., 2000, Multitasking in signal transduction by a promiscuous human Ins(3,4,5,6)P_4 1-kinase/Ins(1,3,4)P_3 5/6-kinase. *Biochem. J.* **351:** 551–555.

York, J.D., Odom, A.R., Murphy, R., Ives, E.B., and Wente, S.R., 1999, A phospholipase C-dependent inositol polyphosphate kinase pathway required for efficient messenger RNA export. *Science* **285:** 96–100.

Yoshida, K.T., Wada, T., Koyama, H., Mizobuchi-Fukuoka, R., and Naito, S., 1999, Temporal and spatial patterns of accumulation of the transcript of *myo*-inositol-1-phosphate synthase and phytin-containing particles during seed development in rice. *Plant Physiol.* 119**:** 65–72.

Chapter 5

Inositol in Bacteria and Archaea

Mary F. Roberts
Department of Chemistry, Boston College, Chestnut Hill, MA 02467, USA

1. INTRODUCTION

Phosphorylated *myo*-inositol is an important moiety in biochemistry as the basis for signal transduction pathways in eukaryotes. Inositol also occurs in bacteria and archaea, not as part of a signaling pathway, but with diverse and unique uses. *myo*-Inositol occurs as part of mycothiol, a molecule comparable to glutathione in mycobacteria, as part of an unusual osmolyte in hyperthermophiles [*e.g.*, di-*myo*-inositol-1,1′-phosphate (DIP)], and as the lipid headgroup anchor for a series of glycosylated lipids in mycobacteria that are critical in the interaction of pathogenic mycobacteria with mammalian cells.

myo-Inositol biosynthesis occurs by the same steps in all organisms, and comparisons of bacterial/archaeal enzymes to their eukaryotic counterparts often yield interesting surprises. The inositol moiety can be converted to a soluble molecule with specific roles in cells or it can be fixed into membrane components. When unique inositol-containing products are generated in pathogens, these can be targets for drug development. Bacteria have also evolved enzymes that specifically degrade inositol-containing compounds: phosphatidylinositol (PI)-specific phospholipase C would hydrolyze the PI in target membranes, while phytases hydrolyze inositol hexakisphosphate (phytate) to supply cells with inorganic phosphate.

This review aims to present some of the more interesting molecules containing inositol synthesized by bacterial and archaeal cells, and to describe what is known about their biosynthesis. A brief review of bacterial and archaeal inositol biosynthesis and catabolic activities is also provided.

A. Lahiri Majumder and B. B. Biswas (eds.), Biology of Inositols and Phosphoinositides, 103–134
© 2006 *Springer. Printed in the Netherlands.*

2. IDENTITY OF INOSITOL-CONTAINING COMPOUNDS FOUND IN CELLS

In bacteria and archaea, inositol-containing molecules are not ubiquitous but restricted to certain classes of organisms. Inositol solutes involved in osmotic balance have been detected in hyperthermophilic archaea and Thermotoga sp. Several archaea have also been shown to contain phospholipids containing D-*myo*-inositol 1-phosphate. However, mycobacteria contain the largest variety of inositol compounds including mycothiol, a glutathione analogue, as well as an array of diacylglycerol (DAG)-based glycosylinositols and glycosylinositol phosphorylceramides (GIPCs). Structures of these unusual metabolites are presented in Table 1.

2.1 Soluble inositol solutes – what does and does not occur in microorganisms

myo-Inositol occurs and accumulates in many eukaryotic cells. In mammals, it often functions as an osmolyte, a role well established in the CNS (Fisher *et al.*, 2002). There are no reports of it accumulating in bacteria at detectable concentrations. Instead *myo*-inositol is converted to specialized soluble phosphate esters with unique roles in protecting cells against stress and to various phospholipids, many of which are unique to bacteria and aid in infectivity of an organism.

Inositol phosphates are key components of eukaryotic signaling pathways that do not occur in bacteria and archaea. Thus, molecules such as inositol-1-phosphate or polyphosphorylated inositols (IP_2, IP_3) have not been detected in bacteria and archaea. Nonetheless, these organisms have evolved pathways to degrade phosphoinositides as a way of scavenging inorganic phosphate. *myo*-Inositol hexakisphosphate, or phytate, is an abundant plant constituent and is the main storage form of phosphate in seeds. It is also found in other eukaryotes where its metabolism is a basic component of cellular housekeeping (see Raboy, 2003, for review). While there are no reports of phytate synthesis in bacteria, many microorganisms have evolved a class of specialized phosphatases to degrade this compound and thus provide the cells with inorganic phosphate. Bacterial phytases have industrial uses. For example, in the baking industry, lactic acid bacteria possessing phytases enhance demineralization of flour. Phytate may also play other roles in cells. For example, IP_6 appears to mediate iron transport in *Pseudomonas aeruginosa* (Hirst *et al.*, 1999), although mechanistic details are not clear at this time.

2.1.1 DIP, DIP-isomers, and mannosyl-DIP derivatives

DIP is a complex solute that was first noted as a major solute in *Pyrococcus woesei* (Scholz *et al.*, 1992) and *Methanococcus igneus* (Ciulla *et al.*, 1994). In *P. woesei*, DIP accumulates in quantities comparable to intracellular

Table 1. Structures of inositol compounds found in (or utilized by) bacteria and archaea

Solute	**Roles**
myo-Inositol 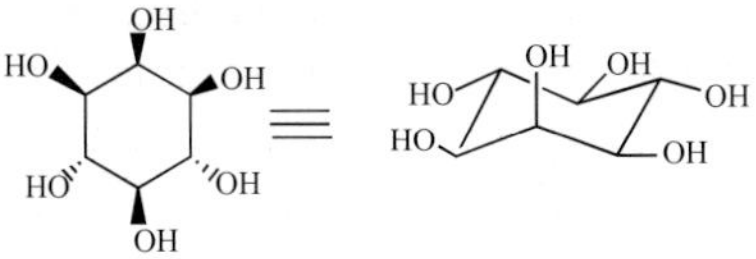	Free *myo*-inositol is not observed to accumulate in bacteria or archaea; it is fixed in unusual solute or lipids
Inositol hexakisphosphate (phytate) 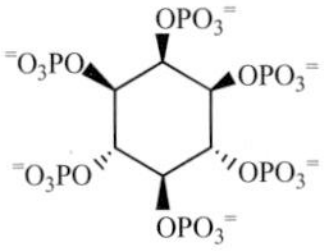	Synthesized by plants; good source of inorganic phosphate for bacteria who have evolved efficient phytases
Di-L,L-*myo*-1,1′-inositol phosphate (DIP) 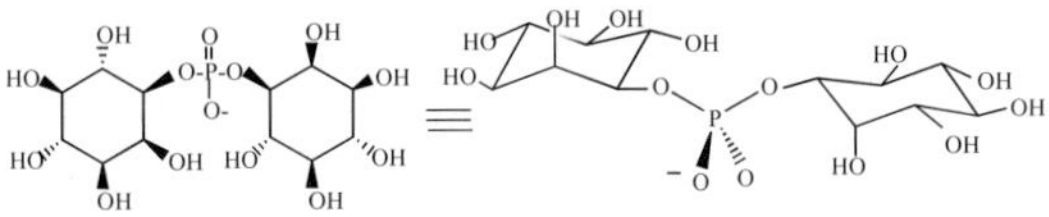	Synthesized in hyperthermophic archaea and Thermotoga sp. in response to osmotic and thermal stress
Di-2-*O*-β-mannosyl-di-L,L-*myo*-1,1-phosphate (mannosyl-DIP) 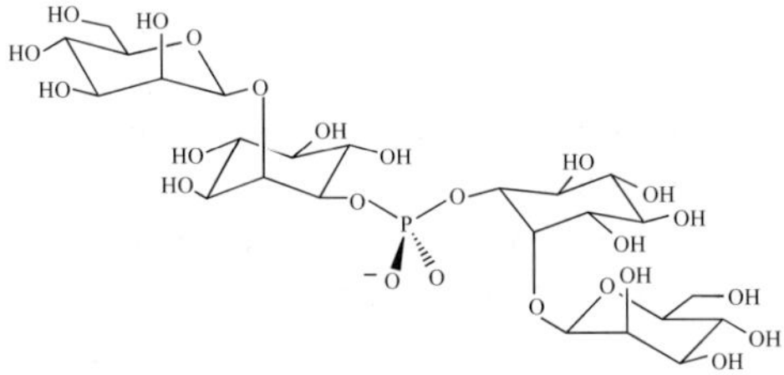	Synthesized in Thermotoga sp. in response to osmotic and thermal stress
Phosphatidylinositol (PI) 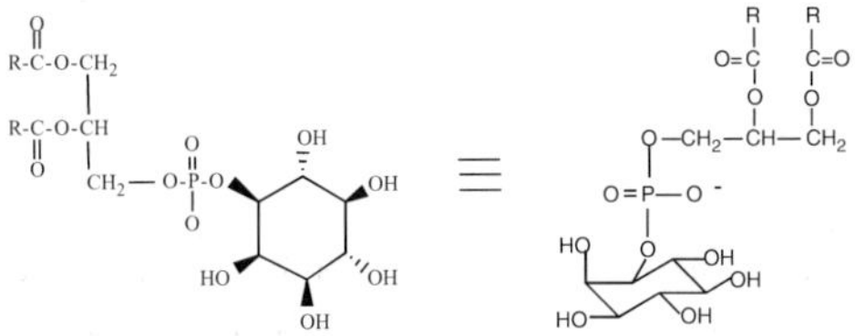	Minor membrane component in most bacteria; base of GPI-anchored proteins and unusual cell wall structures in bacteria

(*continued*)

K^+ (500–600 mM). DIP in *M. igneus* is chiral and composed of L-inositol-1-phosphate (L-I-1-P) units (Chen *et al.*, 1998); in *Thermotoga neapolitana*, a second set of ^1H and ^{13}C resonances is consistent with a DIP molecule formed from both L- and D-I-1-P moieties (Martins *et al.*, 1996). DIP functions as an osmolyte since the intracellular concentration increases with increasing

Table 1. (continued)

Solute	Roles
Glycosylinositol phosphorylceramides: *e.g.*, Manα1 $\rightarrow$ 3 Man α1$\rightarrow$ 6 $GlcNH_2$ α1 $\rightarrow$ 2 Ins1$\rightarrow$ P$\rightarrow$1Cer Manα1 $\rightarrow$3 Manα1$\rightarrow$Ins1$\rightarrow$P$\rightarrow$1Cer	Unusual bacterial phospholipids based on inositol ceramide
Mycothiol 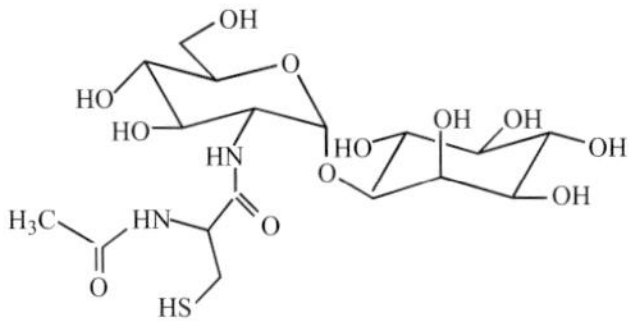	Reducing agent, much like glutathione, in mycobacteria
Lipoarabinomannan 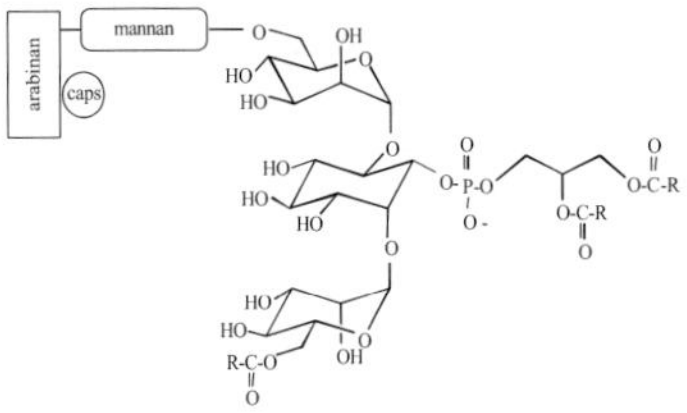	PI-anchored antigen in mycobacteria
2-*O*-(3′-Hydroxy)phytanyl-3-*O*-phytanyl-sn-glycerol phospho-*myo*-inositol (hydroxyarchaetidyl-*myo*-inositol) 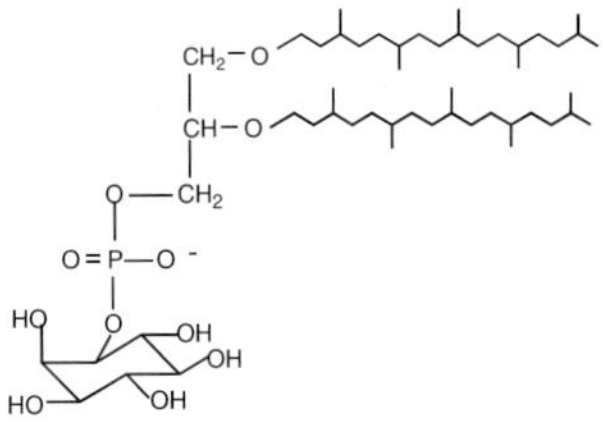	Membrane component of archaea

external NaCl in these cells. However, DIP is detected only when hyperthermophilic Archaea and Thermotogales cells are grown above 75 °C (Chen et al., 1998; Ciulla *et al.*, 1994; Martin *et al.*, 1999; Martins *et al.*, 1996). Two related biosynthetic pathways have been proposed for DIP (Chen *et al.*, 1998; Scholz *et al.*, 1998). The more likely four-step synthesis (Figure 1) includes: (i) conversion of D-glucose-6-phosphate to L-I-1-P catalyzed by the enzyme inositol-1-phosphate synthase (*myo*-inositol-1-phosphate synthase, MIPS); (ii) generation of *myo*-inositol from the I-1-P by an inositol monophosphatase (IMPase); (iii) activation of the I-1-P with CTP to form CDP-inositol (CDP-inositol cytidylyltransferase); and (iv) condensation of CDP-inositol with *myo*-inositol

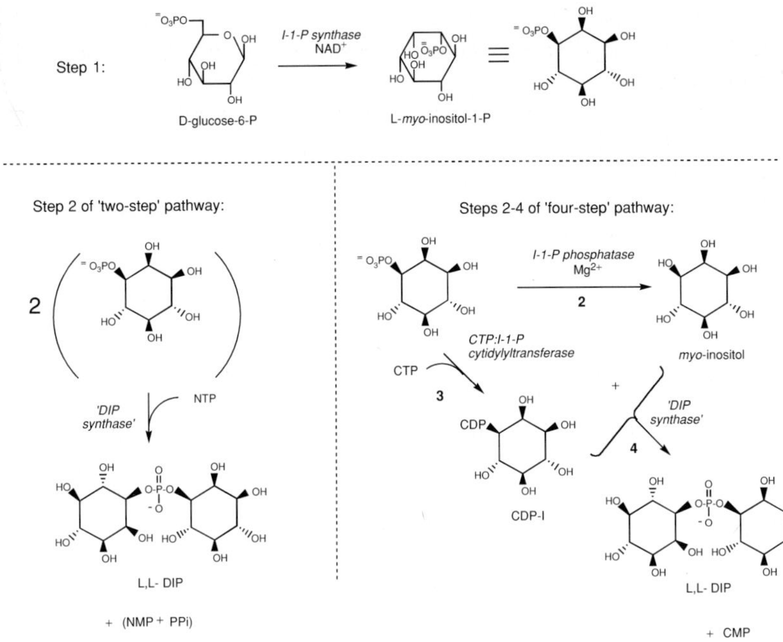

Figure 1. Proposed biosynthetic pathways for DIP.

(DIP synthase) to form DIP (Chen *et al.*, 1998). The two-step mechanism does not utilize *myo*-inositol (Scholz *et al.*, 1998), although direct condensation of two L-I-1-P molecules to form a phosphodiester, even with NTP, is an unprecedented activity for a single enzyme. Furthermore, the two-step mechanism cannot explain forming DIP with both L- and D-I-1-P units as is observed in *T. neapolitana* (Martins *et al.*, 1996). *Archaeoglobus fulgidus* is one of the archaea that accumulates DIP when grown above 75 °C (Martins *et al.*, 1997). In this archaeon, there is no other known use of inositol (*e.g.*, no inositol-containing lipids), so that production of L-I-1-P from D-glucose-6-phosphate (via MIPS) directs carbon resources to DIP only. Since MIPS is the first and committed step in DIP production, it is likely that MIPS protein expression is coupled to growth temperature. A structure of this archaeal enzyme with NAD^+ and Pi bound has been completed recently (Stieglitz *et al.*, 2005) and may shed light on other mechanisms for regulating MIPS activity and DIP production.

T. neapolitana also adds mannose moieties onto the inositol rings, and this new phosphodiester, di-2-*O*-β-mannosyl-di-*myo*-inositol-1,1′-phosphate (Table 1) also appears to accumulate at high growth temperatures. However, DIP is still the major solute accumulated in *T. neapolitana* at supraoptimal growth temperatures (Martins *et al.*, 1996). The addition of mannose to DIP produces a very unique phosphodiester, yet we know very little about why it is synthesized and accumulated. The involvement of these inositol solutes in thermoprotection as well as osmoadaptation could involve modulation of water structure. Perhaps the rigid inositol rings serve to stabilize water hydrogen bonding networks at high temperatures.

2.1.2 Mycothiol

Inositol is an important component of mycothiol [2-(*N*-acetylcysteinyl)amido-2-deoxy-α-D-glucopyranosyl-(1→1)-*myo*-insoitol] (Newton *et al.*, 1995; Spies and Steenkamp, 1994), the major low molecular weight solute in mycobacteria. Mycothiol (for structure see Table 1) is synthesized by attachment of *N*-acetyl-glucosamine to *myo*-inositol, deacetylation of the glucosamine moiety and ligation to L-cysteine, followed by transacetylation of the cysteinyl residue by acetyl-CoA to produce mycothiol (for review see Newton and Fahey, 2002). Mycothiol is thought to protect *Mycobacterium tuberculosis* from inactivation by the host during infections via its antioxidant activity (similar to glutathione) as well as its ability to detoxify a variety of toxic thiol-reactive compounds. Consistent with this protective role, a mutant lacking the Rv1170 gene (which codes for the deacetylase activity needed for mycothiol production) had increased sensitivity to the toxic oxidant cumene hydroperoxide and to the antibiotic rifampin (Buchmeier *et al.*, 2003).

2.2 Inositol-containing lipids

The major use of inositol in bacteria is in generating PIs, versatile membrane components that can be derivatized to anchor proteins or complex carbohydrates to the cell surface. In some microorganisms, the complex carbohydrate structure attached to the PI anchor in the cell envelope is part of the way the microorganism recognizes or binds to target cell components.

2.2.1 Phosphatidylinositol

Gram-negative bacteria normally do not contain much, if any, PI in their membranes. When it does occur, it often correlates with unusual properties of a microorganism. For example, microorganisms that have ice-nucleation activity in supercooled water have been shown to accumulate PI in their membranes (*e.g.*, 0.1–1.0% of total phospholipids in *Escheria coli* K-12 Ice+ strains) (Kozloff *et al.*, 1991a). Corresponding Ice− *E. coli* strains also contained PI, but at 2–30% of the level found in the Ice+ *E. coli* strains. Treatment of these cells with a PI-specific phospholipase C (PI-PLC), which cleaves PI to diacylglcyerol and inositol-1-phosphate, destroyed the ice-nucleating ability. Thus, the functioning of the ice gene apparently increased both the PI synthase activity and the PI content of Ice+ strains from low endogenous levels. These results strongly indicate that PI plays an important role in ice nucleation at −4 °C or above, although it is not thought to be a direct interaction with water. Rather, it has been proposed

(Kozloff *et al.*, 1991a) that PI serves to anchor the appropriate nucleation protein to the cell membrane. The ice-nucleation gene product appears to be attached to the cell membrane via a glycosylphosphatidylinositol (GPI) anchor. The protein is coupled to mannose residues via an *N*-glycan bond to the amide nitrogen; the mannose residues are attached to PI (Kozloff *et al.*, 1991b). This GPI-linked protein structure is the critical element in the class A nucleating structure. The PI mannoside moiety has also been identified as a mycobacterial adhesin mediating binding to non-phagocytic cells (Hoppe *et al.*, 1997).

The distribution of PI in other bacteria appears confined to actinomycetes (*Mycobacterium, Corynebacterium, Nocardia, Micromonospora, Streptomyces, and Propionobacterium*) (Brennan and Ballou, 1968; Brennan and Lehane, 1971; Goren, 1984; Kataoka and Nojima, 1967; Tabaud *et al.*, 1971; Yano *et al.*, 1969), myxobacteria (Elsbach and Weiss, 1988), and *Treponema* (Belisle *et al.*, 1994). The linkages become quite complex, for example the triacylphosphatidylinositol dimannosides (Ac(3)PIM(2)) in *Corynebacterium amycolatum, Corynebacterium jeikeium,* and *Corynebacterium urealyticum* (Yague *et al.*, 2003) and glycosylphosphoinositol mannosides found in Mycobacteria (*e.g.*, Manα1$\rightarrow$2Ins1-P-1Cer and Manα1$\rightarrow$3Manα1$\rightarrow$2Ins1-P-1Cer). For review of cell wall structure and function in *Mycobacterium tuberculosis*, see Brennan (2003).

2.2.2 Inositol-sphingolipids

Mycobacteria have unusual acidic GIPCs as well as GPI-anchors in its cell envelope. The sphingolipids in mycopathogens (Toledo *et al.*, 2001) often have the same linkage and structure found in GPIs of Mycobacteria. Unusual core linkages in *Sporothrix schenckii* (Manα1$\rightarrow$6Ins) and *Paracoccidiodes brasiliensis* (Levery *et al.*, 1998) (Manα1$\rightarrow$2Ins) have also been seen. The synthesis of these GIPCs that are not found in mammalian cells is required for viability of fungi. Indeed, inositol phosphorylceramide (IPC) synthase, the first step in GIPC biosynthesis, could be a potential drug target (Nagiec *et al.*, 1997).

2.2.3 Archaeal inositol-containing lipids

Inositol-containing lipids appear to be more common in archaea than in bacteria; they have been best characterized in methanogens. Archaeal lipids have ether linkages rather than ester linkages; the alkyl chains are appended to the 2- and 3-positions of glycerol rather than the 1- and 2-positions modified in eukaryotic and bacterial lipids. *Methanosarcina barkeri* contains 2-*O*-(3′-hydroxy)phytanyl-3-*O*-phytanyl-sn-glycerol phospho-*myo*-inositol, otherwise known as hydroxyarchaetidyl-*myo*-inositol (Table 1) (Nishihara and Koga, 1991), which can be further modified. A related phosphoglycolipid has also been detected in *Aeropyrum pernix* (Morii *et al.*, 1999). The stereochemical configuration of the phospho-*myo*-inositol residue of glucosaminyl archaetidylinositol was determined to be 1-D-*myo*-inositol 1-phosphate (Nishihara *et al.*, 1992). Since MIPS generates L-I-1-P, the synthesis must use an activated pyrophosphate ester that is attacked by *myo*-inositol; L-I-1-P cannot be the direct precursor. The inositol headgroups of both diether and tetraether polar lipids appear oriented to the cytoplasmic surface of the membrane in *Methanobacterium thermoautotrophicum* (Morii and Koga, 1994); this is similar to mammalian cells where PI molecules are on the cytoplasmic leaflet.

2.3 Cell wall components (lipomannan and lipoarabinomannan from mycobacterium)

In mycobacterial cell envelopes, there are specific cell wall components, lipoarabinomannan (LAM), containing inositol. LAMs (Table 1) have three structural domains: two homopolysaccharides (D-mannan and D-arabinan) constitute the carbohydrate backbone; the mannan core is terminated by a GPI-anchor and arabinan capped by GPI. These phosphatidylmannosides (PIMs) are mediators of adhesion. Tuberculosis is caused by *Mycobacterium tuberculosis*, a facultative intracellular pathogen of alveolar macrophages in the lung. Two human pulmonary surfactant proteins play key roles in the pathogenicity of the organism: hSP-A and hSP-D. hSP-A promotes attachment of *M. tuberculosis* to phagocytes (Downing *et al.*, 1995), while hSP-D reduces uptake of macrophages (Ferguson *et al.*, 1999). hSP-D appears to bind to mannose-capped LAM (Ferguson *et al.*, 1999). hSP-A binds tightly to both pathogenic and non-pathogenic Mycobacterium sp. through specific interactions with the major lipoglycans (mannosylated LAM and lipomannan) (Sidobre *et al.*, 2000). The PI anchor (along with terminal mannose residues) is required for the high affinity of LAMs to the surfactant proteins. PI synthesis in mycobacteria has been reviewed by Salman *et al.*, 1999.

3. ENZYMES OF *MYO*-INOSITOL BIOSYNTHESIS

Inositol is synthesized by the same route in all cells. D-Glucose-6-phosphate (D-G-6-P) is converted to L-*myo*-inositol-1-phosphate (L-I-1-P), which can also be termed D-*myo*-inositol-3-phosphate. The L-I-1-P is then hydrolyzed to *myo*-inositol by a relatively specific phosphatase. As we will see, enzymes that do these reactions in bacteria and archaea can have quite different characteristics from the eukaryotic enzymes.

3.1 myo-Inositol-1-phosphate synthase

MIPS is the rate-limiting step in *myo*-inositol biosynthesis. Regulation of the enzyme from yeast has been examined in detail since there are multiple ways to modulate this activity in cells (Majumder *et al.*, 1997); the eukaryotic enzyme is also of interest since the mammalian homologue may be a potential drug target for bipolar disease (Agam *et al.*, 2002). The well-studied yeast MIPS is a homotetramer (60 kDa subunits) that is activated by ammonium ions (Majumder *et al.*, 1997). It converts D-G-6-P to L-I-1-P with NAD^+ as a cofactor (Loewus, 1977; Loewus *et al.*, 1980; Tian *et al.*, 1999). There are three distinct steps of the reaction with intermediates tightly bound to the protein: (i) D-G-6-P is oxidized to 5-keto-D-G-6-P concomitant with NAD^+ reduction to NADH; (ii) after enolization of the 5-keto-D-G-6-P, an aldol condensation reaction occurs to form the new carbon–carbon bond and yield 2-inosose-1-phosphate; (iii) the inosose compound is reduced to L-I-1-P by NADH regenerating NAD^+ at the active site. The *ino1* gene and its product, MIPS, are found in most eukaryotes but scattered in prokaryotes (Bachawat and Mande, 2000). Sequence homology searches readily identify a putative *ino1* gene product in hyperthermophilic Archaea, *Aquifex aeolicus*, and Thermotoga species; several of these organisms use DIP for osmotic balance or use inositol-containing lipids in their membranes. MIPS activities have been identified in high (G+C) Gram-positive mesophilic bacteria such as Mycobacteria and Streptomyces (Bachawarat and Mande, 1999, 2000), both of which use inositol in cell wall or antibiotic production (in Streptomyces, MIPS is needed for generating *myo*-inositol that is used in synthesizing antibiotics, Walker, 1995). Sequence analyses suggest bacteria recruited the *ino1* gene from archaea (Nesbo *et al.*, 2001), for example it is likely that there was lateral *ino* gene transfer to *T. martima* from *P. horikoshii*. However, it appears that two distinct bacterial lineages appear to have acquired *ino1* from different archaea (Majumdar *et al.*, 2003). Another chapter in this volume provides a detailed review of MIPS structure and mechanism (Chapter 7).

3.1.1 *Mycobacterium tuberculosis* MIPS

myo-Inositol is critical for synthesis of the cell wall lipoglycans of *M. tuberculosis*, and the MIPS commits resources to inositol biosynthesis. Recently, an *ino1*-deficient mutant of *M. tuberculosis* was constructed (Movahedzadeh *et al.*, 2004). The mutant was only viable with high *myo*-inositol in the medium. When this mutant was grown and then incubated in inositol-free medium, levels of mycothiol were reduced while PI mannoside, lipomannan, and LAM levels were not altered. Its infectivity of macrophages was attenuated as well.

The crystal structure of *M. tuberculosis* MIPS has been solved to 1.95 Å (Norman *et al.*, 2002). The most striking difference from the yeast MIPS structure (Stein and Geiger, 2002) was the presence of a metal ion, identified as Zn^{2+}, in the active site in the vicinity of the NAD^+. The observation of a metal ion was surprising because EDTA does not appear to affect the activity of this enzyme. Mutagenesis of four active site residues (D197A, K284A, D310A, and K346A) generated inactive enzyme as detected by the lack of growth when plasmids bearing one of these mutant *ino1* genes was introduced into *M. tuberculosis* containing a defective *ino1* gene (Movahedzadeh *et al.*, 2004). Asp310 was a ligand of the Zn^{2+} ion observed in the crystal structure. Structural studies of D310N caused a loss of the Zn^{2+} ion and a conformational change in the NAD^{+} cofactor. This strongly implies that the Zn^{2+} is critical for activity.

3.1.2 *Archaeoglobus fulgidus MIPS*

An *ino1* gene, identified in this hyperthermophilic archaeon by sequence homology to yeast MIPS, was cloned and overexpressed in *E. coli* (Chen *et al.*, 2000). The archaeal MIPS subunit is 44 kDa, considerably smaller than the yeast enzyme (60 kDa). Like the yeast and mycobacterial enzymes, *A. fulgidus* MIPS is a tetramer. Not surprisingly, it is very thermostable: at 90 °C, K_m is 0.12 mM for G-6-P, 5.1 μM for NAD^{+}, and $k_{cat} = 9.6\ s^{-1}$. Use of D-[5-^{13}C] G-6-P has clearly shown that the product is L-I-1-P (Chen *et al.*, 2000). What appears unique about this MIPS compared to the other ones examined thus far is that it absolutely requires metal ions for activity, with Zn^{2+} and Mn^{2+} optimum but Mg^{2+} also effective. EDTA inhibits the enzyme and halts the reaction after oxidation of G-6-P by NAD^{+} to form 5-keto-glucose-6-phosphate and NADH. The latter two compounds are generated in 1:1 stoichiometry with the enzyme subunits (Chen *et al.*, 2000). This suggests that this archaeal MIPS is a class II aldolase (Figure 2). A structure of the *A. fulgidus* MIPS crystallized without any added NAD^{+} or divalent metal ions has recently been described (Stieglitz *et al.*, 2005). That enzyme has NAD^{+} and Pi bound to each subunit and two of the subunits have a density that has characteristics more consistent with K^{+} as the metal ion. Subunits with this "crystallographic" ion, which occupies a position similar to the putative Zn^{2+} in the *M. tuberculosis* MIPS, have altered NAD^{+} conformations from those subunits that do not have a metal ion present. While this structure will provide a good comparison to both the yeast and *M. tuberculosis* enzymes, what is really needed is the *A. fulgidus* MIPS structure with an activating metal ion, since this type of cation is absolutely necessary for catalytic activity.

Not all MIPS enzymes from thermophiles may have divalent metal ion-dependent activities. While crude protein extracts of *A. fulgidus* exhibited a MIPS activity that absolutely required divalent metal ions and was totally inhibited by 1 mM EDTA, the production of I-1-P from G-6-P (carried out ≥75 °C) by protein extracts from *M. igneus* and *T. maritima* was observed in the presence of 1 mM EDTA. This observation (Chen and Roberts, unpublished results) suggests that either the MIPS from these particular archaea is not divalent metal ion-dependent or that the metal ions bind very tightly and are not easily removed. None of the MIPS activities from thermophiles are affected by the addition of NH_4^{+}. Since the divalent cation in *A. fulgidus* MIPS has been suggested to aid in the aldol condensation and since this is supposedly done by NH_4^{+} in the yeast enzyme, it is of interest how this step occurs in the other archaea and bacteria.

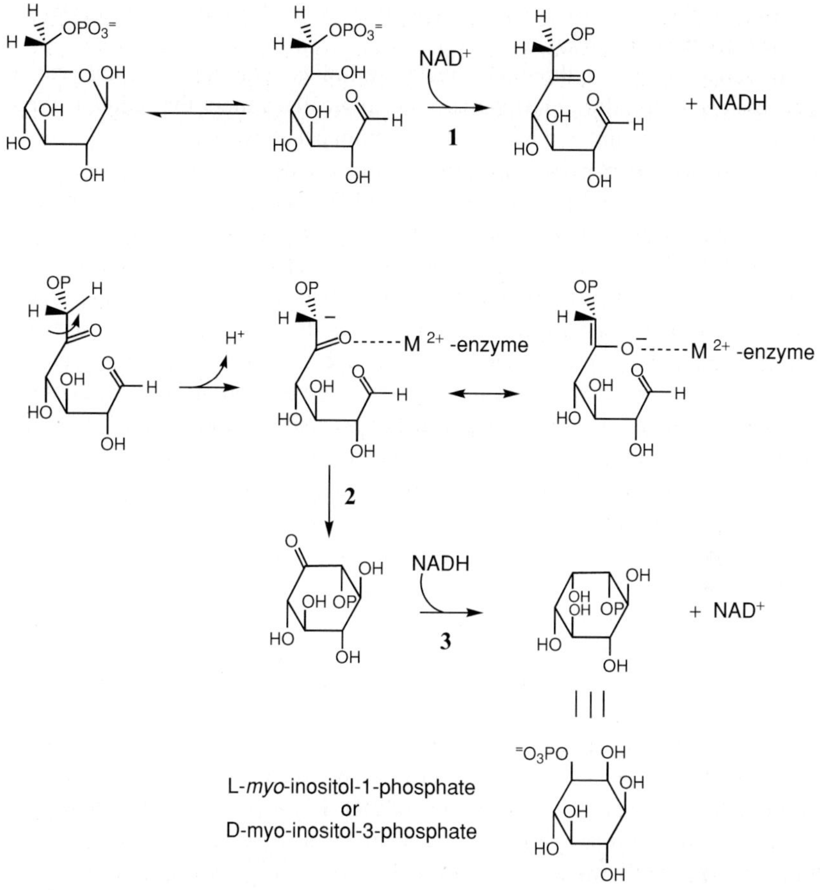

Figure 2. Sequence of MIPS reactions for the enzyme from *A. fulgidus*.

3.2 Inositol monophosphatase

In the bacteria that use PI derivatives (*e.g.*, lipomannan biosynthesis in Mycobacteria), IMPase activities are required. However, in some bacteria that do not use significant PI in their membranes, IMPase activities have been identified and characterized. The biological function of an IMPase in a non-PI containing organism must be something other than generating *myo*-inositol. A good example is the *E. coli suhB* gene product (SuhB), which has high sequence homology to mammalian IMPases and exhibits IMPase activity (PI is not a significant *E. coli* phospholipid). In archaea, IMPase activities are required for DIP biosynthesis (Chen *et al.*, 1998). However, homologues occur in all sequenced archaea whether or not they accumulate DIP or have PI analogues in their membranes. As will be discussed, one of the more unusual features of these IMPases is their dual fructose 1,6-bisphosphatase (FBPase) activity (Stec *et al.*, 2000).

In general there are three phosphatase families: alkaline, acid, and protein phosphatases. Alkaline phosphatases are typically dimers that contain three metal ions per subunit and have a pH optimum pH above 8. Acid phosphatases exhibit an optimum pH$<$7 and are usually divided into three classes: low molecular weight acid phosphatases ($<$20 kDa), high molecular weight acid phosphatases (50–60 kDa), and purple acid phosphatases (which contain an Fe–Fe or Fe–Zn center at the active site). Phosphatases specific for I-1-P appear to be most similar (in kinetic characteristics but not in mechanism) to the alkaline phosphatases, but their structures define a superfamily that also includes inositol polyphosphate 1-phosphatase, fructose 1, 6-bisphosphatase, and Hal2. The members of this superfamily share a common structural core of 5 α-helices and 11 β-strands. Many are Li^+-sensitive (York *et al.*, 1995), and more recent structures of archaeal IMPase proteins suggest the Li^+ -sensitivity is related to the disposition of a flexible loop near the active site (Stieglitz *et al.*, 2002).

3.2.1 *E. coli* SuhB

The *suhB* gene of *E. coli* was identified by isolating mutants that suppressed a variety of temperature-sensitive defects including specific mutations in protein secretion, DNA synthesis (Chang *et al.*, 1991), and the heat shock response (Yano *et al.*, 1990). The SuhB protein has significant sequence similarity to human IMPase and was shown to exhibit Li^+ -sensitive IMPase activity (Matsuhisa *et al.*, 1995) similar to the mammalian enzyme (see Table 2). Since there are essentially no *myo*-inositol-containing phospholipids in most *E. coli*, the occurrence of an IMPase activity strongly suggests that this enzyme must either use another specific substrate or possess another type of "activity." Insight into the function of SuhB and other bacterial homologues has been provided by analysis of the suppressor behavior of the *suhB* gene. *E. coli* mutations in *rpoH* and *dnaB* cause the cells to grow only at 30 °C but not at 42 °C. If *suhB* is also mutated, the cells now grow only at 42 °C. Introducing the SuhB product back into the double mutants via a plasmid allows the cells to grow at 30 °C once again. Thus, mutant *suhB* is a suppressor of mutant *rpoH* or *dnaB* phenotypes. The *E. coli* mutant with *suhB* missing or inactivated is cold-sensitive with defects in protein synthesis (Inada and Nakamura, 1995). Interestingly, RNA cleavage mutations restore minimal protein synthesis to *suhB* mutants (Inada and Nakamura, 1995). Thus, it has been suggested that that the SuhB protein/IMPase can affect a number of genes at the translational level (Chang *et al.*, 1991).

One possibility is that the double-strand RNA-processing activity of RNase III, an endoribonuclease involved in the rate-limiting first cleavage step of mRNA degradation (Chang *et al.*, 1991), is potentially lethal to *E. coli*, and the

Table 2. Characteristics of IMPase activities from bacteria and archaea

Source[a]	M_n	Assay T (°C)	K_m (I-1-P) mM[b]	Unusual substrates	K_D (Mg^{2+}) mM	IC_{50} (Li^+) mM	k_{cat} (s^{-1})	References
Ec	M_1[c]	37	0.068	–	0.5	2	4.4	Chen and Roberts (2000)
Mt	M_2	37	0.18	–	6.0[d]	0.9	3.6	Nigou *et al.* (2002)
Rl		37?	0.23	–				Janczarek and Skorupska (2001)
Af	M_2	85	0.11	PNPP, G-1-P, 2′-AMP, FBP	15[e]	290	4.3	Stieglitz *et al.* (2002)
Mi	M_2	85	0.3	PNPP, G-1-P, 2′-AMP, FBP	6–8	160		
Mj	M_2	85	0.09	2′-AMP best substrate, FBP	3–4	200	4.2	Chen and Roberts (1998)
Tm	M_4	95	0.13	D-I-1-P is 20 times better than L-I-1-P	5–10	100	210	Chen and Roberts (1999)

[a]Ec, *E. coli*; Mt, *Mycobacterium tuberculosis*; Rl, *Rhizobium leguminosa;* Af, *Archaeoglobus fulgidus*; Mi, *Methanococcus igneus*; Mj, *M. jannaschii;* Tm, *Thermotoga maritima. Other abbreviations: PNPP, p-nitrophenylphosphate; G-1-P, glucose-1-phosphate; FBP, fructose-1,6-bisphosphate.*
[b]For D-I-1-P as the substrate unless otherwise noted.
[c]SuhB aggregates, but this can be suppressed with KCl >0.2 M.
[d]The optimum Mg^{2+} concentration is provided instead of a K_D.
[e]The apparent K_D for Mg^{2+} depends on the substrate; this is an average value.

normal function of SuhB is to modulate the lethal action of RNase III. SuhB could alter mRNA stability by (1) modulating RNase III activity or (2) binding to target RNA molecules and protecting them from degradation by RNase III. The first possibility is based on the observation that the RNase III activity was stimulated when the protein was phosphorylated, and SuhB acting as a protein phosphatase could dephosphorylate RNase III. For this to be the mechanism, the phosphatase activity of SuhB must be intact. However, work by Chen and Roberts (2000) demonstrated that the phosphatase activity is absolutely unrelated to the SuhB function in the *dnaB*121 mutant. SuhB functioning in RNA protection appears a more plausible explanation at this point. Suppose high temperature (42 °C) destabilizes double-strand RNA (dsRNA) structure, while at lower temperatures, RNase III, specific for dsRNA, forms a complex with its target RNA and degrades the RNA. SuhB binding to dsRNA could protect it from RNase III degradation. This implies that at the lower temperature, wild type SuhB is required to compensate for the lethal RNase III activity, and loss of SuhB would always generate a cold-sensitive phenotype. Since dsRNA is destabilized at the higher growth temperature, RNase III cannot degrade it and the cells are able to grow.

The observations with the *dnaB* (*dnaB*121) mutation can be understood in this context (Figure 3). At 42 °C, the mutant *dnaB* gene product is not able to form the functional hexamer (Yano *et al.*, 1990) that participates in initiation and elongation of *E. coli* replication by interacting with dnaC and other *E. coli* proteins (Inada and Nakamura, 1996). Without functional dnaB protein, DNA replication is blocked and cells cannot grow. At 30 °C, the mutant dnaB product can still form functional hexamer and wild type SuhB protects the dsRNA from degradation by RNase III. Under these conditions, the cells can grow. The *dnaB* and *suhB* double mutant, although capable of forming functional dnaB (mutant) hexamer at 30 °C, cannot grow because of unregulated RNase III activity at this temperature. At 42 °C, RNase III activity is reduced because of the smaller population of dsRNA; the presence of mutant SuhB must make the mutant dnaB functional at this temperature, although the mechanism for this is unknown.

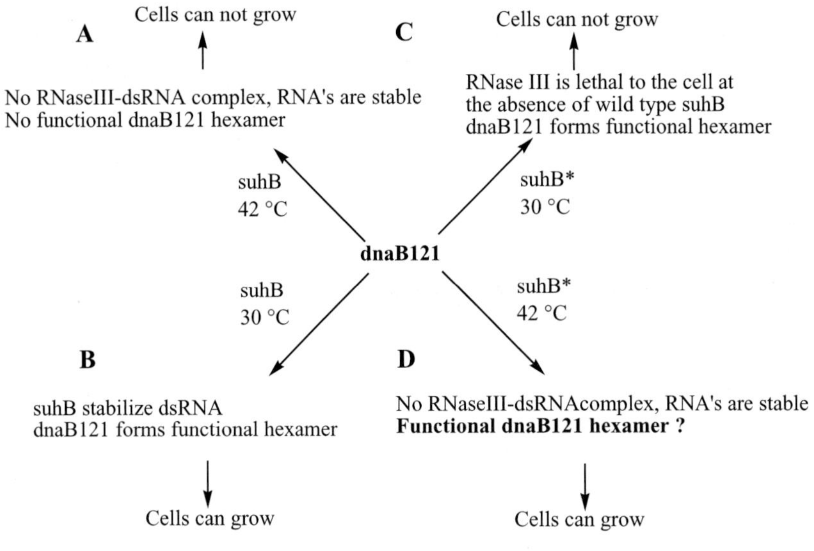

Figure 3. Suppressor effect of mutant *suhB* on the *dnaB121* mutant of *E. coli*.

3.2.2 *Mycobacterium tuberculosis* IMPase

Since inositol lipids are so prevalent in the cell membranes of mycobacteria, significant IMPase activity is needed. There are four ORFs in the *M. tuberculosis* genome with the IMPase signature DPIDGT and WDXAAG found in the superfamily of lithium-sensitive phosphatases (Neuwald *et al.*, 1991), of which three are close to other known genes: *suhB*, *impA*, and *csyQ* (Nigou *et al.*, 2002). The one with closest homology to human IMPase was identified as the *suhB* gene product. Cloning, expression, purification, and characterization of the IMPase protein showed it was very similar to the *E. coli* protein in terms of physical characteristics (Table 2). Similar to *E. coli* SuhB, the *M. tuberculosis* IMPase migrated anomalously on gels and exhibited a native molecular mass suggesting it was a trimer. Kinetic characteristics for this IMPase were very similar to those for *E. coli* SuhB. Substrate specificity studies indicated that I-1-P was the preferred substrate; glycerol-2-phosphate, 2′-AMP, mannitol-1-phosphate were not as effectively hydrolyzed. Although *M. tuberculosis*, as a human pathogen, has an optimal growth temperature of 37 °C, the IMPase is quite stable with an optimum assay temperature of 80 °C. Site-directed mutagenesis indicated that Glu83, Asp104, Asp107, Asp-235, and Trp234, which aligned with human IMPase active site residues, were necessary for catalysis. The residue Ile68, which is comparable to Val70 in HAL2 (a yeast PAPase), was shown to affect Li^+-sensitivity: L81A exhibited a 10-fold increase in the IC_{50} for Li^+.

How critical is the IMPase in mycobacteria? In a related but non-pathogenic organism, *M. smegmatis*, a mutant constructed to have a defective IMPase had altered cell envelope permeability (Parish *et al.*, 1997). Certainly, mutating the IMPase gene should affect the distribution of LAMs since the PI moiety is the basic scaffold, and in the *M. smegmatis* mutant, there was a decrease in PIM2 consistent with altered cell permeability. The same mutant also indirectly provides information on mycothiol production. Mycothiol has an inositol unit that could be incorporated via I-1-P or from *myo*-inositol. The IMPase mutants show dramatically decreased mycothiol levels, consistent with *myo*-inositol as the direct precursor of the inositol unit in mycothiol.

3.2.3 *Rhizobium leguminosarum* bv. Trifoli pssB gene product

A protein of 284 residues, the *pssB* gene product, has been shown to have IMPase activity (Janczarek and Skorupska, 2001). The phosphatase activity of PssB (Table 2) is suggested to have a regulatory function in exopolysaccharide (EPS) synthesis. Mutation of *pssB* caused EPS overproduction, while introducing *pssB* in the wild type strain reduced the levels of EPS. These alterations in EPS production could be correlated with a non-nitrogen-fixing phenotype of rhizobia.

3.2.4 Archaeal IMPase homologues

Hyperthermophilic archaea possess genes whose translated products have considerable sequence homology to human IMPase. Recombinant IMPase activities from *Methanococcus jannaschii*, *Archaeoglobus fulgidus*, and the closely related bacterium *Thermotoga maritima* have been expressed in *E. coli* (Chen and Roberts, 1998, 1999; Stieglitz *et al.*, 2002). The occurrence of an IMPase in archaea that accumulate DIP would be expected. However, IMPase homologues appear to be ubiquitous in archaeal genomes, so that this protein is likely to have another role. In fact, in most archaeal genomes, the genes homologous to

IMPase are annotated as coding for SuhB homologues. MJ0109 in *M. jannaschii*, the archaeal protein first identified by sequence homology as an IMPase, exhibits slightly broader substrate specificity than human or *E. coli* IMPases (*e.g.*, *p*-nitrophenylphosphate is a substrate), but is poorly inhibited by Li^+ (Table 2) (Chen and Roberts, 1998). The structure of this archaeal IMPase solved by molecular replacement using the human IMPase structure (Johnson *et al.*, 2001; Stec *et al.*, 2000), clearly belongs to the IMPase superfamily (Figure 4A) but it actually has some features reminiscent of pig kidney FBPase, a member of the IMPase superfamily that is a tetramer (a dimer of dimers). In particular, active site loop 1 in the archaeal IMPase (MJ0109) was quite different from the corresponding loop in the human I-1-Pase (Figure 4B) (Johnson *et al.*, 2001). The disposition of this loop was much closer to that in pig kidney FBPase (Stec *et al.*, 2000).

Indeed, MJ0109 can catalyze the hydrolysis of the phosphate at C-1 of fructose 1,6-bisphosphate (FBP) generating inorganic phosphate and fructose-6-phosphate (F-6-P) (Stec *et al.*, 2000). G-6-P and F-6-P are not substrates for MJ0109. The FBPase-specific activity exhibited by MJ0109 is comparable to the IMPase-specific activity, making MJ0109 a dual phosphatase that can process substrates in completely different pathways (Stec *et al.*, 2000). FBPase activity gives the protein a potential role in gluconeogenesis, although another protein in *M. jannaschii* with FBPase and not IMPase activity has been partially purified and cloned (Rashid *et al.*, 2002). The IMPase activity links the protein to inositol biosynthesis and stress responses, although *M. jannaschii* does not accumulate DIP and has no *ino1* homologue. The IMPases from *T. maritima* and *A. fulgidus*, both of which accumulate DIP, have also been shown to be bifunctional IMPase/FBPase enzymes (Stec *et al.*, 2000). These IMPases from hyperthermophiles are functionally quite different from *E. coli* SuhB, which cannot hydrolyze FBP. Detailed analyses of the crystal structures of MJ0109 (Johnson *et al.*, 2001) and the homologue from *A. fulgidus*, AF2372 (Stieglitz *et al.*, 2002), with substrates or products and metal ions bound suggest these phosphatases work using a three metal ion-assisted catalytic mechanism. *In vivo* regulation mechanisms for these enzymes are not known, although the IMPase from *A. fulgidus* has two closely spaced cysteine sidechains, Cys150 and Cys186, that inactive the enzyme when they are oxidized to form an intramolecular disulfide (Stieglitz, 2003). Such a mode of phosphatase regulation occurs in chloroplast FBPase enzymes (Balmer *et al.*, 2001; Jacquot *et al.*, 1997).

3.2.5 Glucose-1-phosphatase has specific phytase activity

The glucose-1-phosphatase from *E. coli* hydrolyzes phytate as well as glucose-1-phosphate. This enzyme is a member of the histidine acid phosphatase family (Lee *et al.*, 2003). Glucose-1-phosphatase cleaves only the 3-phosphate from phytate, and no further hydrolysis takes place. It is the only phosphatase

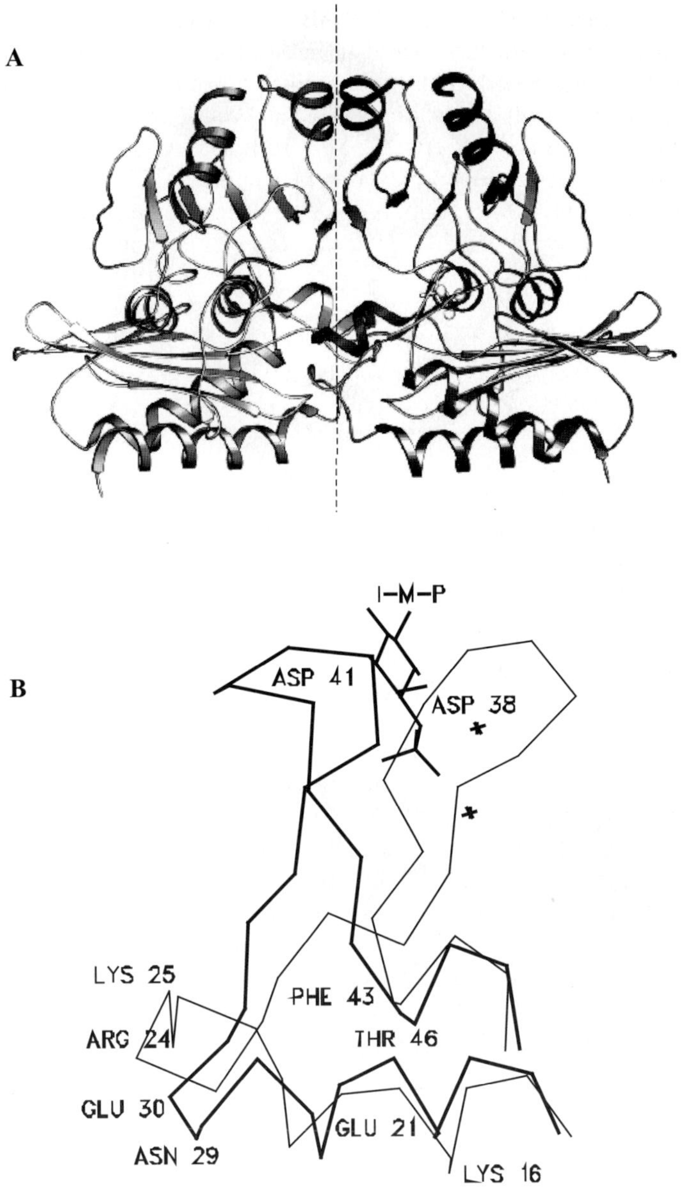

Figure 4. (A) Ribbon diagram showing the structure of the MJ0109 IMPase/FBPase dimer. (B) Comparison of the catalytic loop 1 in MJ0109 (thin line) compared to human IMPase (thick line) with I-1-P and metal ions (+) shown. Reproduced with permission from (Johnson *et al.*, 2001).

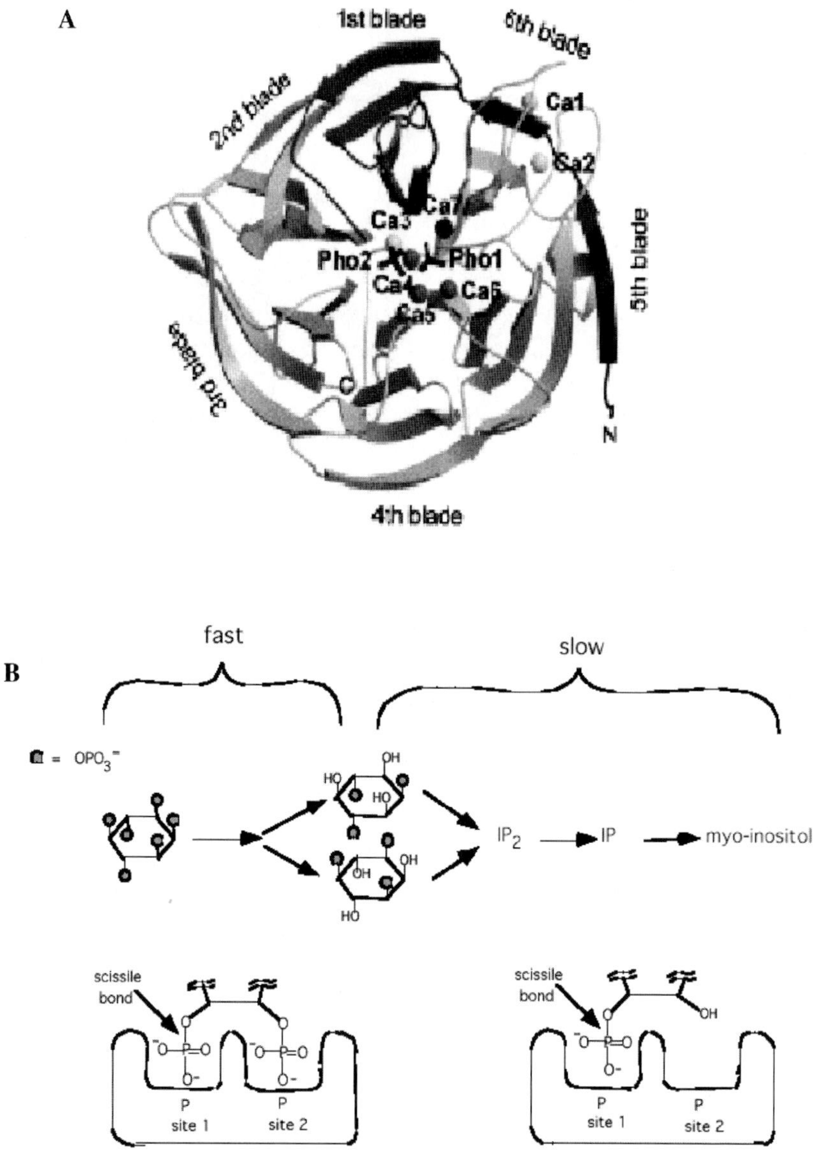

Figure 5. (A) Phytase crystal structure and (B) model for two phosphate binding sites (from Shin *et al.*, 2001).

known to possess this unique specificity. The enzyme, secreted into the host cell, may be involved in the pathogenic inositol phosphate signal transduction pathways.

3.2.6 Phytases

Although phytate is not accumulated in microorganisms, its hydrolysis can provide a large amount of inorganic phosphate to bacterial cells. While a number of bacterial phytases (*e.g.*, from *E. coli*, Bacillus sp.) have been studied in depth, new bacterial phytases are characterized frequently, in part for their biotechnology usefulness. For example, a phytase recently been purified from *Pseudomonas syringae MOK1* (45 kDa) has an optimal activity at pH 5.5 and is inhibited by EDTA. It exhibits a high specificity for phytate and little or no activity for other inositol phosphates (Cho *et al.*, 2003). Strong phytase activity associated with a number of lactic acid bacteria has commercial importance as this enzyme is a useful additive that aids in demineralization of whole grain flours (Vohra and Satyanarayana, 2003).

The stepwise degradation of phytate has been shown to proceed via initial removal of the 3-phosphate to form D-I(1,2,4,5,6)P_5, then dephosphorylation to I(2,4,5,6)P_4, then to I(2,4,6)P_3 or I(2,5,6)P_3. A second pathway is suggested by the observation that I(1,2,4,5,6)P_5 can be degraded to I(1,2,5,6)P_4 to finally I(1,2,6)P_3. Further degradation to I-2-P can occur after prolonged incubation times (Greiner *et al.*, 2002). Insight into the mechanism for phytate hydrolysis has been provided by crystal structures (Shin *et al.*, 2001). The thermostable, calcium-dependent Bacillus phytase adopts a β-propeller fold. The structure shows two phosphates and four Ca^{2+} in the active site (Figure 5A). The non-equivalent phosphate sites represent a "cleavage site" and an "affinity site" that increases the binding affinity for substrates containing adjacent phosphate groups (Figure 5B). The two phosphate binding sites explain the puzzling formation of the alternately dephosphorylated *myo*-inositol triphosphates from phytate and the much slower hydrolysis of *myo*-inositol monophosphates. A structure of *E. coli* phytase with phytate (Lim *et al.*, 2000) suggested that the mechanism of that particular phytase is similar to histidine acid phosphatase, although no pronounced sequence homology was found.

4. BIOSYNTHESIS AND DEGRADATION OF PI

The detection of PI in bacterial and archaeal systems implies that PI synthase, which is an integral membrane protein, is present. Once PI is available, kinases could add phosphates to generate various phosphoinositides, although to date this chemistry has only been demonstrated to occur in eukaryotes. Bacterial catabolic activities specific to PI have also been identified. In most cases these are secreted soluble proteins that can cleave PI and in some cases GPI-anchored proteins to DAG. These phospholipase activities often play a role in infectivity of the bacteria.

4.1 PI Synthase

The enzyme that converts CDP-DAG and *myo*-inositol to PI belongs to the CDP-alcohol phosphatidyltransferase class-I family. In yeast, the enzyme is located in the outer mitochondrial membranes and microsomes (Nikawa and Yamashita, 1997). In *M. tuberculosis*, the *pgsA* gene was identified as encoding a PI synthase enzyme by construction of a conditional mutant. The strain would not grow under non-permissive conditions. There was also a noted loss of cell viability when PI and PI dimannoside dropped to 30% and 50% of what occurs in wild type cells. Therefore PI and the machinery to make it are essential for survival of mycobacteria (Jackson *et al.*, 2000).

Using a eukaryotic PI synthase as query, one can easily identify homologues in mycobacteria as well as in some archaea. The bulk of each PI synthase sequence contains the local, conserved region found in enzymes catalyzing the transfer of the phosphoalcohol moiety from CDP-alcohol, such as phosphatidylserine synthase, cholinephosphotransferase, and phosphatidylglycerolphosphate synthase (Figure 6). The *M. bovis* PI synthase gene codes for a 217-residue membrane protein with an N-terminal hydrophobic segment likely to span the membrane and the region from 58 to 201 constituting the CDP-alcohol transferase domain. Similar sequences in *Streptomyces coelicolor* and even in archaeal genes (*e.g.*, *Pyrococcus furiosus*, *Methanosarcina acetivorans*, *Methanobacterium thermoautotrophicum*) can be identified. There is a very strong alignment of the initial half of the CDP-alcohol transferase domain among these proteins. Whether or not the archaeal genes code for PI synthase activity or another activity that has a CDP-alcohol domain (*e.g.*, the DIP synthase proposed in the biosynthesis of DIP (Chen *et al.*, 1998) awaits heterologous expression of these genes, so far a very difficult task for archaeal membrane proteins.

4.2 Enzymes of inositol phospholipid degradation: PI-specific phospholipase C

Bacterial PI-PLC enzymes have quite a different role from the eukaryotic enzymes involved in signal transduction. A variety of Gram-positive bacteria secrete PI-PLC that is highly specific for non-phosphorylated PI and does not require Ca^{2+} (required by all the mammalian PI-PLC enzymes). The PI-PLC enzyme in a subset of these organisms can also cleave the glycerol-phosphate bond of glycosyl-PI-anchored proteins (Ferguson *et al.*, 1985; Sharom and Lehto, 2002). Secreted PI-PLCs often contribute to the virulence of a microorganism. In *M. tuberculosis*, the gene for PI-PLC is upregulated during the initial 24 h of macrophage infection (Raynaud *et al.*, 2002). Often the PLC enzymes aid in the microorganism avoiding antibacterial host factors as it negotiates

```
My.b.   1   MSKLPFLSRAAFARITTPIARGLLRVGLTPDVVTILGTTASVAGALTLFP  50
Ms.a.   4                       PFARS---VPLSPNTLTLLGFAVSVAAGVA-FA  32
Mb.t.  13             RPVIRRFIDPIAD---RIALPADYITLTGFLVACAASAG-YA  50
Pyr.f.  2   LSNLRPLAKKPLEKIAEPFSK----LGITPNQLTMVGFFLSLLASYEYYL  47
                  +  + ++  P++      + +  + +T++G++++  ++   +

My.b.  51   MGKLFAGACVVWFFVLFDMLDGAMARERGGGTRFGAVLDATCDRISDGAV 100
Ms.a.  33   LGKPFEGGFLILFSGVFDILDGGVARAKGRITPFGGVLDSVCDRYSDGLM  82
Mb.t.  51   SGSLITGAALLAASGFIDVLDGAVARRRFRPTAFGGFLDSTLDRLSDGII 100
Pyr.f.48    NNQVF-GSLILLLGAFLDALDGSLARLTGRVTKFGGFLDSTMDRLSDAAI  96
              + + G+++++++++D+LDG++AR      T FG++LD++ DR+SD+++

My.b. 101   FCGLL------WWIAF-HMRDRPLVI-----ATLICLVTSQVISYIKARA 138
Ms.a.  83   FLGIM------AGAIYGRLSFAPFLGVEGWLWAGFALIGSFLVSYTRARA 126
Mb.t. 101   IIGIT------------AGGFTGLL-----TGLLALHSGLMVSYVRARA 132
Pyr.f. 97   IFGIALGELVNWKVAF-----------------LALIGSYMVSYTRCRA 128
            ++G++                              ++L   +++SY ++RA

My.b. 139   EASGLRG-DG----GFIERPERLIIVLTGAGVSDFPFVPWPPALSVGMWL 183
Ms.a. 127   ESAGCRKLSV----GIAERTERMVILALGA-LSGF--------LGWALVL 163
Mb.t. 133   ESLGIEC-AV----GIAERAERIIIILAGSLAGYLIHPW--FMDAAIIVL 175
Pyr.f.129   E---LAG-SGTLAVGIAERGERLLILVI-AGLFGI--------IDIGVYL 165
            E    +            G   ER ER++I+     +         +          ++L

My.b.   184 LAVASVITCVQRL---HTVWTSPGAID         207
Ms.a.   164 IAVFSHITMIQRV                       176
Mb.t.   176 AALGYFTMI-QRM---IYVW                191
Pyr.f.  166 VAILSWITFLQRV---Y                   179
            +A+ +  +  QR+
```

Figure 6. Alignment of archaeal PI synthase homologues with the *M. bovis* PI synthase (My.b.): Ms.a., *Methanosarcina acetivorans*; Mb.t., *Methanobacterium thermoautotrophicum*; Py.f., *Pyrococcus furiosus*. The transmembrane sequence of the *M. bovis* enzyme is in italics while the CDP-alcohol transferase domain is underlined.

different compartments of cells (*e.g.*, *Bacillus cereus*, *B. thuringiensis*, *B. anthracis*). In *B. anthracis*, the *plcR* gene, which regulates expression of phospholipases and other lytic factors, is expressed under anaerobic conditions (Klichko *et al.*, 2003).

For all the PI-PLC enzymes examined thus far, PI hydrolysis occurs in two sequential reactions as shown in Figure 7 (Griffith and Ryan 1999): (i) an intramolecular phosphotransferase reaction at a phospholipid/aggregate surface to produce DAG and water-soluble 1,2-cyclic phosphate, cIP (or cIP_x in the case of a phosphoinositide substrate), followed by (ii) a phosphodiesterase reaction where the cIP is hydrolyzed to inositol-1-phosphate (or IP_x). The second reaction can be studied separately; it occurs with a soluble monomeric substrate and is much slower (both lower k_{cat} and higher K_m) than the first step (Volwerk *et al.*, 1990; Zhou *et al.*, 1997). It is an excellent probe for allosteric effects of membranes on catalysis (Zhou *et al.*, 1997; Wu *et al.*, 1997).

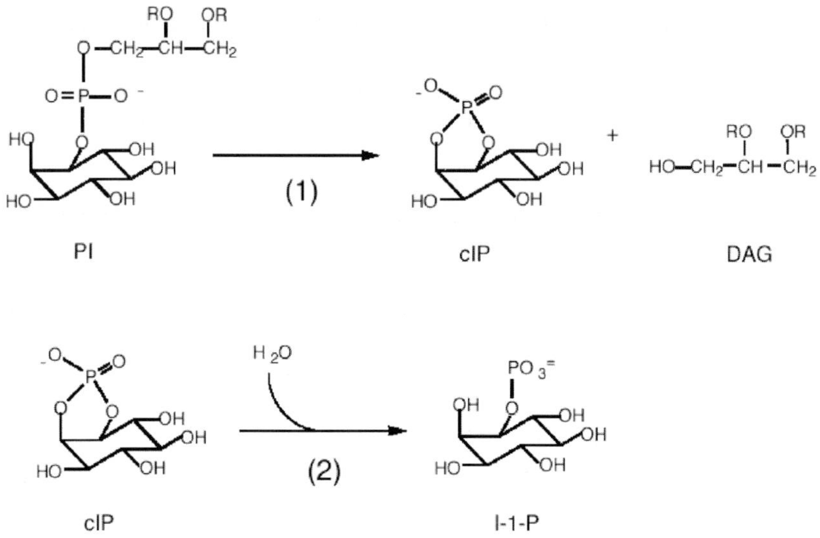

Figure 7. Reactions catalyzed by PI-PLC enzymes.

Crystal structures exist of two bacterial PI-PLC enzymes, the protein from *B. cereus* (Heinz *et al.*, 1995), which can cleave GPI-anchors, and the PI-PLC from *Listeria monocytogenes* (Moser *et al.*, 1997), which is not able to effectively release GPI-anchored proteins. While the sequence homology of these two proteins is limited, the structures are very similar. The bacterial PI-PLC proteins are folded into a distorted TIM-barrel, where the parallel β-strands form an inner circular and closed barrel with α-helices located on the outside between neighboring β-strands, that is structurally very similar to the catalytic domain of PLCδ_1, the only mammalian PI-PLC for which there is a structure (Essen *et al.*, 1996; Heinz *et al.*, 1998). The availability of structures and results of mutagenesis provide details on the catalytic mechanism for this type of enzyme (for review and more extensive references see Mihai *et al.* (2003)).

In the bacterial PI-PLC structures, the top of the barrel rim has several hydrophobic residues that are fully exposed to solvent and poorly defined in the crystal structures (implying significant mobility). The active site of PI-PLC is accessible and well-hydrated, and these mobile elements at the top of the barrel offer a different motif for interactions of the protein with phospholipid interfaces. The PI-PLC from *B. thuringiensis* (nearly identical in sequence to the enzyme from *B. cereus* whose crystal structure was determined) exhibits the property of interfacial activation, where enhanced activity is observed when the substrate PI is present in an interface compared to monomeric substrate (Lewis *et al.*, 1993). However, other non-substrate lipids such as phosphatidylcholine (PC), phosphatidic acid (PA), and other anionic lipids have an effect on the activity of PI-PLC toward both substrates PI and water-soluble cIP (Zhou *et al.*, 1997). In particular, the presence of PC enhances the catalytic activity of the enzyme toward PI as well as cIP. Two residues at the top of the barrel – Trp47 in short helix B and Trp242 in a flexible loop (Figure 8) – are required for the protein to bind to activating PC surfaces (Feng *et al.*, 2002). Each tryptophan interacts with a PC molecule (Zhang *et al.*, 2004), and in so doing must alter the active site so that PI cleavage and cIP hydrolysis occur with enhanced k_{cat} and decreased apparent K_m (Zhou and Roberts, 1998; Zhou *et al.*, 1997). Removal of both tryptophan residues generates an enzyme that no longer effectively binds to membranes. The PI-PLC from *L. monocytogenes* also has aromatic residues fully exposed to solvent at the top of the barrel (Moser *et al.*, 1997) that are likely to be involved in surface binding of the enzyme. However, as discussed below, the interaction of this region of the protein with a net neutral amphiphile is different for the Listeria protein, and the differences from the Bacillus PI-PLC may be an important key to the role of PI-PLC in *L. monocytogenes* infectivity or mammalian cells.

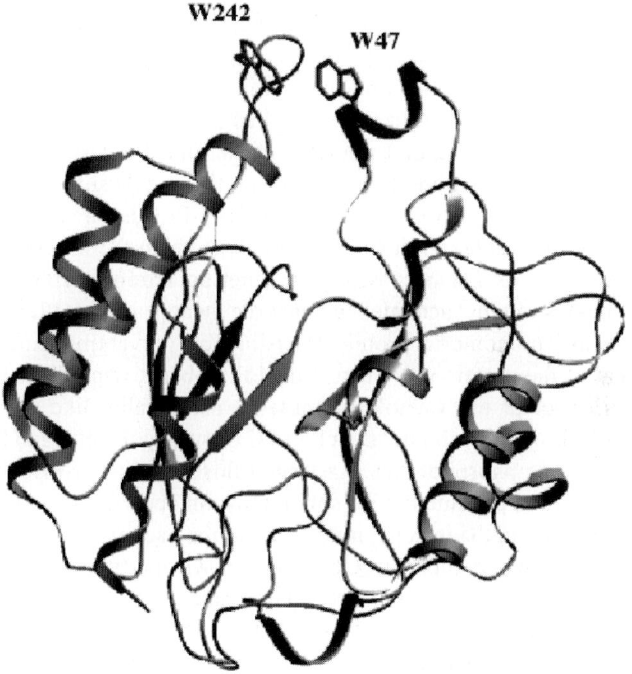

Figure 8. Structure of *B. cereus* PI-PLC.

4.3 Role of inositol enzymes in infection and virulence: *Listeria monocytogenes* PI-PLC

In bacteria, the biological role of secreted PI-PLC enyzmes, along with other non-specific phospholipases, is to aid in survival of the microorganism.

Bacterial PI-PLC enzymes that cleave GPI-anchored proteins, such as the PI-PLC secreted by *B. thuringiensis*, *B. cereus*, and *Staphylococcus aureus*, could target those membrane components in the extracellular leaflet of the plasma membrane, which is rich in the zwitterionic lipids PC and sphingomyelin. However, secreted PI-PLC enzymes whose targets are PI and not GPI-anchors, must follow a different path to aid bacterial survival since PI is not found in the external leaflet of most organisms. *Listeria monocytogenes* is a foodborne pathogen of humans and animals that can cause serious infections in immunocompromised individuals, pregnant women, and the elderly. The substrate for this PI-PLC is the PI (not GPI-anchors) that is on the inner monolayer of the plasma membrane of mammalian cells. When *L. monocytogenes* is internalized into macrophages, both listeriolysin O and PI-PLC expression are upregulated. PI-PLC presumably gains access to host PI by means of phagosome permeabilization and destruction caused by listeriolysin O (Sibelius *et al.*, 1999; Wadsworth and Goldfine, 2002). When this PI-PLC hydrolyzes PI and generates DAG, host PKCβ is activated. This along with elevated intracellular calcium plays an important role in escape of the bacterium from the phagosome.

While the *L. monocytogenes* PI-PLC has a similar structure to the *B. cereus* enzyme, its regulation of activity by amphiphiles and ionic strength is quite different. The Listeria enzyme, like the Bacillus PI-PLC, binds very tightly to vesicles of negatively charged phospholipids, and the surface interaction appears to cause aggregation of the enzyme on the vesicle surface. Unlike the *B. thuringiensis* PI-PLC, the *L. monocytogenes* PI-PLC has a very weak affinity for PC vesicles, although it can bind to short-chain PC micelles. Neutral amphiphiles such as PC or a detergent such as Triton X-100 as well as increased ionic strength "activate" the enzyme by appearing to shift the equilibrium toward monomeric protein, possibly by preventing anionic lipid aggregation of the protein (W. Chen and M.F. Roberts, unpublished results). The inner leaflet of the vacuole membrane, presumably like the external leaflet of the plasma membrane, is rich in PC or sphingomyelin and has a low content of anionic phospholipids. Assuming this bacterial PI-PLC has weak affinity for PC (and presumably for sphingomyelin and PE as well) under the moderate ionic strengths in cells, it will stay in the vacuolar fluid and be easily dispersed into the cytoplasm. Once there it will partition with the negatively charged components of the target membrane. As long as there are some zwitterionic/neutral lipids around, it is likely the enzyme can bind and partially insert into the bilayer in a way that allows it to effectively hydrolyze PI and generate DAG.

5. PERSPECTIVES FOR THE FUTURE

While inositol compounds are not ubiquitous in bacteria and archaea, they do occur and often have critical roles in those organisms. Inositol-containing phospholipids do occur, but in these two domains of life as opposed to Eukarya, they are not components of signaling pathways. Often PI (or the sphingolipid counterpart) serves as a membrane anchor of complex glycolipids and surface proteins. These lipid units are absolutely critical to the integrity and infectivity of mycobacteria. Soluble inositol compounds are also not very common, but those that have been detected have rather unusual roles in microorganisms. Intracellular concentrations of DIP and related solutes are modulated by external NaCl, indicating they are part of the osmotic response of the cells. Their biosynthesis and accumulation is also modulated by the growth temperature of the organisms. Mycothiol is used as a redox agent in mycobacteria – in essence it is a more complex glutathione analogue. Clearly, there are many areas that are ripe for further investigation.

The occurrence of PI anchors in mycobacteria is important for growth and infectivity of these cells, although the inositol unit itself does not play a direct role in the infectivity. Rather the membrane-anchored inositol unit is a structural scaffold. If a unique step in inositol biosynthesis can be inhibited, new drugs might be available against these organisms. The *M. tuberculosis* MIPS is not likely to be a good target since its structure is so similar to the yeast (and presumably human) enzyme. IMPases are likely to be difficult targets since there may be several activities that can carry out this chemistry in *M. tuberculosis*. One of the more promising targets has been suggested to be IPC synthase, since these particular inositol-containing lipids are unique to mycobacteria. Several natural products (*e.g.*, khafrefungin, Nakamura *et al.*, 2003, and galbonolide, Sakoh *et al.*, 2004) that target IPC synthase have been identified, but improving these as inhibitors of IPC synthase and ensuring that they have no effect on mammalian enzymes is likely to require a structure of an IPC synthase. The deacetylase activity needed for mycothiol production might also be a target since mycothiol helps the bacterium fend off oxidative stresses of the host. A recent structure of the deacetylase from *M. tuberculosis* that is absolutely critical for mycothiol production, while providing detailed information of the mechanism for this enzyme, also indicates it may be a representative of a wider family of enzymes, including eukaryotic ones, that work on *N*-acetylglucosamine residues (McCarthy *et al.*, 2004). However, the availability of the structure may allow researchers to develop very specific inhibitors of this enzyme.

DIP and its derivatives are interesting compounds for hyperthermophilic cells to synthesize in response to salt and/or temperature stress. Significant cell resources are necessary to make this molecule, particularly when the intracellular concentration is >0.5 M. With this in mind, one can pose two major questions. (i) What is there about the DIP molecule that makes it a particularly useful solute at high ionic strength and high temperature? (ii) How is DIP accumulation at supraoptimal temperatures regulated? For the first of these, studies will be needed to see how DIP affects water structure around proteins or other macromolecules. The second awaits the development of genetic systems for hyperthermophiles, a non-trivial task at this point.

Inositol enzymology is also ripe for future studies. The recent MIPS structures have certainly aided in our appreciation of the chemistry this enzyme enables. However, they also point out that a significant conformational change is likely to be necessary for the cyclization step of the reaction. Structures with linear G-6-P analogues or transition intermediate mimics fail to show how C1 and C6 are moved close enough for the aldol condensation reaction to occur. While it is not clear how to generate and characterize enzyme intermediate complexes on an atomic level, doing so would open up our understanding of this interesting enzyme. Whether or not metal ions are needed for all MIPS is another intriguing question at this point, since recent structural work with the yeast enzyme suggests it too has a metal ion in the vicinity of the NAD^+.

Continued structural characterization of inositol enzymes could shed light on some of their more uncommon roles. For example, we already know that SuhB does not need its IMPase activity to suppress diverse temperature-sensitive mutants in *E. coli* (Chen and Roberts, 2000). What are its targets in *E. coli*– RNAse III, mRNA molecules, or something else? The availability of a structure for this enzyme for comparison to all the other members of the IMPase superfamily might supply the key to unraveling this other biological activity. And perhaps more intriguingly, we can ask if other bacterial IMPase enzymes exhibit behavior like SuhB such that their inactivation gives rise to a cold-sensitive phenotype, or is this a property solely of the *E. coli* enzyme? The archaeal IMPase/FBPase duality also serves as an interesting entry into how specific activities in this class of phosphatases, members of the IMPase superfamily, evolved. Any details on PI synthases would aid in our understanding of this step in PI biosynthesis. The possible identity of archaeal PI synthase sequences could also provide new targets for expression and characterization that might be better suited for crystallization and structural studies.

ACKNOWLEDGMENTS

The author would like to thank the DOE Department of Energy Biosciences Division DE-FG02-91ER20025 and NIH GM60418 for support of some of the research discussed in this review article.

REFERENCES

Agam, G., Shamir, A., Shaltiel, G., and Greenberg, M.L., 2002, *myo*-Inositol-1-phosphate (MIP) synthase: A possible target for antibipolar drugs. *Bipolar Disord.* **4**(Suppl 1): 15–20.

Bachawarat, N., and Mande, S.C., 1999, Identification of the INO1 gene of *Mycobacterium tuberculosis* H37Rv reveals a novel class of inositol-1-phosphate synthase enzyme. *J. Mol. Biol.* **291:** 531–536.

Bachawarat, N., and Mande, S.C., 2000, Complex evolution of the inositol-1-phosphate synthase gene among archaea and eubacteria. *Trends Genet.* **16:** 111–113.

Balmer, Y., Stritt-Etter, A., Hirasawa, M., Jacquot, J.P., Keryer, E., Knaff, D.B., and Schurmann, P., 2001, Oxidation-reduction and activation properties of chloroplast fructose 1,6-bisphosphatase with mutated regulatory site. *Biochemistry* **40:** 15444–15450.

Belisle, J.T., Brandt, M.E., Radolf, J.D., and Norgard, M.V., 1994, Fatty acids of *Treponema pallidum* and *Borrelia burgdorferi* lipoproteins. *J. Bacteriol.* **176:** 2151–2157.

Brennan, P.J., 2003, Structure, function, and biogenesis of the cell wall of *Mycobacterium tuberculosis*. *Tuberculosis* **83:** 91–97.

Brennan, P.J., and Ballou, C.E., 1968, Phosphatidylmyoinositol monomannoside in *Propionibacterium shermanii*. *Biochem. Biophys. Res. Commun.* **30:** 69–75.

Brennan, P.J., and Lehane, D.P., 1971, The phospholipids of corynebacteria. *Lipids* **6:** 401–409.

Buchmeier, N.A., Newton, G.L., Koledin, T., and Fahey, R.C., 2003, Association of mycothiol with protection of *Mycobacterium tuberculosis* from toxic oxidants and antibiotics. *Mol. Microbiol.* **47:** 1723–1732.

Chang, S.F., Ng, D., Baird, L., and Georgopoulos, C., 1991, Analysis of an *Escherichia coli dnaB* temperature-sensitive insertion mutation and its cold-sensitive extragenic suppressor. *J. Biol. Chem.* **266:** 3654–3660.

Chen, L., and Roberts, M.F., 1998, Cloning and expression of the inositol monophosphatase gene from *Methanococcus jannaschii* and characterization of the enzyme. *Appl. Environ. Microbiol.* **64:** 2609–2615.

Chen, L., and Roberts, M.F., 1999, Characterization of a tetrameric inositol monophosphatase from the hyperthermophilic bacterium *Thermotoga maritima*. *Appl. Environ. Microbiol.* **65**: 4559–4567.

Chen, L., and Roberts, M.F., 2000, Overexpression, purification, and analysis of complementation behavior of *E. coli* SuhB protein: Comparison with bacterial and archaeal inositol monophosphatases. *Biochemistry* **39:** 4145–4153.

Chen, L., Spiliotis, E., and Roberts, M.F., 1998, Biosynthesis of Di-*myo*-inositol-1,1′-phosphate, a novel osmolyte in hyperthermophilic archaea. *J. Bacteriol.* **180:** 3785–3792.

Chen, L., Zhou, C., Yang, H., and Roberts, M.F., 2000, Inositol-1-phosphate synthase from *Archaeoglobus fulgidus* is a class II aldolase. *Biochemistry* **39:** 12415–12423.

Chi, H., Tiller, G.E., Dasouki, M.J., Romano, P.R., Wang, J., O'Keefe, R.J., Puzas, J.E., Rosier, R.N., and Reynolds, P.R., 1999, Multiple inositol polyphosphate phosphatase: evolution as a distinct group within the histidine phosphatase family and chromosomal localization of the human and mouse genes to chromosomes 10–23 and 19. *Genomics* **56:** 324–336.

Cho, J.S., Lee, C.W., Kang, S.H., Lee, J.C., Bok, J.D., Moon, Y.S., Lee, H.G., Kim, S.C., and Choi, Y.J., 2003, Purification and characterization of a phytase from *Pseudomonas syringae* MOK1. *Curr. Microbiol.* **47:** 290–294.

Ciulla, R.A., Burggraf, S., Stetter, K O., and Roberts, M.F., 1994, Occurrence and role of di-*myo*-inositol-1,1′-phosphate in *Methanococcus igneus*. *Appl. Environ. Microbiol.* **60:** 3660–3664.

Downing, J.F., Pasula, R., Wright, J.R., Twigg, H.L., III, and Martin, W.J., II, 1995 Surfactant protein A promotes attachment of *Mycobacterium tuberculosis* to alveolar macrophages during infection with human immunodeficiency virus. *Proc. Natl. Acad. Sci. USA* **92:** 4848–4852.

Elsbach, P., and Weiss, J., 1988, Phagocytosis of bacteria and phospholipid degradation. *Biochim. Biophys. Acta* **947:** 29–52.

Essen, L.O., Perisic, O., Cheung, R., Katan, M., and Williams, R.L., 1996, Crystal structure of a mammalian phosphoinositide-specific phospholipase C δ. *Nature* **380:** 595–602.

Feng, J., Wehbi, H., and Roberts, M.F., 2002, Role of tryptophan residues in interfacial binding of phosphatidylinositol-specific phospholipase C. *J. Biol. Chem.* **277:** 19867–19875.

Ferguson, J.S., Voelker, D.R., McCormack, F.X., and Schlessinger, L.S., 1999, Surfactant protein D binds to *Mycobacterium tuberculosis* bacilli and lipoarabinomannan via carbohydrate-lectin interactions resulting in reduced phagocytosis of the bacteria by macrophages. *J. Immunol.* **163:** 312–321.

Ferguson, M.A., Low, M.G., and Cross, G.A., 1985, Glycosyl-sn-1,2-dimyristylphosphatidylinositol is covalently linked to *Trypanosoma brucei* variant surface glycoprotein. *J. Biol. Chem.* **260:** 14547–14555.

Fisher, S.K., Novak, J.E., and Agranoff, W., 2002, Inositol and higher inositol phosphates in neural tissues: Homeostasis, metabolism and functional significance. *J. Neurochem.* **82:** 736–754.

Goren, M.B., 1984. Biosynthesis and structures of phospholipids and sulfatides. In: Kubica, G.P., and Wayne, L.G. (eds.), The Mycobacteria: A Sourcebook. Marcel Dekker, Inc., New York, pp. 379–415.

Greiner, R., Farouk, A., Alminger, M.L., and Carlsson, N.G., 2002, The pathway of dephosphorylation of *myo*-inositol hexakisphosphate by phytate-degrading enzymes of different Bacillus sp. *Can. J. Microbiol.* **48:** 986–994.

Griffith, O.H., and Ryan, M., 1999, Bacterial phosphatidylinositol-specific phospholipase C: Structure, function, and interaction with lipids. *Biochim. Biophys. Acta* **1441:** 237–254.

Heinz, D.W., Essen, L.O., and Williams, R.L., 1998, Structural and mechanistic comparison of prokaryotic and eukaryotic phosphoinositide-specific phospholipases C. *J. Mol. Biol.* **275:** 635–650.

Heinz, D.W., Ryan, M., Bullock, T.L., and Griffith, O.H., 1995, Crystal structure of the phosphatidylinositol-specific phospholipase C from *Bacillus cereus* in complex with *myo*-inositol. *EMBO J.* **14:** 3855–3863.

Hirst, P.H., Riley, A.M., Mills, S.J., Spiers, I.D., Poyner, D.R., Freeman, S., Potter, B.V., Smith, A.W., 1999, Inositol polyphosphate-mediated iron transport in *Pseudomonas aeruginosa. J. Appl. Microbiol.* **86:** 537–543.

Hoppe, H.C., de Wet, B.J., Cywes, C., Daffe, M., and Ehlers, M.R., 1997, Identification of phosphatidylinositol mannoside as a mycobacterial adhesin mediating both direct and opsonic binding to nonphagocytic mammalian cells. *Infect. Immun.* **65:** 3896—3905.

Inada, T., and Nakamura, Y., 1995, Lethal double-stranded RNA processing activity of ribonuclease III in the absence of suhB protein of *Escherichia coli*. *Biochimie* **77:** 294–302.

Inada, T., and Nakamura, Y., 1996, Autogenous control of the suhB gene expression of *Escherichia coli*. *Biochimie* **78:** 209–212.

Jackson, M., Crick, D.C., and Brennan, P.J., 2000, Phosphatidylinositol is an essential phospholipids of mycobacteria. *J. Biol. Chem.* **275:** 30092–30099.

Jacquot, J.P., Lopez-Jaramillo, J., Miginiac-Maslow, M., Lemaire, S., Cherfils, J., Chueca, A., and Lopez, J., 1997, Cysteine-153 is required for redox regulation of pea chloroplast fructose-1,6-bisphosphatase. *FEBS Lett.* **401:** 143–147.

Janczarek, M., and Skorupska, A., 2001, The *Rhizobium leguminosarum* bv. Trifolii *pssB* gene product is an inositol monophosphatase that influences exopolysaccharide synthesis. *Arch. Microbiol.* **175:** 143–151.

Johnson, K.A., Chen, L., Yang, H., Roberts, M.F., and Stec, B., 2001, Crystal structure and catalytic mechanism of the MJ0109 gene product: A bifunctional enzyme with inositol monophosphatase and fructose 1,6-bisphosphatase activities. *Biochemistry* **40:** 618–630.

Kataoka, T., and Nojima, S., 1967, The phospholipid compositions of some Actinomycetes. *Biochim. Biophys. Acta* **144:** 681–683.

Klichko, V.I., Miller, J., Wu, A., Popv, S.G., and Alibekk, K., 2003, Anaerobic induction of *Bacillus anthracis* hemolytic activity. *Biochem. Biophys. Res. Commun.* **303:** 855–862.

Kozloff, L.M., Turner, M.A., and Arellanno, F., 1991a, Formation of bacterial membrane ice-nucleating lipoglcoprotein complexes. *J. Bacteriol.* **173:** 6528–6536.

Kozloff, L.M., Turner, M.A., Arellanno, F., and Lute, M., 1991b, Phosphatidylinositol, a phospholipid of ice-nucleating bacteria. *J. Bacteriol.* **173:** 2053–2060.

Lee, D.C., Cottrill, M.A., Forsberg, C.W., and Jia, Z., 2003, Functional insights revealed by the crystal structures of *Escherichia coli* glucose-1-phosphatase. *J. Biol. Chem.* **278:** 31412–31418.

Levery, S.B., Toledo, M.S., Straus, A.H., and Takahashi, H.K., 1998, Structure elucidation of sphingolipids from the mycopathogen *Paracoccidiodes brasiliensis*: An immunodominant β-galactofuranose residue is carried by a novel glycosylinositol phosphorylceramide antigen. *Biochemistry* **37:** 8764–8775.

Lewis, K., Garigapati, V., Zhou, C., and Roberts, M.F., 1993, Substrate requirements of bacterial phosphatidylinositol-specific phospholipase C. *Biochemistry* **32:** 8836–8841.

Lim, D., Golovan, S., Forsberg, C.W., and Jia, Z., 2000, Crystal structures of *Escherichia coli* phytase and its complex with phytate. *Nat. Struct. Biol.* **7:** 108–113.

Loewus, M.W., 1977, Hydrogen isotope effects in the cyclization of D-glucose 6-phosphate by *myo*-inositol-1-phosphate synthase. *J. Biol. Chem.* **252:** 7221–7223.

Loewus, M.W., Loewus, F.A., Brillinger, G.U., Otsuka, H., and Floss, H.G., 1980, Stereochemistry of the *myo*-inositol-1-phosphate synthase reaction. *J. Biol. Chem.* **255:** 11710–11712.

Majumdar, A.L., Chatterjee, A., Dastidar, K.G., and Majee, M., 2003, Diversification and evolution of L-*myo*-inositol 1-phosphate synthase. *FEBS Lett.* **553:** 3–10.

Majumder, A.L., Johnson, M.D., and Henry, S.A., 1997, 1L-*myo*-inositol-1-phosphate synthase. *Biochim. Biophys. Acta* **1348:** 245–256.

Martin, D.D., Ciulla, R.A., and Roberts, M.F., 1999, Osmoadaptation in archaea. *Appl. Environ. Microbiol.* **65:** 1815–1825.

Martins, L.O., Carreto, L.S., Da Costa, M.S., and Santos, H., 1996, New compatible solutes related to di-*myo*-inositol-phosphate in members of the order Thermotogales. *J. Bacteriol.* **178:** 5644–5651.

Martins, L.O., Huber, R., Huber, H., Stetter, K.O., da Costa, M.S., and Santos, H., 1997, Organic solutes in hyperthermophilic *Archaea. Appl. Environ. Microbiol.* **63:** 896–902.

Matsuhisa, A., Suzuki, N., Noda, T., and Shiba, K., 1995, Inositol monophosphatase activity from the *Escherichia coli* suhB gene product. *J. Bacteriol.* **177:** 200–205.

McCarthy, A.A., Peterson, N.A., Knijff, R., and Baker, E.N., 2004, Crystal structure of MshB from *Mycobacterium tuberculosis*, a deacetylase involved in mycothiol biosynthesis. *J. Mol. Biol.* **335:** 1131–1141.

Mihai, C., Kravchuk, A.V., Tsai, M.D., and Bruzik, K.S., 2003, Application of Bronsted-type LFER in the study of the phospholipase C mechanism. *J. Amer. Chem. Soc.* **125:** 3236–3242.

Morii, H., Yagi, H., Akutsu, H., Nomura, N., Sako, Y., and Koga, Y., 1999, A novel phosphoglycolipid archaetidyl(glucosyl)inositol with two sesterterpanyl chains from the aerobic hyperthermophilic archaeon *Aeropyrum pernix* K1.

Morii, H., and Koga, Y., 1994, Asymmetrical topology of diether- and tetraether-type polar lipids in membranes of *Methanobacterium thermoautotrophicum* cells. *J. Biol. Chem.* **269:** 10492–10497.

Moser, J., Gerstel, B., Meyer, J.E., Chakraborty, T., Wehland, J., and Heinz, D.W., 1997, Crystal structure of the phosphatidylinositol-specific phospholipase C from the human pathogen *Listeria monocytogenes*. *J. Mol. Biol.* **273:** 269–282.

Movahedzadeh, F., Smith, D.A., Norman, R.A., Dinadayala, P., Murray-Rust, J., Russell, D.G., Kendall, S.L., Rison, S.C., McAlister, M.S., Bancroft, G.J., McDonald, N.Q., Daffe, M., Av-Gay, Y., and Stoker, N.G., 2004, The *Mycobacterium tuberculosis ino1* gene is essential for growth and virulence. *Mol. Microbiol.* **51:** 1003–1014.

Nagiec, M.M., Nagiec, E.E., Baltisberger, J.A., Wells, G.B., Lester, R.L., and Dickson, R.C., 1997, Sphingolipid synthesis as a target for antifungal drugs. Complementation of the insoitol phosphorylceramide synthase defect in a mutant strain of *Saccharomyces cerevisiae* by the *AUR1* gene. *J. Biol. Chem.* **272:** 9809–9817.

Nakamura, M., Mori, Y., Okuyama, K., Tanikawa, K., Yasuda, S., Hanada, K., and Kobayashi, S., 2003, Chemistry and biology of khafrefungin. Large-scale synthesis, design, and structure-activity relationship of khafrefungin, an antifungal agent. *Org. Biomol. Chem.* **1:** 3362–3376.

Nesbo, C.L., L'Haridon, S., Stetter, K.O., and Doolittle, W.F., 2001, Phylogenetic analyses of two "archaeal" genes in *Thermotoga maritima* reveal multiple transfers between archaea and bacteria. *Mol. Biol. Evol.* **18:** 362–375.

Neuwald, A.F., York, J.D., and Majerus, P.W., 1991, Diverse proteins homologous to inositol monophosphatase. *FEBS Lett.* **294:** 16–18.

Newton, G.L., Bewley, C.A., Dwyer, T.J., Horn, R., Abaromowitz, Y., Cohen, G., Davies, J., Faulkner, D.J., and Fahey, R.C., 1995, The structure of U17 isolated from *Streptomyces clavuligerus* and its properties as an antioxidant thiol. *Eur. J. Biochem.* **230:** 821–825.

Newton, G.L., and Fahey, R.C., 2002, Mycothiol biochemistry. *Arch. Microbiol.* **178:** 388–394.

Nigou, J., Dover, L.G., and Besra, G.S., 2002, Purification and biochemical characterization of *Mycobacterium tuberculosis* SuhB, an inositol monophosphatase involved in inositol biosynthesis. *Biochemistry* **41:** 4392–4398.

Nikawa, J., and Yamashita, S., 1997, Phosphatidylinositol synthase from yeast. *Biochim. Biophys. Acta* **1348:** 173–178.

Nishihara, M., and Koga, Y., 1991, Hydroxyarchaetidylserine and hydroxyarchaetidyl-*myo*-inositol in *Methanosarcina barkeri*: Polar lipids with a new ether core portion. *Biochim. Biophys. Acta* **1082:** 211–217.

Nishihara, M., Utagawa, M., Akutsu, H., and Koga, Y., 1992, Archaea contain a novel diether phosphoglycolipid with a polar head group identical to the conserved core of eucaryal glycosyl phosphatidylinositol. *J. Biol. Chem.* **267:** 12432–12435.

Norman, R.A., McAlister, M.S., Murray-Rust, J., Movahedzadeh, F., Stoker, N.G., and McDonald, N.Q., 2002, Crystal structure of inositol 1-phosphate synthase from *Mycobacterium tuberculosis*, a key enzyme in phosphatidylinositol synthesis. *Structure* **10:** 393–402.

Parish, T., Liu, J., Nikaido, H., and Stoker, N.G., 1997, A *Mycobacterium smegmatis* mutant with a defective inositol monophosphate phosphatase gene homolog has altered cell envelope permeability. *J. Bacteriol.* **179:** 7827–7833.

Raboy, V., 2003, *myo*-Inositol-1,2,3,4,5,6-hexakisphosphate. *Phytochemistry* **64:** 1033–1043.

Rashid, N., Imanaka, H., Kanai, T., Fukui, T., Atomi, H., and Imanaka, T., 2002, A novel candidate for the true fructose-1,6-bisphosphatase in archaea. *J. Biol. Chem.* **277:** 30649–30655.

Raynaud, C., Guilhot, C., Rauzier, J., Bordat, Y., Pelicic, V., Manganelli, R., Smith, I., Gicquel, B., and Jackson, M., 2002, Phospholipases C are involved in the virulence of *Mycobacterium tuberculosis*. *Mol. Microbiol.* **45:** 203–217.

Sakoh, H., Sugimoto, Y., Imamura, H., Sakuraba, S., Jona, H., Bamba-Nagano, R., Yamada, K., Hashizume, T., and Morishima, H., 2004, Novel galbonolide derivatives as IPC synthase inhibitors: Design, synthesis and in vitro antifungal activities. *Bioorg. Med. Chem. Lett.* **14:** 143–145.

Salman, M., Lonsdale, J.T., Besra, G.S., and Brennan, P.J., 1999, Phosphatidylinositol synthesis in mycobacteria. *Biochim. Biophys. Acta* **1436:** 437–450.

Scholz, S., Sonnenbichler, J., Schafer, W., and Hensel, R., 1992, Di-*myo*-inositol-1,1′-phosphate: A new inositol phosphate isolated from *Pyrococcus woesei*. *FEBS Lett.* **306:** 239–242.

Scholz, S., Wolff, S., and Hensel, R., 1998, The biosynthesis pathway of di-*myo*-inositol-1,1′-phosphate in *Pyrococcus woesei*. *FEMS Microbiol. Lett.* **168:** 37–42.

Sharom, F.J., and Lehto, M.T., 2002, Glycosylphosphatidylinositol-anchored proteins: Structure, function, and cleavage by phosphatidylinositol-specific phospholipase C. *Biochem. Cell. Biol.* **80:** 535–549.

Shin, S., Ha, N.C., Oh, B.C., Oh, T.K., and Oh, B.H., 2001, Enzyme mechanism and catalytic property of beta propeller phytase. *Structure* **9:** 851–858.

Sibelius, U., Schulz, E.C., Rose, F., Hattar, K., Jacobs, T., Weiss, S., Chakraborty, T., Seeger, W., and Grimminger, F., 1999, Role of *Listeria monocytogenes* exotoxins listeriolysin and phosphatidylinositol-specific phospholipase C in activation of human neutrophils. *Infect. Immun.* **67:** 1125–1130

Sidobre, S., Nigou, J., Puzo, G., and Riviere, M., 2000, Lipoglycans are putative ligands for the human pulmonary surfactant protein A attachment to mycobacteria. *J. Biol. Chem.* **275:** 2415–2422.

Spies, H.S.C., and Steenkamp, D.J., 1994, Thiols of intracellular pathogens. Identification of ovothiol A in *Leishmania donovani* and structural analysis of a novel thiol from *Mycobacterium bovis*. *Eur. J. Biochem.* **224:** 203–213.

Stec, B., Yang, H., Johnson, K.A., Chen, L., and Roberts, M.F., 2000, MJ0109 is an enzyme that is both an inositol monophosphatase and the 'missing' archaeal fructose-1,6-bisphosphatase. *Nat. Struct. Biol.* **7:** 1046–1050.

Stein, A.J., and Geiger, J.H., 2002, The crystal structure and mechanism of 1-L-*myo*-inositol-1-phosphate synthase. *J. Biol. Chem.* **277:** 9484–9491.

Stieglitz, K.A., Johnson, K.A., Yang, H., Roberts, M.F., Seaton, B.A., Head, J.F., and Stec, B., 2002, Crystal structure of a dual activity IMPase/FBPase (AF2372) from *Archaeoglobus fulgidus*. The story of a mobile loop. *J. Biol. Chem.* **277:** 22863–22874.

Stieglitz, K.A., Yang, H., Roberts, M.F., and Stec, B., 2005, Reaching for mechanistic consensus across life kingdoms: Structure and insights into catalysis of the inositol-1-phosphate synthase (MIPS) from *Archaeoglobus fulgidus*. *Biochemistry* **44:** 213–224.

Stieglitz, K.A., Seaton, B.A., Head, J.F., Stec, B., and Roberts, M.F., 2003, Unexpected similarity in regulation between an archaeal inositol monophosphatase/fructose bisphosphatase and chloroplast fructose bisphosphatase. *Protein Sci.* **12:** 760–767.

Tabaud, H., Tisnovska, H., and Vilkas, E., 1971, Phospholipids and glycolipids of a *Micromonospora* strain. *Biochimie* **53:** 55–61.

Tian, F., Migaud, M.E., and Frost, J.W., 1999, Stereochemistry of the *myo*-inositol-1-phosphate synthase reaction. *J. Biol. Chem.* **255:** 11710–11712.

Toledo, M.S., Levery, S.B., Glushka, J., Straus, A.H., and Takahashi, H.K., 2001, Structure elucidation of sphingolipids from the mycopathogen *Sporothrix schenckii*: Identification of novel glycosylinositol phosphorylceramides with core Manα1$\rightarrow$6Ins linkage. *Biochem. Biophys. Res. Commun.* **280:** 19–24.

Vohra, A., and Satyanarayana, T., 2003, Phytases: Microbial sources, production, purification, and potential biotechnological applications. *Crit. Rev. Biotechnol.* **23:** 29–60.

Volwerk, J.J., Shashidhar, M.S., Kuppe, A., and Griffith, O.H., 1990, Phosphatidylinositol-specific phospholipase C from *Bacillus cereus* combines intrinsic phosphotransferase and cyclic phosphodiesterase activities: A ^{31}P NMR study. *Biochemistry* **29:** 8056–8062.

Wadsworth, S.J., and Goldfine, H., 2002, Mobilization of protein kinase C in macrophages induced by *Listeria monocytogenes* affects its internalization and escape from the phagosome. *Infect. Immun.* **70:** 4650–4660.

Walker, J.B., 1995, Enzymatic synthesis of aminocyclitol moieties of aminoglycoside antibiotics from inositol by *Streptomyces* spp.: Detection of glutamine-aminocyclitol aminotransferase and diaminocyclitol aminotransferase activities in a spectinomycin producer. *J. Bacteriol.* **177:** 818–822.

Wu, Y., Perisic, O., Williams, R.L., Katan, M., and Roberts, M.F., 1997, Phosphoinositide-specific phospholipase C δ1 activity toward micellar substrates, inositol 1,2-cyclic phosphate, and other water-soluble substrates: A sequential mechanism and allosteric activation. *Biochemistry* **36:** 11223–11233.

Yague, G., Segovia, M., and Valero-Guillen, P.L., 2003, Phospholipid composition of several clinically relevant Corynebacterium species as determined by mass spectrometry: An unusual fatty acyl moiety is present in inositol-containing phospholipids of *Corynebacterium urealyticum*. *Microbiology* **149:** 1675–1685.

Yano, I., Furukawa, Y., and Kusunose, M., 1969, Phospholipids of *Nocardia coeliaca*. *J. Bacteriol.* **98:** 124–130.

Yano, R., Nagai, H., Shiba, K., and Yura, T., 1990, A mutation that enhances synthesis of σ^{32} and suppresses temperature-sensitive growth of the *rpoH15* mutant of *Escherichia coli*. *J. Bacteriol.* **172:** 2124–2130.

York, J.D., Ponder, J.W., and Majerus, P.W., 1995, Definition of a metal-dependent/Li^+-inhibited phosphomonoesterase protein family based upon a conserved three-dimensional core structure. *Proc. Natl. Acad. Sci. USA* **92:** 5149–5153.

Zhang, X., Wehbi, H., and Roberts, M.F., 2004, Crosslinking phosphatidylinositol-specific phospholipase C traps two activating phosphatidylcholine molecules on the enzyme. *J. Biol. Chem.* **279:** 20490–20500.

Zhou, C., and Roberts, M.F., 1998, Nonessential activation and competitive inhibition of bacterial phosphatidylinositol-specific phospholipase C by short-chain phospholipids and analogues. *Biochemistry* **37:** 16430–16439.

Zhou, C., Wu, Y., and Roberts, M.F., 1997, Activation of phosphatidylinositol-specific phospholipase C toward inositol 1,2-(cyclic)-phosphate. *Biochemistry* **36:** 347–355.

Chapter 6

Regulation of 1D-*myo*-Inositol-3-Phosphate Synthase in Yeast

Lilia R. Nunez and Susan A. Henry
Department of Molecular Biology and Genetics, Cornell University, Ithaca, NY 14853, U.S.A.

1. INTRODUCTION

The structural gene for 1D-*myo*-inositol-3-phosphate synthase[1] (IP synthase), *INO1*, is among the most highly regulated genes in the model eukaryote, *Saccharomyces cerevisiae*. *INO1* is regulated on a transcriptional level, not only in response to the phospholipid precursor inositol, but also to other phospholipid precursors such as choline (reviewed in Carman and Henry, 1989, 1999; Greenberg and Lopes, 1996; Henry and Patton-Vogt, 1998). In addition, *INO1* transcript abundance is modulated in response to growth phase, and availability of nutrients such as nitrogen and glucose (Carman and Henry, 1999; Henry and Patton-Vogt, 1998). Transcription of the *INO1* gene also responds to signals emanating from several signal transduction pathways and is exquisitely sensitive to perturbations in the general RNA polymerase II (polII) transcription machinery (Chang *et al.*, 2002; Cox *et al.*, 1997; Henry and Patton-Vogt, 1998; Ouyang *et al.*, 1999; Shirra and Arndt, 1999; Shirra *et al.*, 2001).

Although regulation of the *INO1* gene has been studied in yeast by a number of laboratories over a period spanning several decades, much of its complex regulation remains to be elucidated. Analysis of the transcriptional

[1]This enzyme was formerly known as *myo*-inositol-1-phosphate synthase. The recommendation for preferred nomenclature recently changed. Please refer to the website of the Nomenclature Committee of the International Union of Biochemistry and Molecular Biology at http://www.chem.qmul.ac.uk/iubmb/enzyme/.

A. Lahiri Majumder and B. B. Biswas (eds.), Biology of Inositols and Phosphoinositides, 135–156
© 2006 *Springer. Printed in the Netherlands.*

regulation of *INO1* has provided unique insights into the regulation of phospholipid metabolism in yeast and, in so doing, promises to reveal further insights in this model organism into the interplay of signal transduction pathways with general metabolism. In this chapter, the regulation of IP synthase in yeast will be discussed in the context of the complex regulation of phospholipid metabolism, starting with initial studies at the enzymatic level. The cloning of the *INO1* gene of yeast and the subsequent genetic and molecular dissection of the regulatory apparatus controlling it will be the major focus of this chapter.

2. STUDIES OF THE REGULATION OF IP SYNTHASE AT THE ENZYMATIC LEVEL

IP synthase catalyses the rate-limiting step in the synthesis of phosphatidylinositol (PtdIns) (Figure 1) in yeast cells grown in the absence of inositol (Chen and Charalampous, 1963, 1965; Culbertson *et al.*, 1976a; Donahue and Henry, 1981b; Majumder *et al.*, 1997). This cytosolic enzyme is found ubiquitously in eukaryotic organisms and catalyses the formation of *myo*-inositol-3-phosphate from glucose 6-phosphate (reviewed in Majumder *et al.*, 1997, 2003).

The conversion of glucose 6-phosphate to *myo*-inositol-3-phosphate is a three-step reaction including an oxidation step requiring NAD^+ as a hydrogen acceptor, a cyclization step and a reduction step requiring NADH as a hydrogen donor (reviewed in Majumder *et al.*, 1997). Thus, NAD^+ is required as a co-factor, but is regenerated in the course of the three-step reaction. The enzyme has been purified from a variety of organisms and, in each case, the active enzyme is a multimer of identical subunits (Donahue and Henry, 1981b; Maeda and Eisenberg, 1980; Majumder *et al.*, 1997; Pittner *et al.*, 1979).

The purification and characterization of yeast IP synthase were first reported by Donahue and Henry (1981b). The yeast enzyme is a 240 kDa tetramer, composed of four identical subunits of 62 kDa, encoded by the *INO1* gene (Dean-Johnson and Henry, 1989; Donahue and Henry, 1981b). The purified enzyme has a pH optimum of 7.0 and is absolutely dependent upon NAD^+. Yeast IP synthase is also stimulated over 20-fold by NH_4Cl and inhibited by 2-deoxyglucose 6-phosphate. The characteristics of the purified enzyme are very similar to those reported for the enzyme in yeast crude extracts (Culbertson *et al.*, 1976a).

Assessment of IP synthase activity in yeast crude extracts revealed that the enzyme activity is repressed in cells grown in the presence of inositol (Culbertson *et al.*, 1976b). Immunological analysis, using antibody prepared in response to purified yeast IP synthase, demonstrated that the IP synthase subunit is largely absent in crude extracts of cells grown in the presence of inositol (Donahue and Henry, 1981b), suggesting that regulation of IP synthase activity involves lowering of the expression and/or stability of the subunit.

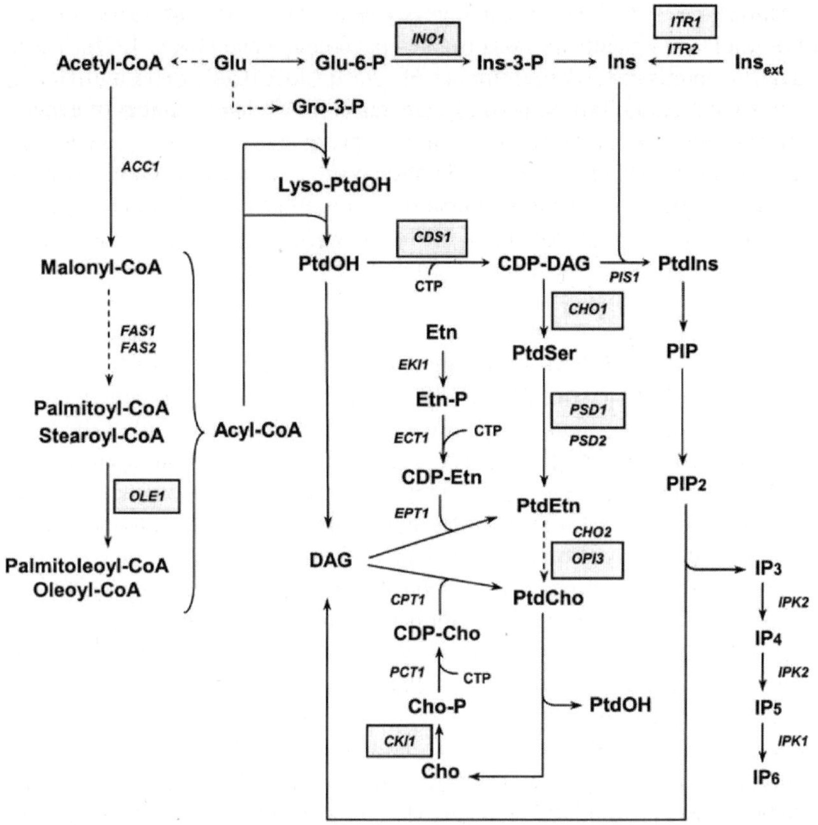

Figure 1. Schematic representation of phospholipid biosynthesis in *S. cerevisiae*. Structural genes are shown in italics. Genes in which expression is coregulated by the presence of inositol and choline in the growth medium (Jesch *et al.*, 2005) and Ino2p and Ino4p are bound to their promoters (Lee *et al.*, 2002) are shown in boxes. Solid arrows indicate direct enzymatic conversions. Dashed arrows indicate conversions that require more than one enzymatic step. Cho, choline; Cho-P, choline phosphate; DAG, diacylglycerol; CDP-DAG, cytidinediphosphate diacylglycerol; CDP-Cho, cytidinediphosphate choline; Etn, ethanolamine; Glu, glucose; Glu-6-P, glucose-6-phosphate; Gro-3-P, glycerol-3-phosphate; Ins, inositol; Ins_{ext}, inositol externally added; Ins-3-P, inositol-3-phosphate; $InsP_3$, inositol trisphosphate; $InsP_4$, inositol tetrakisphosphate; $InsP_5$, inositol pentakisphosphate; $InsP_6$, inositol hexakisphosphate; Lyso-PtdOH, lysophosphatidic acid; PtdIns, phosphatidylinositol; PIP, phosphatidylinositol phosphate; PIP_2, phosphatidylinositol biphosphate; PtdCho, phosphatidylcholine; PtdEtn, phosphatidylethanolamine; PtdOH, phosphatidic acid; PtdSer, phosphatidylserine (Carman and Henry, 1999; Daum *et al.*, 1998; Greenberg and Lopes, 1996; Henry and Patton-Vogt, 1998; Ives *et al.*, 2000; Odom *et al.*, 2000; Paltauf *et al.*, 1992; Saiardi *et al.*, 2001; Shirra *et al.*, 2001).

During the 1970s, a collection of yeast mutants unable to grow in the absence of inositol (Ino$^-$ phenotype) was isolated (Culbertson and Henry, 1975). These mutants represented a large number of genetic loci (Culbertson and Henry, 1975; Donahue and Henry, 1981a). The reason for the large number of genetic loci, capable of conferring an Ino$^-$ phenotype, remained an open question for many years (reviewed in Henry and Patton-Vogt, 1998). However, it gradually became clear that many of these mutations also conferred pleiotropic defects in the global transcriptional machinery and/or signaling pathways. These topics will be discussed in subsequent sections of this chapter. Assays of crude extracts of a selection of the Ino$^-$ mutants isolated by Culbertson and Henry (1975) revealed that IP synthase activity was absent or dramatically reduced (Culbertson *et al.*, 1976b). Following the purification of the yeast IP synthase and the production of specific yeast IP synthase antibody, it was demonstrated that the IP synthase subunit was absent or severely reduced in all of the Ino$^-$ mutants except those with mutations at the *INO1* locus (Donahue and Henry, 1981a; Majumder *et al.*, 1981). Many *ino1* mutants, in contrast, exhibited cross-reacting material when extracts derived from them were tested with anti-IP synthase antibody (Donahue and Henry, 1981b). Furthermore, *ino1* mutants exhibit a complex pattern of interallelic genetic complementation characteristic of genetic loci encoding a subunit of a multimeric protein (Culbertson and Henry, 1975; Majumder *et al.*, 1981). On this basis, the *INO1* gene was first identified as the structural gene for the IP synthase subunit (Donahue and Henry, 1981b). Subsequently, the *INO1* gene was cloned by complementation of the Ino$^-$ phenotype of an *ino1* mutant (Klig and Henry, 1984) and sequenced (Dean-Johnson and Henry, 1989). Analysis of the N-terminal amino acid sequence of purified IP synthase subunit and comparison to the sequence of the protein encoded by the *INO1* gene confirmed its identity as the structural gene for the subunit of IP synthase (Dean-Johnson and Henry, 1989). The yeast *INO1* gene was the first IP synthase structural gene to be isolated and characterized from any organism.

Mutations at the *INO2* and *INO4* loci were later shown to confer pleiotropic defects in lipid metabolism (Loewy and Henry, 1984) in addition to their inability to express IP synthase. Cloning of *INO4* (Klig *et al.*, 1988b) and *INO2* (Nikoloff *et al.*, 1992) by complementation of the Ino$^-$ phenotypes of the *ino2* and *ino4* mutants revealed that these two genes encoded proteins contained in the basic helix-loop-helix (HLH) motif found in a number of eukaryotic DNA binding protcins. This set the stage for further analysis of the coordinate transcriptional regulation of *INO1* and other genes of phospholipid metabolism which will be discussed in the next section of this chapter.

Mutants constitutive for IP synthase activity were identified on the basis of a phenotype associated with *o*ver*p*roduction of *i*nositol (Opi$^-$ phenotype) and the subsequent excretion of inositol which can be detected in a bioassay (Greenberg *et al.*, 1982b). Mutants at the *OPI1* locus were shown to overexpress IP synthase activity and cross-reacting material, regardless of the presence of exogenous inositol, and exhibited other changes in phospholipid metabolism (Greenberg *et al.*, 1982a,b; Klig *et al.*, 1985). Other mutants with Opi^- phenotypes proved to have complex pleiotropic defects in lipid metabolism (Greenberg *et al.*, 1983; Klig *et al.*, 1988a; Letts and Henry, 1985; McGraw and Henry, 1989; Patton-Vogt *et al.*, 1997; Summers *et al.*, 1988); a fact that ultimately proved to be related to the coordinate regulation controlling many steps of phospholipid metabolism in yeast (Henry and Patton-Vogt, 1998) (Figure 1), the topic of a subsequent section of this chapter.

3. TRANSCRIPTIONAL REGULATION OF *INO1*, STRUCTURAL GENE FOR IP SYNTHASE

Following identification of the structural gene for IP synthase, *INO1*, in yeast (Donahue and Henry, 1981b) and its subsequent cloning (Klig and Henry, 1984), it became possible to analyze its regulation at the transcriptional level. Hirsch (1987), Hirsch and Henry (1986) demonstrated that supplementation of the growth medium of wild type cells with inositol leads to a 10-fold or greater decrease in *INO1* transcript abundance. This effect of inositol on *INO1* transcript abundance, mirrors its affect on IP synthase activity (Culbertson *et al.*, 1976a,b) and on IP synthase cross-reacting material in crude extracts (Donahue and Henry, 1981b). The provision of choline in addition to inositol leads to a further two- to three-fold decrease in *INO1* transcript abundance (Hirsch, 1987). Furthermore, *ino2* and *ino4* mutants express very low levels of *INO1* transcript while *opi1* mutant cells express high constitutive levels (Hirsch and Henry, 1986). Thus, it was clear that much of the regulation of IP synthase activity and IP synthase subunit abundance is due to transcriptional regulation of the *INO1* gene. A number of other enzymes of phospholipid biosynthesis exhibit a similar pattern of repression at the level of enzymatic activity or protein abundance in response to inositol and choline (Homann *et al.*, 1987b), and were likewise affected by *ino2*, *ino4*, and *opi1* mutations (Bailis *et al.*, 1987; Kiyono *et al.*, 1987; Klig and Henry, 1984; Kodaki and Yamashita, 1987, 1989; Kodaki *et al.*, 1991a) (reviewed in Carman and Henry, 1989). Thus, as structural genes encoding these enzymes were cloned and sequenced, a common pattern of regulation emerged, and the search for common promoter elements got underway. A repeated element, which we termed UAS_{INO} (Bachhawat *et al.*, 1995; Carman and Henry, 1989) and others have called the inositol-choline responsive element (ICRE) (Schüller *et al.*, 1995), was identified in the promoter of the *INO1* gene (Hirsch, 1987) and the promoters of many other genes encoding enzymes of phospholipid biosynthesis (Greenberg and Lopes, 1996). The consensus sequence for UAS_{INO} is 5′ CATGTGAAAT 3′ (Bachhawat *et al.*, 1995; Schüller *et al.*, 1995) (reviewed in Carman and Henry, 1989; Greenberg and Lopes, 1996). Extensive mutational analysis of this element has been conducted in several laboratories (Bachhawat *et al.*, 1995; Kodaki *et al.*, 1991b; Schüller *et al.*,

1995). The core of the binding sequence contains the consensus for the binding of bHLH proteins (CANNTG). Any mutation deviating from the UAS_{INO} core consensus of 5′ CATGTG 3′ results in reduced expression (Bachhawat *et al.*, 1995).

The basic pattern of transcriptional regulation of UAS_{INO}-containing genes is as follows: they are repressed in logarithmically grown cells when inositol is present in the growth medium. Transcription of these genes is further repressed when choline is added to medium already containing inositol, but addition of choline in the absence of inositol has no effect (Carman and Henry, 1989; Greenberg and Lopes, 1996; Henry and Patton-Vogt, 1998). This overall pattern of regulation has now been confirmed by microarray analysis, revealing the complete set of inositol and choline regulated genes in yeast (Jesch *et al.*, 2005; Santiago and Mamoun, 2003). The *INO1* gene is the most highly regulated of the UAS_{INO}-containing genes.

The UAS_{INO} element is the binding site for the bHLH transcription factors Ino2p and Ino4p (Ambroziak and Henry, 1994; Bachhawat *et al.*, 1995; Lopes *et al.*, 1991; Nikoloff and Henry, 1994; Schwank *et al.*, 1995). Ino2p and Ino4p interact to form a heterodimer which binds to UAS_{INO} (Ambroziak and Henry, 1994; Lopes and Henry, 1991; Nikoloff and Henry, 1994; Schwank *et al.*, 1995). Any mutation in the core bHLH binding site (5′ CATGTG 3′) reduces the efficiency of Ino2p/Ino4p binding to UAS_{INO} (Bachhawat *et al.*, 1995). Ino2p contains two N-terminal activation domains, necessary for transcriptional activation of UAS_{INO}-containing genes (Dietz *et al.*, 2003; Schwank *et al.*, 1995). The *INO2* gene itself contains a UAS_{INO} element in its promoter and exhibits autoregulation (Ashburner and Lopes, 1995a; Schüller *et al.*, 1992). The regulation of *INO2* is highly complex and involves several interacting mechanisms including transcriptional and translational mechanisms (Ashburner and Lopes, 1995b). The transcriptional regulation of *INO2* requires the participation of Opi1p, Ume6p, and Sin3p (Kaadige and Lopes, 2003). Whereas, *ino2* and *ino4* loss of function mutants have a recessive Ino^- phenotype (Culbertson and Henry, 1975; Loewy and Henry, 1984) due to inability to express *INO1* (Hirsch and Henry, 1986), $INO2^c$ mutants exhibit dominant constitutive *INO1* expression (Gardenour *et al.*, 2004). In addition to its interaction with Ino4p, Ino2p has been reported to interact with several other proteins involved in transcriptional regulation, including Sua7p (TFIIB) (Dietz *et al.*, 2003) and Opi1p (Wagner *et al.*, 2001), the negative regulator of UAS_{INO}-containing genes (Greenberg and Lopes, 1996; White *et al.*, 1991).

The *OPI1* gene was cloned and sequenced and Opi1p was found to contain polyglutamine tracts and a leucine zipper motif (White *et al.*, 1991). Opi1p functions through the UAS_{INO} element (Bachhawat *et al.*, 1995) but it is not thought to interact directly with UAS_{INO} (Graves and Henry, 2000). *opi1Δ* loss of function mutants express *INO1* and other UAS_{INO}-containing genes constitutively (Carman and Henry, 1989; Hirsch and Henry, 1986) while overexpression of Opi1p results in an Ino$^-$ phenotype due to an inability to express *INO1* (Wagner *et al.*, 1999). In addition to interacting with Ino2p, Wagner *et al.* (2001) reported that Opi1p interacts with the Sin3p repressor which affects the expression of a large number of genes in yeast.

Recently, evidence for a role of Opi1p as a sensor of lipid metabolism has emerged (Loewen *et al.*, 2003, 2004). Opi1p interacts with Scs2p (Gavin *et al.*, 2002), an integral membrane protein (Kagiwada *et al.*, 1998) which targets it to the endoplasmic reticulum (ER) (Loewen *et al.*, 2003). Opi1p also interacts with phosphatidic acid (PtdOH) (Loewen *et al.*, 2004), a precursor of many lipids (Figure 1), including PtdIns. When PtdOH levels are high, in the absence of inositol, Opi1p remains bound to Scs2p in the ER. However, when inositol is present, the increased rate of synthesis of PtdIns leads to a reduction of PtdOH levels and Opi1p translocates to the nucleus, resulting in rapid repression of *INO1* (Loewen *et al.*, 2004). Opi1p has also been shown to be a target for phosphorylation by protein kinase C (PKC) and protein kinase A *in vitro* (Sreenivas and Carman, 2003; Sreenivas *et al.*, 2001).

Much remains to be learned about the mechanism by which *INO1* expression is regulated in response to change in growth and environmental conditions. Ino$^-$ or Opi$^-$ phenotypes have been reported for mutants defective in a wide range of functions related to general transcription. For example, partial deletion of the C-terminal domain (CTD) of the PolII large subunit, Rpo21p (Rpb1p), results in an Ino$^-$ phenotype (Nonet and Young, 1989; Scafe *et al.*, 1990b). This is due to an inability to fully derepress the *INO1* gene (Scafe *et al.*, 1990a). Mutations in other genes encoding proteins related to general transcription, including Sin3p, Ume6p, Rpd3p (Elkhaimi *et al.*, 2000; Hudak *et al.*, 1994; Jackson and Lopes, 1996), Dep1p (Lamping *et al.*, 1995), histone 2A (Hirschhorn *et al.*, 1992), and the TATA binding protein (Arndt *et al.*, 1995), also give rise to Ino$^-$ and Opi$^-$ phenotypes and altered *INO1* expression. Others include Rpg9p, encoding a small subunit of PolII (Furter-Graves *et al.*, 1994), Srb2p, a component of the RNA PolII holoenzyme and the mediator subcomplex (Nonet and Young, 1989), Rpb2p, the second largest RNA PolII subunit (Pappas and Hampsey, 2000; Scafe *et al.*, 1990c), and components of the SWI/SNF global transcriptional activator complex (Peterson and Herskowitz, 1992; Peterson *et al.*, 1991). Mutations at other loci that result in changes in *INO1* expression are reviewed in Henry and Patton-Vogt (1998) and illustrated in Figure 2. In addition, mutations in genes encoding components of several signal transduction pathways (Figure 2) also have effects on *INO1* expression and in some cases confer Opi$^-$ or Ino$^-$ phenotypes. The role of signal transduction pathways in *INO1* expression will be discussed in a subsequent section of this chapter.

The mechanism by which sublethal or conditional mutations in so many genes result in inositol auxotrophy still remains to be explained. However, it is clear that *INO1* transcription is exquisitely sensitive to even minor perturbations of the global transcriptional apparatus (reviewed in Henry and Patton-Vogt, 1998). This fact no doubt provides at least a partial explanation for the

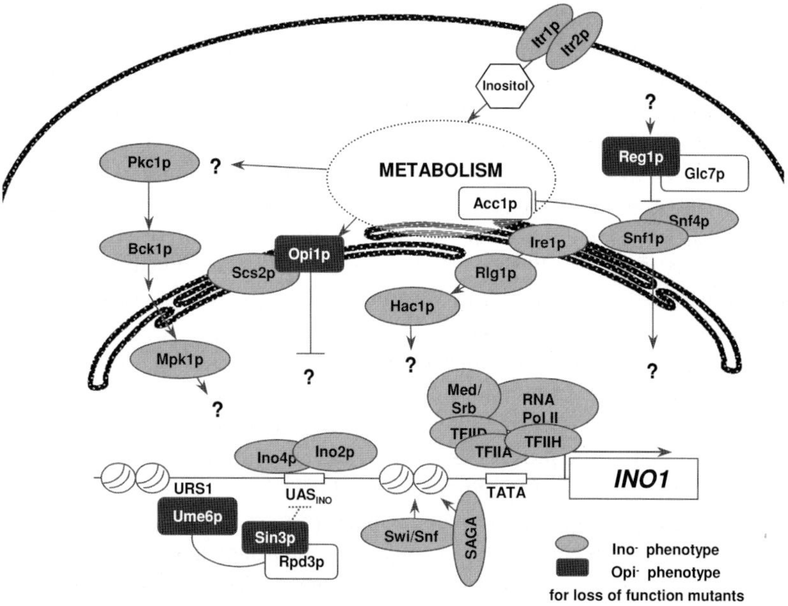

Figure 2. Gene products known to affect *INO1* transcription. The details are discussed in the text and/or reviewed in (Carman and Henry, 1999; Henry and Patton-Vogt, 1998).

observation that mutations at such a diverse array of genetic loci are capable of conferring inositol auxotrophy. This also presumably explains why so many loci were identified in the original screens for mutants with Ino$^-$ and Opi$^-$ phenotypes (Culbertson and Henry, 1975; Donahue and Henry, 1981a; Greenberg *et al.*, 1982a). The exquisite sensitivity of *INO1* transcription to so many factors influencing cellular transcription and/or signaling may reflect unique features of the specific transcriptional mechanisms controlling *INO1* activation and repression that remain to be elucidated.

4. INSIGHTS INTO THE REGULATION OF PHOSPHOLIPID METABOLISM IN YEAST DERIVED FROM STUDIES OF THE *INO1* GENE

The fact that the *INO1* gene is more highly regulated than other UAS$_{INO}$-containing genes has enabled the development of reporter constructs derived from its promoter (Lopes *et al.*, 1991, 1993) which serve as reporters for the overall pattern of regulation of phospholipid biosynthesis in yeast. These constructs have facilitated both regulatory studies and mutant screenings (Gardenour *et al.*, 2004; Henry and Patton-Vogt, 1998; Hudak *et al.*, 1994; Lamping *et al.*, 1995; Slekar and Henry, 1995; Swede *et al.*, 1992). Furthermore, because the repressed or basal level of expression of *INO1* is quite low, mutants unable to activate its transcription produce insufficient inositol, and thus are Ino^-, a readily identifiable phenotype (Culbertson and Henry, 1975; Swede *et al.*, 1992). Equally identifiable, in a simple plate assay, is the Opi^- phenotype associated with overproduction of inositol due to overexpression of *INO1* (Greenberg *et al.*, 1982a,b). These phenotypes have enabled the identification, as described previously, of a broad array of mutants with defects in the regulation or expression of *INO1*. Many of the mutants defective in *INO1* expression or regulation have pleiotropic phenotypes that have produced insights into the overall mechanism of regulation of phospholipid metabolism in yeast. For example, in addition to being unable to derepress *INO1*, the *ino2* and *ino4* mutants proved to have very abnormal phospholipid compositions, including decreased levels of phosphatidylcholine (PtdCho) (Loewy and Henry, 1984). The overall rate of synthesis of PtdCho via methylation of phosphatidylethanolamine (PtdEtn) (Figure 1) was found to be lower in *ino2* and *ino4* mutants than in wild type cells (Loewy and Henry, 1984). This ultimately proved to be due to the fact that virtually all of the genes encoding enzymes in the pathway leading to the *de novo* synthesis of PtdCho from PtdOH *via* methylation of PtdEtn (Figure 1) contain one or more UAS_{INO} elements in their promoters and, thus, require Ino2p and Ino4p for maximal expression (Carman and Henry, 1989; Greenberg and Lopes, 1996). *opi1* mutants similarly exhibit constitutive expression of these same enzymes and have abnormal phospholipid compositions (Klig *et al.*, 1985).

More surprisingly, a series of mutants with defects in structural genes encoding enzymes of phospholipid biosynthesis were also found to have altered regulation of *INO1* and other UAS_{INO}-containing genes (reviewed in Henry and Patton-Vogt, 1998). Mutants such as *cho2* and *opi3* (Figure 1), which are blocked in PtdCho biosynthesis *via* methylation of PtdEtn, exhibit a conditional Opi^- phenotype (Greenberg *et al.*, 1983; McGraw and Henry, 1989; Summers *et al.*, 1988). Specifically, the *cho2* and *opi3* mutants are unable to repress *INO1* in the presence of inositol and exhibit an Opi^- phenotype in the absence of inositol, unless they are supplied with an intermediate that bypasses the specific metabolic lesion caused by the mutation in question (see Figure 1). Metabolites supplied exogenously [*i.e.*, monomethylethanolamine (MME), dimethylethanolamine (DME), and choline for *cho2* or DME and choline for *opi3*], restore PtdCho synthesis and simultaneously restore responsiveness to inositol mediated regulation (Griac *et al.*, 1996; McGraw and Henry, 1989; Summers *et al.*, 1988). *INO1* expression is similarly aberrant in mutants defective in each of the steps leading sequentially from PtdOH to CDP-DAG to PtdCho (*i.e.*, PtdOH $\rightarrow$ CDP-DAG $\rightarrow$ PtdSer $\rightarrow$ PtdEtn $\rightarrow$ PtdMME $\rightarrow$ PtdDME $\rightarrow$ PtdCho; Figure 1) (reviewed in Henry and Patton-Vogt, 1998). These observations lead to the proposal that a metabolite produced during the course of phospholipid metabolism results in a signal leading to repression/derepression of *INO1* and other UAS_{INO}-containing genes. It was originally proposed that this signal was produced by PtdCho, the end product of the above reaction sequence (Griac and Henry, 1996; Griac *et al.*, 1996; Swede, 1994). However, discovery of misregulation of *INO1* in a *sec14^{ts}* strain that also carried *cki1Δ*, a bypass suppressor of its growth defects (Cleves *et al.*, 1991a,b) mutant, provided evidence that the signal for derepression of *INO1* is generated at or near the beginning of the above reaction sequence (*i.e.*, PtdOH or a closely related metabolite) (Henry and Patton-Vogt, 1998; Patton-Vogt *et al.*, 1997; Sreenivas *et al.*, 1998).

The *SEC14* gene encodes a PtdIns/PtdCho transfer protein that is essential for growth and secretion (Aitken *et al.*, 1990; Bankaitis *et al.*, 1989). Bankaitis and colleagues demonstrated that mutations blocking the CDP-choline pathway for PtdCho biosynthesis (*i.e.*, the steps catalyzed by the *CKI1, PCT1,* and *CPT1* gene products; Figure 1) suppress *sec14* growth and secretion defects *via* a bypass mechanism (Cleves *et al.*, 1991a,b). *sec14^{ts}* strains also exhibit an altered phospholipid metabolism and composition (McGee *et al.*, 1994). Subsequently, it was observed that *sec14* strains carrying CDP-choline pathway bypass suppressors, such as *cki1Δ*, exhibit a severe Opi$^-$ phenotype (Patton-Vogt *et al.*, 1997). Moreover, *sec14^{ts} cki1Δ* strains also excrete choline (Opc$^-$ for *o*ver*p*roduction of *c*holine), a phenotype due to elevated phospholipase D (PLD) mediated turnover of PtdCho (Patton-Vogt *et al.*, 1997; Sreenivas *et al.*, 1998).

PLD mediated turnover of PtdCho, which leads to the production of one molecule of PtdOH for every molecule of choline produced, is induced immediately upon elevation of *sec14^{ts} cki1Δ* strains to the *sec14^{ts}* semi-restrictive and restrictive temperature, an event which also correlates with the derepression of *INO1* (Patton-Vogt *et al.*, 1997). Free choline had previously been demonstrated not to be the signal leading to *INO1* derepression (Griac *et al.*, 1996). Thus, it was proposed that PtdOH or a related metabolite is responsible for generating the signal for *INO1* derepression (Henry and Patton-Vogt, 1998). This hypothesis is consistent with the phenotypes of all the mutants in the phospholipid biosynthetic pathways from PtdOH *via* PtdEtn methylation to PtdCho (Figure 1) described to date (Griac, 1997; Griac *et al.*, 1996; Klig *et al.*, 1988a; Letts and Henry, 1985; McGraw and Henry, 1989; Patton-Vogt *et al.*, 1997; Shen and Dowhan, 1996, 1997; Summers *et al.*, 1988). The full rationalization for this model is contained in the review article by Henry and Patton-Vogt (1998).

One of the predictions of the above hypothesis is that PLD is necessary for *INO1* derepression in the *sec14^{ts} cki1* genetic background. Deletion of the *SPO14/PLD1* gene in a *sec14^{ts} cki1Δ* strain (Sreenivas *et al.*, 1998) resulted in elimination of the Opc$^-$ phenotype as well as the Opi$^-$ phenotype characteristic of the *sec14^{ts} cki1Δ* parent strain (Sreenivas *et al.*, 1998). Thus, PLD mediated turnover of PtdCho is responsible for the Opi$^-$ and Opc$^-$ phenotypes observed in the *sec14^{ts} cki1Δ* strain. Moreover, compared to the *sec14^{ts} cki1Δ* parent, the *sec14^{ts} cki1Δ pld1Δ* triple mutant cannot survive elevation to the restrictive temperature of 37 °C. Thus, the *SPO14/PLD1* gene product is also necessary for bypass suppression of the *sec14* defect by *cki1Δ* and other CDP-choline pathway suppressors (Sreenivas *et al.*, 1998). Bankaitis and colleagues have shown that Pld1p is required for bypass of the *sec14* growth and secretory phenotypes by all known bypass suppressors (Xie *et al.*, 1998). Thus, it appears that PtdOH produces a signal controlling derepression of *INO1* and other UAS_{INO}-containing genes. The recent discovery that Opi1p binds PtdOH in the ER and translocates to the nucleus when PtdOH is consumed upon addition of inositol (Loewen *et al.*, 2004) has defined the mechanism by which this regulatory loop functions.

5. ROLE OF SIGNAL TRANSDUCTION PATHWAYS AND MEMBRANE TRAFFICKING IN *INO1* EXPRESSION AND REGULATION

The roles of various signal transduction pathways in the regulation of phospholipid biosynthesis has been probed taking advantage of the high sensitivity of *INO1* transcription to environmental and metabolic changes, coupled with genetic analysis of mutants exhibiting Ino^- and Opi^- phenotypes. Observations from such analyses suggest that combinatorial inputs of several signal transduction pathways contribute to the complex regulation of *INO1* and other UAS_{INO}-containing genes (Carman and Henry, 1999).

The first signal transduction pathway to be shown to influence *INO1* expression was the Unfolded Protein Response (UPR) pathway which regulates ER homeostasis, including ER membrane biogenesis (reviewed in Patil and Walter, 2001). This signaling pathway responds to stimuli generated by unfolded proteins accumulating in the ER lumen (Cox and Walter, 1996; Cox *et al.*, 1993; Mori *et al.*, 1992, 1993). The activation of the UPR pathway promotes the transcription of genes encoding protein folding chaperons such as Kar2p (BiP) (Kohno *et al.*, 1993; Mori *et al.*, 1992, 1993; Nikawa and Yamashita, 1992) *via* the *HAC1*, *IRE1*, and *RLG1* gene products. Under ER stress, the Ire1p kinase, located in the ER, is activated by auto phosphorylation and catalyses the splicing of *HAC1* mRNA (Cox and Walter, 1996; Mori *et al.*, 2000), which is then ligated by Rlg1p (Sidrauski *et al.*, 1996). Since only the spliced form of *HAC1* mRNA is effectively translated (Chapman and Walter, 1997; Kawahara *et al.*, 1997), this mechanism controls the production of Hac1p which then binds to *U*nfolded *P*rotein *R*esponse *E*lements (UPRE) found in promoters of chaperone genes such as *KAR2* (BiP).

Mutants in the UPR pathway, *hac1Δ*, *ire1Δ*, and *rlg1-100* are inositol auxotrophs and levels of *INO1* expression in *ire1Δ* and *hac1Δ* cells are reduced compared to wild type strains (Cox *et al.*, 1993, 1997; Nikawa and Yamashita, 1992; Sidrauski *et al.*, 1996). The *opi1Δ* mutation suppresses the Ino^- phenotype and restores *INO1* expression in *ire1Δ* and *hac1Δ* cells (Cox *et al.*, 1997). The UPR is also induced in wild type cells grown in the absence of inositol, leading Cox *et al.* (1997) to propose that activation of the UPR is involved in the mechanism of *INO1* activation.

Chang *et al.* (2002) studied *INO1* and UPRE transcriptional activation in *sec14^{ts} cki1Δ* cells. As described in the previous section of this chapter, the *SEC14* gene encodes a PtdCho/PtdIns transporter. Mutations (*cki1Δ, pct1Δ,* or *cpt1Δ*) (Figure 1) which inactivate the CDP-choline (Kennedy) pathway for PtdCho biosynthesis enable cells possessing even a complete deletion of *SEC14* to survive and grow (Cleves *et al.*, 1991b). Chang *et al.* (2002) showed that both UPRE and *INO1* expression were highly induced in *sec14^{ts} cki1Δ* cells elevated to the *sec14^{ts}* semi-permissive or restrictive temperatures. However, introduction of *ire1Δ* or *hac1Δ* mutations into the *sec14^{ts} cki1Δ* genetic background did not eliminate the high level of *INO1* expression. Furthermore, *INO1* was still regulated in response to inositol in the *sec14^{ts} cki1Δ ire1Δ* and *sec14^{ts} cki1Δ hac1Δ* strains.

Consistent with results reported by Cox *et al.* (1997), Chang (2001) also showed that *INO1* is expressed in *hac1Δ* and *ire1Δ* cells when the cells are first shifted to inositol-free media. However, this initial level of *INO1* expression is not maintained. The residual *INO1* expression in *hac1Δ* and *ire1Δ* cells is repressed in the presence of inositol (Chang *et al.*, 2002) indicating that the UPR pathway is not responsible for transducing the signal that results in *INO1* repression in response to inositol. Moreover, the UPR is not required for high level activation of *INO1* observed in *sec14^{ts} cki1Δ* cells, as demonstrated by the Opi^{-} phenotype and high *INO1* expression levels observed in *sec14^{ts} cki1Δ ire1Δ* or *hac1Δ* triple mutants. Thus, the UPR pathway does not appear to be directly responsible for mediating the inositol-sensitive signal controlling expression of *INO1* and other UAS_{INO}-containing genes. The UPR does, however, appear to be required to maintain the overall level of *INO1* expression, especially under stress conditions sensed by the UPR. (Chang, 2001; Chang *et al.*, 2002).

The glucose response pathway also mediates signals that influence the expression of phospholipid biosynthetic genes. This signaling pathway controls the transcription of glucose-repressible genes in response to the availability of glucose in the media (reviewed in Carlson, 1999). The glucose response regulatory cascade controlling catabolite repression includes the glucose hexokinase Hxk2p, the Glc7p-Reg1p protein phosphatase complex, the Snf1p-Snf4p protein kinase complex, as well as the Mig1p, Ssn6p, and Tup1p regulatory proteins. Glc7p-Reg1p complex is required for repression of genes, such as *SUC2,* responsive to glucose levels. The Glc7p-Reg1p complex exerts its effects by catalyzing dephosphorylation of Snf1p, thus, inactivating the Snf1p-Snf4p kinase. Mutants of the glucose response signaling pathway present phenotypes associated with *INO1* misregulation. For example, *snf1Δ* and *snf4Δ* are inositol auxotrophs (Hirschhorn *et al.*, 1992; Shirra and Arndt, 1999) and exhibit low levels of *INO1* expression (Shirra *et al.*, 2001). In contrast, *reg1* mutations result in an Opi^{-} phenotype and exhibit increased expression of *INO1* (Ouyang *et al.*, 1999; Ruiz-Noriega, 2000; Shirra and Arndt, 1999; Shirra *et al.*, 2001).

Yeast Snf1p is functionally related to mammalian AMP-activated protein kinase. Acetyl-CoA carboxylase (Acc1p), which catalyses the rate-limiting step in fatty acid biosynthesis (Figure 1), is a target of the Snf1p-Snf4p kinase in yeast and the AMP activated kinase in mammals (Davies *et al.*, 1992; Woods *et al.*, 1994). A dominant partial loss of function allele of *ACC1* was shown to restore *INO1* expression in a *snf1Δ* mutant, thereby demonstrating that Acc1p is a target of Snf1p-Snf4p relevant to *INO1* expression (Shirra *et al.*, 2001). Snf1p is also required for histone H3 Ser-10 phosphorylation, which in turn is required for histone H3 Lys-14 Gcn5-mediated acetylation at the promoter of *INO1*. Phosphorylation and acetylation of histone H3 is necessary for proper *INO1* expression (Lo *et al.*, 2001). Although *INO1* does not show classic glucose repression, the availability of glucose and the nature of other available carbon sources affect its expression (Ruiz-Noriega, 2000). In cells grown under conditions that activate the Snf1p-Snf4p kinase such as low glucose or alternative carbon sources, *INO1* expression is increased, and in some cases, some upregulation is observed even in the presence of inositol (Ruiz-Noriega, 2000).

The PKC signaling pathway, known in the budding yeast as the cell integrity pathway, is one of the five mitogen-activated protein kinase (MAPK) cascades that have been identified in *S. cerevisiae* (Errede *et al.*, 1995; Gustin *et al.*, 1998; Heinisch *et al.*, 1999; Levin and Errede, 1995). The signaling proteins in this pathway include the PKC homolog Pkc1p, the MAP kinase kinase kinase (MEKK) Bck1p, the redundant Thr/Tyr kinases Mkk1p and Mkk2p, and the MAP kinase Mpk1p, also called Slt2p (Heinisch *et al.*, 1999; Irie *et al.*, 1993; Lee and Levin, 1992; Lee *et al.*, 1993; Levin *et al.*, 1994). In mammalian cells, PKC has been implicated in the control of cellular proliferation and differentiation (reviewed in Nishizuka, 1984a, 1986), and transmits signals to the nucleus *via* one or more MAPK cascades, which may include Raf-1, Ras, MEKs, and ERKs. Activated ERKs can, in turn, activate transcription factors such as myc, myb, max, fos, and jun, enabling the expression of genes encoding factors involved in cell proliferation and tumor invasion (Chang and Karin, 2001; Treisman, 1996). In mammalian cells, PKC has also been shown to be activated by diacylglycerol (DAG), which is produced by phospholipase C (PLC) acting upon phosphatidylinositol 4,5-biphosphate (PIP_2). The hydrolysis of PIP_2 by PLC also produces inositol 1,4,5-triphosphate ($InsP_3$), which in turn results in Ca^{++} release (reviewed in Nishizuka, 1984b).

The yeast *PKC1* gene was originally identified in a genome-wide screen of *S. cerevisiae* using probes derived from cDNAs encoding isoenzymes of rat PKC (Levin *et al.*, 1990). Yeast Pkc1p has substrate specificity and primary sequence similar to some mammalian PKC isoforms (Levin *et al.*, 1990; Watanabe *et al.*, 1994). Since yeast Pkc1p is similar to mammalian PKC, it was expected that DAG and Ca^{++} would be involved in its activation. However, Pkc1p purified from yeast does not respond to Ca^{++}, DAG, phorbol esters, or phospholipids *in vitro* (Antonsson *et al.*, 1994; Watanabe *et al.*, 1994). Thus, the question as to whether Pkc1p in yeast is activated by products of phospholipid hydrolysis remains to be answered.

Yeast mutants defective in the cell integrity (PKC) signaling pathway exhibit several phenotypes associated with cell wall defects, including tendency to lyse at the restrictive temperature of 37 °C, a phenotype that is rescued by adding osmotic support, usually 1 M sorbitol, to the media (Levin and Errede, 1995). Recent studies in our laboratory have revealed that mutants defective in the cell integrity pathway also exhibit an Ino^- phenotype (Nunez and Henry, unpublished results). Mutations in *PKC1*, *BCK1,* and *MPK1* genes all result in inositol auxotrophy, indicative of defects in *INO1* expression. Our preliminary results confirm that *mpk1Δ* mutations result in low *INO1* expression under certain growth conditions. The *opi1Δ* mutation, which confers a high level of constitutive *INO1* expression (Greenberg *et al.*, 1982b), was introduced into each of the three mutant backgrounds listed above. The presence of the *opi1Δ* mutation suppressed the inositol auxotrophy of *pck1Δ, bck1Δ,* and *mpk1Δ* mutants, and conferred an Opi^- phenotype in the double mutant strains (Nunez and Henry, unpublished results). Thus, we propose that the activation of *INO1* responds to molecular stimuli generated by the PKC signal transduction pathway.

In mammalian cells, products derived from the turnover of inositol-containing phospholipids have been shown to play diverse roles in cellular signaling. Although it is as yet not known whether $InsP_3$ produced by phospholipase C (Plc1p) regulates Ca^{++} homeostasis in yeast, inositol polyphosphates have recently received attention in yeast as potential transcriptional regulators. Two independent groups reported that Arg82p (renamed Ipk2p), a yeast transcriptional regulator, has inositol polyphosphate kinase activity (Ives *et al.*, 2000; Odom *et al.*, 2000; Saiardi *et al.*, 1999). Ipk1p, another novel inositol polyphosphate was also characterized (Ives *et al.*, 2000). Ipk2p and Ipk1p phosphorylate $InsP_3$ produced by PLC, thus resulting in inositol hexakisphosphate ($InsP_6$; phytic acid) biosynthesis. The kinase activities of Ipk2p and Ipk1p and the resulting formation of $InsP_6$ are required for mRNA export (Saiardi *et al.*, 2000; York *et al.*, 1999). $InsP_6$ appears to influence ATP-dependent chromatin remodeling, but the exact mechanism is still unknown (Shen *et al.*, 2003; Steger *et al.*, 2003). A role for inositol phosphates in regulation of transcription has also been suggested (Odom *et al.*, 2000). More recently, it has been shown that *INO1* mRNA levels are reduced to 18% and 41% in the *ipk2Δ* and *ipk1Δ* mutant, respectively (Shen *et al.*, 2003). Thus, it appears likely that the inositol phosphates may play a role in regulating expression *INO1*, which, in turn, encodes an enzyme catalyzing a crucial rate-limiting step early in the pathway which leads to their formation (Figure 1).

6. CONCLUDING REMARKS

In the period of nearly three decades since the isolation of the first *Saccharomyces cerevisiae* inositol auxotrophs (Culbertson and Henry, 1975), great progress has been made in describing the genetic regulation of IP synthase in yeast. However, each advance, from the purification and characterization of the enzyme and the subsequent cloning and sequencing of its structural gene, *INO1*, to the recent progress in elucidating the regulatory mechanisms controlling *INO1* expression, has raised still more questions. Analysis of the regulation of *INO1* has provided unexpected insights into complex mechanisms controlling phospholipid metabolism in relationship to cellular signaling pathways. Despite this progress, it is clear that research on the mechanisms controlling *INO1* expression and its relationship to fundamental cellular processes has many more insights to reveal. This work was supported by a grant of the National Institutes of Health to S.A.H. (GM019629).

REFERENCES

Aitken, J.F., vanHeusden, G.P., Temkin, M., and Dowhan, W., 1990, The gene encoding the phosphatidylinositol transfer protein is essential for cell growth. *J. Biol. Chem.* **265:** 4711–4717.

Ambroziak, J., and Henry, S.A., 1994, *INO2* and *INO4* gene products, positive regulators of phospholipid biosynthesis in *Saccharomyces cerevisiae*, form a complex that binds to the *INO1* promoter. *J. Biol. Chem.* **269:** 15344–15349.

Antonsson, B., Montessuit, S., Friedli, L., Payton, M.A., and Paravicini, G., 1994, Protein kinase C in yeast. Characteristics of the *Saccharomyces cerevisiae* PKC1 gene product. *J Biol. Chem.* **269:** 16821–16828.

Arndt, K.M., Ricupero-Hovasse, S., and Winston, F., 1995, TBP mutants defective in activated transcription *in vivo*. *EMBO J.* **14:** 1490–1497.

Ashburner, B.P., and Lopes, J.M., 1995a, Autoregulated expression of the yeast *INO2* and *INO4* helix-loop-helix activator genes effects cooperative regulation on their target genes. *Mol. Cell. Biol.* **15:** 1709–1715.

Ashburner, B.P., and Lopes, J.M., 1995b, Regulation of yeast phospholipid biosynthesis involves two superimposed mechanisms. *Proc. Natl. Acad. Sci. U.S.A.* **92:** 9722–9726.

Bachhawat, N., Ouyang, Q., and Henry, S.A., 1995, Functional characterization of an inositol-sensitive upstream activation sequence in yeast: A *cis*-regulatory element responsible for inositol-choline mediated regulation of phospholipid biosynthesis. *J. Biol. Chem.* **270:** 25087–25095.

Bailis, A.M., Poole, M.A., Carman, G.M., and Henry, S.A., 1987, The membrane-associated enzyme phosphatidylserine synthase is regulated at the level of mRNA abundance. *Mol. Cell. Biol.* **7:** 167–176.

Bankaitis, V.A., Malehorn, D.E., Emr, S.D., and Greene, R., 1989, The *Saccharomyces cerevisiae SEC14* gene encodes a cytosolic factor that is required for transport of secretory proteins from the yeast Golgi complex. *J. Cell Biol.* **108:** 1271–1281.

Carlson, M., 1999, Glucose repression in yeast. *Curr. Opin. Microbiol.* **2:** 202–207.

Carman, G.M., and Henry, S.A., 1989, Phospholipid biosynthesis in yeast. *Ann. Rev. Biochem.* **58:** 635–669.

Carman, G.M., and Henry, S.A., 1999, Phospholipid biosynthesis in the yeast *Saccharomyces cerevisiae* and interrelationship with other metabolic processes. *Prog. Lipid Res.* **38:** 361–399.

Chang, H.J., 2001, Role of the unfolded protein response pathway in phospholipid biosynthesis and membrane trafficking in *Saccharomyces cerevisiae.* Department of Biological Sciences, Carnegie Mellon University.

Chang, H.J., Jones, E.W., and Henry, S.A., 2002, Role of the unfolded protein response pathway in regulation of *INO1* and in the *sec14* bypass mechanism in *Saccharomyces cerevisiae*. *Genetics* **162:** 27–43.

Chang, L., and Karin, M., 2001, Mammalian MAP kinase signalling cascades. *Nature* **410:** 37–40.

Chapman, R.E., and Walter, P., 1997, Translational attenuation mediated by an mRNA intron. *Curr. Biol.* **7:** 850–859.

Chen, I.W., and Charalampous, F.C., 1963, A soluble enzyme system from yeast which catalyzes the biosynthesis of inositol from glucose. *Biochem. Biophys. Res. Commun.* **12:** 62–67.

Chen, I.W., and Charalampous, F.C., 1965, Biochemical studies on inositol. 8. Purification and properties of the enzyme system which converts glucose 6-phosphate to inositol. *J. Biol. Chem.* **240:** 3507–3512.

Cleves, A., McGee, T., and Bankaitis, V., 1991a, Phospholipid transfer proteins: A biological debut. *Trends Cell. Biol.* **1:** 30–34.

Cleves, A.E., McGee, T., Whitters, E.A., Champion, K.M., Aitken, J.R., Dowhan, W., Goebl, M., and Bankaitis, V.A., 1991b, Mutations in the CDP-choline pathway for phospholipid biosynthesis bypass the requirement for an essential phospholipid transfer protein. *Cell* **64:** 789–800.

Cox, J.S., Chapman, R.E., and Walter, P., 1997, The unfolded protein response coordinates the production of endoplasmic reticulum protein and endoplasmic reticulum membrane. *Mol. Biol. Cell* **8:** 1805–1814.

Cox, J.S., Shamu, C.E., and Walter, P., 1993, Transcriptional induction of genes encoding endoplasmic reticulum resident proteins requires a transmembrane protein kinase. *Cell* **73:** 1197–1206.

Cox, J.S., and Walter, P., 1996, A novel mechanism for regulating activity of a transcription factor that controls the unfolded protein response. *Cell* **87:** 391–404.

Culbertson, M.R., Donahue, T.F., and Henry, S.A., 1976a, Control of inositol biosynthesis in *Saccharomyces cerevisiae*: Properties of a repressible enzyme system in extracts of wild type (Ino^+) cells. *J. Bacteriol.* **126:** 232–242.

Culbertson, M.R., Donahue, T.F., and Henry, S.A., 1976b, Control of inositol biosynthesis in *Saccharomyces cerevisiae*: Inositol-phosphate synthetase mutants. *J. Bacteriol.* **126:** 243–250.

Culbertson, M.R., and Henry, S.A., 1975, Inositol-requiring mutants of *Saccharomyces cerevisiae*. *Genetics* **80:** 23–40.

Daum, G., Lees, N.D., Bard, M., and Dickson, R., 1998, Biochemistry, cell biology and molecular biology of lipids of *Saccharomyces cerevisiae*. *Yeast* **14:** 1471–1510.

Davies, S.P., Carling, D., Munday, M.R., and Hardie, D.G., 1992, Diurnal rhythm of phosphorylation of rat liver acetyl-CoA carboxylase by the AMP-activated protein kinase, demonstrated using freeze-clamping. Effects of high fat diets. *Eur. J. Biochem.* **203:** 615–623.

Dean-Johnson, M., and Henry, S.A., 1989, Biosynthesis of inositol in yeast: Primary structure of *myo*-inositol 1-phosphate synthase locus and functional characterization of its structural gene, the *INO1* locus. *J. Biol. Chem.* **264:** 1274–1283.

Dietz, M., Heyken, W.T., Hoppen, J., Geburtig, S., and Schuller, H.J., 2003, TFIIB and subunits of the SAGA complex are involved in transcriptional activation of phospholipid biosynthetic genes by the regulatory protein Ino2 in the yeast *Saccharomyces cerevisiae*. *Mol. Microbiol.* **48:** 1119–1130.

Donahue, T.F., and Henry, S.A., 1981a, Inositol mutants of *Saccharomyces cerevisiae*: Mapping the *ino1* locus and characterizing alleles of the *ino1*, *ino2* and *ino4* loci. *Genetics* **98:** 491–503.

Donahue, T.F., and Henry, S.A., 1981b, *myo*-Inositol-1-phosphate synthase: Characteristics of the enzyme and identification of its structural gene in yeast. *J. Biol. Chem.* **256:** 7077–7085.

Elkhaimi, M., Kaadige, M.R., Kamath, D., Jackson, J.C., Biliran, H., Jr. and Lopes, J.M., 2000, Combinatorial regulation of phospholipid biosynthetic gene expression by the *UME6, SIN3* and *RPD3* genes. *Nucleic Acids Res.* **28:** 3260–3167.

Errede, B., Cade, R.M., Yashar, B.M., Kamada, Y., Levin, D.E., Irie, K., and Matsumoto, K., 1995, Dynamics and organization of MAP kinase signal pathways. *Mol. Reprod. Dev.* **42:** 477–485.

Furter-Graves, E.M., Hall, B.D., and Furter, R., 1994, Role of a small RNA pol II subunit in TATA to transcription start site spacing. *Nucleic Acids Res.* **22:** 4932–4936.

Gardenour, K.R., Levy, J., and Lopes, J.M., 2004, Identification of novel dominant INO2c mutants with an Opi-phenotype. *Mol. Microbiol.* **52:** 1271–1280.

Gavin, A.C., Bosche, M., Krause, R., Grandi, P., Marzioch, M., Bauer, A., Schultz, J., Rick, J.M., Michon, A.M., Cruciat, C.M., Remor, M., Hofert, C., Schelder, M., Brajenovic, M., Ruffner, H., Merino, A., Klein, K., Hudak, M., Dickson, D., Rudi, T., Gnau, V., Bauch, A., Bastuck, S., Huhse, B., Leutwein, C., Heurtier, M.A., Copley, R.R., Edelmann, A., Querfurth, E., Rybin, V., Drewes, G., Raida, M., Bouwmeester, T., Bork, P., Seraphin, B., Kuster, B., Neubauer, G., and Superti-Furga, G., 2002, Functional organization of the yeast proteome by systematic analysis of protein complexes. *Nature* **415:** 141–147.

Graves, J.A., and Henry, S.A., 2000, Regulation of the yeast *INO1* gene: The products of the *INO2, INO4,* and *OPI1* regulatory genes are not required for repression in response to inositol. *Genetics* **154:** 1485–1495.

Greenberg, M., Goldwasser, P., and Henry, S., 1982a, Characterization of a yeast regulatory mutant constitutive for inositol-1-phosphate synthase. *Mol. Gen. Genet.* **186:** 157–163.

Greenberg, M.L., Klig, L.S., Letts, V.A., Loewy, B.S., and Henry, S.A., 1983, Yeast mutant defective in phosphatidylcholine synthesis. *J. Bacteriol.* **153:** 791–799.

Greenberg, M.L., and Lopes, J.M., 1996, Genetic regulation of phospholipid biosynthesis in yeast. *Microbiol. Rev.* **60:** 1–20.

Greenberg, M.L., Reiner, B., and Henry, S.A., 1982b, Regulatory mutations of inositol biosynthesis in yeast: Isolation of inositol-excreting mutants. *Genetics* **100:** 19–33.

Griac, P., 1997, Regulation of yeast phospholipid biosynthetic genes in phosphatidylserine decarboxylase mutants. *J. Bacteriol.* **179:** 5843–5848.

Griac, P., and Henry, S.A., 1996, Phosphatidylcholine biosynthesis in *Saccharomyces cerevisiae*: Effects on regulation of phospholipid synthesis and respiratory competence. In: Op den Kamp, J.A.F. (ed.), NATO ASI Series: Molecular Dynamics of Biological Membranes. Springer, Verlag, pp. 339–346.

Griac, P., Swede, M.J., and Henry, S.A., 1996, The role of phosphatidylcholine biosynthesis in the regulation of the *INO1* gene of yeast. *J. Biol. Chem.* **271:** 25692–25698.

Gustin, M.C., Albertyn, J., Alexander, M., and Davenport, K., 1998, MAP kinase pathways in the yeast *Saccharomyces cerevisiae*. *Microbiol. Mol. Biol. Rev.* **62:** 1264–1300.

Heinisch, J.J., Lorberg, A., Schmitz, H.P., and Jacoby, J.J., 1999, The protein kinase C-mediated MAP kinase pathway involved in the maintenance of cellular integrity in *Saccharomyces cerevisiae*. *Mol. Microbiol.* **32:** 671–680.

Henry, S.A., and Patton-Vogt, J.L., 1998, Genetic regulation of phospholipid metabolism: Yeast as a model eukaryote. In: Moldave, K. (ed.), Progress in Nucleic Acid Research and Molecular Biology. Academic Press Inc., San Diego, CA, USA, pp. 133–179.

Hirsch, J.P., 1987, *cis-* and *trans*-acting regulation of the *INO1* gene of *Saccharomyces cerevisiae*. Ph.D. thesis, Albert Einstein College of Medicine.

Hirsch, J.P., and Henry, S.A., 1986, Expression of the *Saccharomyces cerevisiae* inositol-1-phosphate synthase (*INO1*) gene is regulated by factors that affect phospholipid synthesis. *Mol. Cell. Biol.* **6:** 3320–3328.

Hirschhorn, J.N., Brown, S.A., Clark, C.D., and Winston, F., 1992, Evidence that SNF2/SWI2 and SNF5 activate transcription in yeast by altering chromatin structure. *Genes Dev.* **6:** 2288–2298.

Homann, M.J., Bailis, A.M., Henry, S.A., and Carman, G.M., 1987b, Coordinate regulation of phospholipid biosynthesis by serine in *Saccharomyces cerevisiae*. *J. Bacteriol.* **169:** 3276–3280.

Hudak, K.A., Lopes, J.M., and Henry, S.A., 1994, A pleiotropic phospholipid biosynthetic regulatory mutation in *Saccharomyces cerevisiae* is allelic to *sin3* (*sdi1, ume4, rpd1*). *Genetics* **136:** 475–483.

Irie, K., Takase, M., Lee, K.S., Levin, D.E., Araki, H., Matsumoto, K., and Oshima, Y., 1993, MKK1 and MKK2, which encode *Saccharomyces cerevisiae* mitogen-activated protein kinase-kinase homologs, function in the pathway mediated by protein kinase C. *Mol. Cell. Biol.* **13:** 3076–3083.

Ives, E.B., Nichols, J., Wente, S.R., and York, J.D., 2000, Biochemical and functional characterization of inositol 1,3,4,5,6-pentakisphosphate 2-kinases. *J. Biol. Chem.* **275:** 36575–36583.

Jackson, J.C., and Lopes, J.M., 1996, The yeast *UME6* gene is required for both negative and positive transcriptional regulation of phospholipid biosynthetic gene expression. *Nucleic Acids Res.* **24:** 1322–1329.

Jesch, S.A., Zhao, X., Wells, M.T., and Henry, S.A., 2005, Genome-wide analysis reveals inositol, not choline, as the major effector of Ino2p-Ino4p and unfolded protein response target gene expression in yeast. *J. Biol. Chem.* **280:** 9106–9118.

Kaadige, M.R., and Lopes, J.M., 2003, Opi1p, Ume6p and Sin3p control expression from the promoter of the INO2 regulatory gene via a novel regulatory cascade. *Mol. Microbiol.* **48:** 823–832.

Kagiwada, S., Hosaka, K., Murata, M., Nikawa, J., and Takatsuki, A., 1998, The *Saccharomyces cerevisiae* SCS2 gene product, a homolog of a synaptobrevin-associated protein, is an integral membrane protein of the endoplasmic reticulum and is required for inositol metabolism. *J. Bacteriol.* **180:** 1700–1708.

Kawahara, T., Yanagi, H., Yura, T., and Mori, K., 1997, Endoplasmic reticulum stress-induced mRNA splicing permits synthesis of transcription factor Hac1p/Ern4p that activates the unfolded protein response. *Mol. Biol. Cell* **8:** 1845–1862.

Kiyono, K., Miura, K., Kushima, Y., Hikiji, T., Fukushima, M., Shibuya, I., and Ohta, A., 1987, Primary structure and product characterization of the *Saccharomyces cerevisiae CHO1* gene that encodes phosphatidylserine synthase. *J. Biochem.* **102:** 1089–1100.

Klig, L.S., and Henry, S.S., 1984, Isolation of the yeast *INO1* gene: Located on an autonomously replicating plasmid, the gene is fully regulated. *Proc. Natl. Acad. Sci. U.S.A.* **81:** 3816–3820.

Klig, L.S., Homann, M.J., Carman, G.M., and Henry, S.A., 1985, Coordinate regulation of phospholipid biosynthesis in *Saccharomyces cerevisiae*: Pleiotropically constitutive *opi1* mutant. *J. Bacteriol.* **162:** 1135–1141.

Klig, L.S., Homann, M.J., Kohlwein, S.D., Kelley, M.J., Henry, S.A., and Carman, G.M., 1988a, *Saccharomyces cerevisiae* mutant with a partial defect in the synthesis of CDP-diacylglycerol and altered regulation of phospholipid biosynthesis. *J. Bacteriol.* **170:** 1878–1886.

Klig, L.S., Hoshizaki, D.K., and Henry, S.A., 1988b, Isolation of the yeast INO4 gene, a positive regulator of phospholipid biosynthesis. *Curr. Genet.* **13:** 7.

Kodaki, T., Hosaka, K., Nikawa, J.-I., and Yamashita, S., 1991a, Identification of the upstream activation sequences responsible for the expression and regulation of the *PEM1* and *PEM2* genes encoding the enzymes of the phosphatidylethanolamine methylation pathway in *Saccharomyces cerevisiae*. *J. Biochem.* **109:** 276–287.

Kodaki, T., Nikawa, J., Hosaka, K., and Yamashita, S., 1991b, Functional analysis of the regulatory region of the yeast phosphatidylserine synthase gene, *PSS*. *J. Biochem.* **173:** 7992–7995.

Kodaki, T., and Yamashita, S., 1987, Yeast phosphatidylethanolamine methylation pathway. *J. Biol. Chem.* **262:** 15428–15435.

Kodaki, T., and Yamashita, S., 1989, Characterization of the methyltransferases in the yeast phosphatidylethanolamine methylation pathway by selective gene disruption. *Eur. J. Biochem.* **185:** 243–251.

Kohno, K., Normington, K., Sambrook, J., Gething, M.-J., and Mori, K., 1993, The promoter region of the yeast *KAR2* (BiP) gene contains a regulatory domain that responds to the presence of unfolded proteins in the endoplasmic reticulum. *Mol. Cell. Biol.* **13:** 877–890.

Lamping, E., Paltauf, F., Henry, S.A., and Kohlwein, S.D., 1995, Isolation and characterization of a mutant of *Saccharomyces cerevisiae* with pleiotropic deficiencies in transcriptional activation and repression. *Genetics* **137:** 55–65.

Lee, K.S., Irie, K., Gotoh, Y., Watanabe, Y., Araki, H., Nishida, E., Matsumoto, K., and Levin, D.E., 1993, A yeast mitogen-activated protein kinase homolog (Mpk1p) mediates signalling by protein kinase C. *Mol. Cell. Biol.* **13:** 3067–3075.

Lee, K.S., and Levin, D.E., 1992, Dominant mutations in a gene encoding a putative protein kinase (B*CK1*) bypass the requirement for a *Saccharomyces cerevisiae* protein kinase C homolog. *Mol. Cell. Biol.* **12:** 172–182.

Lee, T.I., Rinaldi, N.J., Robert, F., Odom, D.T., Bar-Joseph, Z., Gerber, G.K., Hannett, N.M., Harbison, C.T., Thompson, C.M., Simon, I., Zeitlinger, J., Jennings, E.G., Murray, H.L., Gordon, D.B., Ren, B., Wyrick, J.J., Tagne, J.B., Volkert, T.L., Fraenkel, E., Gifford, D.K., and Young, R.A., 2002, Transcriptional regulatory networks in *Saccharomyces cerevisiae. Science* **298:** 799–804.

Letts, V.A., and Henry, S.A., 1985, Regulation of phospholipid synthesis in phosphatidylserine synthase-deficient (*cho1*) mutants of *Saccharomyces cerevisiae. J. Bacteriol.* **163:** 560–567.

Levin, D., Fields, F.O., Kunisawa, R., Bishop, J.M., and Thorner, J., 1990, A candidate protein kinase C gene, *PKC1*, is required for the *S. cerevisiae* cell cycle. *Cell* **62:** 312–224.

Levin, D.E., Bowers, B., Chen, C.Y., Kamada, Y., and Watanabe, M., 1994, Dissecting the protein kinase C/MAP kinase signalling pathway of *Saccharomyces cerevisiae. Cell. Mol. Biol. Res.* **40:** 229–239.

Levin, D.E., and Errede, B., 1995, The proliferation of MAP kinase signaling pathways in yeast. *Curr. Opin. Cell. Biol.* **7:** 197–202.

Lo, W.-S., Duggan, L., Tolga Emre, N.C., Belotserkovskya, R., Lane, W.S., Shiekhattar, R., and Berger, S.L., 2001, Snfl – a histone kinase that works in concert with the histone acetyltransferase Gcn5 to regulate transcription. *Science* **293:** 1142–1146.

Loewen, C.J., Roy, A., and Levine, T.P., 2003, A conserved ER targeting motif in three families of lipid binding proteins and in Opi1p binds VAP. *EMBO J.* **22:** 2025–2035.

Loewen, C.J.R., Gaspar, M.L., Jesch, S.A., Delon, C., Ktistakis, N.T., Henry, S.A., and Levine, T.P., 2004, Phospholipid metabolism regulated by a transcription factor sensing phosphatidic acid. *Science* **304:** 1644–1647.

Loewy, B.S., and Henry, S.A., 1984, The *INO2* and *INO4* loci of *Saccharomyces cerevisiae* are pleiotropic regulatory genes. *Mol. Cell. Biol.* **4:** 2479–2485.

Lopes, J.M., and Henry, S.A., 1991, Interaction of *trans* and *cis* regulatory elements in the *INO1* promoter of *Saccharomyces cerevisiae. Nucleic Acids Res.* **19:** 3987–3994.

Lopes, J.M., Hirsch, J.P., Chorgo, P.A., Schulze, K.L., and Henry, S.A., 1991, Analysis of sequences in the *INO1* promoter that are involved in its regulation by phospholipid precursors. *Nucleic Acids Res.* **19:** 1687–1693.

Lopes, J.M., Schulze, K.L., Yates, J.W., Hirsch, J.P., and Henry, S.A., 1993, The *INO1* promoter of *Saccharomyces cerevisiae* includes an upstream repressor sequence (URS1) common to a diverse set of yeast genes. *J. Bacteriol.* **175:** 4235–4238.

Maeda, T., and Eisenberg, F., Jr., 1980, Purification, structure, and catalytic properties of L-*myo*-inositol-1-phosphate synthase from rat testis. *J. Biol. Chem.* **255:** 8458–8464.

Majumder, A., Duttagupta, S., Goldwasser, P., Donahue, T., and Henry, S., 1981, The mechanism of interallelic complementation at the *INO1* locus in yeast: Immunological analysis of mutants. *Mol. Gen. Genet.* **184:** 347–354.

Majumder, A.L., Chatterjee, A., Dastidar, K.G., and Majee, M., 2003, Diversification and evolution of L-myo-inositol 1-phosphate synthase. *FEBS Lett.* **553:** 3–10.

Majumder, A.L., Johnson, M.D., and Henry, S.A., 1997, 1L-*myo*-inositol 1-phosphate synthase. *Biochim. Biophys. Acta* **1348:** 245–256.

McGee, T.P., Skinner, H.B., Whitters, E.A., Henry, S.A., and Bankaitis, V.A., 1994, A phosphatidylinositol transfer protein controls the phosphatidylcholine content of yeast Golgi membranes. *J. Cell Biol.* **124:** 273–287.

McGraw, P., and Henry, S.A., 1989, Mutations in the *Saccharomyces cerevisiae opi3* gene: Effects on phospholipid methylation, growth and cross-pathway regulation of inositol synthesis. *Genetics* **122:** 317–330.

Mori, K., Ma, W., Gething, M.-J., and Sambrook, J., 1993, A transmembrane protein with a $cdc2^{+}$/CDC28-related kinase activity is required for signaling from the ER to the nucleus. *Cell* **74:** 743–756.

Mori, K., Ogawa, N., Kawahara, T., Yanagi, H., and Yura, T., 2000, mRNA splicing-mediated C-terminal replacement of transcription factor Hac1p is required for efficient activation of the unfolded protein response. *Proc. Natl. Acad. Sci.* **97:** 4660–4665.

Mori, K., Sant, A., Kohno, K., Normington, K., Gething, M.-J., and Sambrook, J.F., 1992, A 22 bp *cis*-acting element is necessary and sufficient for the induction of the yeast *KAR2* (BiP) gene by unfolded proteins. *EMBO J.* **11:** 2583–2593.

Nikawa, J.-I., and Yamashita, S., 1992, *IRE1* encodes a putative protein kinase containing a membrane-spanning domain and is required for inositol prototrophy in *Saccharomyces cerevisiae*. *Mol. Microbiol.* **6:** 1441–1446.

Nikoloff, D.M., and Henry, S.A., 1994, Functional characterization of the *INO2* gene of *Saccharomyces cerevisiae*. *J. Biol. Chem.* **269:** 7402–7411.

Nikoloff, D.M., McGraw, P., and Henry, S.A., 1992, The *INO2* gene of *Saccharomyces cerevisiae* encodes a helix-loop-helix protein that is required for activation of phospholipid synthesis. *Nucleic Acids Res.* **20:** 3253.

Nishizuka, Y., 1984a, The role of protein kinase C in cell surface signal transduction and tumour promotion. *Nature* **308:** 693–698.

Nishizuka, Y., 1984b, Turnover of inositol phospholipids and signal transduction. *Science* **225:** 1365–1370.

Nishizuka, Y., 1986, Studies and perspectives of protein kinase C. *Science* **233:** 305–312.

Nonet, M., and Young, R., 1989, Intragenic and extragenic suppressors of mutations in the heptapeptide repeat domain of *Saccharomyces cerevisiae* RNA polymerase II. *Genetics* **123:** 715–724.

Odom, A.R., Stahlberg, A., Wente, S.R., and York, J.D., 2000, A role for nuclear inositol 1,4,5-trisphosphate kinase in transcriptional control. *Science* **287:** 2026–2029.

Ouyang, Q., Ruiz-Noriega, M., and Henry, S.A., 1999, The *REG1* gene product is required for repression of *INO1* and other inositol-sensitive upstream activating sequence-containing genes of yeast. *Genetics* **152:** 89–100.

Paltauf, F., Kohlwein, S., and Henry, S.A., 1992, Regulation and compartmentalization of lipid synthesis in yeast. In: Pringle, J. (ed.), The Molecular and Cellular Biology of the Yeast Saccharomyces. Cold Spring Harbor Laboratory Press, Plainview, NY, USA, pp. 415–500.

Pappas, D.L., Jr., and Hampsey, M., 2000, Functional interaction between Ssu72 and the Rpb2 subunit of RNA polymerase II in *Saccharomyces cerevisiae*. *Mol. Cell. Biol.* **20:** 8343–8351.

Patil, C., and Walter, P., 2001, Intracellular signaling from the endoplasmic reticulum to the nucleus: The unfolded protein response in yeast and mammals. *Curr. Opin. Cell Biol.* **13:** 349–356.

Patton-Vogt, J.L., Griac, P., Sreenivas, A., Bruno, V., Dowd, S., Swede, M.J., and Henry, S.A., 1997, Role of the yeast phosphatidylinositol/phosphatidylcholine transfer protein (Sec14p) in phosphatidylcholine turnover and *INO1* regulation. *J. Biol. Chem.* **272:** 20873–20883.

Peterson, C.L., and Herskowitz, I., 1992, Characterization of the yeast *SWI1, SWI2*, and *SWI3* genes, which encode a global activator of transcription. *Cell* **68:** 573–583.

Peterson, C.L., Kruger, W., and Herskowitz, I., 1991, A functional interaction between the C-terminal domain of RNA polymerase II and the negative regulator SIN1. *Cell* **64:** 1135–1143.

Pittner, F., Tovorova, J.J., Karnitskaya, E.Y., Khoklov, A.S., and Hoffmann-Ostenhof, O., 1979, *Myo*-inositol 1-phosphate synthase from *Streptomyces griseus* (studies on the biosynthesis of cyclitols, XXXVIII). *Mol. Cell. Biochem.* **25:** 43.

Ruiz-Noriega, M., 2000, Signal transduction and phospholipid biosynthesis in yeast: The role of the glucose response pathway. Department of Biological Sciences, Carnegie Mellon University, p. 157.

Saiardi, A., Caffrey, J.J., Snyder, S.H., and Shears, S.B., 2000, Inositol polyphosphate multikinase (ArgRIII) determines nuclear mRNA export in *Saccharomyces cerevisiae*. *FEBS Lett.* **468:** 28–32.

Saiardi, A., Erdjument-Bromage, H., Snowman, A.M., Tempst, P., and Snyder, S.H., 1999, Synthesis of diphosphoinositol pentakisphosphate by a newly identified family of higher inositol polyphosphate kinases. *Curr. Biol.* **9:** 1323–1326.

Saiardi, A., Nagata, E., Luo, H.R., Snowman, A.M., and Snyder, S.H., 2001, Identification and characterization of a novel inositol hexakisphosphate kinase. *J. Biol. Chem.* **276:** 39179–39185.

Santiago, T.C., and Mamoun, C.B., 2003, Genome expression analysis in yeast reveals novel transcriptional regulation by inositol and choline and new regulatory functions for Opi1p, Ino2p, and Ino4p. *J. Biol. Chem.* **278:** 38723–38730.

Scafe, C., Chao, D., Lopes, J., Hirsch, J.P., Henry, S., and Young, R.A., 1990a, RNA polymerase II C-terminal repeat influences response to transcriptional enhancer signals. *Nature* **347:** 491–494.

Scafe, C., Martin, C., Nonet, M., Podos, S., Okamura, S., and Young, R.A., 1990b, Conditional mutations occur predominantly in highly conserved residues of RNA polymerase II subunits. *Mol. Cell. Biol.* **10:** 1270–1275.

Scafe, C., Nonet, M., and Young, R.A., 1990c, RNA polymerase II mutants defective in transcription of a subset of genes. *Mol. Cell. Biol.* **10:** 1010–1016.

Schüller, H.-J., Richter, K., Hoffmann, B., Ebbert, R., and Schweizer, E., 1995, DNA binding site of the yeast heteromeric Ino2p/Ino4p basic helix-loop-helix transcription factor: Structural requirements as defined by saturation mutagenesis. *FEBS Lett.* **370:** 149–152.

Schüller, H.J., Schorr, R., Hoffman, B., and Schweizer, E., 1992, Regulatory gene *INO4* of yeast phospholipid biosynthesis is positively autoregulated and functions as a transactivator of fatty acid synthase genes *FAS1* and *FAS2* from *Saccharomyces cerevisiae. Nucleic Acids Res.* **20:** 5955–5961.

Schwank, S., Ebbert, R., Rautenstrauss, K., Schweizer, E., and Schuller, H.-J., 1995, Yeast transcriptional activator *INO2* interacts as an Ino2p/Ino4p basic helix-loop-helix heteromeric complex with the inositol/choline-responsive element necessary for expression of phospholipid biosynthetic genes in *Saccharomyces cerevisiae. Nucleic Acids Res.* **23:** 230–237.

Shen, H., and Dowhan, W., 1996, Reducation of CDP-diacylglycerol synthase activity results in the excretion of inositol by *Saccharomyces cerevisiae. J. Biol. Chem.* **271:** 29043–29048.

Shen, H., and Dowhan, W., 1997, Regulation of phospholipid biosynthetic enzymes by the level of CDP-diacylglycerol synthase activity. *J. Biol. Chem.* **272:** 11215–11220.

Shen, X., Xiao, H., Ranallo, R., Wu, W.H., and Wu, C., 2003, Modulation of ATP-dependent chromatin-remodeling complexes by inositol polyphosphates. *Science* **299:** 112–114.

Shirra, M.K., and Arndt, K.M., 1999, Evidence for the involvement of the Glc7-Reg1 phosphatase and the Snf1-Snf4 kinase in the regulation of *INO1* transcription in *Saccharomyces cerevisiae. Genetics* **152:** 73–87.

Shirra, M.K., Patton-Vogt, J., Ulrich, A., Liuta-Tehlivets, O., Kohlwein, S.D., Henry, S.A., and Arndt, K.M., 2001, Inhibition of acetyl coenzyme A carboxylase activity restores expression of the *INO1* gene in a *snf1* mutant strain of *Saccharomyces cerevisiae. Mol. Cell. Biol.* **21:** 5710–5722.

Sidrauski, C., Cox, J.S., and Walter, P., 1996, tRNA Ligase is required for regulated mRNA splicing in the unfolded protein response. *Cell* **87:** 405–413.

Slekar, K.H., and Henry, S.A., 1995, *SIN3* works through two different promoter elements to regulate *INO1* gene expression in yeast. *Nucleic Acids Res.* **23:** 1964–1969.

Sreenivas, A., and Carman, G.M., 2003, Phosphorylation of the yeast phospholipid synthesis regulatory protein Opi1p by protein kinase A. *J. Biol. Chem.* **278:** 20673–20680.

Sreenivas, A., Patton-Vogt, J.L., Bruno, V., Griac, P., and Henry, S.A., 1998, A role for phospholipase D (Pld1p) in growth, secretion, and regulation of membrane lipid synthesis in yeast. *J. Biol. Chem.* **273:** 16635–16638.

Sreenivas, A., Villa-Garcia, M.J., Henry, S.A., and Carman, G.M., 2001, Phosphorylation of the yeast phospholipid synthesis regulatory protein Opi1p by protein kinase C. *J. Biol. Chem.* **276:** 29915–29923.

Steger, D.J., Haswell, E.S., Miller, A.L., Wente, S.R., and O'Shea, E.K., 2003, Regulation of chromatin remodeling by inositol polyphosphates. *Science* **299:** 114–116.

Summers, E.F., Letts, V.A., McGraw, P., and Henry, S.A., 1988, *Saccharomyces cerevisiae cho2* mutants are deficient in phospholipid methylation and cross-pathway regulation of inositol synthesis. *Genetics* **120:** 909–922.

Swede, M.J., 1994, Isolation and characterization of novel regulatory mutants of phospholipid biosynthesis in *Saccharomyces cerevisiae*. Department of Biological Sciences, Carnegie Mellon University.

Swede, M.J., Hudak, K.A., Lopes, J.M., and Henry, S.A., 1992, Strategies for generating phospholipid synthesis mutants in yeast. In Dennis, E.A. (ed.), Methods in Enzymology: Phospholipid Biosynthesis. Academic Press, Inc., San Diego, CA, USA, pp. 21–34.

Treisman, R., 1996, Regulation of transcription by MAP kinase cascades. *Curr. Opin. Cell Biol.* **8:** 205–215.

Wagner, C., Blank, M., Strohman, B., and Schuller, H.J., 1999, Overproduction of the Opi1 repressor inhibits transcriptional activation of structural genes required for phospholipid biosynthesis in the yeast *Saccharomyces cerevisiae*. *Yeast* **15:** 843–854.

Wagner, C., Dietz, M., Wittmann, J., Albrecht, A., and Schuller, H.-J., 2001, The negative regulator Opi1 of phospholipid biosynthesis in yeast contacts the pleiotropic repressor Sin3 and the transcriptional activator Ino2. *Mol. Microbiol.* **41:** 155–166.

Watanabe, M., Chen, C.Y., and Levin, D.E., 1994, *Saccharomyces cerevisiae PKC1* encodes a protein kinase C (PKC) homolog with a substrate specificity similar to that of mammalian PKC. *J. Biol. Chem.* **269:** 16829–16836.

White, M.J., Hirsch, J.P., and Henry, S.A., 1991, The *OPI1* gene of *Saccharomyces cerevisiae*, a negative regulator of phospholipid biosynthesis, encodes a protein containing polyglutamine tracts and a leucine zipper. *J. Biol. Chem.* **266:** 863–872.

Woods, A., Munday, M.R., Scott, J., Yang, X., Carlson, M., and Carling, D., 1994, Yeast *SNF1* is functionally related to mammalian AMP-activated protein kinase and regulates acetyl-CoA carboxylase *in vivo*. *J. Biol. Chem.* **269:** 19509–19515.

Xie, Z., Fang, M., Rivas, M.P., Faulkner, A.J., Sternweis, P.C., Engebrecht, J., and Bankaitis, V.A., 1998, Phospholipase D activity is required for suppression of yeast phosphatidylinositol transfer protein defects. *Proc. Natl. Acad. Sci U.S.A.* **95:** 12346–12351.

York, J.D., Odom, A.R., Murphy, R., Ives, E.A., and Wente, S.R., 1999, A phospholipase C-dependent inositol polyphosphate kinase pathway required for efficient mRNA export. *Science* **285:** 96–100.

Chapter 7

The Structure and Mechanism of *myo*-Inositol-1-Phosphate Synthase

James H. Geiger and Xiangshu Jin
Chemistry Department, Michigan State University, East Lansing, MI 48824, U.S.A.

Abstract: The first and rate-limiting step in the biosynthesis of *myo*-inositol is the conversion of D-glucose 6-phosphate to 1L-*myo*-inositol 1-phosphate catalyzed by 1L-*myo*-inositol 1-phosphate synthase (MIP synthase). MIP synthase has been identified in a wide variety of organisms from bacteria to humans and is relatively well-conserved throughout evolution. It is probably homotetrameric in most if not all cases and always requires NAD^+ as a cofactor, with NADH being reconverted to NAD^+ in the catalytic cycle. This review focuses on the structure and mechanism of MIP synthase, with a particular emphasis on the mechanistic insights that have come from several recent structures of the enzyme. These include the structure of the enzyme from *Saccharomyces cerevisiae, Archeoglobus fulgidus* and *Mycobacterium tuberculosis*.

Keywords: *Inositol biosynthesis, 1L-myo-inositol 1-phosphate synthase, D-glucose 6-phosphate, enzyme structure, enzyme mechanism.*

1. INTRODUCTION, EARLY ENZYMOLOGY

The *de novo* biosynthesis of all inositols follows a common pathway from glucose 6-phosphate (Loewus and Kelly, 1962; Majumder *et al.*, 1997). In the first step, glucose 6-phosphate is converted to *myo*-inositol 1-phosphate (MIP) by *myo*-inositol 1-phosphate synthase, (EC 5.5.1.4; MIPS), and in the second step, MIP is dephosphorylated by *myo*-inositol 1-phosphate phosphatase to yield *myo*-inositol (Chen and Charalampous, 1965; Eisenberg, 1967). The

A. Lahiri Majumder and B. B. Biswas (eds.), Biology of Inositols and Phosphoinositides, 157–180
© 2006 *Springer. Printed in the Netherlands.*

other inositol-containing compounds are then produced via metabolic processing (Majumder *et al.* 1997; Bohnert and Jensen, 1996; Obendorf, 1997; Peterbauer and Richter, 1998; Peterbauer *et al.*, 1998). Though Fischer first proposed that *myo*-inositol was formed by cyclization of D-glucose (Fischer, 1945), the first demonstration of this pathway came only in 1961, when Loewus and Kelly used isotopic labeling to show that the 1 and 6 carbons of glucose condense to form *myo*-inositol (Loewus and Kelly, 1962). The mechanism proposed by Loewus and Kelly has been confirmed by an amalgam of data collected using enzyme from a variety of sources and is shown in Figure 1. Early work on the mechanism of MIPS has been the subject of several reviews (Hoffmann-Ostenhof and Pittner, 1982; Loewus, 1990; Loewus and Loewus,

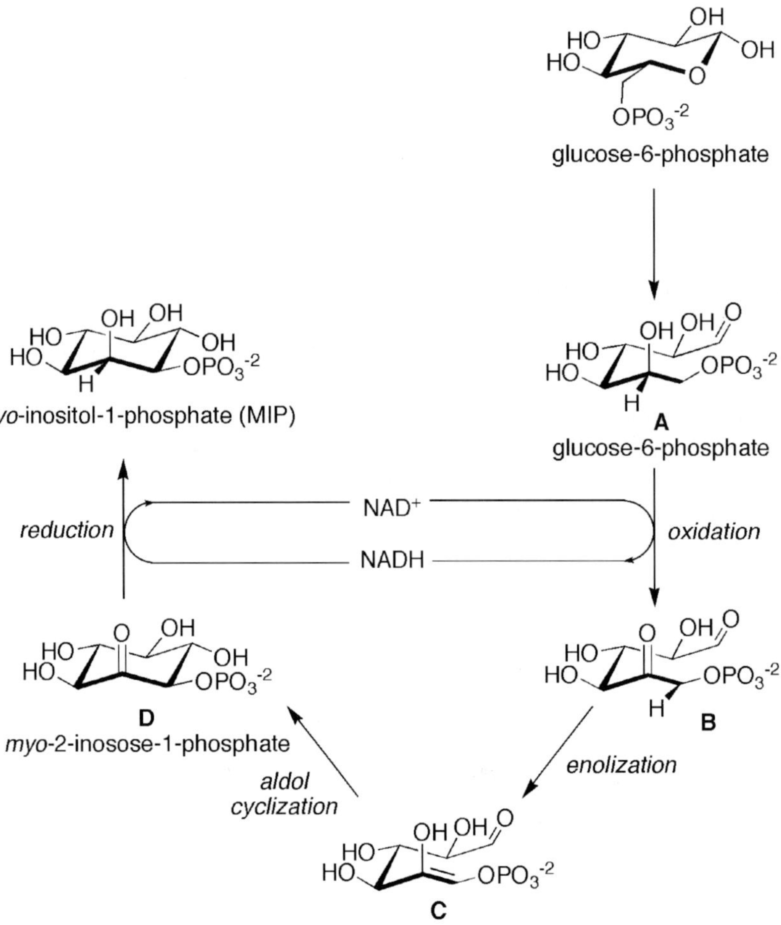

Figure 1. Mechanistic scheme for the conversion of glucose 6-phosphate to MIP by MIPS.

1983; Loewus and Murthy, 2000; Majumder *et al.*, 1997; Parthasarathy and Eisenberg, 1986).

Two possibilities exist for the first step of the reaction. Either the enzyme preferentially binds the ring-opened aldehyde form of glucose 6-phosphate as shown (intermediate **A**) or it binds the ring-closed acetal form and subsequently catalyzes conversion to the ring-opened form. The problem with the former proposal is that the ring-opened form makes up less than 1% of the total glucose 6-phosphate in solution. However, the second proposal would require a complex rearrangement of the substrate in the active site of the enzyme subsequent to binding. This rearrangement would either require a large motion of the phosphate group or a large motion of the rest of the glucose 6-phosphate molecule to accommodate the cyclization (Conrad *et al.*, 2002; Loewus *et al.*, 1980). The next step involves oxidation by NAD^+ of C5 to produce keto intermediate **B**. Proton abstraction of the C6 *pro-R* hydrogen leads to the enol intermediate **C**. Intramolecular aldol cyclization then leads to intermediate **D** and the product results from reduction at C5 to yield the product MIP. Though biochemical data suggested that reduction by NADH involved hydride transfer of the *Pro-S* C4 hydrogen of NADH, this stereochemistry is inconsistent with all of the crystal structures of MIPS determined so far (Byun and Jenness, 1981; Jin and Geiger, 2003; Norman *et al.*, 2002; Stieglitz *et al.*, 2004). An abundance of evidence supports the mechanism described above: (1) NAD^+ is absolutely required for enzymatic activity and enzyme-bound NADH is tightly associated with the enzyme (Barnett and Corina, 1968; Barnett *et al.*, 1973a,b; Byun and Jenness, 1981). In fact, *in situ* formation of NADH has been shown spectroscopically (Peng, 1996). (2) Chemically synthesized intermediate **B** was incubated with *apo*-MIPS and [4-^{3}H] NADH and produced [^{3}H]-*myo*-inositol, confirming that **B** is an intermediate (Kiely and Sherman, 1975). Incubation of MIPS, NAD^+, and **B** can produce glucose-6-phosphate, showing that the first step of the reaction is also reversible (Barnett *et al.*, 1973b). (3)Trapping experiments using tritiated borohydride were consistent with the presence of intermediate **D** (Chen and Eisenberg, 1975). Intermediate **D** was also chemically synthesized, shown to be a competitive inhibitor of MIPS and was also shown to be a substrate for the NADH-bound form of the enzyme, demonstrating definitively that **D** is indeed the last intermediate in the mechanism (Migaud and Frost, 1995). (4) Kinetic isotope effects were detected for hydrogens on both C5 and C6, consistent with the formation of intermediate **B** and indicative of enolization being partially rate-limiting (Byun *et al.*, 1973; Chen and Charalampous, 1967; Loewus, 1977). (5) Identification of the C6 *Pro-R* hydrogen as the hydrogen abstracted from **B** to form the enol was proven by Floss and coworkers using glucose-6-phosphate stereospecifically radiolabeled at C6 (Loewus *et al.*, 1980).

Though the mechanistic scheme described above is generally accepted, many questions remain. (1) The nature of the enolization/aldol condensation.

Two mechanisms predominate in enzymatically catalyzed aldol reactions. In Type 1 aldolases, Schiff base formation occurs between an enzyme lysine residue and a substrate carbonyl group serving to activate deprotonation and enolization. Type II aldolases activate the carbonyl by divalent cation chelation (Horecker *et al.*, 1972). The divalent cation serves as a Lewis acid to activate deprotonation/enolization. Several eukaryotic MIPS enzymes have been shown not to require divalent cations for activity or to be affected by metal chelating agents, such as EDTA, indicating that these enzymes are not Type II divalent cation requiring aldolases (Chen *et al.*, 2000a; Chen *et al.*, 2000b; Maeda and Eisenberg, 1980; Mauck *et al.*, 1980). Efforts to demonstrate the presence of a Schiff base have also proven unsuccessful. Though an earlier report using impure enzyme showed different results (Pittner and Hoffmann Ostenhof, 1976; Pittner and Hoffmann Ostenhof, 1978), attempts at trapping a Schiff base intermediate using borohydride have been unsuccessful and there is no transfer of ^{18}O from water to product as is expected if Schiff base formation was to occur (Maeda and Eisenberg, 1980; Sherman *et al.*, 1977). On the other hand, MIPS from *Archaeoglobus fulgidus* (aMIPS), is very sensitive to EDTA and does require divalent cations for its activity, indicating that it may use a type II-like aldolase mechanism (Chen *et al.*, 2000b).

2. MECHANISTIC INSIGHTS FROM INHIBITION OF MIPS

Several inhibitors of MIPS have been identified and shed much light on the mechanism. These include 2-deoxy-D-glucose 6-phosphate (entry v in Table 1), and 2-deoxy-D-glucitol 6-phosphate (entry vi in Table 1) (Barnett *et al.*, 1973a; Wong and Sherman, 1985; Wong *et al.*, 1982). Table 1 shows most of the known inhibitors of MIPS with their inhibition constants (K_i). The Frost lab has made the most significant contributions to this field by synthesizing a variety of mechanism-based inhibitors in an effort to better understand the enzyme. As shown, the proposed intermediate **D**, *myo*-2-inosose 1-phosphate (entry viii in Table 1), was synthesized and found to be a micromolar inhibitor, consistent with its identification as an intermediate in the reaction (Migaud and Frost, 1995). Intermediate **D** is an intermediate of the NAD^+-bound enzyme because the NADH-bound enzyme would be required for reduction to product. Consistent with this hypothesis, NADH-bound yMIPS readily converts intermediate **D** to product (Migaud and Frost, 1995). 1-deoxy 1-phosphonomethyl-*myo*-2-inosose and dihydroxyacetone 1-phosphate were also found to be competitive inhibitors of the enzyme (entries ix and xi in Table 1, respectively) (Migaud and Frost, 1996), while 2-deoxy-*myo*-inositol 1-phosphate was not an inhibitor. This shows that the pre-oxidized center at C2 is critical for binding this class of inhibitor. A number of acyclic glucose 6-phosphate derivatives that only mimic the acyclic form of glucose 6-phosphate were all found to be inhibitors of MIPS (entries v–vii), while molecules that enforce a cyclic conformation (entries i–iv in Table 1), were not inhibitors of

Table 1. Inhibitors of MIPS and their inhibition constants

entry #	compound	K_i
i.	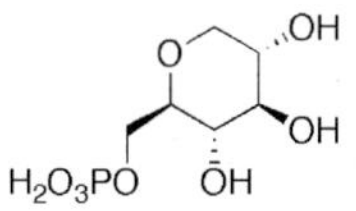	No inhibition
ii.	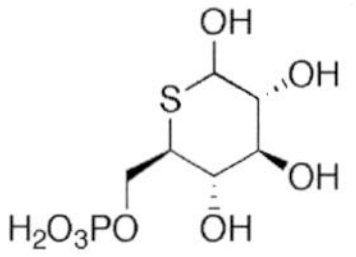	No inhibition
	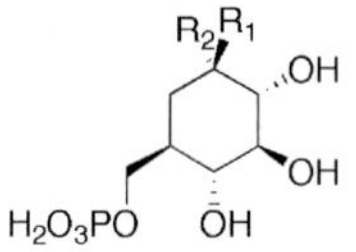	
iii.	R_1 = OH, R_2 = H	No inhibition
iv.	R_1 = H, R_2 = OH	No inhibition
v.	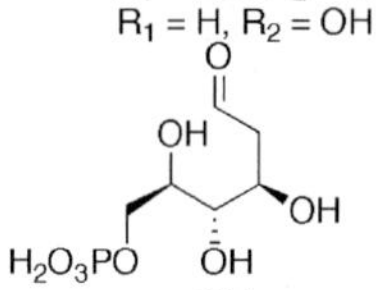	9.1 μM
vi.	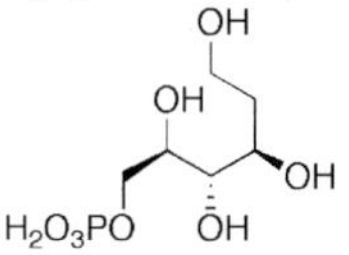	2.3 μM
vii.	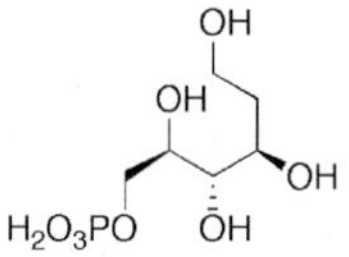	2.3 μM
viii.	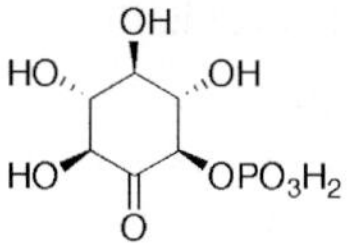	3.6 μM
ix.	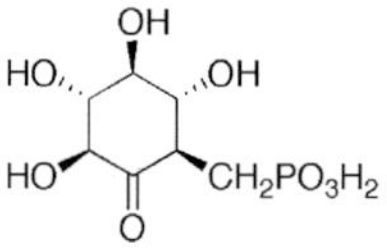	37 μM @7.2 160 μM @6.4 6.4 μM @6.4
x.	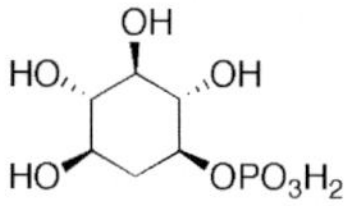	No inhibition
xi.	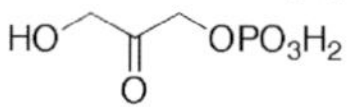	0.7 mM
	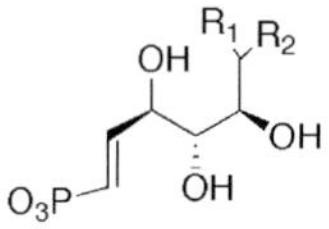	
xii.	R_1 = CHO, R_2 = OH	1.1 mM
xiii.	R_1 = CH_2OH, R_2 = OH	30 μM
xiv.	R_1 = CHO, R_2 = H	100 μM
xv.	R_1 = CH_2OH, R_2 = H	0.67 μM
	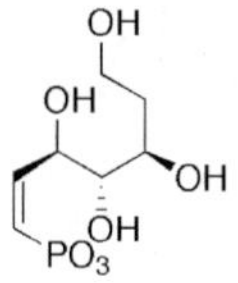	
xvi.	R_1 = CHO, R_2 = OH	No inhibition
xvii.	R_1 = CH_2OH, R_2 = OH	No inhibition

the enzyme. These results are consistent with the idea that MIPS binds only the acyclic form of glucose 6-phosphate, despite the fact that the acyclic form represents less than 1% of the molecules in solution. It is interesting to note that the 2-deoxy derivative of most of these molecules are more potent inhibitors than their more glucose-like parents. The relative orientation of the phosphate group relative to the rest of the molecule was also explored using vinyl phosphonate glucitol derivatives that enforce either a *cis*- or *trans*- orientation to the phosphate group (entry xii-xvii). The fact that only the *trans*-isomers showed enzyme inhibition activity led to the conclusion that the phosphate group is bound in a transoid conformation in the MIPS active site. This would put the phosphate group in an orientation consistent with intramolecular proton extraction from C5 by the phosphate group during enolization.

MIPS has now been identified in at least 65 organisms (Majumder *et al.*, 1997; Majumder *et al.*, 2003). These include Cyanobacteria, eubacteria, Archaebacteria, plants, fungi, and animals (Majumder *et al.*, 1997). MIPS from fungi to humans is remarkably conserved.

3. OVERALL STRUCTURE OF THE MONOMER

The role of the enzyme in catalysis was particularly unclear as no active-site mutants had been identified until very recently (Jin *et al.*, 2004). However, a flurry of structure determinations of MIPS from several organisms has dramatically impacted our understanding of the enzyme's mechanism. Structures of MIPS from *Mycobacterium tuberculosis*, *Archeoglobus fulgidus* and *Saccharomyces cerevisiae* (baker's yeast) have been determined since 2001 (Norman *et al.*, 2002; Steiglitz *et al.*, 2004; Stein and Geiger 2002). The structure of the yeast enzyme has been determined in the *apo*-form, the NAD^+-bound form, the NADH/phosphate/glycerol-bound form, the NAD^+/2-deoxy glucitol phosphate form, and the NAD^+/2-deoxy-D-glucitol 6-(*E*)-vinylhomophosphonate form (Jin and Geiger, 2003; Jin *et al.*, 2004; Stein and Geiger 2000; Stein and Geiger, 2002).

S. cerevisiae MIPS (yMIPS) consists of three domains, a Rossmann fold domain responsible for binding NAD^+ and containing some of the active site residues, a tetramerization/catalytic domain responsible for oligomerization and containing the remaining active site residues, and a central domain primarily responsible for domain orientation and oligomerization (Figure 2) (Stein and Geiger, 2002). The Rossmann fold includes residues 66–266 in yMIPS, the catalytic domain includes residues 327–441, and the central domain includes residues 1–65, which form two long β strands that drape over the top of the

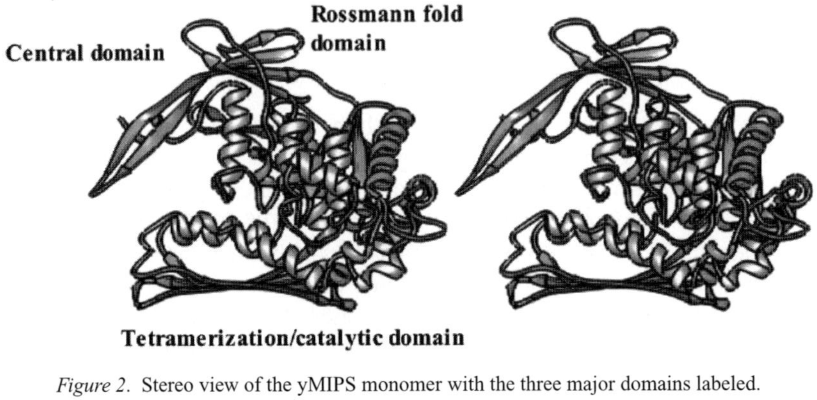

Figure 2. Stereo view of the yMIPS monomer with the three major domains labeled.

enzyme. The central domain also includes a c-terminal region (residues 433–533) and an insertion in the Rossmann fold that includes residues 93–140. Together these three elements, the n-terminus, Rossman fold insertion, and c-terminus fold to produce the central domain. An additional insertion in the Rossmann fold includes residues 149–215 and serves to surround the adenine portion of NAD^+, preventing ready dissociation of the cofactor consistent with the dinucleotide's catalytic role in the enzyme's mechanism. Numerous interactions between the three domains prevent relative motion between the three domains and serve to fix the position of NAD^+ bound to the Rossmann fold domain relative to the catalytic domain. The result is a large rigid core made up of all three domains of the enzyme. Given the relatively high sequence identity between yMIPS and other eukaryotic MIPS throughout their length, it is quite likely that the domain structure of the higher eukaryotic MIPS will be very similar to that of yMIPS (Figure 3).

There are significant differences in domain architecture between yMIPS and *M. tuberculosis* MIPS (mMIPS) and MIPS from *A. fulgidus* (aMIPS). Neither of these enzymes contains the n-terminal 65 amino acids or the large insertion in the Rossmann fold, which subtracts large portions of the central domain seen in yMIPS. However, they both have c-terminal regions similar to yMIPS, which serve to fix the relative orientation between the catalytic domain and Rossmann fold domain. Both mMIPS and aMIPS have similar catalytic and Rossmann fold domains when compared to yMIPS (Norman *et al.*, 2002; Stieglitz *et al.*, 2004).

The overall fold of MIPS is similar to that of diaminopimelic acid dehydrogenase from *Corynebacterium glutamicum* and dihydrodipicolinate reductase from *E. coli*. Though the reactions catalyzed by these enzymes are quite different, they all use either NAD^+ or $NADP^+$ bound to a structurally similar Rossmann fold domain, and all three contain a β sheet domain located underneath the Rossmann fold and directly beneath the nicotinamide moiety that defines part of the active site (Norman *et al.*, 2002; Stein and Geiger, 2002).

4. STRUCTURE OF THE TETRAMER

As shown in Figure 4, yMIPS, aMIPS, and mMIPS all form tetrameric structures. The tetramer can be understood as a dimer of dimers with approximate 222 symmetry (Figure 4). Each monomer in the tetramer makes extensive interactions with two other protomers (Norman *et al.*, 2002; Stein and Geiger, 2002; Stieglitz *et al.*, 2004). The bottom of the catalytic domain β sheet makes an extensive interaction with the equivalent β sheet from a neighboring protomer to form one of the interfaces. The other interface begins at the bottom of the molecule with a strand-to-strand interaction between adjoining protomers. This strand-to-strand interaction creates a continuous β sheet that stretches across both protomers. The interface then continues across the entire back face of the molecule, mostly involving residues from the central domain. While the β sheet interaction buries 6,000 Å^2 of surface area in yMIPS, the more extensive interface in yMIPS stretching from bottom to top of the molecule buries 11,700 Å^2 of surface area (Stein and Geiger, 2002). Since both of these interfaces are relatively hydrophobic, stable dissociation of the tetramer is highly unlikely. Though there are several reports indicating some eukaryotic MIPS enzymes to be trimeric in solution (Gumber *et al.*, 1984; Maeda and Eisenberg, 1980; Ogunyemi *et al.*, 1978; Raychaudhuri *et al.*, 1997), the high

```
                              β1              β2
         10        20        30        40        50        60
  1 MTEDNIAPITSVKVVTDKCTYKDNELLTKYSYENAVV--TKTASGRFD--VTPTVQDYVFKLDLKKPEKL Yeast
  1 M---FIES---FKVESPNVKYTENEIHSVYDYETTEVVHEKTVNGTYQWIVKPKTVKYDFKTDIRVP-KL plant
  1 M---KPTNNSTLEVISPKVQVDDEFITTDYDYQTSHV--KRTADGQLQ--VHPQTTSLKIRTGRHVP-KL fly
  1 M---EAAAQFFVE--SPDVVYGPEAIEAQYEYRTTRV--SREG-GVLK--VHPTSTRFTFRTARQVP-RL human
  1                                                                     M Archae
  1 MSEHQSLPAPEASTE-------------------------------------------------------V mycobacterium

    β3     α1        β4    β5     α2  β6       β7
       70        80        90       100       110       120       130
 67 GIMLIGLGGNNGSTLVASVLANKHNVEFQTKEGVKQPNYFGSMTQCSTLKLGIDAEGNDVYAPFNSLL-PM Yeast
 64 GVMLVGLGGNNGSTLTAGVIANKEGISWATKDKVQQANYFGSLTQASSIRVGS-FNGEEIYAPFKSLL-PM plant
 63 GVMLVGWGGNNGSTLTAALEANRRQLKWRKRTGVQEANWYGSITQASTVFIGSDEDGGDVYVPMKELL-PM fly
 60 GVMLVGWGGNNGSTLTAAVLANRLRLSWPTRSGRKEANYYGSLTQAGTVSLGLDAEGQEVFVPFSAVL-PM human
 19 KVWLVGAYGIVSTTAMVGARAIERGI----------APKIGLVSEL----------------PHFEGIEKY archae
 11 RVAIVG-VGNCASSLVQGVEYYYNAD--------DTSTVPGLMHVRFGP---------------------- mycobacterium

    α3   β8         α4             α5        β9       α6
     140       150       160       170       180       190
137 VSPNDFVV-SGWDINNA----DLYEAMQ-RS----QVLEYDLQQRLKAKMSLVKPLPSIYYPDFI-AANQD Yeast
133 VNPDDVVF-GGWDISDMN----LADAMA---RA----RVLDIDLQKQLRPYMENIVPLPGIFDPD-FIAAN plant
133 VEPDNIIV-DGWDISGLH----LGDAMR---RA----EVLDVALQDQIYDQLAQLRPRPSIYDPD-FIAAN fly
130 VAPNDLVF-DGWDISSLN----LAEAMR---RAKVLDWGLQEQLWPHMEALRPRPSVYIPEFIAA-NQSAR human
 47 APF-SFEF-GGHEIRLL---SNAYEAAK-EHWELNRHFDREILEAVKSDLEGIVARKGTAL--NC-GSGIK archae
 57 YHVRDVKFVAAFDVDAKKVGFDLSDAIF-ASENNTIKIAD-----VAPT--NVIVQRGPTL--DGIGKYYA mycobacterium

    α6 β10         α7                β11                α8
     200       210       220       230       240       250       260
197 -ERANNCINLDEKGNVTTRGKWTHLQRIRRDIQNFKEENALDKVIVLWTANTERYVEVSPGVNDTMENLLQ Yeast
198 -QGSRANH-------VIKGTKKEQVDHIIKDMREFKEKNKVDKVVVLWTANTERYSNVVVGMNDTMENLME plant
198 -QSDRADN-------VIRGTRLEQYEQIRKDIRDFRERSGVDSVIVLWTANTERFADVQPGLNTTSQELIA fly
195 -ADN-----------LIPGSRAQQLEQIRRDIRDFRSSAGLDKVIVLWTANTERFCEVIPGLNDTAENLLR human
109 -ELGDIKTLE-----GEGLSLAEMVSRIEEDIKSFA----DDETVVINVASTEPLPNYSEEYHGSLEGFER archae
118 DTIELSDAE---------------PVDVVQALKEAK----VD--VLVSYLPVGSE---------------- mycobacterium

    α8      α9        β12            α10  β13       α11
     270       280       290       300       310       320       330
267 SIKNDH-EEIAPSTIFAAASILEGVPYINGSP-QNTF-VPGLVQLAEHEGTFIAGDDLKS--GQTKLKSVL Yeast
263 SVDRDE-AEISPSTLYAIACVLEGIPFINGSP-QNTF-VPGLIDMAIRNNVLIGGDDFKS--GQTKMKSVL plant
263 SLEANH-SEVSPSTIFAMASIAEGCTYINGSP-QNTF-VPGLIQLAEEKNVFIAGDDFKS--GQTKIKSVL fly
259 TIELGL--EVSPSTLFAVASILEGCAFLNGSP-QNTL-VPGALELAWQHRVFVGGDDFKS--GQTKVKSVL human
160 MIDEDRKEYASASMLYAYAALKLGLPYANFTP-SPGSAIPALKELAEKKGVPHAGNDGKT--GETLVKTTL archae
152 ----------EADKFYAQCAIDAGVAFVNALP-VFIASDPVWAKKFTDARVPIVGDDIKSQVGATITHRVL mycobacterium
```

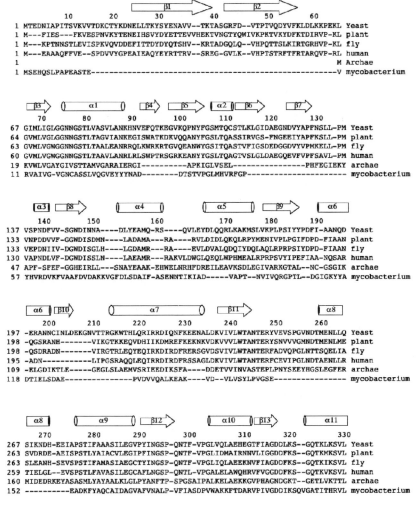

Figure 3. Sequence alignment of MIPS from several sources. Elements of secondary structure as defined in yMIPS are shown above the alignment and numbered. Cylinders are helices and wide arrows β strands.

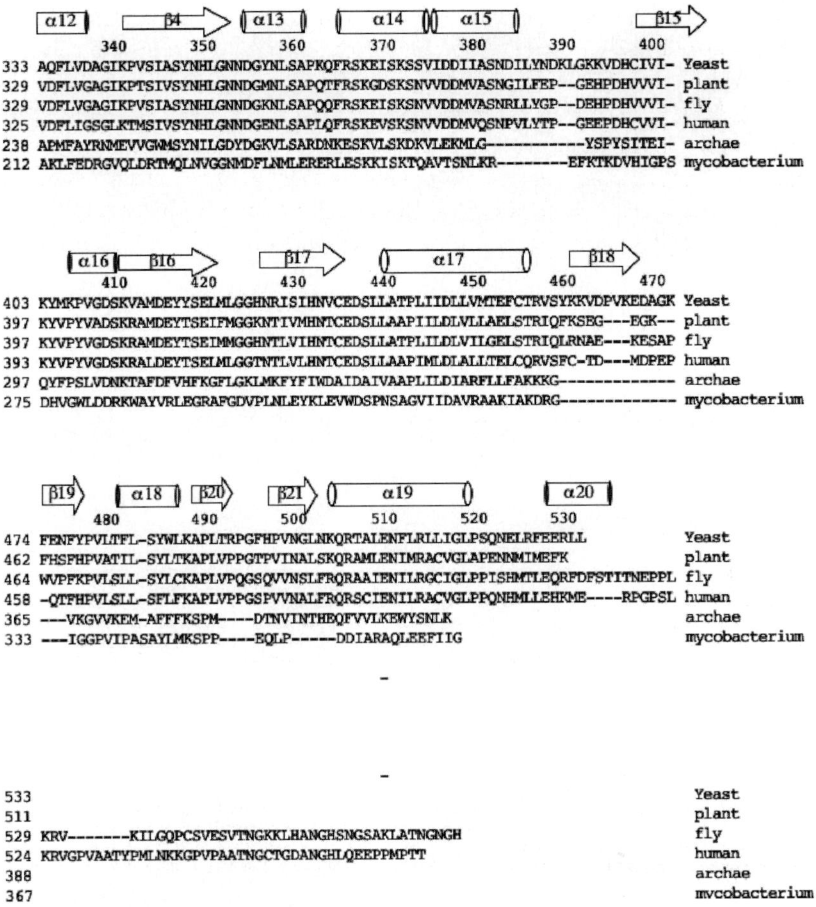

Figure 3. (continued)

sequence identity between MIPS from higher eukaryotes (approximately 50% identity between yMIPS and all known eukaryotic sequences) and the two extensive interfaces seen in yMIPS lead to the conclusion that all eukaryotic MIPS are likely to be tetrameric. Both aMIPS and mMIPS tetramers are relatively similar to the yMIPS tetramer, characterized by the packing of β sheets and strand-to-strand interaction of β strands between protomers. It would appear that the oligomerization state of the enzyme has been conserved throughout evolution. The most notable differences between the structures lay in the central domain region. Both aMIPS and mMIPS lack the n-terminal region and the Rossman fold insertion that largely make up this domain. The result is that both aMIPS and mMIPS have significantly smaller interfaces between protomers in this interface (Norman *et al.*, 2002; Stein and Geiger, 2002; Stieglitz *et al.*, 2004).

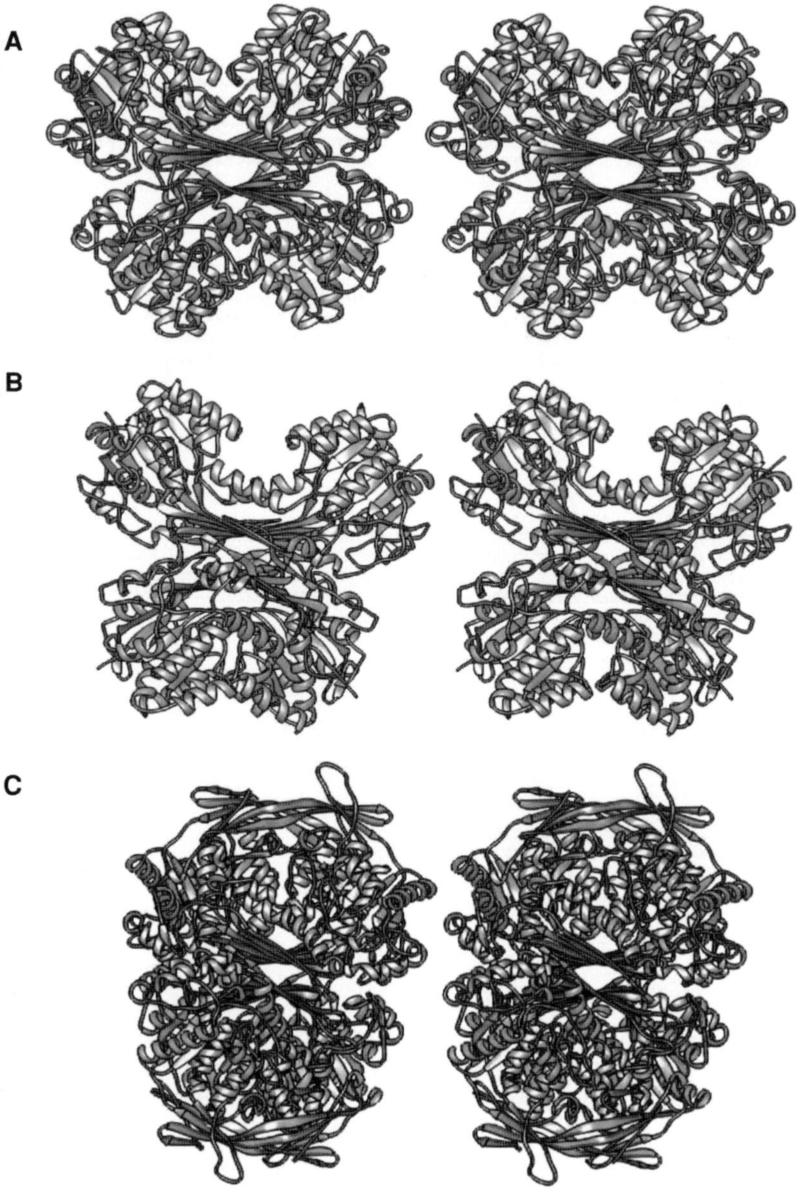

Figure 4. Stereoviews of the aMIPS (a), mMIPS (b), and yMIPS (c) tetramers

5. NAD$^+$ BINDING OF MIPS

Several structures of yMIPS bound to NAD$^+$ or NADH have been determined. In fact, there are five independently determined structures with 12 independent copies (due to more than one molecule in the asymmetric unit) of NAD$^+$ or NADH (Jin and Geiger, 2003; Jin *et al.*, 2004; Stein and Geiger, 2002). Binding of the nucleotide is virtually identical in all of these structures from the adenine to the last phosphodiester moiety. Interactions between the two riboses, adenine, and two phosphodiesters are essentially identical in all of these structures. Table 2 delineates the interactions seen in these structures. The loop between β3 and α1 (Ile71-Gly78) plays a critical role in the binding as it does in all Rossmann Fold dinucleotide-binding domains. The sequence of yMIPS, GLGGNNG, is similar to the characteristic sequence for Rossmann folds seen in this region (GXGXXG). The glycine residues in this sequence

Table 2. Hydrogen bonds between NAD$^+$ and MIPS for the structures that are known

YMIPS	Dist. Å	mMIPS	Distance	aMIPS	Distance
G75N O3*	2.9	G24N	3.5	G10 N O3	3.2
N76N AO2 (H)	2.9	N25N AO2	3.1	I11 N NO2	3.3
N76N NO2 (H)	3.2				
N77N NO2 (H)	2.9	C26 N O1	2.9	V12 N NO2	3.0
N77ND2 AO2	3.2	C26 SG O1	3.0		
		C26 SG N7	3.0		
D148 OD1 O3*	2.6	D69 OD1 O3*	2.8	E57 OE1 AO3	3.1
D148 OD2 O2*	2.6	D69 OD2 O2*	2.7	E57 OE2 O2	3.1
				R59 AO2	2.7
		K74 NZ O3*	2.9		
S184 OG AN1	2.7				
I185 O AN6	3.1	P107 O N6	2.7	T99 O AN6	2.5
		T108 O N6	2.9	A100 N AN6	3.3
A245 O O3*	3.1	L146 O O3*	2.7	A148 O NO3	2.4
		Y116 OH O1	2.7		
T247 N O3*	3.0	V148 N O3*	3.1	T150 N O3*	3.3
T247 OG1 O3*	3.1			T150 OG1 NO3	3.1
T247 OG1 O2*	2.6			T150 OG1 NO2	2.9
		Y157 OH N1	2.7	Y185 OH AN1	3.2
				T228 OG1 NN7	2.7
				G229 N O2P	2.7
		D235 OD1 O2*	2.7	D261 OD2 NO2	3.2
N355 O AO1	2.8				
				D274 NZ NO2	3.0
		D310 O O7	2.7	D332 O NN7	3.2
		S314 OG1 O7	3.1		

Data from Jin and Geiger, 2003; Norman *et. al.*, 2002; Stieglitz *et. al.*, 2004.

produce a short, tight turn that leaves room for the dinucleotide to cross from one side of the sheet to the other. There are several direct interactions between this loop and NAD^+ as shown in Table 2. There is a marked difference between NAD^+ binding in MIPS and nucleotide binding in other Rossman fold enzymes, namely an insertion in the Rossmann fold between β8 and α8 in the Rossmann fold domain (encompassing amino acids 153–214 in yMIPS) that serves to completely engulf the adenine portion of the dinucleotide. There are many interactions between this inserted region and NAD^+, both hydrophobic (W147, I149, and I185 all make hydrophobic interactions with adenine in yMIPS) and hydrogen bonding (S184 and I185 both make hydrogen bonds with NAD^+) (Table 2). The complete burial of the adenine and the direct interactions between it and yMIPS are doubtless responsible for the relatively tight binding of NAD^+. Tighter NAD^+ binding for this enzyme makes sense because NAD^+ is used catalytically and is not turned over with each enzymatic cycle. In contrast to the adenine, ribose, and phosphate parts of NAD^+, there are surprising differences in the position of the nicotinamide moiety in various yMIPS structures. The nicotinamide position varies between two extremes (1) The nicotinamide amide nitrogen is hydrogen bonded to the closest phosphate oxygen, pinching together the nicotinamide ring and phosphate (Jin and Geiger, 2003; Stein and Geiger, 2002). (2) The phosphate and nicotinamide are forced apart by an ion or water molecule that chelates both the nicotinamide amide oxygen moiety and the phosphate oxygen (Figure 5) (Jin and Geiger, 2003; Jin *et al.*, 2004).

The amide/phosphate hydrogen bond is found in many yMIPS structures, including all the yMIPS/NAD^+ complex structures determined in our laboratory. The second extreme is seen in the yMIPS/NADH/phosphate/glycerol structure and in the yMIPS/NAD^+/2-deoxy-glucitol 6-(*E*)-vinylhomophosphonate structure. The feature found between the nicotinamide and phosphate

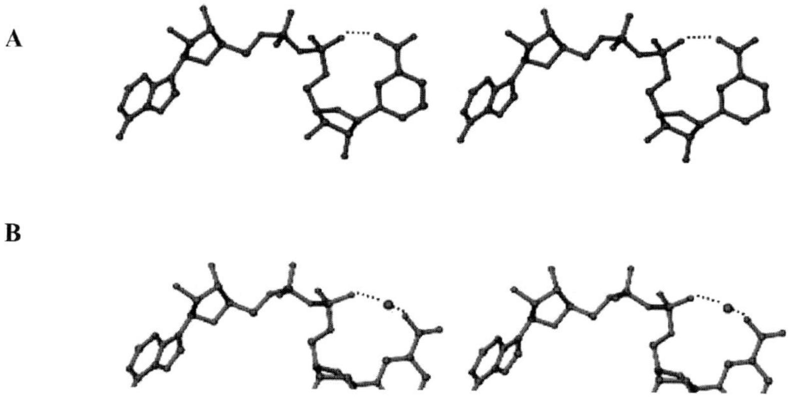

Figure 5. The two conformations of the NAD^+ nicotinamide seen in yMIPS structures. Hydrogen bonding or metal chelation is shown with dotted lines.

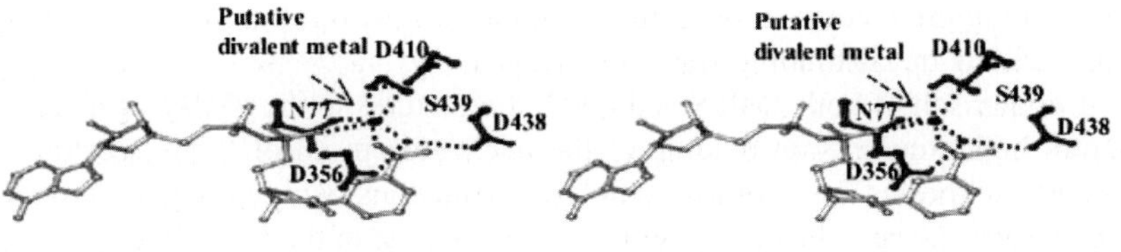

Figure 6. Chelation sphere about the metal in the yMIPS/NAD$^+$/2-deoxyglucitol-6-(E)-vinyl phosphonate complex. Protein amino acids are shown with dark lines and NAD$^+$ is shown with less dark lines.

moieties was postulated to be some sort of cation in both of these structures, based on the relatively short distances between the feature and the chelating groups (Jin and Geiger, 2003; Jin *et al.*, 2004). Treatment of yMIPS/NADH/phosphate/glycerol complex crystals with EDTA resulted in loss of the putative metal ion, indicating the putative metal to be a divalent cation. Close inspection of the coordination about this feature in the yMIPS/NAD$^+$/vinyl phosphonate structure also points to its identification as a metal ion (Figure 6). As shown, the putative metal is hexacoordinate, chelated by the nicotinamide amide, the nicotinamide phosphate oxygen, N77, S439, D410, and a water molecule. Since this water molecule is also coordinated to two carboxylates, D356 and D438, it must donate a lone pair to the putative metal unless one of these acidic groups is protonated. S323 must also coordinate with a lone pair because it is already making a hydrogen bond to D410 with its hydrogen. Add to this the direct coordination of D410, and the result is that four of the six coordinating groups must be donating lone pairs as opposed to hydrogens, inconsistent with identification of this feature as a water molecule. The other two ligands, N77 and the nicotinamide amide can also use their oxygen atoms to donate lone pairs, completing the all oxygen chelation sphere about the metal (Jin *et al.*, 2004). This metal ion is also seen in both the aMIPS and mMIPS structures chelating both the nicotinamide oxygen and phosphate oxygen atoms, but the remainder of the coordination sphere is not identical in the three (Norman *et al.*, 2002; Stieglitz *et al.*, 2004). In mMIPS, the coordination sphere of the putative Zn^{2+} is tetrahedral, characterized by shorter ligand distances (around 2.2 Å) and is completed by a water molecule and S311 (equivalent to S439 in yMIPS) similar to that seen in yMIPS. In aMIPS, this metal is pentacoordinate with D304 (equivalent to D410 in yMIPS), D332 (equivalent to D438 in yMIPS), and a water molecule completing the coordination sphere. Though aMIPS requires divalent cation for its activity, this metal was tentatively identified as a K^+ ion due to the longer ligand lengths (2.4–3.0 Å), occupancies, B factors, and the fact that the protein was treated with EDTA before crystallization. It was also suggested that the monovalent cation might be replacing a divalent cation in this structure. The difference in chelation between the three structures is not currently understood, but in all three cases, the metal appears to play an important role in positioning the nicotinamide in the active site. However, the catalytic importance of this metal-binding site is not known.

Comparison of both aMIPS and mMIPS structures to the yMIPS structures shows high similarity in binding of the nucleotide in all of these structures. Table 2 delineates the hydrogen bonding interactions between MIPS from all three organisms and the dinucleotide. As shown, nine of the 13 hydrogen bonds between yMIPS and NAD^+ are conserved in aMIPS and mMIPS, though many of the residues making these interactions are not. This occurs because many of the interactions are via the main chain and because many of the changes nevertheless preserve the hydrogen bond. Of the 17 and 18 hydrogen bonds between mMIPS and NAD^+ and aMIPS and NAD^+, respectively, only 3 and 2 hydrogen bonds are unique to each enzyme, respectively. This shows that in spite of relatively low sequence similarity between the three enzymes whose structures are known, the enzyme/NAD^+ interface has been remarkably preserved.

6. THE ACTIVE SITE OF MIPS

The active site of MIPS is located between the bottom of the Rossmann fold domain and the β sheet of the catalytic domain. Two helices, α14 and α15 stretch across the front of the active site and α13 forms the side of the active site. NAD^+ forms the top of the active site with the adenine ring more or less centered in the active site cavity. Numerous structures of MIPS have been determined and shed considerable light on the mechanism of the enzyme. Generally, the active site is characterized by a great deal of flexibility. This is particularly true of yMIPS.

The first structure of yMIPS was solved with only partial occupancy of NAD^+ in the active site. Residues 352–409 were not built into the model and were thought to be disordered in the structure (Stein and Geiger, 2002). This encompasses α13, α14, α15, and β15. Subsequent structures were determined both of the apo enzyme and the fully occupied NAD^+-bound enzyme. yMIPS bound to NAD^+ has been crystallized and its structure determined from three crystal forms, a C2 form with two independent molecules in the asymmetric unit, a $P2_12_12$ crystal form with four molecules in the asymmetric unit, and a $P2_1$ form, also with four monomers per asymmetric unit (Jin and Geiger, 2003, Kniewel *et al.*, 2002).

All these structures also contained a disordered region in the vicinity of the active site, but the amount of disorder was substantially less than that seen in the original structure. While residues 352–390 are still disordered in apo yMIPS, the C2 yMIPS/NAD^+ complex, and three of the four molecules of the $P2_1$ NAD^+-bound yMIPS complex, β15 residues (391–408) are relatively well-ordered in most of these structures. A smaller still disordered region is reported in the $P2_12_12$ yMIPS/NAD^+ crystal form. In this structure, only residues 365–375, encompassing α14 are disordered (Kniewel *et al.*, 2002).

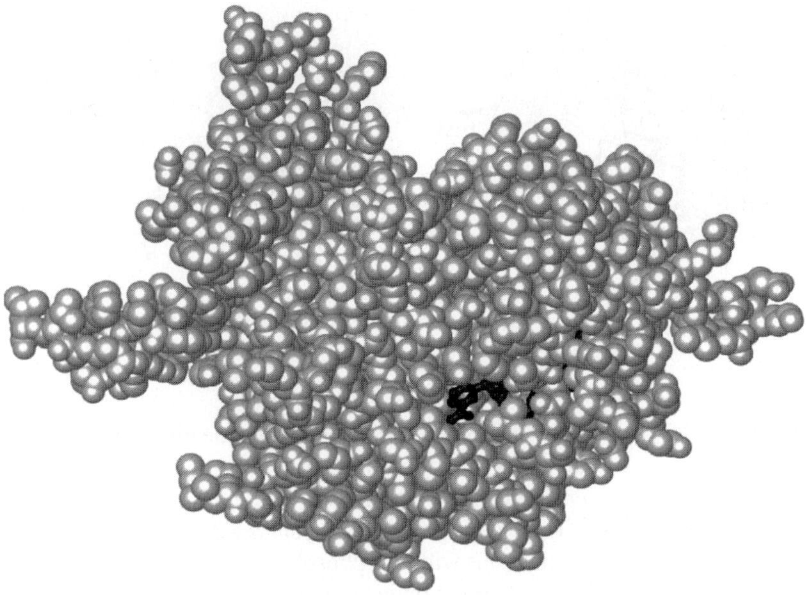

Figure 7. Space filling model of yMIPS/NAD^+ complex. NAD^+ atoms are darkened relative to the protein atoms.

α14 stretches across the front of the active site and when disordered leaves the nicotinamide ring of NAD^+ exposed (Figure 7).

The structure of mMIPS also contains a 30 residue disordered region about the active site that corresponds closely to the disordered region identified in yMIPS (Norman *et al.*, 2002). This region corresponds to yMIP's α14, α15, and β15.

Several structures of yMIPS have also been determined bound to small molecules. These molecules include two inhibitors, phosphate, glycerol, and an undefined small molecule bound to only one of four protomers in the $P2_1$ yMIPS/NAD^+ crystal form (Jin *et al.*, 2004; Jin and Geiger, 2003; Stein and Geiger, 2002). In all these structures, the entire region from 352–408 becomes ordered, completely encapsulating the active site and NAD^+ nicotinamide moiety. It thus appears that a disordered-to-ordered transition of at least the α14 helix occurs upon active site binding leading to a complete burial of both the active site and substrate. Together these structures provide critical insights regarding the mechanism of yMIPS.

When crystals of apo yMIPS are soaked in NADH instead of NAD^+, a surprising result was obtained. The yMIPS active site contained not only NADH but two other small molecules as well. A phosphate ion and a glycerol molecule were both located in the active site resulting in the folding of the entire disordered region and completely encapsulating both phosphate and glycerol within

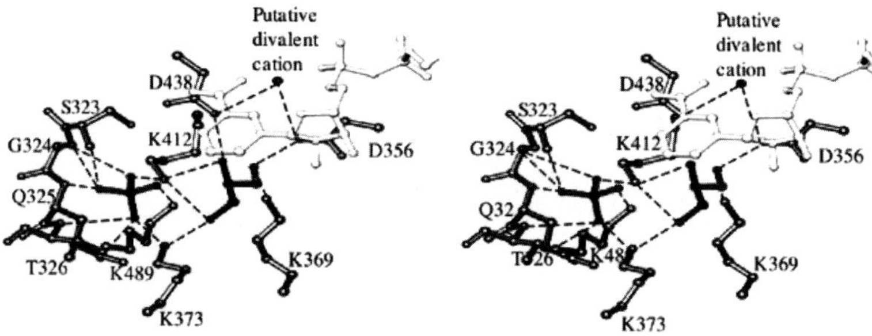

Figure 8. Phosphate and glycerol bound in the active site of yMIPS. Phosphate and glycerol are dark relative to the enzyme side chains.

the active site of the enzyme (Figure 8). The loop connecting β13 and α12 (residues 323–326) surrounds the phosphate and numerous interactions are made between the mainchain amide nitrogens of this loop and the phosphate (Jin and Geiger, 2003). Three highly conserved lysines K373, K412, and K489 also make salt bridge interactions with this phosphate. A significant conformational change of the 323–326 loop relative to both the apo and NAD^+-bound yMIPS structures is required to accommodate the phosphate. The loop is flipped out of the active site, serving to enlarge the size of the active site in this structure. The glycerol moiety sits next to the phosphate and also makes numerous interactions with side chain residues in the active site. The side chains of D438, D356, K369, K373, and K412 all make hydrogen bonds with the glycerol and are all highly conserved residues in MIPS (Jin and Geiger, 2003). A phosphate is also bound within the active site of the aMIPS enzyme and the entire active site is ordered in this structure. Interactions between the phosphate and protein are very similar to that seen in the yMIPS structure. The same three conserved lysines (K274, K278, and K367 in aMIPS correspond to K373, K412, and K489 in yMIPS) interact with the phosphate in aMIPS and similar main chain interactions between the aMIPS loop and the phosphate are also seen (Stieglitz *et al.*, 2004).

Recently, a structure of yMIPS bound to the mechanism-based inhibitor 2-deoxy-D-glucitol 6-(*E*)-vinyl homophosphonate was determined to 2.0 Å resolution (Jin *et al.*, 2004). This structure significantly contributed to the understanding of the mechanism of the enzyme. A *trans* orientation between the phosphorous of the phosphonate moiety and C5 is enforced by incorporation of an olefinic linkage. Surprisingly, instead of binding in a pseudocyclic conformation that brings C1 and C6 in proximity for bond formation during the intramolecular aldol cyclization, the inhibitor is bound in a more extended conformation with C1 and C6 on opposite sides of the molecule (Figure 9). C5 is directly under the nicotinamide C4 and they are 3.8 Å apart, relatively well positioned for hydride transfer from the nicotinamide to C4 as demanded by the mechanism. In agreement with the above structures, the phosphonate of

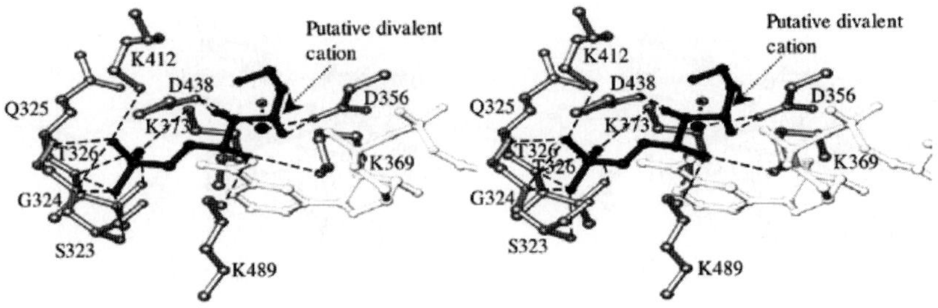

Figure 9. Interactions seen in the active site of the yMIPS/2-deoxy-D-glucitol 6-(*E*)-vinylhomophosphonate complex. Inhibitor is shown with dark lines relative to the enzyme side chains and NAD^+.

the inhibitor is positioned identically to the phosphate in the phosphate-bound structure. Most of the interactions between phosphate and enzyme are preserved in the inhibitor-bound complex. The main chain nitrogens of S323, G324, Q325, and T326 all make hydrogen bonds to the phosphonate. K412 and K373 make salt bridge interactions to the phosphonate as they do to the phosphate in the yMIPS phosphate-bound structure. K489, however, has changed its position and now joins K369 to make interactions with O5 of the inhibitor. All of the oxygen atoms of the inhibitor save O1 make hydrogen bonds with conserved residues in the active site, many of which also make interactions with glycerol in the yMIPS/glycerol complex structure. S323, G324, Q325, T326, D356, K369, K373, K412, D438, and K489 all make hydrogen bonds to the inhibitor and all, save Q325, which makes a main chain interaction with the phosphonate are highly conserved throughout evolution (Figure 3).

7. MODELING SUBSTRATE AND INTERMEDIATE IN THE ACTIVE SITE

Since the inhibitor is bound in a relatively extended conformation not consistent with the cyclization reaction, it was necessary to approximate a catalytically active conformation using molecular modeling simulations. In fact, when the substrate was modeled into the active site in the same conformation as the inhibitor, a steric clash occurs between the 2-hydroxyl group of the substrate Glucose-6-phosphate and L360 (atoms come within 2 Å of one another) (Jin *et al.*, 2004). This indicates that the substrate is not capable of adopting a conformation similar to that of the inhibitor. Modeling of the substrate in the active site was then performed by assuming that the inhibitor faithfully represented the catalytically active conformation of the substrate in its phosphate, C6, C5, and O5 positions. The rest of the substrate was then adjusted to conform to a

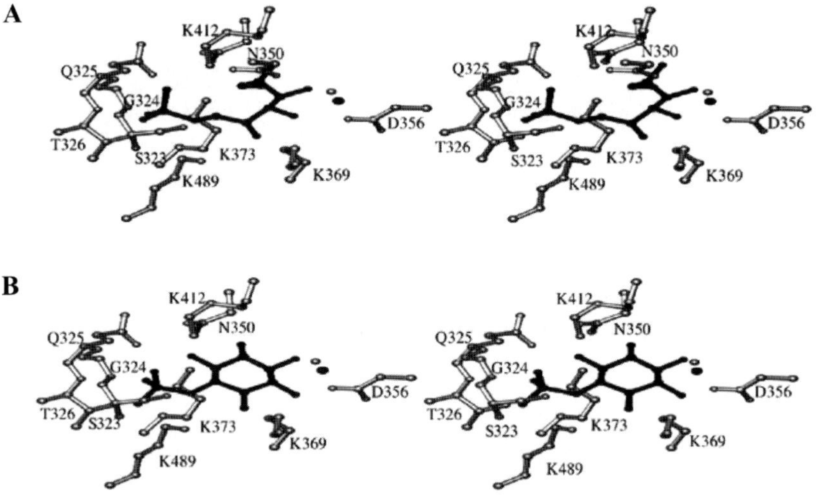

Figure 10. The substrate (a) and last intermediate (b) are modeled in the yMIPS active site in a conformation competent for catalysis. Intermediates and putative metal ion are dark relative to the enzyme side chains.

pseudocyclic chair-like conformation and energy minimization was performed using Insight II. The result of this modeling is shown in Figure 10 (Jin *et al.*, 2004). O1 now makes interactions with both K373 and K412, reflecting the interaction between K369, K489, and O5. In addition, D356 makes a hydrogen bond to O4, and O3 and O2 are almost within hydrogen bonding distance of D438 and N350, respectively. All of these residues are *absolutely* conserved from bacteria to humans.

Myo-2-inosose 1-phosphate, the final intermediate in the reaction pathway, was also modeled into the active site of the enzyme also by overlaying the phosphate, C5 and O5 onto the equivalent atoms of the 2-deoxy-D-glucitol 6-(*E*)-vinyl homophosphonate structure (Figure 10) (Jin *et al.*, 2004). Very similar interactions were seen between the enzyme and this molecule as well, even though there are significant changes in the positions particularly of C1 and O1 (the glucose-based numbering is preserved here for clarity). O1 is still bound tightly to both K373 and K412 in this structure and most of the hydrogen bonds between the hydroxyl groups of the intermediate and the enzyme are also preserved.

8. MECHANISM

A reasonable hypothesis for the detailed mechanism of the reaction was proposed based on a combination of the inhibitor-bound structure and the modeling results. Given that none of the substrate-like cyclic derivatives in Table 1

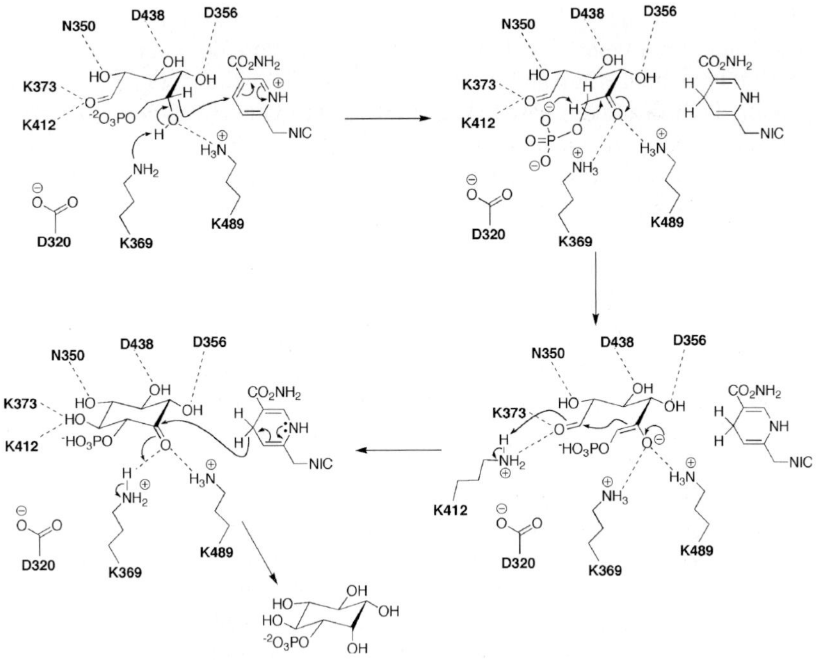

Figure 11. Proposal for the mechanism of yMIPS.

are inhibitors of the enzyme and the nature of the binding of 2-deoxy-D-glucitol 6-(*E*)-vinyl homophosphonate in the active site, it is likely that MIPS binds exclusively the acyclic, aldehyde form of glucose-6-phosphate in a conformation similar to that shown in Figure 10. The hydrogen at C5 is then transferred as a hydride to NAD^+ to form a keto intermediate (Figure 11). Deprotonation of O5 is assisted by K369 and K489, possibly by proton shuttle to D320, which is tightly hydrogen-bound to K369. Enol formation then occurs by deprotonation of the Pro(R) hydrogen from O6. The phosphate oxygen could act as the extracting base in this reaction, though K489 cannot be ruled out as both are reasonably close to C6 in the structure. Though neither group would be expected to have a pKa in a reasonable range to perform this extraction from data collected in free solution, there is precedent for phosphate ester-mediated deprotonation in the dehydroquinate synthase enzyme (Bender *et al.*, 1989; Widlanski *et al.*, 1989) and there are many reports of buried lysines with dramatically altered pKa's (Lee and Houk, 1997; Schmidt and Westheim, 1971). The developing negative charge on O5 is stabilized by K489 and K369, activating enol formation. In the next step, carbon–carbon bond formation occurs via intramolecular aldol cyclization between C1 and C6. Though a metal ion resides in the active site of yMIPS, aMIPS, and mMIPS as described above, this metal is nowhere near O5 or O1, and is in fact essentially buried by its ligands in yMIPS. It is very unlikely to be acting as a Lewis acid in the reaction.

Though several lines of biochemical evidence argue against the formation of a Schiff base for a type 1 aldol mechanism, the structure does not rule out this possibility as both K369 and K489 interact directly with O5 and either could potentially form a Schiff base to C5. However, formation of such a Schiff base would require significant motion of either C5 or the main chain atoms of K369 or K489 as neither lysine can reach C5 in the present structure (Jin *et al.*, 2004). A third possibility, which is consistent with both biochemical and structural data, is that the reaction occurs by a distinct mechanism that does not require either Schiff base formation or Lewis acid activation. In this mechanism, the developing negative charge on O1 is strongly stabilized by both K373 and K412 just as K369 and K489 stabilized the negative charge during enol formation. Product formation then occurs by hydride transfer from nicotinamide back to C5 to form the inositol product. This hypothesis predicts that all four of the active site lysines will be critically important for enzyme activity. Consistent with this prediction, mutation of either K369, K489, or K412 resulted in an inactive enzyme (Jin *et al.*, 2004).

Many significant questions remain to fully characterize the mechanism of MIPS. For example, the above mechanism does not explain why divalent cations are critically important for aMIPS activity, while several lines of evidence indicate that they are dispensable for the catalytic activity of MIPS from eukaryotes (Chen *et al.*, 2000b). Though yMIPS, aMIPS, and mMIPS all contain a metal-binding site between the nicotinamide amide group and the phosphate, this metal appears to be too buried to be directly involved in the mechanism and has been identified in all three structures, not just aMIPS. To reconcile this issue, electrostatic calculations and modeling studies were conducted and a second potential metal-binding site in aMIPS was identified (Stieglitz *et al.*, 2004). This site is close to a putative cation-binding site suggested in the first yMIPS structure, but the two enzymes differ in that N354 in yeast is not conserved in *Archaebacteria fulgidus* and is instead an aspartate (D259 in *A. fulgidus*). Modeling of the substrate glucose-6-phosphate and first intermediate 5-keto-glucose-6-phosphate in the active site of aMIPS indicated that the second metal could assist in Lewis acid stabilization of the negative charge that accumulates on O1 during the aldol cyclization. This would represent a modification of the type II aldolase mechanism that typically involves stabilization of negative charge on the oxygen during enolate formation, though chelation of both oxygens has also been suggested (Dreyer and Schulz, 1996). However, these modeling studies are in agreement with the ring closure geometry proposed for yMIPS and are also in agreement regarding the importance of all four of the critical lysine residues identified in yMIPS to be critical for catalysis (Stieglitz *et al.*, 2004).

This structural model for the mechanism of MIPS is not consistent with the reported structure of the yMIPS/2-deoxyglucitol-6-phosphate complex structure (Stein and Geiger, 2002). In this structure, the inhibitor 2-deoxyglucitol-6-phosphate's phosphate group does not occupy the phosphate-binding site, but instead is modeled into a site on the opposite side of the active site from this position. In fact, the entire inhibitor is flipped in the active site when compared to 2-deoxy-D-glucitol 6-(*E*)-vinyl homophosphonate. In fact, S323 flips into the active site and occludes the phosphate-binding site in this structure. Several lines of evidence indicate that the 2-deoxyglucitol-6-phosphate model does not mimic catalytically active substrate binding:

1. The phosphate position disagrees with three other structures, two yMIPS structures and the aMIPS structure.
2. Residue D356 is in the same conformation in all other MIPS structures save the 2-deoxy-glucitol-6-phosphate complex structure, a conformation that would occlude the 2-deoxy-glucitol-6-phosphate binding mode seen in this structure. This indicates that this mode of binding is inconsistent with all of these structures.
3. The electrostatic potential calculated at neutral pH is positive in the phosphate-binding site defined in the other three structures while it is negative at neutral pH in the phosphate-binding site proposed in the 2-deoxy-glucitol-6-phosphate complex structure.
4. The 2-deoxy-glucitol-6-phosphate complex crystals were grown and stabilized at pH 4.5, a pH at which the enzyme is completely inactive. On the other hand, the 2-deoxy-D-glucitol 6-(*E*)-vinyl homophosphonate complex crystals were produced at pH 5.5, where the enzyme retains about half of its activity.
5. S323 is in the flipped out position in both the aMIPS and mMIPS structures, which is required for phosphate binding. This residue is flipped into the active site and occludes the phosphate-binding site in the 2-deoxy-glucitol-6-phosphate complex structure.

All of this data lead to the conclusion that 2-deoxy-glucitol-6-phosphate is modeled in a catalytically incompetent conformation, possibly due to the low pH used to produce the crystals. The mechanism based on this structure must then also be disguarded (Jin and Geiger, 2003; Jin *et al.*, 2004; Stein and Geiger, 2002).

In conclusion, the combination of structural and biochemical results using enzyme from a variety of sources is beginning to coalesce around a unified mechanism for the critical and complex transformation of glucose 6-phosphate to MIP catalyzed by MIP synthase.

ACKNOWLEDGMENTS

We thank B. Stec for generously providing both the coordinates and manuscript of the aMIPs structure before publication. We also thank B. Borhan for his assistance with some of the figures.

REFERENCES

Barnett, J.E., and Corina, D.L., 1968, The mechanism of glucose 6-phosphate-D-myo-inositol 1-phosphate cyclase of rat testis. The involvement of hydrogen atoms. *Biochem. J.* **108:** 125–129.

Barnett, J.E.G., Rasheed, A., and Corina, D.L., 1973a, Inhibitors of MIP synthase. *Biochem. Soc. Trans.* **1:** 1267.

Barnett, J.E.G., Rasheed, A., and Corina, D.L., 1973b, Partial reactions of D-glucose 6-phosphate-1l-myoinositol 1-phosphate cyclase. *Biochem. J.* **131**: 21–30.

Bender, S.L., Widlanski, T., and Knowles, J.R., 1989, Dehydroquinate synthase – the use of substrate-analogs to probe the early steps of the catalyzed reaction. *Biochemistry* **28:** 7560–7572.

Bohnert, H.J., and Jensen, R.G., 1996, Strategies for engineering water-stress tolerance in plants. *Trends Biotechnol.* **14:** 89–97.

Byun, S.M., and Jenness, R., 1981, Stereospecificity of L-*myo*-inositol-1-phosphate synthase for nicotinamide adenine dinucleotide. *Biochemistry* **20:** 5174–5177.

Byun, S.M., Jenness, R., Ridley, W.P., and Kirkwood, S., 1973, Stereospecificity of D-glucose-6-phosphate – 1l-myo-inositol-1-phosphate cycloaldolase on hydrogen-atoms at C-6. *Biochem. Biophys. Res. Commun.* **54:** 961–967.

Chen, C.H.J., and Eisenberg, F., 1975, Myoinosose-2 1-phosphate – intermediate in myoinositol 1-phosphate synthase reaction. *J. Biol. Chem.* **250:** 2963–2967.

Chen, I.W., and Charalampous, F.C., 1965, Biochemical studies on inositol. 8. Purification and properties of the enzyme system which converts glucose 6-phosphate to inositol. *J. Biol. Chem.* **240:** 3507–3512.

Chen, I.W., and Charalampous, F.C., 1967, Studies on the mechanism of cyclization of glucose 6-phosphate to D-inositol 1-phosphate. *Biochim. Biophys. Acta.* **136:** 568–570.

Chen, L., Zhou, C., Yang, H., and Roberts, M.F., 2000a, Inositol-1-phosphate synthase from *Archaeoglobus fulgidus* is a class II adolase. *FASEB J.* **14:** 77.

Chen, L.J., Zhou, C., Yang, H.Y., and Roberts, M.F., 2000b, Inositol-1-phosphate synthase from *Archaeoglobus fulgidus* is a class II aldolase. *Biochemistry* **39:** 12415–12423.

Conrad, R.M., Grogan, M.J., and Bertozzi, C.R., 2002, Stereloselective synthesis of myo-inositol via ring-closing metathesis: A building block for glycosylphosphatidylinositol (GPI) anchor synthesis. *Org. Lett.* **4:** 1359–1361.

Dreyer, M.K., and Schulz, G.E., 1996, Catalytic mechanism of the metal-dependent fuculose aldolase from *Escherichia coli* as derived from the structure. *J. Mol. Biol.* **259:** 458–466.

Eisenberg, F., Jr., 1967, D-myoinositol 1-phosphate as product of cyclization of glucose 6-phosphate and substrate for a specific phosphatase in rat testis. *J. Biol. Chem.* **242:** 1375–1382.

Fischer, H.O.L., 1945, Chemical and biochemical relationships between hexoses and inositols. *Harvey Lect.* **40:** 156–178.

Gumber, S.C., Loewus, M.W., and Loewus, F.A., 1984, Further-studies on myo-inositol-1-phosphatase from the pollen of lilium-longiflorum thunb. *Plant Physiol.* **76:** 40–44.

Hoffmann-Ostenhof, O., and Pittner, F., 1982, The biosynthesis of myo-inositol and its isomers. *Can. J. Chem.* **60:** 1863–1871.

Horecker, B.L., Tsolas, O., and Lai, C.Y., 1972, Aldolases. In: Boyer, P.D. (ed.), The Enzymes, Vol. 7. Academic Press, New York, pp. 213–258.

Jin, X.S., Foley, K.M., and Geiger, J.H., 2004, The structure of the 1L-myo-inositol-1-phosphate synthase-NAD(+)-2deoxy-D-glucitol 6-(E)-vinylhomophosphonate complex demands a revision of the enzyme mechanism. *J. Biol. Chem.* **279:** 13889–13895.

Jin, X.S., and Geiger, J.H., 2003, Structures of NAD(+)- and NADH-bound 1-L-myo-inositol 1-phosphate synthase. *Acta Crystallogr Section D-Biol. Crystallogr.* **59:** 1154–1164.

Kiely, D.E., and Sherman, W.R., 1975, Chemical model for cyclization step in biosynthesis of L-myo-inositol 1-phosphate. *J. Am. Chem. Soc.* **97:** 6810–6814.

Kniewel, R., Buglino, J.A., Shen, V., Chadya, T., Beckwith, A., and Lima, C.D., 2002, Structural analysis of *Saccharomyces cerevisiae* myo-inositol phosphate synthase. *J. Struct. Funct. Genomics* **2:** 129–134.

Lee, J.K., and Houk, K.N., 1997, A proficient enzyme revisited: The predicted mechanism for orotidine monophosphate decarboxylase. *Science* **276:** 942–945.

Loewus, F.A., 1990, Inositol biosynthesis. In: Morre, D.J., Boss, W.F., and Loewus, F.A. (eds.), Inositol Metabolism in Plants. Wiley-Liss Inc., New York, NY, pp. 13–19.

Loewus, F.A., and Loewus, M.W., 1983, Myo-inositol: Its biosynthesis and metabolism. *Annu. Rev. Plant Physiol.* **34:** 137–161.

Loewus, F.A., and Kelly, S., 1962, Conversion of glucose to inositol in parsley leaves. *Biochem. Biophys. Res. Commun.* **7:** 204–208.

Loewus, F.A., and Murthy, P.P.N., 2000, *Myo*-inositol metabolism in plants. *Plant Sci.* **150:** 1–19.

Loewus, M.W., 1977, Hydrogen isotope effects in the cyclization of D-glucose 6-phosphate by myo-inositol-1-phosphate synthase. *J. Biol. Chem.* **252:** 7221–7223.

Loewus, M.W., Loewus, F.A., Brillinger, G.U., Otsuka, H., and Floss, H.G., 1980, Stereochemistry of the myo-inositol-1-phosphate synthase reaction. *J. Biol. Chem.* **255:** 11710–11712.

Maeda, T., and Eisenberg, F., Jr., 1980, Purification, structure, and catalytic properties of L-myo-inositol-1-phosphate synthase from rat testis. *J. Biol. Chem.* **255:** 8458–8464.

Majumder, A.L., Chatterjee, A., Dastidar, K.G., and Majee, M., 2003, Diversification and evolution of L-myo-inositol 1-phosphate synthase. *FEBS Lett.* **553:** 3–10.

Majumder, A.L., Johnson, M.D., and Henry, S.A., 1997, 1L-myo-inositol-1-phosphate synthase. *Biochim. Biophys. Acta.* **1348:** 245–256.

Mauck, L.A., Wong, Y.H., and Sherman, W.R., 1980, L-myo-Inositol-1-phosphate synthase from bovine testis: Purification to homogeneity and partial characterization. *Biochemistry* **19:** 3623–3629.

Migaud, M.E., and Frost, J.W., 1995, Inhibition of myo-Inositol-1-phophate Synthase by a Reaction Coordinate Intermediate. *J. Am. Chem. Soc.* **117:** 5154–5155.

Migaud, M.E., and Frost, J.W., 1996, Elaboration of a general strategy for inhibition of myo-inositol 1-phosphate synthase: Active site interactions of analogues possessing oxidized reaction centers. *J. Am. Chem. Soc.* **118:** 495–501.

Norman, R.A., Mcalister, M.S.B., Murray-Rust, J, Movahedzadeh, F, Stoker, N.G., and Mcdonald, N. Q., 2002, Crystal structure of Inositol 1-Phosphate Synthase from *Mycobacterium tuberculosis*, a key enzyme in phophatidylinositol synthesis. *Structure* **10:** 393–402.

Obendorf, R.L., 1997, Oligosaccharides and galactosyl cyclitols in seed desiccation tolerance. *Seed Sci. Res.* **7:** 63–74.

Ogunyemi, E.O., Pittner, F., and Hoffmannostenhof, O., 1978, Studies on biosynthesis of cyclitols 36. Purification of *myo*-inositol-1-phosphate synthase of duckweed, Lemna-Gibba, to homogeneity by affinity chromatography on nad-sepharose molecular and catalytic properties of enzyme. *Hoppe-Seylers Zeitschrift Fur Physiologische Chemie* **359:** 613–616.

Parthasarathy, R., and Eisenberg, F., 1986, The inositol phospholipids – a stereochemical view of biological-activity. *Biochem. J.* **235:** 313–322.

Peng, J. (1996) Mechanistic Studies and Inhibition of Dehydroquinate Synthase and Myo-Inositol 1-Phosphate Synthase. Ph.D. Thesis, Michigan State University.

Peterbauer, T., Puschenreiter, M., and Richter, A., 1998a, Metabolism of galactosylononitol in seeds of Vigna umbellata. *Plant Cell Physiol.* **39:** 334–341.

Peterbauer, T., and Richter, A., 1998b, Galactosylononitol and stachyose synthesis in seeds of adzuki bean – purification and characterization of stachyose synthase. *Plant Physiol.* **117:** 165–172.

Pittner, F., and Hoffmann Ostenhof, O., 1976, Studies on the biosynthesis of cyclitols, XXXV. On the mechanism of action of myo-inositol-1-phosphate synthase from rat testicles. *Hoppe Seylers Z. Physiol. Chem.* **357:** 1667–1671.

Pittner, F., and Hoffmann Ostenhof, O., 1978, Studies on the biosynthesis of cyclitols, XXXVII. On mechanism and function of Schiff's base formation as an intermediary reaction step of myo-inositol-1-phosphate synthase from rat testicles. *Hoppe Seylers Z. Physiol. Chem.* **359:** 1395–1400.

Raychaudhuri, A., Hait, N.C., Dasgupta, S., Bhaduri, T. J., Deb, R., and Majumder, A.L., 1997, L-myo-inositol 1-phosphate synthase from plant sources – characteristics of the chloroplastic and cytosolic enzymes. *Plant Physiol.* **115:** 727–736.

Schmidt, D.E., and Westheim, F., 1971, Pk of lysine amino group at active site of acetoacetate decarboxylase. *Biochemistry* **10:** 1249–1253.

Sherman, W.R., Rasheed, A., Mauck, L.A., and Wiecko, J., 1977, Incubations of testis *myo*-inositol-1-phosphate synthase with D-(5-18O)glucose 6-phosphate and with H218O show no evidence of Schiff base formation. *J. Biol. Chem.* **252:** 5672–5676.

Stein, A.J., and Geiger, J.H., 2000, Structural studies of MIP synthase. *Acta Crystallogr. D* **56:** 348–350.

Stein, A.J., and Geiger, J.H., 2002, The crystal structure and mechanism of 1-L-*myo*-inositol-1-phosphate synthase. *J. Biol. Chem.* **277:** 9484–9491.

Stieglitz, K., Honging, Y., Roberts, M.F., and Stec, B., 2004, Reaching for mechanistic consensus across life kingdoms: Structure and insights into catalysis of the inositol-1-phosphate synthase (mIPS) from *Archeoglobus fulgidus*. *Biochemistry*, **44:** 213–224.

Widlanski, T., Bender, S.L., and Knowles, J.R., 1989, Dehydroquinate synthase – a sheep in wolfs clothing. *J. Am. Chem. Soc.* **111:** 2299–2300.

Wong, Y.H., Mauck, L.A., and Sherman, W.R., 1982, L-*Myo*-inositol-1-phosphate synthase from bovine testis. *Methods Enzymol.* **90:** 309–314.

Wong, Y.H.H., and Sherman, W.R., 1985, Anomeric and other substrate-specificity studies with myo-inositol-1-P synthase. *J. Biol. Chem.* **260:** 1083–1090.

Chapter 8

Phosphoinositide Metabolism: Towards an Understanding of Subcellular Signaling

Wendy F. Boss, Amanda J. Davis, Yang Ju Im, Rafaelo M. Galvão, and Imara Y. Perera
Department of Plant Biology, North Carolina State University, Raleigh, NC 27695-7612, USA

1. INTRODUCTION

The polyphosphorylated inositol phospholipids have multiple effects on cellular metabolism including regulating cytoskeletal structure, membrane associated enzymes, ion channels and pumps, vesicle trafficking, and producing second messengers (Cockcroft and De Matteis, 2001; Laude and Prior, 2004; Roth, 1999; Roth, 2004; Simonsen *et al.*, 2001; Takenawa and Itoh, 2001; Yin and Janmey, 2002). These diverse functions emphasize the importance of understanding both the spatial and the temporal regulation of phosphoinositide (PI) metabolism and the need to characterize the subcellular pools (Laude and Prior, 2004; Roth, 2004; Sprong *et al.*, 2001; Yin and Janmey, 2002).

Phosphatidylinositol-4,5-bisphosphate (PtdIns(4,5)P_2) pools can be regulated by PI kinases, PtdIns(4,5)P_2-phosphatases, phospholipase C (PLC), and lipid transfer and binding proteins. The contribution of each to PtdIns(4,5)P_2 pools will depend on the metabolic status of the cell. Enzymes involved in PtdIns(4,5)P_2 biosynthesis are shown in Figure 1. Our goal is to convince the reader of the importance of characterizing the metabolic fluxes within the discrete subcellular phospholipid microdomains that make up the lipid signaling pools.

The chemistry of the lipid head group helps to explain the ubiquitous nature of these polar lipids. Sugar phosphates are the most fundamental structures in biology. They provide the backbone of DNA and RNA and are the building

A. Lahiri Majumder and B. B. Biswas (eds.), Biology of Inositols and Phosphoinositides, 181–204
© 2006 *Springer. Printed in the Netherlands.*

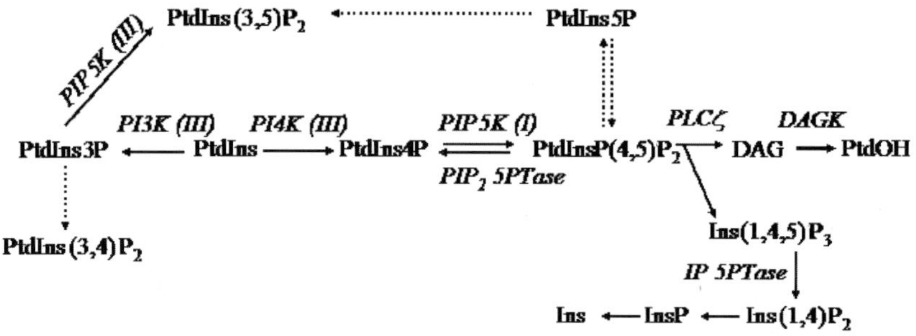

Figure 1. A summary of the PI pathway in plants. The enzymes catalyzing each step are shown in italics. Dashed arrows indicate reactions that have not been reported in plants. *Abbreviations*: DAG, diacylglycerol; DAGK, diacylglycerol kinase; Ins, inositol; InsP, inositol monophosphate; Ins(1,4)P_2, inositol (1,4,) bisphosphate; Ins(1,4,5)P_3, inositol (1,4,5) trisphosphate; IP 5PTase, inositol phosphate 5-phosphatase; PtdIns, phosphatidylinositol; PtdIns3P, phosphatidylinositol-3-phosphate; PtdIns4P, phosphatidylinositol-4-phosphate; PtdIns5P, phosphatidylinositol-5-phosphate; PtdIns(4,5)P_2, phosphatidylinositol-4,5-bisphosphate; PtdIns(3,4)P_2, phosphatidylinositol-3, 4-bisphosphate; PtdIns(3,5)P_2, phosphatidylinositol-3,5-bisphosphate; PI4K, phosphatidylinositol 4-kinase; PIP5K, phosphatidylinositol phosphate 5-kinase; PIP_2 5PTase, phosphatidylinositol(4,5) bisphosphate 5-phosphatase; PLC, phospholipase C.

blocks for carbon metabolism and energy production by the cell. It is likely that glycolipids were part of the original matrix coating RNA as the first cells were formed, but to our knowledge, this remains to be tested (Chen and Szostak, 2004; Hanakahi *et al.*, 2000; Hanczyc *et al.*, 2003). Notably, as organisms evolved, the core proteins that synthesized sugar phosphates became fundamental for life, and inositol phospholipids became a part of the lipid bilayer. In addition, the inositol phosphate head group provides a means for rapidly changing the membrane surface charge. The impact of changing the phosphorylation status of proteins on the biology of the cell is well known. Edmond Fischer and Edwin Krebs were acknowledged in 1992 for their seminal work in this area. However, the impacts of altering lipid phosphorylation on fundamental cellular processes are only beginning to be appreciated (Loewen *et al.*, 2004).

The potential to convey explicit information through the multiple phosphorylation sites of the inositol ring validates the conservation of the PI signaling pathway by biological systems. Furthermore, the stereospecificity of the five hydroxyl groups in the inositol phospholipids that extend from the surface of the bilayer conveys structural identity, which will impact multiple interacting molecules such as metabolites, proteins, and perhaps even nucleic acids. Discrete lipid microdomains are defined by the lipid, protein, and/or nucleic acid scaffold surrounding them. These regions may be part of the bilayer, they may bridge the cytoskeleton/bilayer interface, or they may be scaffolds of molecules that are mobilized within the cell.

It is becoming increasingly clear that the PI lipids are organized within discrete subcellular microdomains. At this stage of our knowledge, however, it is easier to describe microdomains in terms of function since the structures defining them are poorly characterized. We will use the term plasma membrane signaling pool to describe the microdomains of PtdIns(4,5)P_2 associated with the plasma membrane which are rapidly hydrolyzed in response to a stimulus. Our assumption is that some, but not all, plasma membrane PtdIns(4,5)P_2 microdomains are a part of the same signaling pool. A corollary to this is that different stimuli will activate different signaling pools of PtdIns(4,5)P_2 and thereby mediate different stimulus–response pathways.

The ability to sense and respond to the environment is essential for the survival of all living organisms. The PI pathway is involved in sensing environmental stimuli and is one of the most conserved signaling pathways. Components of the PI pathway are present in both prokaryotes and eukaryotes. For example, membranes from the hyperthermophilic archaeon *Pyrococcus furiosus* will phosphorylate phosphatidylinositol (PtdIns) to form PtdIns3P (Figure 2).

Comparative analysis of the enzymes involved in the PI pathway suggests that as organisms evolved, the complexity of the pathway also increased and multiple families of enzymes were produced as shown in Table 1 (Cockcroft and De Matteis, 2001; Drøbak *et al.*, 1999; Mueller-Roeber and Pical, 2002; Roth, 2004; Stevenson *et al.*, 2000). This increase in diversity would increase the potential for generating discrete subcellular lipid domains.

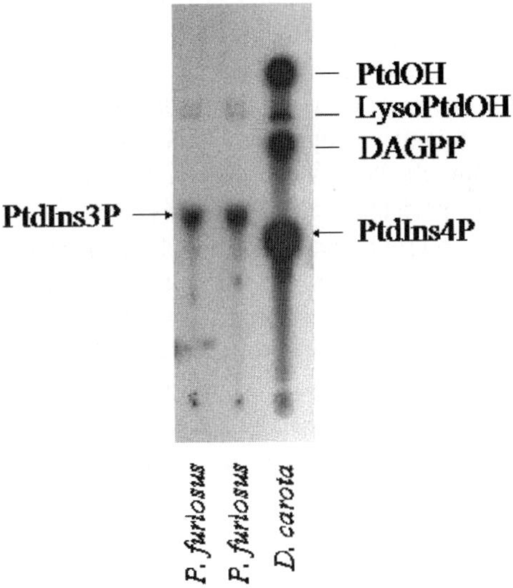

Figure 2. Membranes isolated from *Pyrococcus furiosus* will phosphorylate PtdIns to form PtdIns3P. Microsomal membranes were isolated from *Daucus carota* (wild carrot cells) grown in cell culture and *P. furiosus*. Lipid kinase activity was assayed as previously described using PtdIns as a substrate (Bunney *et al.*, 2000). The lipids were extracted and separated by thin layer chromatography as described by Walsh *et al.* (1991) to separate PtdIns3P and PtdIns4P. The migrations of the lipid standards are indicated (Brglez and Boss, unpublished results).

Table 1. A comparison of the gene families for the major PI pathway enzymes in humans and plants

Gene	Families in humans	Families in plants
PI3K	Type I, II, and III	Type III only
PI4K	Type II and III	Type III and possibly II
PIPK	Type I, II, and III	Type I and III
PLC	β, γ, δ, ϵ, and ζ	ζ

2. LIPID–PROTEIN INTERACTIONS

Insights into PI-mediated signaling domains can be gained from understanding lipid–protein interaction. The inositol phospholipids have been shown to bind to proteins [ATPases, profilin, phospholipase D (PLD), dynamin, and patellin] and to alter their function. For example, PtdIns(4,5)P_2 is required for PLDγ1 and β1 function and stimulates the activity of PLDα1 (Pappan and Wang, 1999; Wang, 1999; Wang, 2001). In a dynamic system such as a living cell this means that if PtdIns(4,5)P_2 is hydrolyzed in response to a stimulus, there would be a decrease in the specific activity of the PLD within the PtdIns(4,5)P_2 microdomain until PtdIns(4,5)P_2 is resynthesized. In this manner, an oscillation of the activity of selected PLD isoforms would be produced in response to sequential changes in PLC activity and PtdInsP kinase activity. This complicates the life of the researcher trying to interpret data on lipid-mediated signaling and emphasizes the need to include time course studies and to simultaneously monitor the suite of lipids and lipid signaling pathways (Li *et al.*, 2004; Meijer and Munnik, 2003; Van Leeuwen *et al.*, 2004; Welti *et al.*, 2002).

Other PI binding proteins include profilins. Profilins are low molecular weight cytoskeletal binding proteins that bind PtdIns(4,5)P_2 and other inositol lipids. Maize profilins have been classified into two groups (Gibbon *et al.*, 1998; Kovar *et al.*, 2000). They are thought to competitively bind actin and regulate actin assembly at different locations within the cell. The class I profilins are more abundant and are proposed to be low affinity G-actin buffers. The class II profilins are in lower abundance but have a higher affinity for G-actin. Both classes of profilins bind PtdIns(4,5)P_2; however, the class I profilins bind with a higher affinity (Kovar *et al.*, 2000). Studies of knockout and T-DNA insertion lines have yet to clearly delineate functions of the profilin isoforms. This is most likely because of the pleotropic effects of the profilins on actin cytoskeleton, vesicle trafficking and signaling, all of which may be mediated through PI pathway intermediates (Drøbak *et al.*, 2004; Huang *et al.*, 2003; Staiger *et al.*, 1997). These studies emphasize the need to understand binding affinities and the *in vivo* dynamics of lipid–protein interactions.

Patellin is a newly identified PI binding protein. Patellin 1 (PATL1) is one of a family of six Arabidopsis proteins containing a Sec14 lipid binding domain and a Golgi dynamics domain (GOLD). The GOLD domain is found in proteins involved in vesicle trafficking and secretion (Anantharaman and Aravind, 2002). The PATL1 Sec14 domain is homologous to the Sec14p domain in yeast, which binds PtdIns and phosphatidylcholine (Bankaitis *et al.*, 1989). However, biochemical characterization of PATL1 indicates that it binds with highest specificity to PtdIns5P and PtdIns(4,5)P_2. Furthermore, it was shown that PATL1 localized to the cell plate during cytokinesis. The domain structure of PATL1 and its lipid binding properties indicate a role for PATL1 in membrane trafficking to the expanding cell plate and reveal a role for PIs in cell plate formation (Peterman *et al.*, 2004). It will be interesting to determine what role, if any, previously identified lipid transfer proteins play in this process (Kapranov *et al.*, 2001; Monks *et al.*, 2001). Further study is needed to determine more precisely the function of the protein–lipid interactions in these processes.

Another family of lipid binding proteins involved in vesicle trafficking is dynamins. Dynamins are a family of large molecular weight (60–110 kDa) GTP binding proteins involved in membrane trafficking to and from various compartments, cell plate formation, and mitochondrial and chloroplast division. Although dynamins are found in both plants and animals, there are distinct differences between them. All the animal dynamins contain a pleckstrin homology (PH) lipid-binding domain for targeting the protein to the membrane. The PH domain consists of approximately 100 amino acids that bind with varied specificity to inositol phospholipids (Lemmon, 2003; Lemmon *et al.*, 1996). Of the 16 dynamins in Arabidopsis, only two (DRP2A and DRP2B) contain a PH domain (Hong *et al.*, 2003). The most extensively characterized of the two, DRP2A, has been localized to the trans Golgi network and has been shown to be involved in trafficking to the vacuole (Jin *et al.*, 2001). The PH domain in DRP2A (previously ADL6) has a unique structure with three regions of 16–24 amino acids inserted between the characteristic β sheets that make up the PH domain. Lipid binding studies of the full length DRP2A protein shows that it binds strongly to PtdIns3P with minimal binding to PtdIns4P. In contrast, the PH domain alone binds PtdIns3P and PtdIns4P equally. Further studies indicated that the lipid binding specificity of the DRP2A PH domain was determined by interactions of the PH domain and the C-terminus of DRP2A suggesting that this is an important site for lipid–protein regulation *in vivo* (Lee *et al.*, 2002). Further investigation is needed into the exact function of DRP2A in membrane trafficking and how the PH domain and PtdIns3P contribute to its function.

The specific binding of a PH domain to lipids depends not only on the structure of the parent protein but also on the protein environment surrounding the membrane lipids it will bind. Studies using GFP-PH domains to identify lipid microdomains have shown that some lipids are inaccessible to PH domain binding and emphasize the need to characterize the binding specificity of the PH domain prior to engaging in *in vivo* studies (Balla *et al.*, 2000; Balla and Varnai, 2002).

3. SUBCELLULAR LIPID DOMAINS

The proteins and other molecules that bind to the inositol lipids will affect their turnover rate and the extent of their impact on cellular metabolism. If so many cytoskeletal proteins bind to the inositol lipids, does this mean that the lipids are associated with the cytoskeleton? PtdIns 3-kinase activity is cytoskeletal associated (Dove *et al.*, 1994) and at least one form of PtdIns 4-kinase, (the type III PtdIns 4-kinase) is cytoskeletal associated (Stevenson *et al.*, 1998).

A*t*PI4Kα1 (Figure 3) is probably the major enzyme contributing to the previously reported F-actin associated PtdIns 4-kinase activity (Tan and Boss, 1992; Xu *et al.*, 1992). While it was also reported that PIP5K activity was associated with the actin-enriched fraction from plants, the specific isoform of the enzyme has not been identified (Tan and Boss, 1992). As more specific antibodies and molecular probes become available, it will be important to monitor lipid protein interaction *in vivo* and *in vitro*.

*At*PI4Kα1 is a low abundance protein and is difficult to detect on immunoblots of membranes from Arabidopsis or *Nicotina tobaccum*. When

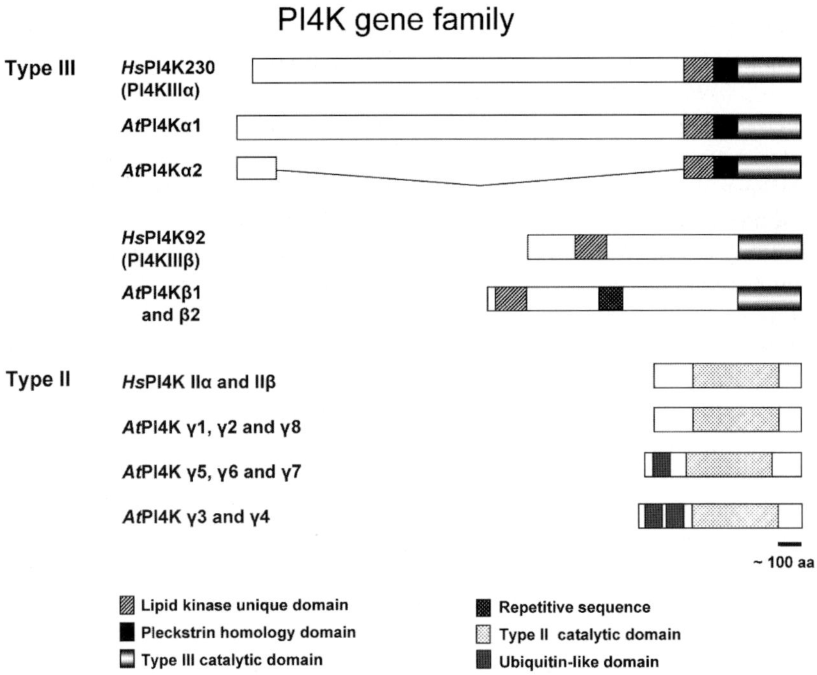

Figure 3. Linear representations the *Arabidopsis thaliana* (*At*) PI4Ks in comparison to human (*Hs*) isoforms. Thus far, only the type III *At*PI4Ks have been shown to be functional enzymes. Conserved domains are indicated and described in the text.

expressed in a heterologous system, however, the protein is readily detectable and has threefold higher specific activity than *At*PI4Kβ1 the other characterized PtdIns 4-kinase isoform (Stevenson-Paulik *et al.*, 2003). Of interest, *At*PI4Kβ1 has a charged repeat domain in the N-terminus that has been proposed to be important for membrane association (Xue *et al.*, 1999). Based on *in vitro* data and heterologous expression in the eukaryotic *Spodoptera frugiperda* (Sf9) insect cells, it is not evident which membrane, if any, is the predominate location. It is possible, that like the yeast and human orthologs, *At*PI4Kβ1 will be involved in membrane secretion from the Golgi apparatus and could potentially supply the plasma membrane PtdIns4P (Stevenson-Paulik *et al.*, 2003; Xue *et al.*, 1999). Like their human orthologs, both *At*PI4Kα1 and *At*PI4Kβ1 can be phosphorylated *in vitro* (Suer *et al.*, 2001; Zhao *et al.*, 2000) (Stevenson-Paulik and Galvão, unpublished results). It will be interesting to see which peptide domains and posttranslational modifications regulate the intracellular distribution of the 4-kinases.

A unique feature of the plant type III PI4Kα is the PH domain that binds to PtdIns4P. Two other PH domains that selectively bind PtdIns4P are the oxysterol binding protein, OSBP, and FAPP1 (phosphatidylinositol-*f*our-phosphate *a*da*p*tor *p*rotein-1) (Dowler *et al.*, 2000; Levine and Munro, 2002). One Arabidopsis PH domain protein, *At*PH1, isolated as a homolog of pleckstrin has been reported (Mikami *et al.*, 1999); however, it binds PtdIns3P and not PtdIns4P (Dowler *et al.*, 2000). The first question asked about the *At*PI4Kα1PH domain was what effect, if any, did it have on enzyme activity? It turns out that *At*PI4Kα1 activity was differentially sensitive to PtdIns4P, the product of the reaction. The specific activity of *At*PI4Kα1 was inhibited 70% by 0.5 mM PtdIns4P in the presence of 1:1 concentration of substrate, PtdIns. The effect of PtdIns4P was not simply due to charge, as AtPI4Kα1 activity was stimulated approximately 50% by equal concentrations of the other negatively charged lipids, PtdIns3P, phosphatidic acid, and phosphatidylserine (Stevenson-Paulik *et al.*, 2003). In contrast, the specific activity of AtPI4Kβ1, which does not have a PH domain, was stimulated twofold by PtdIns4P but not other negatively charged lipids and was inhibited 50% by phosphatidylcholine. Expression of *At*PI4Kα1 without the PH domain compromised PtdIns 4-kinase activity; decreased association with fine actin filaments *in vitro*, and resulted in mis-localization of the kinase *in vivo* in the insect cells. These studies support the idea that the two Arabidopsis type III PI4Kα1 and PI4Kβ1 are responsible for distinct PI pools.

The fact that *At*PI4Kα1 activity is inhibited by PtdIns4P and inhibition is relieved by adding rPH domain suggests a mechanism whereby other PtdIns4P-binding proteins such as profilin (Drøbak *et al.*, 1994), dynamin-like proteins (Kim *et al.*, 2001b), actin depolymerizing factor (Gungabissoon *et al.*, 1998), phospholipid transfer proteins (Cockcroft, 1996; Cockcroft and De Matteis, 2001; Monks *et al.*, 2001), and potentially PIP5K should affect PI4Kα1 activity or subcellular distribution and illustrates the importance of studying specific microdomains within the cell. In addition, the PH domain may be involved in the previously characterized activation of PI4K by eEF1A, a translational elongation factor and actin binding and bundling protein (Yang and Boss, 1994a). eEF1A was shown to bind PLCδ PH domain and increase PLC activity (Chang *et al.*, 2002). The potential role of eEF1A in regulating PI4K, PLC, and PIP5K (Davis *et al.*, unpublished results), in additional to its function in actin bundling, provides a mechanism where the status of the inositol lipids, cytoskeletal structure, and protein translation would be coordinately regulated.

In addition to the cytoskeleton, the nucleus is an area where PI metabolism is prevalent. An important and often overlooked observation is that the plant type III PtdIns 3-kinase activity is present in nuclei and immunodetection of the enzyme indicates co-localization with putative transcription initiation sites (Bunney *et al.*, 2000). Also, PtdIns 3-kinase activity and transcript increase significantly during root nodulation (Hernandez *et al.*, 2004; Hong and Verma, 1994) consistent with a role in membrane trafficking and/or transcriptional regulation. As previously reported (Bunney *et al.*, 2000; Drøbak, 1992; Mueller-Roeber and Pical, 2002; Stevenson *et al.*, 2000), the only family of PtdIns 3-kinases found in plants is the type III Vps34-like family that is associated with vacuolar trafficking in yeast (De Camilli *et al.*, 1996; Herman and Emr, 1990; Simonsen *et al.*, 2001) and may have a similar function in plants (Kim *et al.*, 2001a; Matsuoka *et al.*, 1995). Like *S. cerevisiae*, terrestrial plants do not have the plasma membrane type I and II PtdIns 3-kinases associated with PtdIns(3,4,5)P_3 biosynthesis and with much of the human signaling. It should be noted, however, that in a heterologous system, Arabidopsis PIP5K1 will produce PtdIns(3,4,5)P_3 by phosphorylating PtdIns(3,4)P_2 (Elge *et al.*, 2001).

Early studies indicated that nuclei isolated from protoplasts have a high percentage of [^{3}H]inositol PtdIns(4,5)P_2 relative to the total [^{3}H]inositol phospholipids (Hendrix *et al.*, 1989). In addition, histones cause an 80-fold increase in microsomal PtdIns 4-kinase (Yang and Boss, 1994b) activity suggesting that PtdIns4P and PtdIns(4,5)P_2 along with PtdIns3P (Bunney *et al.*, 2000) may be important components of the nuclear matrix or nuclear membrane. Hydrolysis of the inositol phospholipids by PLC and the formation of diacylglycerol would facilitate membrane fusion and could be an integral part of nuclear membrane formation during karyokinesis. The presence of PtdIns(4,5)P_2 in the nucleus is also consistent with a role for Ins(1,4,5)P_3 derived Ins(1,4,5,6) P_4-mediated transport of RNA from the nucleus (Odom *et al.*, 2000). Whether the lipids or inositol phosphates would interact directly or indirectly with RNA is not known. Analysis of inositol lipid or inositol phosphate binding to RNA is virtually uninvestigated in plants. The role of the PI pathway in plant nucleus is far from resolved and provides a rich area for future studies (Irvine, 2003).

One caveat of the studies of nuclei is that the ER is closely associated with the nuclear envelope. Lipid kinase activity has been associated with ER and the role of the inositol lipids in membrane trafficking is well accepted (Cockcroft and De Matteis, 2001; Godi *et al.*, 2004; Roth, 2004; van Meer and Sprong,

2004). The early biochemical studies of the lipid kinases were done prior to having the molecular and biochemical tools that we have today. Future *in vivo* studies of the enzymes and lipids should help to sort out the functional biology. The genomics and biochemistry of the enzymes involved in inositol lipid metabolism have been reviewed in detail quite recently (Drøbak, 1992; Mueller-Roeber and Pical, 2002; Van Leeuwen *et al.*, 2004).

The question of lipid trafficking in plants has been addressed thus far primarily with inhibitor studies and the expression of lipid binding peptides. These studies have led to intriguing results consistent with a role of PtdIns 3-kinase in trafficking to the vacuole (Kim *et al.*, 2001a; Matsuoka *et al.*, 1995). Future studies should focus on identifying the scaffolds of interacting proteins in order to understand more precisely the mechanisms involved. The role of lipid transfer proteins in plants is still a mystery. While orthologs of the yeast phosphatidylinositol lipid transfer proteins have been identified, their precise function in regulating membrane lipid distribution is not clear (Drøbak *et al.*, 1999; Kapranov *et al.*, 2001; Monks *et al.*, 2001). Knockouts of one isoform of these lipid transfer proteins lead to aberrant root hair morphology, but otherwise normal plant growth suggesting that there is functional redundancy within this multigene family that helps to maintain a normal phenotype (Vincent *et al.*, 2005). More complete biochemical and genetic characterization of the family of the putative lipid transfer proteins will reveal specific functions.

4. REGULATION OF PHOSPHATIDYLINOSITOL-4,5-BISPHOSPHATE POOLS

Regulation of PtdIns (4,5) P_2 biosynthesis begins with the families of PI4Ks. The properties of *At*PI4Kα1 and *At*PI4Kβ1 are described above and more extensively in recent review (Mueller-Roeber and Pical, 2002). These PI4Ks are orthologs of the yeast and mammalian Stt4p/PI4Kα and Pik1p/PI4Kβ, respectively. Although two other type III PI4Ks, *At*PI4Kα2 and β2 can be identified in the genomic database, there are no ESTs reported for these two isoforms and attempts to recover a full length cDNA using PCR have been unsuccessful (Galvão, Stevenson-Paulik *et al.*, unpublished results) suggesting that these may not encode functional enzymes. Curiously, there are several 60–70-kDa polypeptides detected with antibodies raised against the *At*PI4K α1 that were proposed to be encoded by *At*PI4Kα2 (Shank *et al.*, 2001; Stevenson-Paulik *et al.*, 2003), but the lack of ESTs and the failure to obtain PCR products suggest that these bands either represent proteolytic products of *At*PI4Kα1 or β1 or are alternative splice variants encoded by these transcripts. While it is possible that *At*PI4Kα2 and *At*PI4Kβ2 transcripts are expressed in specific cell types under developmental control, they must be in relatively low abundance as, to our knowledge, they have not been detected in any of the publicly available transcript array databases.

Another family of PI4Ks that provides further opportunity for study is the type II family. Mammalian and yeast type II PI4Ks were recently cloned and characterized and are thought to function to supply PtdIns4P to the plasma membrane signaling pool (Balla *et al.*, 2002; Minogue *et al.*, 2001; Nishikawa *et al.*, 1998; Wei *et al.*, 2002). There are eight putative type II PI4Ks sequences in the *Arabidopsis* genome (Mueller-Roeber and Pical, 2002); however, to date, none of them have been shown to encode functional lipid kinases. At the amino acid sequence level, the lipid kinase domain from the putative plant type II PI4Ks are 26–30 % identical to the conserved type II lipid kinase domains from either mammalian or yeast enzymes. However, unlike the yeast or mammalian homologs, five of the plant proteins contain N-terminal *ub*iquitin-*l*ike (UBL) domain. Most of UBL domain-containing proteins share the ability to interact with 26S proteasome but their functions vary depending on the other domains present (Eskildsen *et al.*, 2003; Hartmann-Petersen *et al.*, 2003). If these plant proteins are functional lipid kinases, there may be connections between the controlled protein degradation and the PI pathway, that is, they might be regulated by ubiquitin/26S proteasome. Alternatively, the plant type II PI4Ks may not have lipid kinase activity, but rather be like the target of rapamycin (TOR) protein kinases that contain a lipid kinase-like domain but lack lipid kinase activity (Harris and Lawrence, 2003; Helliwell *et al.*, 1994).

Two intriguing conundrums in plant PI signaling are, which PI4K isoform(s) supplies the plasma membrane pool of PtdIns4P and how is the plasma membrane PtdIns(4,5)P_2 signaling pool regulated? Answers to these questions are integral to understanding the dynamics of the plasma membrane PI signaling pathway. Because PI4Kβ1 is product stimulated and because the relative ratio of PtdIns4P to PtdIns(4,5)P_2 is high in plants, PI4Kβ1 might be the major source of PtdIns4P. It remains to be seen whether PI4Kβ1 is trafficked to the plasma membrane and supplies the plasma membrane signaling pool or if that pool is determined by the newly identified family of putative PI4Ks, the type II enzymes.

Although several labs have shown that whole cell PtdIns(4,5)P_2 is rapidly turning over (Munnik *et al.*, 1998), the plasma membrane signaling pool is poorly characterized. A profile of the Arabidopsis PIPKs and PLC is given in Figure 4.

Evidence suggests that under certain physiological conditions, the plasma membrane signaling pool can become limiting. For example, if cells are growing in nutrient depleted medium, the plasma membrane PIP kinase activity increases prior to the production of Ins(1,4,5)P_3 in response to stimulation (Heilmann *et al.*, 2001). An interesting characteristic of plants is that the ratio of PtdIns4P to PtdIns(4,5)P_2 is on the order of 10 or 20 to 1 (Boss, 1989; Gross and Boss, 1993; Meijer and Munnik, 2003; Mueller-Roeber and Pical, 2002; Stevenson *et al.*, 2000), compared to values of 1:1 or 1:2 found in the most reactive animal cells such as brain and iris muscle cells. These observations led many to suggest that PIPKs limit the production of PtdIns(4,5)P_2 and the flux

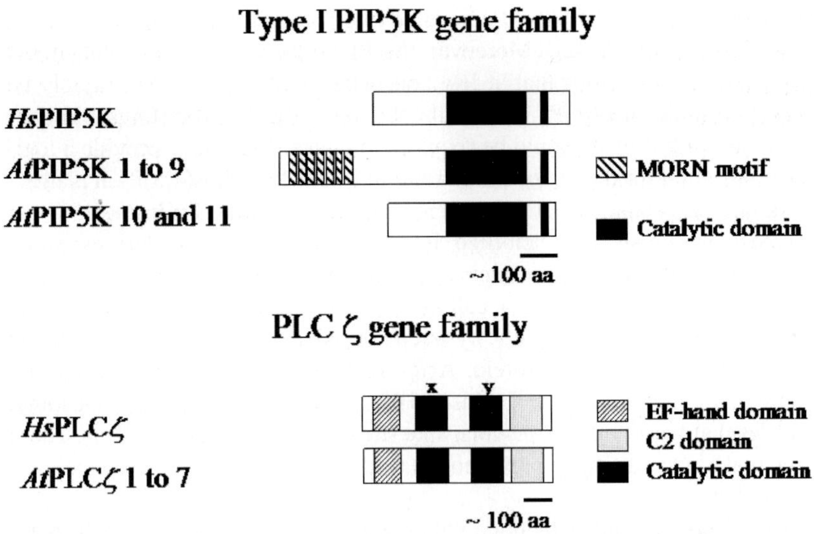

Figure 4. Linear representation of the Arabidopsis PIPKs and PLCs. The Arabidopsis PtdInsP 5-kinases are most similar to the human type I PtdInsP 5-kinases. There are 11 putative type I *At*PtdInsP 5-kinases in Arabidopsis arranged in two subfamilies based on size. Subfamily B contains *At*PIPK1-9, all of which contain membrane occupation and recognition nexus (MORN) repeats. *At*PIPK10-11 are in Subfamily A with molecular weights less than that of the members of subfamily B and contain no MORN repeats. The Arabidopsis phosphoinositide specific phospholipase C family is most similar to the animal PLCζ. There are seven functional PI-PLCs in Arabidopsis (Hunt *et al.*, 2004). All isoforms contain EF-hand motifs, the X and Y catalytic domains characteristic of PI-PLCs and a C2 lipid-binding domain.

through the plant PI pathway. Evidence supporting this hypothesis was reported recently by Perera *et al.* (2002) who showed that when the human inositol polyphosphate 5-phosphatase (InsP 5-ptase) was expressed in plants, there was increased Ins(1,4,5)P_3 turnover and a decrease in PtdIns(4,5)P_2. The human InsP 5-ptase is localized to the plasma membrane and specifically hydrolyzes Ins(1,4,5)P_3 and Ins(1,3,4,5)P_4 but not PtdIns(4,5)P_2. As a result of expressing the human InsP 5-ptase the rate of flux through the plasma membrane PI pool increased dramatically. Although the plasma membrane PtdInsP 5-kinase activity increased threefold, there was a decrease in net PtdIns(4,5)P_2 suggesting that synthesis was not keeping up with the hydrolysis. These data indicate that PtdIns(4,5)P_2 biosynthesis was limiting in this transgenic system; however, they do not explain the mechanism nor do they explain the differences in the inositol phospholipid ratios generally reported for plants and animals.

Insights into the reasons for the high ratio of PtdIns4P to PtdIns(4,5)P_2 in plants have come from comparative biochemical studies of recombinant proteins. The *Km* of the human PIPKβ1 for PtdIns4P is close to 70-fold lower than that found for *At*PIPK1 [1 μM compared to 69 μM, respectively (Davis *et al.*, 2004; Kunz *et al.*, 2002)]. Moreover, the V_{max} of the Arabidopsis enzyme is sixfold less. This means that at low concentrations, PtdIns4P will rapidly be converted to PtdIns(4,5)P_2 in animal cells, but a much higher threshold concentration of PtdIns4P would be required in plants. These data provide a logical explanation for the relative high ratio of PtdIns4P:PtdIns(4,5)P_2 in plants.

In animal cells, selectively increasing intracellular PtdIns(4,5)P_2 can decrease apoptosis and stabilize intracellular membranes. For example, increasing PIP5K1α expression 2.6-fold inhibited caspase activity and decreased apoptosis in HeLa cells (Mejillano *et al.*, 2001). Furthermore, increasing PtdIns(4,5)P_2 either by overexpressing PIP5K or a constitutively active PIP5K-regulatory protein, Arf6-GTP, resulted in internalization and trapping of membranes in PtdIns(4,5)P_2-positive, actin-coated endosomal vesicles that fused to form vacuolar-like structures (Brown *et al.*, 2001). These data make a compelling argument that altering PIP5K can affect cellular metabolism.

Could similar studies be done with plant enzymes? Only 1 of the 14 putative plant PIPKs, *At*PIPK1, has been well characterized (Davis *et al.*, 2004; Elge *et al.*, 2001; Mueller-Roeber and Pical, 2002; Westergren *et al.*, 2001). *AtPIPK1* encodes a 752 amino acid protein with a predicted molecular weight of 86 kDa. Studies using promoter-Gus fusions indicate high levels of *AtPIPK1* expression in procambial tissue of leaves flowers and roots (Elge *et al.*, 2001). Using recombinant protein, Westergren *et al.* (2001) have shown that the enzyme activity is downregulated by protein phosphorylation and that the substrate preference was for PtdIns4P although PtdIns3P and PtdIns5P were also phosphorylated, but little else is known about the regulation of the plant PIPKs. The substrate specificity and enzyme activity of the human type I and II PIPKs are regulated by an activation loop (Anderson *et al.*, 1999; Doughman *et al.*, 2003; Kunz *et al.*, 2002). Interchanging the activation loops of PIPKs predictively alters the site of inositol phosphorylation converting a type I PtdIns4P 5-kinase into a type II PtdIns5P 4-kinase and vice versa (Kunz *et al.*, 2000). Furthermore, using site-directed mutagenesis, demonstrated that a conserved Glu (362) in the activation loop of the human type I PIPKβ is essential for efficient PtdIns4P 5-kinase function. Similarly, the activation loop of the human type IIβ enzymes has a conserved Ala (381) that is essential for PtdIns5P 4-kinase function (Kunz *et al.*, 2002). Because the activation loop plays such a critical role in the human PIPKs, it is possible that evolutionary changes in the activation loop of the plant PIPKs are responsible, in part, for the differences in the specific activities.

The Arabdopsis PtdInsP kinase isoforms, *At*PIPK1-11, all have the conserved Glu of the human type I activation loop suggesting that the plant proteins are type I PIPKs. Biochemical data are more equivocal, but also indicate that PtdIns4P is the preferred substrate for PIPK1 (Davis *et al.*, 2004; Elge *et al.*, 2001; Westergren *et al.*, 2001). In addition, biochemical data using recombinant *At*PIP5K10 indicates that it too is a type I PtdIns4P 5-kinase (Perera *et al.*, 2005). *At*PIPK10 and *At*PIPK11 are both members of subfamily A of PIPKs in contrast to *At*PIPK1-9, subfamily B (Mueller-Roeber and Pical, 2002). *At*PIPK10 (At4g01190) and *At*PIPK11 (At1g01460) both lack putative membrane occupation and recognition nexus (MORN) repeats in the N-terminus suggesting that they will reside in distinct subcellular locations. Transcript profiles have indicated that *At*PIPK11 is prevalent in pollen (Becker *et al.*, 2003) so they may also represent a tissue-specific isoform of the enzyme. Whether this is the isoform that is involved in Rac regulation of pollen tube growth remains to be seen (Kost *et al.*, 1999). Regulation of plant PIPKs by Rac is an important finding that identifies a signaling complex for generating PtdIns(4,5)P_2 microdomains and a mechanism for altering PIPK activity *in vivo*.

Based on predicted amino acid sequences and biochemical data there is no evidence of type II PIPKs in Arabidopsis. The type II PIPKs are PtdInsP 4-kinases that preferentially phosphorylate PtdIns5P. Putative type III kinases (*At*Fab1a-d) have been identified in the Arabidopsis genome. The predicted amino acid sequences of the *At*Fabs are similar to the human type III kinases and the yeast FAB enzymes that have a PtdIns3P binding domain (FYVE domain) and that phosphorylate PtdIns3P to form PtdIns(3,5)P_2. At least one isoform, *At*Fab1a, has been shown to be a functional PtdIns3P 5-kinase (Drøbak, unpublished results).

Although *in vitro* kinase assays have led to important insights into the regulation of the pathway, it is important to appreciate the potential limitations of these assays that are usually conducted under optimal conditions. Until we can characterize the enzymes within their subcellular microenvironments, we will not truly understand their impact. The effects of altering Rac expression on PIPK activity, actin cytoskeleton and pollen tube growth is one example (Kost *et al.*, 1999). Other examples of how microdomains can affect the product produced by endogenous enzymes comes from recent studies of yeast lacking PTEN, a protein that dephosphorylates PtdIns(4,5)P_2. The PTEN mutant produced PtdIns(3,4,5)P_3, a lipid not normally present in the cells (Mitra *et al.*, 2004). Presumably as a result of the cell not being able to dephosphorylate PtdIns(4,5)P_2 there was a sufficient increase in inositol lipids to favor additional phosphorylation. Elge *et al.* (2001) showed that when expressed in insect cells and given excess substrate *At*PIPK1 produced PtdIns(3,4,5)P_3. Whether this would actually happen under normal or stressed conditions might vary with different species (DeWald *et al.*, 2001; Dove *et al.*, 1997; Pical *et al.*, 1999). It is not difficult to imagine that alterations of PI flux during a sustained stress or as a result of genetic manipulation could result in PtdIns(3,4,5)P_3 production in plants.

Plasma membrane PtdIns(4,5)P_2 levels will be regulated both by synthesis and catabolism. Hydrolysis by C-type phospholipases is the best studied PtdIns(4,5)P_2 catabolic pathway (Cote and Crain, 1993; Hunt *et al.*, 2004). PLC activity in plants was first described by Sommarin's lab (Melin *et al.*, 1992; Melin *et al.*, 1987). Analysis of tissue-specific expression profiles and the calcium sensitivity of the seven active *Arabidopsis* isoforms are reported by Hunt *et al.* (2004). All the data thus far indicate that the plant PLCs are calcium stimulated and belong to the ζ family of enzymes (Hernandez-Sotomayor *et al.*, 1999; Hunt *et al.*, 2004). That is, they are calcium regulated but lack the PH domain of the δ family of PLCs (Mueller-Roeber and Pical, 2002). The PH domain of the PLC δ binds to PtdIns(4,5)P_2 and Ins(1,4,5)P_3 and in doing so facilitates the binding of the enzyme to the membrane lipids until sufficient Ins(1,4,5)P_3 is produced to displace it (Cifuentes *et al.*, 1994). In this manner PLC δ comes on and off the membrane in response to stimuli. It is not known what regulates the subcellular distribution of the plant PtdIns(4,5)P_2 PLCs. Most of the activity is membrane associated and localized increases in cytosolic Ca^{2+} could selectively activate the specific isoforms (Hunt *et al.*, 2004). The specific activities reported in early studies of isolated membranes were similar for plants and animals (Sandelius and Sommarin, 1990). If this proves to be true when purified recombinant proteins are compared, it would support the hypothesis that under normal conditions, plant PLCs do not limit the flux through the PI pathway.

It is important to remember that there are no G-protein regulated (β family) or tyrosine kinase-regulated (γ family) PLCs in Arabidopsis. Of all the enzymes in the plant PI pathway, the PLCs are the least complex and provide the best support for the notion that the plant PI pathway is not as highly evolved as the human brain PI signaling pathway. A corollary to this hypothesis is that the ζ family of PLCs should be the seminal isoforms from which the others evolved. Attempts have been made to alter the expression at least two of these *in planta* (Hunt *et al.*, 2003; Mills *et al.*, 2004; Sanchez and Chua, 2001). While the antisense plants had no reported guard cell phenotype (Sanchez and Chua, 2001), plants expressing the sense construct of PLC 1 and 2 under the control of a guard cell promoter had a wilted phenotype in response to water stress and had reduced, not increased levels, of PLC transcript and reduced levels of guard cell PLC activity. Hunt *et al.* (2003) suggested that the decreased response of the guard cells to water stress and abscisic acid (ABA) was consistent with the lack Ins(1,4,5)P_3-mediated signaling. Comparative analyses of pharmacological and genetic manipulations of PLC activity of guard cells have revealed the complexity of the response (Mills *et al.*, 2004). Until we know more about the subcellular localization of the proteins and how their activity is regulated, it is difficult to know what pleiotropic effects result from changes in PLC activity.

Another potential fate of the PtdIns(4,5)P_2 plasma membrane signaling pool is that it can be dephosphorylated by phosphatases. The InsP 5-ptase and the effects of altering gene expression have been described in Chapter 3. The family of 5PTases most likely to affect the plasma membrane signaling pool are the type II, which preferentially hydrolyze inositol phospholipid substrates. Mutations in these genes have been associated with several diseases in humans (Pendaries *et al.*, 2003; Tronchere *et al.*, 2003). There are at least nine putative inositol phospholipid PTases in Arabidopsis all of which contain a predicted suppressor of actin (SAC, Novick *et al.*, 1989) domain and are predicted to hydrolyze PtdInsPs (Zhong and Ye, 2003). Some mutations in these SAC domain-containing proteins affect cell wall biosynthesis and have been characterized as fragile fiber mutants (Zhong and Ye, 2003). The lack of some SAC-like proteins during development (Despres *et al.*, 2003) suggests an important role for PtdIns(4,5)P_2 during embryo development and is consistent with increased PtdInsP 5-kinase activity and PtdIns(4,5)P_2 in embryogenic cell cultures (Chen *et al.*, 1991; Ek-Ramos *et al.*, 2003; Wheeler and Boss, 1987). Mutations in tissue- and cell-specific isoforms of the lipid phosphatases that result in increased microdomains of PtdIns(4,5)P_2 will no doubt lead to important insights. One exciting example of cellular and subcellular specificity has been reported for another member of this family, SAC9 (Williams *et al.*, 2005). The SAC9 mutant has flaccid guard cells and is a dwarf plant suggesting that SAC9 is regulating the guard cell plasma membrane signaling pool of PtdIns(4,5)P_2. PtdIns(4,5)P_2 levels in SAC9 mutant plants increase 40-fold. Ins(1,4,5)P_3 levels also increase indicating a shunt of PtdIns(4,5)P_2 into the PLC signaling pool. While one might at first approximation assume that the increased PtdIns(4,5)P_2 and Ins(1,4,5)P_3 might distribute equally throughout the cell, apparently it does not as reflected in the differences in phenotype of the SAC9 mutant and the fragile fiber mutants. The flaccid stomata of the SAC9 mutant is the most intriguing evidence that altering the flux through the plasma membrane PtdIns(4,5)P_2 signaling pathways will affect growth and development. More complete analysis of the impact of altering PtdIns(4,5)P_2 5PTases and PLC on other compensatory pathways and measurements of metabolic flux will surely produce the insights needed to interpret these very intriguing results.

As illustrated by the examples above, the key to understanding the PI pathway is to comprehend the nature of the regulation of discrete subcellular microdomains of the lipids. Any changes in a specific plasma membrane signaling pool, whether a transient oscillation in PtdIns(4,5)P_2 or a more sustained change in rate of flux through the pathway in response to a stimulus, will result in changes in cytoskeletal structure, membrane enzyme activity, and pump or channel activity within the microdomains where the lipid resides.

Evidence for the rapid and transient production of Ins(1,4,5)P_3 comes from many studies beginning with those of Satter's group (Morse *et al.*, 1987). Once Rich Crain's group demonstrated the validity of using the commercially available brain binding assay for monitoring InsP(1,4,5)P_3 in plants, the field expanded rapidly (Cote and Crain, 1993; Cote *et al.*, 1996). Prior to this time, many were mislead by the large amounts of [^{3}H]inositol labeled metabolites that could be recovered from plants, many of which co-migrated with Ins(1,4,5)P_3 on the simple gravity-mediated ion exchange chromatography columns used. The rapid and selective Ins(1,4,5)P_3 assays enabled more extensive time course studies that revealed sustained elevations of Ins(1,4,5)P_3 in addition to the rapid transient increases in response to stimuli (DeWald *et al.*, 2001; Drøbak and Watkins, 2000; Perera *et al.*, 1999; Perera *et al.*, 2001). The shorter transients were predicted to be part of a ubiquitous "wake up call" which is probably transmitted by several signaling pathways in sessile organisms such as plants and the long-term response was predicted to reflect a specific Ins(1,4,5)P_3-mediated event (Perera *et al.*, 1999).

We do not know what roles Ins(1,4,5)P_3 plays in regulating plant growth and development. Of interest, plant cells appear to be able to survive quite well with almost no detectable steady state Ins(1,4,5)P_3 (Perera *et al.*, 2002; Perera *et al.*, unpublished results). It is important to remember that this is not a static system and, as mentioned above, the flux through the pathway will increase as a result of the increased rate of hydrolysis of Ins(1,4,5)P_3 and the attenuation of Ins(1,4,5)P_3 signal. Thus, transgenic plants expressing the human InsP 5-ptase gene may prove to be a useful tool for assessing the impact of the plasma membrane generated Ins(1,4,5)P_3 and PtdIns(4,5)P_2 turnover on plant growth and development. Understanding the source of the Ins(1,4,5)P_3, how these signals intersect and what determines which interacting signaling networks or interactomes will be activated is a challenge for future studies.

5. FUTURE CHALLENGES

Living cells are constantly engaged in sensing, interpreting, and responding to environmental cues. To fully appreciate the dynamics of the signaling microdomains it is essential to think in terms of flux. This is difficult for most of us as our vision is limited to a great extent by the measurement methods we use. For example, we often analyze transcript levels, enzyme activities, or cellular structures at fixed moments in time. Only in death does one reach a truly static state. The challenge for scientists is to envision the dynamics of the living cell and appreciate the constant flux within the cellular milieu. While modern microscopy is making great advances, the ability to investigate microdomains and metabolic fluxes of the PI pathway is limited in part by our lack of nanoprobes. We know from *in vivo* labeling studies that synthesis (phosphorylation) of PtdIns and PtdInsP occurs within seconds indicating rapid turnover of the phosphate group. We also know that the lipid kinases are associated with many of the membranes of the cell, the cytoskeleton, and the nucleus (Drøbak *et al.*, 1999; Mueller-Roeber and Pical, 2002); however, we have yet to fully characterize the subcellular distribution of the specific isoforms of the plant PI metabolizing enzymes.

As more genetic manipulations of the PI pathway are encountered, extensive and more complete analyses of the inositol metabolites as well as the phospholipids will be required. Fortunately, the methods have been refined and standards are more readily available than in 1985; so there is the potential to characterize the metabolic fluxes in signaling pools (Welti *et al.*, 2002; Wenk *et al.*, 2003) (Chapters 1, 3, and 4 in this book).

In addition, through evolving technologies, we are gaining the tools to do these studies *in vitro* and *in vivo* by using molecular genetics and recombinant proteins. With these new techniques, however, the cautions and comments of one of the great biochemists of the 20th century, Efraim Racker are still appropriate: *"As a biochemist I have stressed the need for obtaining pure proteins more than once; I have also stressed the need to distrust data obtained with crude systems. As a physiologist I want to stress the need to pay attention to data obtained with native membrane or even with intact cells and to distrust data obtained with pure proteins. . . . the only way we can find out whether the whole is the sum of the parts is by putting the pieces together again and learning how they work. It is the task of the physiologist to help the biochemist by pointing out what is missing."* (Racker, 1965)

In vitro analyses of protein–lipid interactions are essential to understand the mechanisms involved, and as Racker indicates, these must be analyzed in the context of the whole organism. An example of the importance of Racker's advice comes from studies of a potential PI pathway inhibitor neomycin. In earlier studies, adding PtdIns(4,5)P_2 increased the specific activity of reconstituted the dog kidney ATPase (Lipsky and Lietman, 1980). If neomycin was added to the PtdIns(4,5)P_2 micelles prior to adding them to the ATPase, neomycin inhibited ATPase activation as one might anticipate; however, if PtdIns(4,5)P_2 was added to the ATPase and then neomycin was added, there was no effect of neomycin on the PtdIns(4,5)P_2-mediated activation. That is, in the reconstituted system, PtdIns(4,5)P_2 bound the ATPase with such a high affinity that neomycin could not competitively bind and displace the lipid. Although many have documented the fact that neomycin at 1:1 concentrations will bind to PtdIns(4,5)P_2 micelles and prevent PLC hydrolysis *in vitro* (Gungabissoon *et al.*, 1998; Staxen *et al.*, 1999), it is unlikely that neomycin would preferentially bind PtdIns(4,5)P_2 already associated with cellular proteins including PLC. Why is it that so many use neomycin to inhibit the PI pathway *in vivo* and see an effect? There are several explanations. One may be that neomycin, which is a positively charged aminoglycoside will bind to the cell wall and will prevent the penetration of elicitors (Cho *et al.*, 1995). Another is that neomycin, once it enters the cell, will inhibit protein synthesis and thereby affect cellular metabolism.

The insights from the studies using neomycin are mentioned to emphasize the need for multifaceted approach to studying these pathways and to highlight the need to characterize individual pools of PtdIns(4,5)P_2. Importantly, genetic manipulations also are not sufficient to predictively alter a pathway as multicellular organisms will induce compensatory mechanisms to survive. Conflicting results from inhibitor and molecular genetic studies may seem insurmountable (Mills et al., 2004), and yet, by combining technologies and doing comprehensive analysis at the cellular and subcellular level, these conflicts will eventually be resolved and scientists will weave together the interactome of the signaling networks. Advances in nanotechnology will surely lead to new levels of understanding as probes are developed to visualize fluxes in lipid microdomains *in vivo*. Our ultimate goal should be to reach a point where as scientists we can predictively alter plant growth responses by manipulating the flux through a selective signaling pathway.

ACKNOWLEDGMENTS

This work was supported in part by a National Science Foundation grant to WFB, a NASA grant to IYP, and a fellowship from the National Council for Scientific and Technological Development (CNPq), Ministry of Science and Technology (MCT), Brazil to RMG. The authors would like to thank Mary Williams of Harvey Mudd College, Claremont, CA for sharing her unpublished work and Amy Grunden of North Carolina State University for *Pyrococcus furiosus* membranes and Irena Brglez for doing the kinase assays and maintaining the laboratory. This chapter was written as a perspective and not a comprehensive review. Several reviews have been written recently and are cited in the chapter.

REFERENCES

Anantharaman, V., and Aravind, L., 2002, The GOLD domain, a novel protein module involved in Golgi function and secretion. *Genome Biol.* **3:** research0023.0021–0023.0027.

Anderson, R.A., Boronenkov, I.V., Doughman, S.D., Kunz, J., and Loijens, J.C., 1999, Phosphatidylinositol phosphate kinases, a multifaceted family of signaling enzymes. *J. Biol. Chem.* **274:** 9907–9910.

Balla, T., Bondeva, T., and Varnai, P., 2000, How accurately can we image inositol lipids in living cells? *Trends Pharmacol. Sci.* **21:** 238–241.

Balla, A., Tuymetova, G., Barshishat, M., Geiszt, M., and Balla, T., 2002, Characterization of type II phosphatidylinositol 4-kinase isoforms reveals association of the enzymes with endosomal vesicular compartments. *J. Biol. Chem.* **277:** 20041–20050.

Balla, T., and Varnai, P., 2002, Visualizing cellular phosphoinositide pools with GFP-fused protein-modules. *Sci. STKE* **2002:** PL3.

Bankaitis, V.A., Malehorn, D.E., Emr, S.D., and Greene, R., 1989, The saccharomyces-cerevisiae sec14 gene encodes a cytosolic factor that is required for transport of secretory proteins from the yeast Golgi-complex. *J. Cell Biol.* **108:** 1271–1281.

Becker, J.D., Boavida, L.C., Carneiro, J., Haury, M., and Feijo, J.A., 2003, Transcriptional profiling of Arabidopsis tissues reveals the unique characteristics of the pollen transcriptome. *Plant Physiol.* **133:** 713–725.

Boss, W.F., 1989, Phosphoinositide metabolism: Its relation to signal transduction in plants, in Boss, W.F., Morre, D.J. (ed): Second Messengers in Plant Growth and Development. New York, Alan R. Liss, pp 29–56.

Brown, F.D., Rozelle, A.L., Yin, H.L., Balla, T., and Donaldson, J.G., 2001, Phosphatidylinositol 4,5-bisphosphate and Arf6-regulated membrane traffic. *J. Cell Biol.* **154:** 1007–1017.

Bunney, T.D., Watkins, P.A., Beven, A.F., Shaw, P.J., Hernandez, L.E., Lomonossoff, G.P., Shanks, M., Peart, J., and Drøbak, B.K., 2000, Association of phosphatidylinositol 3-kinase with nuclear transcription sites in higher plants. *Plant Cell* **12:** 1679–1688.

Chang, J.-S., Seok, H., Kwon, T.-K., Min, D.S., Ahn, B.-H., Lee, Y.H., Suh, J.-W., Kim, J.-W., Iwashita, S., Omori, A., Ichinose, S., Numata, O., Seo, J.-K., Oh, Y.-S., and Suh, P.-G., 2002, Interaction of elongation factor-1alpha and pleckstrin homology domain of phospholipase c-gamma1 with activating its activity. *J. Biol. Chem.* **277:** 19697–19702.

Chen, I.A., and Szostak, J.W., 2004, Membrane growth can generate a transmembrane pH gradient in fatty acid vesicles. *Proc. Natl. Acad. Sci. U.S.A.* **101:** 7965–7970.

Cho, M.H., Tan, Z., Erneux, C., Shears, S.B., and Boss, W.F., 1995, The effects of mastoparan on the carrot cell plasma membrane polyphosphoinositide phospholipase C. *Plant Physiol.* **107:** 845–856.

Cifuentes, M.E., Delaney, T., and Rebecchi, M.J., 1994, D-*myo*-Inositol 1,4,5-trisphosphate inhibits binding of phospholipase C-delta 1 to bilayer membranes. *J. Biol. Chem.* **269:** 1945–1948.

Cockcroft, S., 1996, Phospholipid signaling in leukocytes. *Curr. Opin. Hematol.* **3:** 48–54.

Cockcroft, S., and De Matteis, M.A., 2001, Inositol lipids as spatial regulators of membrane traffic. *J. Membr. Biol.* **180:** 187–194.

Cote, G.G., and Crain, R.C., 1993, Biochemistry of phosphoinositides. *Annu. Rev. Plant Physiol. Plant Mol. Biol.* **44:** 333–356.

Davis, A.J., Perera, I.Y., and Boss, W.F., 2004, Cyclodextrins enhance recombinant phosphatidylinositol phosphate kinase activity. *J. Lipid. Res.* **45:** 1783–1789.

De Camilli, P., Emr, S.D., McPherson, P.S., and Novick, P., 1996, Phosphoinositides as regulators in membrane traffic. *Science* **271:** 1533–1539.

Despres, B., Bouissonnie, F., Wu, H., Gomord, V., Guilleminot, J., Grellet, F., Berger, F., Delseny, M., and Devic, M., 2003, Three SAC1-like genes show overlapping patterns of expression in Arabidopsis but are remarkably silent during embryo development. *Plant J.* **34:** 293–306.

DeWald, D.B., Torabinejad, J., Jones, C.A., Shope, J.C., Cangelosi, A.R., Thompson, J.E., Prestwich, G.D., and Hama, H., 2001, Rapid accumulation of phosphatidylinositol 4,5-bisphosphate and inositol 1,4,5-trisphosphate correlates with calcium mobilization in salt-stressed Arabidopsis. *Plant Physiol.* **126:** 759–769.

Doughman, R.L., Firestone, A.J., and Anderson, R.A., 2003, Phosphatidylinositol phosphate kinases put $PI4,5P_2$ in its place. *J. Membr. Biol.* **194:** 77–89.

Dove, S.K., Cooke, F.T., Douglas, M.R., Sayers, L.G., Parker, P.J., and Michell, R.H., 1997, Osmotic stress activates phosphatidylinositol-3,5-bisphosphate synthesis. *Nature* **390:** 187–192.

Dove, S.K., Lloyd, C.W., and Drøbak, B.K., 1994, Identification of a phosphatidylinositol 3-hydroxy kinase in plant cells: Association with the cytoskeleton. *Biochem. J.* **303:** 347–350.

Dowler, S., Currie, R.A., Campbell, D., Deak, M., Kular, G., Downes, C.P., and Alessi, D., 2000, Identification of pleckstrin-homology-domain-containing proteins with novel phosphoinositide-binding specificities. *Biochem. J.* **351:** 19–31.

Drøbak, B.K., 1992, The plant phosphoinositide system. *Biochem. J.* **288:** 697–712.

Drøbak, B.K., Dewey, R.E., Boss, W.F., 1999, Phosphoinositide kinases and the synthesis of polyphosphoinositides in higher plant cells, in Jeon KW (ed): International Review of Cytology. New York, Academic Press, vol **189,** pp 95–130.

Drøbak, B.K., Franklin-Tong, V.E., and Staiger, C.J., 2004, The role of the actin cytoskeleton in plant cell signaling. *New Phytol.* **163:** 13–30.

Drøbak, B.K., and Watkins, P.A., 2000, Inositol(1,4,5)trisphosphate production in plant cells: an early response to salinity and hyperosmotic stress. *FEBS Lett.* **481:** 240–244.

Drøbak, B.K., Watkins, P.A.C., Valenta, R., Dove, S.K., Lloyd, C.W., Staiger, C.J., 1994, Inhibition of plant plasma membrane phosphoinositide phospholipase C by the actin-binding protein, profilin. *Plant J.* **6:** 389–400.

Ek-Ramos, M.J., Racagni-Di Palma, G., and Hernandez-Sotomayor, S.M.T., 2003, Changes in phosphatidylinositol and phosphatidylinositol monophosphate kinase activities during the induction of somatic embryogenesis in *Coffea arabica*. *Physiol. Plant.* **119:** 270–277.

Elge, S., Brearley, C., Xia, H.J., Kehr, J., Xue, H.W., and Mueller-Roeber, B., 2001, An Arabidopsis inositol phospholipid kinase strongly expressed in procambial cells: Synthesis of Ptdlns(4,5) P_2 and Ptdlns(3,4,5)P_3 in insect cells by 5-phosphorylation of precursors. *Plant J.* **26:** 561–571.

Eskildsen, S., Justesen, J., Schierup, M.H., and Hartmann, R., 2003, Characterization of the 2′-5′-oligoadenylate synthetase ubiquitin-like family. *Nucleic Acids Res.* **31:** 3166–3173.

Gibbon, B.C., Zonia, L.E., Kovar, D.R., Hussey, P.J., and Staiger, C.J., 1998, Pollen profilin function depends on interaction with proline-rich motifs. *Plant Cell* **10:** 981–993.

Godi, A., Di Campli, A., Konstantakopoulos, A., Di Tullio, G., Alessi, D.R., Kular, G.S., Daniele, T., Marra, P., Lucocq, J.M., and De Matteis, M.A., 2004, FAPPs control Golgi-to-cell-surface membrane traffic by binding to ARF and PtdIns(4)P. *Nat. Cell Biol.* **6:** 393–404.

Gross, W., Boss, W.F., 1993, Inositol phospholipids and signal transduction, in D.P.S. Verma (ed): Control of Plant Gene Expression. Boca Raton, CRC Press, Inc, pp 17–32.

Gungabissoon, R.A., Jiang, C.-J., Drøbak, B.K., Maciver, S.K., and Hussey, P.J., 1998, Interaction of maize actin-depolymerising factor with actin and phosphoinositides and its inhibition of plant phospholipase C. *Plant J.* **16:** 689–696.

Hanakahi, L.A., Bartlet-Jones, M., Chappell, C., Pappin, D., and West, S.C., 2000, Binding of inositol phosphate to DNA-PK and stimulation of double-strand break repair. *Cell* **102:** 721–729.

Hanczyc, M.M., Fujikawa, S.M., and Szostak, J.W., 2003, Experimental models of primitive cellular compartments: Encapsulation, growth, and division. *Science* **302:** 618–622.

Harris, T.E., and Lawrence, J.C., Jr., 2003, TOR signaling. *Sci. STKE* **2003:** re15.

Hartmann-Petersen, R., Hendil, K.B., and Gordon, C., 2003, Ubiquitin binding proteins protect ubiquitin conjugates from disassembly. *FEBS Lett.* **535:** 77–81.

Heilmann, I., Perera, I.Y., Gross, W., and Boss, W.F., 2001, Plasma membrane phosphatidylinositol 4,5-bisphosphate decreases with time in culture. *Plant Physiol.* **126:** 1507–1518.

Helliwell, S.B., Wagner, P., Kunz, J., Deuter-Reinhard, M., Henriquez, R., and Hall, M.N., 1994, TOR1 and TOR2 are structurally and functionally similar but not identical phosphatidylinositol kinase homologues in yeast. *Mol. Biol. Cell* **5:** 105–118.

Hendrix, W., Assefa, H., and Boss, W.F., 1989, The polyphosphoinositides, phosphatidylinositol monophosphate and phosphatidylinositol bisphosphate are present in nuclei isolated from carrot protoplasts. *Protoplasma* **151:** 62–72.

Herman, P.K., and Emr, S.D., 1990, Characterization of VPS34, a gene required for vacuolar protein sorting and vacuole segregation in *Saccharomyces cerevisiae*. *Mol. Cell Biol.* **10:** 6742–6754.

Hernandez, L.E., Escobar, C., Drøbak, B.K., Bisseling, T., and Brewin, N.J., 2004, Novel expression patterns of phosphatidylinositol 3-hydroxy kinase in nodulated *Medicago* spp. plants. *J. Exp. Bot.* **55:** 957–959.

Hernandez-Sotomayor, S.M.T., De Los Santos-Briones, C., Munoz-Sanchez, J.A., and Loyola-Vargas, V.M., 1999, Kinetic analysis of phospholipase C from *Catharanthus roseus* transformed roots using different assays. *Plant Physiol.* **120:** 1075–1082.

Hong, Z., Bednarek, S.Y., Blumwald, E., Hwang, I., Jurgens, G., Menzel, D., Osteryoung, K.W., Raikhel, N.V., Shinozaki, K., Tsutsumi, N., and Verma, D.P.S., 2003, A unified nomenclature for Arabidopsis dynamin-related large GTPases based on homology and possible functions. *Plant Mol. Biol.* **53:** 261–265.

Hong, Z., and Verma, D.P., 1994, A phosphatidylinositol 3-kinase is induced during soybean nodule organogenesis and is associated with membrane proliferation. *Proc. Natl. Acad. Sci. U.S.A.* **91:** 9617–9621.

Huang, S.J., Blanchoin, L., Kovar, D.R., and Staiger, C.J., 2003, Arabidopsis capping protein (AtCP) is a heterodimer that regulates assembly at the barbed ends of actin filaments. *J. Biol. Chem.* **278:** 44832–44842.

Hunt, L., Mills, L.N., Pical, C., Leckie, C.P., Aitken, F.L., Kopka, J., Mueller-Roeber, B., McAinsh, M.R., Hetherington, A.M., and Gray, J.E., 2003, Phospholipase C is required for the control of stomatal aperture by ABA. *Plant J.* **34:** 47–55.

Hunt, L., Otterhag, L., Lee, J.C., Lasheen, T., Hunt, J., Seki, M., Shinozaki, K., Sommarin, M., Gilmour, D.J., Pical, C., and Gray, J.E., 2004, Gene-specific expression and calcium activation of *Arabidopsis thaliana* phospholipase C isoforms. *New Phytol.* **162:** 643–654.

Irvine, R.F., 2003, Nuclear lipid signalling. *Nat. Rev. Mol. Cell Biol.* **4:** 349–360.

Jin, J.B., Kim, Y.A., Kim, S.J., Lee, S.H., Kim, D.H., Cheong, G.W., and Hwang, I., 2001, A new dynamin-like protein, ADL6, is involved in trafficking from the trans-Golgi network to the central vacuole in Arabidopsis. *Plant Cell* **13:** 1511–1525.

Kapranov, P., Routt, S.M., Bankaitis, V.A., de Bruijn, F.J., and Szczyglowski, K., 2001, Nodule-specific regulation of phosphatidylinositol transfer protein expression in *Lotus japonicus*. *Plant Cell* **13:** 1369–1382.

Kim, D.H., Eu, Y.-J., Yoo, C.M., Kim, Y.-W., Pih, K.T., Jin, J.B., Kim, S.J., Stenmark, H., and Hwang, I., 2001a, Trafficking of phosphatidylinositol 3-phosphate from the trans-Golgi network to the lumen of the central vacuole in plant cells. *Plant Cell* **13:** 287–301.

Kim, Y.W., Park, D.S., Park, S.C., Kim, S.H., Cheong, G.W., and Hwang, I., 2001b, Arabidopsis dynamin-like 2 that binds specifically to phosphatidylinositol 4-phosphate assembles into a high-molecular weight complex in vivo and in vitro. *Plant Physiol.* **127:** 1243–1255.

Kost, B., Lemichez, E., Spielhofer, P., Hong, Y., Tolias, K., Carpenter, C., and Chua, N.H., 1999, Rac homologues and compartmentalized phosphatidylinositol 4,5-bisphosphate act in a common pathway to regulate polar pollen tube growth. *J. Cell Biol.* **145:** 317–330.

Kovar, D.R., Drøbak, B.K., and Staiger, C.J., 2000, Maize profilin isoforms are functionally distinct. *Plant Cell* **12:** 583–598.

Kunz, J., Fuelling, A., Kolbe, L., and Anderson, R.A., 2002, Stereo-specific substrate recognition by phosphatidylinositol phosphate kinases is swapped by changing a single amino acid residue. *J. Biol. Chem.* **277:** 5611–5619.

Kunz, J., Wilson, M.P., Kisseleva, M., Hurley, J.H., Majerus, P.W., and Anderson, R.A., 2000, The activation loop of phosphatidylinositol phosphate kinases determines signaling specificity. *Mol. Cell* **5:** 1–11.

Laude, A.J., and Prior, I.A., 2004, Plasma membrane microdomains: Organization, function and trafficking. *Mol. Membr. Biol.* **21:** 193–205.

Lee, S.H., Jin, J.B., Song, J.H., Min, M.K., Park, D.S., Kim, Y.W., and Hwang, I.H., 2002, The intermolecular interaction between the PH domain and the C-terminal domain of Arabidopsis dynamin-like 6 determines lipid binding specificity. *J. Biol. Chem.* **277:** 31842–31849.

Lemmon, M.A., 2003, Phosphoinositide recognition domains. *Traffic* **4:** 201–213.

Lemmon, M.A., Ferguson, K.M., and Schlessinger, J., 1996, PH domains: Diverse sequences with a common fold recruit signaling molecules to the cell surface. *Cell* **85:** 621–624.

Levine, T.P., and Munro, S., 2002, Targeting of Golgi-specific pleckstrin homology domains involves both PtdIns 4-kinase-dependent and -independent components. *Curr. Biol.* **12:** 695–704.

Li, W., Li, M., Zhang, W., Welti, R., and Wang, X., 2004, The plasma membrane-bound phospholipase Ddelta enhances freezing tolerance in *Arabidopsis thaliana*. *Nat. Biotechnol.* **22:** 427–433.

Lipsky, J.J., and Lietman, P.S., 1980, Neomycin inhibition of adenosine triphosphatase: Evidence for a neomycin–phospholipid interaction. *Antimicrob. Agents Chemother.* **18:** 532–535.

Loewen, C.J., Gaspar, M.L., Jesch, S.A., Delon, C., Ktistakis, N.T., Henry, S.A., and Levine, T.P., 2004, Phospholipid metabolism regulated by a transcription factor sensing phosphatidic acid. *Science* **304:** 1644–1647.

Matsuoka, K., Bassham, D.C., Raikhel, N.V., and Nakamura, K., 1995, Different sensitivity to wortmannin of two vacuolar sorting signals indicates the presence of distinct sorting machineries in tobacco cells. *J. Cell Biol.* **130:** 1307–1318.

Meijer, H.J.G., and Munnik, T., 2003, Phospholipid-based signaling in plants. *Annu. Rev. Plant Biol.* **54:** 265–306.

Mejillano, M., Yamamoto, M., Rozelle, A.L., Sun, H.Q., Wang, X.D., and Yin, H.L., 2001, Regulation of apoptosis by phosphatidylinositol 4,5-bisphosphate inhibition of caspases, and caspase inactivation of phosphatidylinositol phosphate 5-kinases. *J. Biol. Chem.* **276:** 1865–1872.

Melin, P.M., Pical, C., Jergil, B., and Sommarin, M., 1992, Polyphosphoinositide phospholipase C in wheat root plasma membranes. Partial purification and characterization. *Biochim. Biophys. Acta* **1123:** 163–169.

Melin, P.M., Sommarin, M., Sandelius, A.S., and Jergil, B., 1987, Identification of Ca^{2+}-stimulated polyphosphoinositide phospholipase C in isolated plant plasma membranes. *FEBS Lett.* **223:** 87–91.

Mikami, K., Takahashi, S., Katagiri, T., Yamaguchi-Shinozaki, K., and Shinozaki, K., 1999, Isolation of an *Arabidopsis thaliana* cDNA encoding a pleckstrin homology domain protein, a putative homologue of human pleckstrin. *J. Exp. Bot.* **334:** 729–730.

Mills, L.N., Hunt, L., Leckie, C.P., Aitken, F.L., Wentworth, M., McAinsh, M.R., Gray, J.E., and Hetherington, A.M., 2004, The effects of manipulating phospholipase C on guard cell ABA-signalling. *J. Exp. Bot.* **55:** 199–204.

Minogue, S., Anderson, J.S., Waugh, M.G., dos Santos, M., Corless, S., Cramer, R., and Hsuan, J.J., 2001, Cloning of a human type II phosphatidylinositol 4-kinase reveals a novel lipid kinase family. *J. Biol. Chem.* **276:** 16635–16640.

Mitra, P., Zhang, Y., Rameh, L.E., Ivshina, M.P., McCollum, D., Nunnari, J.J., Hendricks, G.M., Kerr, M.L., Field, S.J., Cantley, L.C., and Ross, A.H., 2004, A novel phosphatidylinositol(3,4,5)P_3 pathway in fission yeast. *J. Cell Biol.* **166:** 205–211.

Monks, D.E., Aghoram, K., Courtney, P.D., DeWald, D.B., and Dewey, R.E., 2001, Hyperosmotic stress induces the rapid phosphorylation of a soybean phosphatidylinositol transfer protein homolog through activation of the protein kinases SPK1 and SPK2. *Plant Cell* **13:**1205–1219.

Morse, M.J., Crain, R.C., and Satter, R.L., 1987, Light-stimulated inositol phospholipid turnover in *Samanea saman* leaf pulvini. *Proc. Natl. Acad. Sci. U.S.A.* **84:** 7075–7078.

Mueller-Roeber, B., and Pical, C., 2002, Inositol phospholipid metabolism in Arabidopsis. Characterized and putative isoforms of inositol phospholipid kinase and phosphoinositide-specific phospholipase C. *Plant Physiol.* **130:** 22–46.

Munnik, T., Irvine, R.F., and Musgrave, A., 1998, Phospholipid signalling in plants. *Biochim. Biophys. Acta* **1389:** 222–272.

Nishikawa, K., Toker, A., Wong, K., Marignani, P.A., Johannes, F.J., and Cantley, L.C., 1998, Association of protein kinase Cmu with type II phosphatidylinositol 4-kinase and type I phosphatidylinositol-4-phosphate 5-kinase. *J. Biol. Chem.* **273:** 23126–23133.

Novick, P., Osmond, B.C., and Botstein, D., 1989, Suppressors of yeast actin mutations. *Genetics* **121:** 659–674.

Odom, A.R., Stahlberg, A., Wente, S.R., and York, J.D., 2000, A role for nuclear inositol 1,4, 5-trisphosphate kinase in transcriptional control. *Science* **287:** 2026–2029.

Pappan, K., and Wang, X., 1999, Molecular and biochemical properties and physiological roles of plant phospholipase D. *Biochim. Biophys. Acta* **1439:**151–166.

Pendaries, C., Tronchere, H., Plantavid, M., and Payrastre, B., 2003, Phosphoinositide signaling disorders in human diseases. *FEBS Lett.* **546:** 25–31.

Perera, I.Y., Davis, A.J., Galanopoulou, D., Im, Y.J., Boss, W.F., 2005, Characterization and comparative analysis of Arabidopsis phosphatidylinositol phosphate 5-kinase 10 reveals differences in Arabidopsis and human phosphatidylinositol phosphate kinases. *FEBS Lett.* **579:** 3427–32.

Perera, I.Y., Heilmann, I., and Boss, W.F., 1999, Transient and sustained increases in inositol 1,4,5-trisphosphate precede the differential growth response in gravistimulated maize pulvini. *Proc. Natl. Acad. Sci. U.S.A.* **96:** 5838–5843.

Perera, I.Y., Heilmann, I., Chang, S.C., Boss, W.F., and Kaufman, P.B., 2001, A role for inositol 1,4,5-trisphosphate in gravitropic signaling and the retention of cold-perceived gravistimulation of oat shoot pulvini. *Plant Physiol.* **125:** 1499–1507.

Perera, I.Y., Love, J., Heilmann, I., Thompson, W.F., and Boss, W.F., 2002, Up-regulation of phosphoinositide metabolism in tobacco cells constitutively expressing the human type I inositol polyphosphate 5-phosphatase. *Plant Physiol.* **129:** 1795–1806.

Peterman, T.K., Ohol, Y., McReynolds, L., and Luna, E.J., 2004, Patellin1, a novel sec14-like protein, localizes to the cell plate and binds phosphoinositides. *Plant Physiol.* **136:** 3080–3094.

Pical, C., Westergren, T., Dove, S.K., Larsson, C., and Sommarin, M., 1999, Salinity and hyperosmotic stress induce rapid increases in phosphatidylinositol 4,5-bisphosphate, diacylglycerol pyrophosphate, and phosphatidylcholine in *Arabidopsis thaliana* cells. *J. Biol. Chem.* **274:** 38232–38240.

Racker, E., 1985, Reconstitutions of transporters, receptors, and pathological states. Orlando, Academic Press.

Roth, M.G., 1999, Lipid regulators of membrane traffic through the Golgi complex. *Trends Cell Biol.* **9:** 174–179.

Roth, M.G., 2004, Phosphoinositides in constitutive membrane traffic. *Physiol. Rev.* **84:** 699–730.

Sanchez, J.P., and Chua, N.H., 2001, Arabidopsis PLC1 is required for secondary responses to abscisic acid signals. *Plant Cell* **13:** 1143–1154.

Sandelius, A.S., Sommarin, M., 1990, Membrane-localized reactions involved in poly-phosphoinositide turnover in plants, in Morre, D.J., Boss, W.F., Loewus, F. (ed): Inositol Metabolism in Plants. New York, Wiley-Liss, pp 139–161.

Shank, K.J., Su, P., Brglez, I., Boss, W.F., Dewey, R.E., and Boston, R.S., 2001, Induction of lipid metabolic enzymes during the endoplasmic reticulum stress response in plants. *Plant Physiol.* **126:** 267–277.

Simonsen, A., Wurmser, A.E., Emr, S.D., and Stenmark, H., 2001, The role of phosphoinositides in membrane transport. *Curr. Opin. Cell. Biol.* **13:** 485–492.

Sprong, H., van der Sluijs, P., and van Meer, G., 2001, How proteins move lipids and lipids move proteins. *Nat. Rev. Mol. Cell Biol.* **2:** 504–513.

Staiger, C.J., Gibbon, B.C., Kovar, D.R., and Zonia, L.E., 1997, Profilin and actin-depolymerizing factor: Modulators of actin organization in plants. *Trends Plant Sci.* **7:** 275–281.

Staxen, I., Pical, C., Montgomery, L.T., Gray, J.E., Hetherington, A.M., and McAinsh, M.R., 1999, Abscisic acid induces oscillations in guard-cell cytosolic free calcium that involve phosphoinositide-specific phospholipase C. *Proc. Natl. Acad. Sci. U.S.A.* **96:** 1779–1784.

Stevenson, J.M., Perera, I.Y., and Boss, W.F., 1998, A phosphatidylinositol 4-kinase pleckstrin homology domain that binds phosphatidylinositol 4-monophosphate. *J. Biol. Chem.* **273:** 22761–22767.

Stevenson, J.M., Perera, I.Y., Heilmann, I., Persson, S., and Boss, W.F., 2000, Inositol signaling and plant growth. *Trends Plant Sci.* **5:** 252–258.

Stevenson-Paulik, J., Love, J., and Boss, W.F., 2003, Differential regulation of two Arabidopsis type III phosphatidylinositol 4-kinase isoforms. A regulatory role for the pleckstrin homology domain. *Plant Physiol.* **132:** 1053–1064.

Suer, S., Sickmann, A., Meyer, H.E., Herberg, F.W., and Heilmeyer, L.M., Jr., 2001, Human phosphatidylinositol 4-kinase isoform PI4K92. Expression of the recombinant enzyme and determination of multiple phosphorylation sites. *Eur. J. Biochem.* **268:** 2099–2106.

Takenawa, T., and Itoh, T., 2001, Phosphoinositides, key molecules for regulation of actin cytoskeletal organization and membrane traffic from the plasma membrane. *Biochim. Biophys. Acta* **1533:** 190–206.

Tan, Z., and Boss, W.F., 1992, Association of phosphatidylinositol kinase, phosphatidylinositol monophosphate kinase, and diacylglycerol kinase with the cytoskeleton and F-actin fractions of carrot (*Daucus carota* L.) cells grown in suspension culture. *Plant Physiol.* **100:** 2116–2120.

Tronchere, H., Buj-Bello, A., Mandel, J.L., and Payrastre, B., 2003, Implication of phosphoinositide phosphatases in genetic diseases: The case of myotubularin. *Cell Mol. Life Sci.* **60:** 2084–2099.

Van Leeuwen, W., Okresz, L., Bogre, L., and Munnik, T., 2004, Learning the lipid language of plant signalling. *Trends Plant Sci.* **9:** 378–384.

van Meer, G., and Sprong, H., 2004, Membrane lipids and vesicular traffic. *Curr. Opin. Cell Biol.* **16:** 373–378.

Vincent, P., Chua, M., Nogue, F., Fairbrother, A., Mekeel, H., Xu, Y., Allen, N., Bibikova, T.N., Gilroy, S., Bankaitis, V.A., 2005, A Sec14p-nodulin domain phosphatidylinositol transfer protein polarizes membrane growth of Arabidopsis thaliana root hairs. *J Cell Biol.* **168:** 801–12.

Walsh, J.P., Caldwell, K.K., and Majerus, P.W., 1991, Formation of phosphatidylinositol 3-phosphate by isomerization from phosphatidylinositol 4-phosphate. *Proc. Natl. Acad. Sci. U.S.A.* **88:** 9184–9187.

Wang, X., 1999, The role of phospholipase D in signaling cascades. *Plant Physiol.* **120:** 645–651.

Wang, X., 2001, Plant phospholipases. *Annu. Rev. Plant Physiol. Plant Mol. Biol.* **52:** 211–231.

Wei, Y.J., Sun, H.Q., Yamamoto, M., Wlodarski, P., Kunii, K., Martinez, M., Barylko, B., Albanesi, J.P., and Yin, H.L., 2002, Type II phosphatidylinositol 4-kinase beta is a cytosolic and peripheral membrane protein that is recruited to the plasma membrane and activated by Rac-GTP. *J. Biol. Chem.* **277:** 46586–46593.

Welti, R., Li, W., Li, M., Sang, Y., Biesiada, H., Zhou, H.-E., Rajashekar, C.B., Williams, T.D., and Wang, X., 2002, Profiling membrane lipids in plant stress responses. Role of phospholipase Dalpha in freezing-induced lipid changes in Arabidopsis. *J. Biol. Chem.* **277:** 31994–32002.

Wenk, M.R., Lucast, L., Di Paolo, G., Romanelli, A.J., Suchy, S.F., Nussbaum, R.L., Cline, G.W., Shulman, G.I., McMurray, W., and De Camilli, P., 2003, Phosphoinositide profiling in complex lipid mixtures using electrospray ionization mass spectrometry. *Nat. Biotechnol.* **21:** 813–817.

Westergren, T., Dove, S.K., Sommarin, M., and Pical, C., 2001, AtPIP5K1, an *Arabidopsis thaliana* phosphatidylinositol phosphate kinase, synthesizes PtdIns(3,4)P_2 and PtdIns(4,5)P_2 in vitro and is inhibited by phosphorylation. *Biochem. J.* **359:** 583–589.

Wheeler, J.J., and Boss, W.F., 1987, Polyphosphoinositides are present in plasma membranes isolated from fusogenic carrot cells. *Plant Physiol.* **85:** 389–392.

Williams, M.E., Torabinejad, J., Cohick ,E., Parker, K., Drake, E.J., Thompson, J.E., Hortter, M., Dewald, D.B., 2005, Mutations in the Arabidopsis phosphoinositide phosphatase gene SAC9 lead to over accumulation of PtdIns(4,5)P_2 and constitutive expression of the stress-response pathway. *Plant Physiol.* **138:** 686–700.

Xu, P., Lloyd, C.W., Staiger, C.J., and Drøbak, B.K., 1992, Association of phosphatidylinositol 4-kinase with the plant cytoskeleton. *Plant Cell* **4:** 941–951.

Xue, H.W., Pical, C., Brearley, C., Elge, S., and Muller-Rober, B., 1999, A plant 126-kDa phosphatidylinositol 4-kinase with a novel repeat structure. Cloning and functional expression in baculovirus-infected insect cells. *J. Biol. Chem.* **274:** 5738–5745.

Yang, W., and Boss, W.F., 1994a, Regulation of phosphatidylinositol 4-kinase by the protein activator PIK-A49. Activation requires phosphorylation of PIK-A49. *J. Biol. Chem.* **269:** 3852–3857.

Yang, W., and Boss, W.F., 1994b, Regulation of the plasma membrane type III phosphatidylinositol 4-kinase by positively charged compounds. *Arch. Biochem. Biophys.* **313:** 112–119.

Yin, H.L., and Janmey, P.A., 2002, Phosphoinositide regulation of the actin cytoskeleton. *Annu. Rev. Physiol.* **2:** 2.

Zhao, X.H., Bondeva, T., and Balla, T., 2000, Characterization of recombinant phosphatidylinositol 4-kinase beta reveals auto- and heterophosphorylation of the enzyme. *J. Biol. Chem.* **275:** 14642–14648.

Zhong, R., and Ye, Z.-H., 2003, The SAC domain-containing protein gene family in Arabidopsis. *Plant Physiol.* **132:** 544–555.

Chapter 9

Cracking the Green Paradigm: Functional Coding of Phosphoinositide Signals in Plant Stress Responses

Laura Zonia and Teun Munnik
Section of Plant Physiology, Swammerdam Institute for Life Sciences, University of Amsterdam, Kruislaan 318, NL-1098 SM, Amsterdam, The Netherlands

1. INTRODUCTION

Plant form and function represent unique adaptations to life on earth. Although many fundamental elements of biochemistry and cell biology are shared among mammals, plants, fungi and protists, each has the potential to express these elements differently. Thus it should not be surprising that while plant cells contain many of the same elements of phosphoinositide signaling systems as other organisms, the functional coding of these elements may differ because both the information signaled as well as the responses elicited are different. The goal of this review is to provide a critical overview of what is currently known about phosphoinositide signaling in plants and in plant stress responses, to focus on a specific system in which differential coding of phosphatidylinositol signals has been elucidated, and to indicate directions for future research.

2. RECENT DISCOVERIES ARE REWRITING THE BOOK ON PHOSPHOINOSITIDE SIGNALING DURING CELLULAR HOMEOSTASIS

To delineate the phosphoinositide signaling pathways that are induced upon stress, it is necessary to briefly discuss what is known about homeostatic cellular signaling and to highlight recent discoveries that are critically relevant for

A. Lahiri Majumder and B. B. Biswas (eds.), Biology of Inositols and Phosphoinositides, 207–238
© 2006 *Springer. Printed in the Netherlands.*

plant signaling. There have been numerous excellent reviews of the current understanding of plant phosphoinositide and lipid signaling pathways in recent years (Munnik *et al.*, 1998; Stevenson *et al.*, 2000; Irvine and Schell, 2001; Munnik, 2001; Munnik and Meijer, 2001; Müller-Roeber and Pical, 2002; Wang *et al.*, 2002; Meijer and Munnik, 2003; Ryu, 2004; Van Leeuwen *et al.*, 2004; Wang, 2004; see also Perera and Boss, this volume). In spite of this considerable progress, detailed signaling pathways in unstimulated cells have not yet been completely elucidated, in part because of the low incorporation of radioisotope into phosphoinositide isomers in plants compared with mammalian cells. This apparent low abundance of phosphoinositides may reflect a rapid turnover rate of specific signals that are essential for cellular homeostasis. Additionally, specific phosphoinositides may be sequestered into cellular microdomains with high local concentrations that become diluted after whole-cell extraction and analysis. Recent exciting discoveries about phosphoinositide signaling in other organisms could potentially impact the understanding of plant phosphoinositide signaling in the next few years.

2.1. A novel route for synthesis of PtdIns(3,4,5)P_3 has been identified in yeast

One of the isoforms synthesized by the mammalian PI 3-kinase (PI3K) pathway is PtdIns(3,4,5)P_3, which is a critical signal involved in growth regulation. However, PtdIns(3,4,5)P_3 has not been detected in plants or yeast and it has been thought that they lack this phosphoinositide signal. The class of PI3K enzymes that catalyze synthesis of PtdIns(3,4,5)P_3 from PtdIns(4,5)P_2 in mammalian cells is also lacking in plants and yeast, which contain only the class III PI3K (Vps34p) that phosphorylates PtdIns to PtdIns(3)P. PtdIns(3,4,5)P_3 is dephosphorylated by the D-3 phosphatase PTEN (Maehama and Dixon, 1998), which is deleted or inactivated in many carcinomous cells (Simpson and Parsons, 2001; Sulis and Parsons, 2003; Parsons, 2004). PTEN also functions as a protein tyrosine phosphatase and an inositol 1,3,4,5,6-pentakisphosphate 3-phosphatase in mammalian cells (Caffrey *et al.*, 2001).

Recent compelling work shows that PtdIns(3,4,5)P_3 may be undetectable in yeast cells because of homeostatic regulatory pathways. Mitra *et al.* (2004) have identified a PTEN homologue in *Schizosaccharomyces pombe*, and when they disrupted this gene the cells accumulated PtdIns(3,4,5)P_3 to levels comparable with those found in mammalian cells. They went on to show that synthesis of PtdIns(3,4,5)P_3 required PI3K (Vps34p) and PIP5K (Its3p). *Arabidopsis* also contains a PTEN homologue (Luan *et al.*, 2001) that is expressed exclusively in pollen grains during the late stage of development (Gupta *et al.*, 2002). Disruption of *AtPTEN1* induces pollen cell death after mitosis (Gupta *et al.*, 2002). The *Arabidopsis* genome also contains both Vps34p and PIP5K1 (Welters *et al.*, 1994; Mikami *et al.*, 1998; Elge *et al.*, 2001). In vitro, AtPIP5K can phosphorylate PtdIns3P and PtdIns4P to

$PtdIns(3,4)P_2$ and $PtdIns(4,5)P_2$ (Westergren *et al.*, 2001). When transfected into insect cells, AtPIP5K directed the synthesis of both $PtdIns(4,5)P_2$ and $PtdIns(3,4,5)P_3$ (Elge *et al.*, 2001). Thus, it may be possible that the alternate route for synthesis of $PtdIns(3,4,5)P_3$ documented in yeast cells also exists in plants.

2.2. Plant PLC signaling: Differences and similarities to the mammalian paradigm

Most studies in plants have focused on demonstrating that the 'cannonical' mammalian pathway – defined as PLC hydrolysis of $PtdIns(4,5)P_2$ to produce DAG, which activates protein kinase C (PKC), and $Ins(1,4,5)P_3$, which mobilizes cytosolic Ca^{2+} ($[Ca^{2+}]_{cyt}$) from intracellular stores – also exists in plants (for reviews see Munnik *et al.*, 1998a; Stevenson *et al.*, 2000; Müller-Roeber and Pical, 2002; Meijer and Munnik, 2003). However, even in mammalian cells the understanding of PLC signaling is undergoing revision as new details emerge. Plants contain most of the components of this signaling pathway, with notable exceptions being PKC and a confirmed IP_3 receptor, but the functional coding of these signals has yet to be conclusively demonstrated. Recent advances in the understanding of phosphoinositide signaling in *Saccharomyces*, *Dictyostelium*, and plants suggests that signaling through the PLC pathway in plants may be quite different from the 'cannonical' pathway described for mammalian cells. These discoveries and their significance for plant cell signaling will be discussed in the following sections.

2.2.1. Are plant PLCs functional homologs of the novel PLCζ family?

Mammalian cells contain 5 classes of PLC isozymes based on their protein sequence, molecular organization and regulation mechanisms –β, γ, δ, ε, and the newly discovered ζ isoform that is expressed in mouse sperm (Rhee, 2001; Kurokawa *et al.*, 2004; Larman *et al.*, 2004; Swann *et al.*, 2004). *Saccharomyces* and *Dictyostelium* contain a single PLCδ isoform, while *Arabidopsis* has 9 putative PLC isoforms (2 of which are thought to be inactive) that have been classified as PLCδ by phylogenetic analysis (Müller-Roeber and Pical, 2002; Hunt *et al.*, 2004). However, plant PLCδ isoforms critically differ from the mammalian PLCδ in that they lack the pleckstrin homology (PH) domain that avidly binds $PtdIns(4,5)P_2$. Instead, plant PLC isoforms only contain the second loop of the conserved EF-hand domain and the C2 lipid-binding domain (Müller-Roeber and Pical, 2002). Based on protein domain organization, plant PLC isoforms are more similar to the recently characterized sperm-specific PLCζ, which also lacks the PH domain (Cox *et al.*, 2002; Saunders *et al.*, 2002). Sperm PLCζ can induce $[Ca^{2+}]_{cyt}$ oscillations in egg cells after fertilization, yet it has only been detected in sperm cell soluble fractions and after fertilization in the egg cell pronucleus (Saunders *et al.*, 2002; Larman *et al.*, 2004). Some plant PLC isoforms are constitutively expressed in specific tissues while others are differentially induced in response to stress or phytohormones (Müller-Roeber and Pical, 2002; Hunt *et al.*, 2004; Lin *et al.*, 2004). Plant PLC activity has been detected in soluble fractions, requiring millimolar Ca^{2+} concentrations and preferring PtdIns as a substrate, and in the plasma membrane, requiring 0.1–10 μM Ca^{2+} and preferring PtdIns4P and PtdIns(4,5)P_2 as substrates (Müller-Roeber and Pical, 2002). The substrate preference could have interesting ramifications for plant signaling, as estimates have reported that PtdIns represents ca. 5% of total phospholipids, and the ratio of PtdIns:PtdInsP:PtdInsP_2 is on the order of 385:35:1 (Hetherington and Drøbak, 1992; Munnik *et al.*, 1994).

2.2.2. PtdIns(4,5)P_2: More than a substrate for PLC

PtdIns(4,5)P_2 does not only act as a substrate for PLC (and potentially PI3K) but is also emerging as an important signaling lipid itself. It recruits, organizes and activates cytoskeletal and protein complexes at the plasma membrane (Martin, 1998; Janmey *et al.*, 1999; Lemmon, 2003; Yin and Janmey, 2003; Carroll *et al.*, 2004; Van Leeuwen *et al.*, 2004), and has an important role in membrane trafficking (Itoh and Takenawa, 2004; Roth, 2004). PtdIns(4,5)P_2 also stimulates PLD activity, thereby affecting alternate signaling pathways (Wang, 2000; Cockcroft, 2001; McDermott *et al.*, 2004). Thus, PtdIns(4,5)P_2 has the potential to link multiple signaling pathways and integrate several downstream cellular responses. Unfortunately, there have been relatively few studies in plants that focus on these alternative functions of PtdIns(4,5)P_2 signaling and how they may impact cellular responses (Drøbak *et al.*, 1994; Kost *et al.*, 1999; Kovar *et al.*, 2000).

2.2.3. Not DAG but PtdOH may be the primary lipid second messenger in plants

In mammalian cells, the DAG generated by hydrolysis of PtdIns(4,5)P_2 functions as a lipid signal that activates PKC. However, as mentioned above, PKC has not been detected in plants and gene homologs are lacking from the *Arabidopsis* genome. DAG signaling is terminated by phosphorylation via DAG kinase (DGK), which produces phosphatidic acid (PtdOH) (Kanoh *et al.*, 2002; Cipres *et al.*, 2003; Luo *et al.*, 2004). The results obtained to date for plants indicate that DAG likely does not function as the primary lipid second messenger of the PLC pathway. Increasingly, the evidence indicates that the fate of DAG in plants is to be channelled into lipid metabolic pathways or converted to PtdOH, which has been characterized as an important signal elicited as a general response to stress (Munnik *et al.*, 1998b; Miège and Maréchal, 1999; Munnik, 2001; Munnik and Meijer, 2001; Wang, 2001; Meijer and Munnik,

2003; Wang, 2004). PtdOH can be produced via 2 different pathways, PLC-DGK and/or PLD, and may be functionally differentiated by the pathway that produces it (Den Hartog *et al.*, 2001, 2003; Arisz *et al.*, 2003; Meijer and Munnik, 2003; Testerink and Munnik, 2005). The pathway that generates PtdOH can be determined experimentally by utilizing a differential labelling strategy, in addition to the use of primary alcohols (which are competitive substrates for PLD activity) as reporters to detect PLD activity (Munnik *et al.*, 1995, 1998b; Munnik, 2001). PtdOH signaling may also be differentiated by tissue-specific expression of different phospholipase isoenzymes in addition to spatial and/or developmental constraints. In plant, yeast and mammalian cells, PtdOH has important roles in growth control, vesicle trafficking, regulation of lipid and protein kinases, and cytoskeletal organization (English, 1996; Cockcroft, 2001; Munnik, 2001; Shen *et al.*, 2001; Meijer and Munnik, 2003; Foster and Xu, 2003; Freyberg *et al.*, 2003; Ktistakis *et al.*, 2003; Anthony *et al.*, 2004; Duman *et al.*, 2004; Nakanishi *et al.*, 2004; Testerink and Munnik, 2004; Testerink *et al.*, 2004; Testerink and Munnik, 2005). In plants and yeast, PtdOH signaling activity is terminated via phosphorylation by PtdOH kinase (PAK) to form the novel lipid diacylglycerol pyrophosphate (DGPP) (Munnik *et al.*, 1996; Munnik, 2001).

2.2.4. Ins(1,4,5)P_3 can be generated by 2 different pathways

Alternate routes exist for the formation of Ins(1,4,5)P_3 in addition to hydrolysis of PtdIns(4,5)P_2 by PLC. *Dictyostelium* has only a single PLC gene, yet cells with a deletion of this gene have normal levels of Ins(1,4,5)P_3 (Bominaar *et al.*, 1991; Drayer *et al.*, 1994). It was shown that the primary route for synthesis of Ins(1,4,5)P_3 in these cells was via dephosphorylation of Ins(1,3,4,5,6)P_5 by an Ins(1,3,4,5,6)P_5 3/6-bisphosphatase, which is similar to the mammalian multiple inositol polyphosphate phosphatase (MIPP) (Craxton *et al.*, 1995; Van Dijken *et al.*, 1995, 1997; Van Haastert and Van Dijken, 1997). MIPP catalyzes this reaction in rat liver extracts (Craxton *et al.*, 1995; Van Dijken *et al.*, 1995). Whether this pathway functions in plants must be determined, but it is possible that not all Ins(1,4,5)P_3 measured in plant cells is generated through the PLC pathway. It may be derived, at least in part, by dephosphorylation of higher-order inositol polyphosphates in certain tissues or during certain developmental conditions.

2.2.5. Ins(1,4,5)P_3 may importantly function as a precursor for InsP_4, InsP_5 and InsP_6 isomers

PLC signaling in mammalian cells generates Ins(1,4,5)P_3, which in turn binds to intracellular receptors that mobilize $[Ca^{2+}]_{cyt}$. Although microinjection of Ins(1,4,5)P_3 has been demonstrated to induce an increase in $[Ca^{2+}]_{cyt}$ in plant cells (Gilroy *et al.*, 1990; Franklin-Tong *et al.*, 1996; Malhó, 1998), a confirmed IP_3 receptor homologue has not been identified in plant cells or in the *Arabidopsis* genome. In some cases, it has been implicitly assumed that Ins(1,4,5)P_3 is the only active inositol polyphosphate signal in plants and that its signaling activity is terminated by dephosphorylation. However, recent evidence suggests that in fact Ins(1,4,5)P_3 may be further phosphorylated to $InsP_4$, $InsP_5$ and $InsP_6$, because $InsP_6$ is a much stronger elicitor than Ins(1,4,5)P_3 in mobilizing $[Ca^{2+}]_{cyt}$ from the endoplasmic reticulum in stomatal guard cells (Lemtiri-Chlieh *et al.*, 2000, 2003).

The importance of pathways converting Ins(1,4,5)P_3 to $InsP_6$ have been characterized in yeast cells, where PLC activation is not associated with increases in Ins(1,4,5)P_3 but with increases in $InsP_6$ (Perera *et al.*, 2004; York *et al.*, 2001). Hypoosmotic stress activates hydrolysis of PtdIns(4,5)P_2 and the Ins(1,4,5)P_3 that is produced is rapidly phosphorylated to $InsP_6$ (Perera *et al.*, 2004). Yeast mutants lacking Arg82p or Ipk1p still hydrolyze PtdIns(4,5)P_2 in response to hypoosmotic stress but do not accumulate $InsP_6$ (Perera *et al.*, 2004). In fact, the only inositol polyphosphate detected in wild-type yeast cells under standard conditions was $InsP_6$ (York *et al.*, 2001). PLC activity was conditionally required throughout the cell cycle, and cells lacking PLC failed to produce $InsP_6$ while cells over-expressing PLC had detectable levels of Ins(1,4,5)P_3 and increased levels of $InsP_4$, $InsP_5$ and $InsP_6$ (York *et al.*, 1999, 2001). The Ins(1,4,5)P_3 produced by PLC hydrolysis of PtdIns(4,5)P_2 was not functioning as a signal itself, but instead was rapidly phosphorylated to $InsP_4$, $InsP_5$ and $InsP_6$. $InsP_6$ was shown to have a critical function in transcriptional regulation and mRNA export from the nucleus (York *et al.*, 1999, 2001; Ives *et al.*, 2000; Odom et al, 2000).

In *Arabidopsis*, Ins(1,4,5)P_3 can be phosphorylated to $InsP_6$ by the activities of an $InsP_3$/$InsP_4$/$InsP_5$ 3/5/6-kinase and an $InsP_5$ 2-kinase (Stevenson-Paulik *et al.*, 2002). An *Arabidopsis* 3/6 dual-specificity inositol polyphosphate kinase (AtIpk2b) has been cloned and characterized, and it can utilize Ins(1,4,5)P_3, Ins(1,4,5,6)P_4 or Ins(1,3,4,5)P_4 and produce Ins(1,3,4,5,6)P_5 (Xia *et al.*, 2003). $InsP_4$ isomers are known to have important roles in ion channel regulation (Ca^{2+} channels and Ca^{2+}-activated Cl^- channels) (Irvine and Schell, 2001; Ho *et al.*, 2002; Ho and Shears, 2002; Zonia *et al.*, 2002). Furthermore, higher-order inositol polyphosphates and inositol pyrophosphates have important roles in diverse cellular processes including chemotaxis in *Dictyostelium* (Luo *et al.*, 2003), endocytic trafficking in yeast (Saiardi *et al.*, 2002), protein kinase activity and integration of signaling networks in mammalian and yeast cells (Menniti *et al.*, 1993; Shears, 1996; Ho *et al.*, 2000; Shears, 2004). Taken together, these results suggest that Ins(1,4,5)P_3 may not be a primary inositol polyphosphate signal produced following PLC hydrolysis of PtdIns(4,5)P_2 (or potentially PtdInsP_3 or PtdIns), but instead may be utilized for the production of other important phosphoinositide signals in plant cells.

2.2.6. Does Ca^{2+} signaling have a general function as a simple binary switch or does it carry frequency encoded information?

$[Ca^{2+}]_{cyt}$ is an important component of cellular signaling cascades, and in mammalian cells its mobilization is often coupled to the generation of $Ins(1,4,5)P_3$. Ca^{2+} dynamics are known to exhibit different spectral frequency profiles and take the form of spikes, waves, or prolonged oscillations that encode specific information, called the 'Ca^{2+} signature'. In plants, such a Ca^{2+} signature has been implicated during the process of cell volume changes in guard cells (Staxén *et al.*, 1999; Allen *et al.*, 2000, 2001). However, a recent critical evaluation of the data and literature has shown that in many cases, similar putative Ca^{2+} signatures induce different cellular responses, while in other cases very different putative Ca^{2+} signatures induce very similar cellular responses (Scrase-Field and Knight, 2003). In other cell systems, the Ca^{2+} dynamics can exhibit great variability while the cellular responses show great stability (Scrase-Field and Knight, 2003). These conflicting reports have led to the suggestion that stomatal guard cells may be a special case, and that a more global role for $[Ca^{2+}]_{cyt}$ signaling in plant cells may be to function as a simple binary switch (on/off) (Scrase-Field and Knight, 2003).

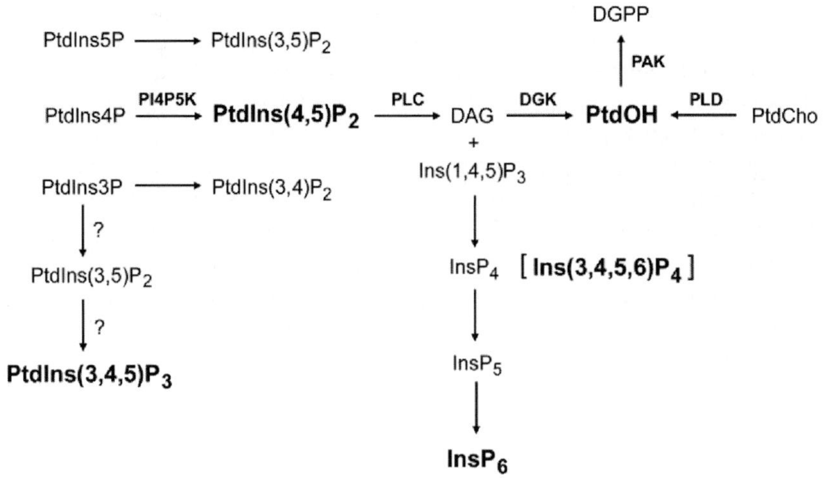

Figure 1. Diagram of some of the main phosphoinositide and phospholipid signaling pathways, with emphasis (in **bold text**) on recently discovered pathways that may have special relevance for plant cell signaling. Enzymes involved in the major lipid pathways that are discussed in the text are shown. The alternate pathway generating $PtdIns(3,4,5)P_3$ is included based on data from *S. pombe*. The significance of $Ins(3,4,5,6)P_4$ is discussed in Section 4.2.1.

2.2.7. Towards a new paradigm of phosphoinositide signaling in plant cells

In summary, recent advances in the understanding of phosphoinositides and phospholipids are leading to new insights in plant cell signaling. It is important to note that with respect to the PLC pathway, evidence is lacking to definitively assign a function for DAG, and a confirmed IP_3 receptor has still not been identified in plant cells or in the *Arabidopsis* genome. Instead, a picture is emerging that PtdOH and higher-order inositol polyphosphates are the key second messengers generated by PLC signaling. In addition, as only a fraction of possible phosphoinositide pathways have been studied in plants, it is likely that there will be exciting discoveries in the near future. A diagram illustrating a subset of the important phosphoinositide and phospholipid pathways that have been characterized in plants, protists, and mammals is presented in Figure 1, with an emphasis on the recently discovered pathways discussed in this section that may have special relevance for plant cell signaling.

3. PHOSPHOINOSITIDE SIGNALING RESPONSES TO STRESS

Plants continuously monitor their local environments and decode information that ultimately impacts all aspects of their metabolism, growth and reproduction. For this reason, the responses elicited by both biotic and abiotic stresses are among the best-characterized phosphoinositide signaling networks described in plants. Environmental stress also impacts the regulation and expression of genes involved in these signaling networks. Lin *et al.* (2004) performed DNA chip analysis of 79 independent clones representing 82 isozymes involved in phosphoinositide and lipid signaling networks and showed that the expression profile trends of different enzyme families were similar following treatments with abscisic acid (ABA), auxin (IAA), Ca^{2+} and mannitol. Importantly, many isoforms of the families PLC, PLD, PIPK and inositol polyphosphatase (IPPase) exhibited constitutive expression in unstimulated tissues and were differentially regulated by drought, osmotic shock, salt stress, cold and hormones (Lin *et al.*, 2004). This study provides an important new view of the dynamic complexity of phosphoinositide signaling responses to stress in plants.

3.1. Drought, hyperosmosis and salt stress elicit a number of phosphoinositide and phospholipid signals

The loss of productivity of arable land due to desertification has induced intensive research into understanding plant responses to drought and salt stress. Osmotic stress causes changes in cell volume that can impact multiple levels of cellular organization and function including plasma membrane curvature and deformation, association of the plasma membrane with the cell wall, interactions of the cytoskeleton and associated protein complexes with the plasma membrane, intracellular components that are sensitive to macromolecular crowding and/or ionic strength, and the partitioning of water across the plasma membrane and within intracellular compartments. Cell volume regulation has been intensively studied in mammalian cells and it is controlled by a number of phospholipid and ionic signals, including PI4P5K and PtdIns(4,5)P_2, PLD and PtdOH, PLA_2 and lyso-PtdOH, mechanosensitive Ca^{2+} channels and Ca^{2+}-activated Cl^- and K^+ currents (for review, Wehner *et al.*, 2003).

Plant cells rapidly respond to hyperosmotic stress with a battery of signals, including PtdIns5P, PtdIns(3,5)P_2, PtdIns(4,5)P_2, PtdOH, lyso-PtdOH, DGPP, and Ins(1,4,5)P_3 (for reviews see Munnik and Meijer, 2001; Xiong *et al.*, 2002; Meijer and Munnik, 2003). These signals have been identified in a number of different plant species and tissues, either singly or in combination, and appear to be correlated with the severity of stress (see below). Osmotic stress also activates MAPK and other protein kinase cascades, and in specific cases these kinase cascades may be downstream from PtdOH signaling (Munnik *et al.*, 1999; Lee *et al.*, 2001, 2003; Munnik and Meijer, 2001; Anthony *et al.*, 2004; Testerink *et al.*, 2004). The number of different signals elicited by hyperosmotic stress indicates how critical the intracellular hydrostatic pressure and osmotic status are for plant cell function, and the complexity of pathways that function during the mobilization of recovery and survival strategies. Cellular targets for these signals include vesicle trafficking, cytoskeletal organization, activation of protein kinase and phosphatase cascades, small G-proteins, ion channels and ion fluxes. This section will provide an overview of some of the most important research in this field.

In a series of investigations using a range of KCl from 0–500 mM in the green algae *Chlamydomonas*, it was shown that different phospholipid signaling pathways were stimulated by different degrees of severity of stress (Meijer *et al.*, 1999, 2001a, 2001b, 2002; Munnik *et al.*, 1999, 2000; Munnik and Meijer, 2001). PLD was stimulated and PtdOH levels increased at 2 different concentration ranges, from 25–75 mM KCl and again at 200–500 mM KCl. PLC and DGK stimulation, together with increased levels of PtdIns(4,5)P_2, PtdOH and DGPP, occurred only at the higher concentration range. Thus, PtdOH generated in response to a strong hyperosmotic stress appears to result from activation of both PLC-DGK and PLD pathways. Increases in isomers phosphorylated at the D-3 position, notably PtdIns(3,5)P_2, occurred between 150–300 mM KCl. PLA_2 stimulation and increased lyso-PtdOH occurred between 200–400 mM KCl. Hyperosmotic stress also induced increases in PtdIns3P, PtdIns4P, PtdIns5P, PtdIns(3,5)P_2 and PtdIns (4,5)P_2 (Meijer *et al.*, 1999, 2001b; Munnik and Meijer, 2001). Increases in PtdIns(4,5)P_2 occurred at the expense of PtdIns4P, not of PtdIns5P. The data suggested that PtdIns5P is likely to be a precursor for PtdIns(3,5)P_2 (Meijer *et al.*, 2001b).

$PtdIns(3,5)P_2$ levels also increase in response to hyperosmotic stress in *Nicotiana tabacum* pollen tubes (Zonia and Munnik, 2004). A recent report characterizes the *Arabidopsis* phosphatase (At5PTase11) that specifically terminates $PtdIns(3,5)P_2$ signaling, and shows that the gene is regulated by ABA and auxin (Ercetin and Gillaspy, 2004). In *Saccharomyces*, $PtdIns(3,5)P_2$ is involved in endocytic trafficking and regulation of the vacuole (Shaw *et al.*, 2003; Whitley *et al.*, 2003; Rudge *et al.*, 2004).

Hyperosmotic stress induces increases in the levels of $PtdIns(4,5)P_2$ in plant cells (Einspahr *et al.*, 1988; Cho et al, 1993; Heilman *et al.*, 1999; Pical *et al.*, 1999; DeWald *et al.*, 2001; Takahashi *et al.*, 2001; Zonia and Munnik, 2004), and in some cases this has been correlated with an increase in $Ins(1,4,5)P_3$ (Drøbak and Watkins, 2000; DeWald *et al.*, 2001; Takahashi *et al.*, 2001). It had been thought that increases in PLC activity and $Ins(1,4,5)P_3$ might have a role in mobilizing $[Ca^{2+}]_{cyt}$, as drought and salt stress are known to elevate $[Ca^{2+}]_{cyt}$ as one of the first cellular responses (Knight *et al.*, 1997; Kiegle *et al.*, 2000). However, as discussed in Section 2.2.5., recent results suggest that higher-order inositol polyphosphate signals may function in the mobilization of $[Ca^{2+}]_{cyt}$ (Lemtiri-Chlieh *et al.*, 2000, 2003), and so the understanding of these responses to hyperosmotic stress in plant cells may undergo revision in the near future.

ABA is a phytohormone involved in plant dessication responses that promotes stomatal closure and inhibits stomatal opening, thereby minimizing water loss via transpiration. It evokes multiple responses in guard cells, inhibiting the plasma membrane H^+-ATPase and activating an anion channel, a non-selective Ca^{2+}-permeable channel, and K^+ efflux (Blatt, 2000; Schroeder *et al.*, 2001). ABA-induced responses have been correlated with PLC activity and increased levels of $Ins(1,4,5)P_3$ in guard cells (Lee *et al.*, 1996; Staxén *et al.*, 1999) and in *Arabidopsis* seedlings (Burnette *et al.*, 2003). In contrast, no increase in $Ins(1,4,5)P_3$ was observed in response to ABA in *Arabidopsis* suspension cells, although $Ins(1,4,5)P_3$ levels did increase within seconds in response to hyperosmotic stress (Takahashi *et al.*, 2001). Some clarification of these conflicting reports has now been obtained. Recent work using transgenic plants with altered levels of PLC activity show that PLC signaling is required for secondary ABA responses and functions in the cascade of events that inhibit stomatal opening, but it is not involved in the primary responses that promote ABA-induced stomatal closure (Sanchez and Chua, 2001; Hunt *et al.*, 2003, Mills *et al.*, 2004). ABA also stimulates PI3K and PI4K activities, and inhibition of PI3K and PI4K lead to decreased levels of PtdInsP isoforms and inhibits ABA-induced stomatal closing (Jung *et al.*, 2002).

ABA-induced stomatal closure stimulates PLD activity and induces increases in PtdOH (Ritchie and Gilroy, 1998, 2000; Jacob *et al.*, 1999). PLDα and PLDγ were identified in *Arabidopsis* guard cells, and antisense suppression of PLDα impairs stomatal closure induced by water deficits (Sang *et al.*, 2001b). Induction of PtdOH signaling also correlates with downstream events that occur during stomatal closure, including activation of cADP-ribose and Rab GTPase pathways (Jacob *et al.*, 1999; Hallouin *et al.*, 2002). Furthermore, there is evidence that PtdOH signaling is key to all ABA-induced cellular responses in *Arabidopsis* suspension cells, in that it was found to be upstream of activation of plasma membrane anion channels (Hallouin *et al.*, 2002).

A yeast *Sec14* phosphatidylinositol transfer protein homologue (Ssh1p) has been identified in soybean that potentiates the activities of PI3K and PI4K (Monks *et al.*, 2001). Hyperosmotic stress induces the phosphorylation of Ssh1p by protein kinases SPK1 and/or SPK2, and the phosphorylated Ssh1p has a much weaker association with the plasma membrane (Monks *et al.*, 2001). These results suggest that hyperosmotic stress diverts the normal homeostatic pathways for inositol phospholipid distribution in cell membranes, which may function in reorganizing vesicle trafficking in response to hyperosmotic stress.

3.2. Pathogen stress and defence responses

Plants are prey to many insects, fungi, and pathogenic bacteria, which cause considerable losses in agricultural productivity throughout the world. This has generated an intensive effort to elucidate plant responses to biotic stress. Some plant species have evolved counteractive mechanisms of defense involving hypersensitive cell death or systemic acquired resistance (for review, Nimchuk *et al.*, 2003). The first step in defence responses involves recognition of specific pathogen-derived effector molecules (elicitors) that activate specific signaling cascades and gene activation programs (Thomma *et al.*, 2001a, 2001b). A battery of cellular response cascades have been identified and include reactive oxygen species (ROS), nitric oxide, Ca^{2+} flux, activation of plasma-membrane NADPH oxidase, changes in activity of plasma membrane ion channels and H^+-ATPase, G-proteins, cyclic GMP and cyclic ADP-ribose, MAPK and CDPK cascades, salicylic and jasmonic acid, and ethylene (Xing *et al.*, 1997; Zimmermann *et al.*, 1997; Blumwald *et al.*, 1998; Durner *et al.*, 1998; Blume *et al.*, 2000; Nürnberger and Scheel, 2001; Taylor *et al.*, 2001; Wendehenne *et al.*, 2001; Delledonne *et al.*, 2002; Cessna *et al.*, 2003; Kroj *et al.*, 2003; Overmyer *et al.*, 2003; Rojo *et al.*, 2003). Plant defense responses are pathogen-dependent, as specific subsets of these cascades are activated in response to specific effector molecules (Thomma *et al.*, 2001a, 2001b).

Early work reported that fungal elicitor induced increases in PtdIns(4,5)P_2 and Ins(1,4,5)P_3 in pea epicotyl (Toyoda *et al.*, 1993). In tobacco suspension cells, increases in Ins(1,4)P_2 and Ins(1,4,5)P_3 were detected, although no significant change in the level of PtdIns(4,5)P_2 was observed (Kamada and Muto, 1994). Surprisingly, decreased levels of Ins(1,4,5)P_3 were reported in *Arabidopsis* suspension cells infected with the bacterial pathogen *Pseudomonas syringae*, suggesting that this method of elicitation may produce a different cellular response (Shigaki and Bhattacharyya, 2000).

Exposure to elicitors and initiation of defence responses activates PLC, PLD, and PLA_2 (Van der Luit *et al.*, 2000; Wang *et al.*, 2000; Laxalt and Munnik, 2002; Scherer *et al.*, 2002; Song and Goodman, 2002; Zhang *et al.*, 2003; De Jong *et al.*, 2004; Kasparovsky et al, 2004), and induces increased expression of genes coding for specific PLD isoforms and lipid transfer proteins (Laxalt *et al.*, 2001; Park *et al.*, 2002; Zabela *et al.*, 2002; Jung *et al.*, 2003). Elicitors induce rapid increases in the levels of PtdOH and DGPP. Differential labelling experiments showed that the increased PtdOH was primarily generated through the PLC-DGK pathway (Van der Luit *et al.*, 2000; Laxalt and Munnik, 2002; Den Hartog *et al.*, 2003; De Jong *et al.*, 2004). PtdOH has previously been shown to be important for the production of superoxide in *Arabidopsis* (Sang *et al.*, 2001a). Recent work has now proven that PtdOH acts upstream of ROS signaling (De Jong *et al.*, 2004; Park *et al.*, 2004), and upstream of wound-induced MAPK cascades (Lee *et al.*, 2001), placing PtdOH signaling as one of the first events during activation of elicitor-dependent defence responses. ROS signaling appeared to be responding to PtdOH generated by the PLC-DGK pathway (De Jong *et al.*, 2004), while the PLD pathway appeared to be involved in activation of MAPK cascades (Lee *et al.*, 2001). Legume plant cells also exhibit rapid increases in PtdOH when challenged with Nod factor from the symbiotic bacteria *Rhizobium* (Den Hartog *et al.*, 2001; Laxalt and Munnik, 2002). Nod factor was shown to increase PtdOH levels by activating both the PLC-DGK and PLD pathways (Den Hartog *et al.*, 2001, 2003).

3.3. Temperature stress

Plant responses to external temperature are essential for optimization of metabolic activity, the timing of flowering and seed set, and during the process of vernalization. Chilling and freezing adversely affect agricultural productivity, yet some plants develop cold tolerance if the temperature is decreased gradually (Pearce, 1999). For this reason, investigations of temperature stress in plants have largely focused on characterizing signaling responses to cold. Cold stress induces increases in phosphoinositides and phospholipids and activates a number of proteins involved in phosphoinositide signaling pathways.

Cold shock decreases the levels of membrane structural lipids PC, PE, PG but increases the levels of PtdOH and lysophospholipids (Welti *et al.*, 2002). *Arabidopsis* mutants deficient in PLDα expression have increased tolerance to freezing (Welti *et al.*, 2002). In contrast, deletion of the plasma membrane-bound PLDδ in *Arabidopsis* causes increased sensitivity to freezing, while overexpression of PLDδ increased freezing tolerance (Li *et al.*, 2004). Increased levels of PtdOH were detected within 10 min after cold treatment of *Nicotiana tabacum* suspension cells, concurrently with decreases in PtdInsP and $PtdInsP_2$ (Gawer *et al.*, 1999). Similarly, in *Arabidopsis* suspension cells, cold induces a rapid increase in the activities of PLC and PLD pathways, increasing the levels of PtdOH (Ruelland *et al.*, 2002). Using lipid profiling analysis and differential labelling strategies, it was estimated that the PLC-DGK pathway generates 80% of this PtdOH and the PLD pathway produces 20% (Ruelland *et al.*, 2002).

Map-based cloning of an *Arabidodopsis* mutant that is less tolerant of cold stress enabled the identification of inositol polyphosphate 1-phosphatase (Xiong *et al.*, 2001; Viswanathan and Zhu, 2002). The mutant plants had higher levels of Ins(1,4,5)P_3 than wild-type plants, indicating a role for inositol polyphosphate signals in the response to cold (Xiong *et al.*, 2001; Viswanathan and Zhu, 2002). In yeast cells, deletion of an inositol polyphosphate 5-phosphatase (INP51) that exhibits activity against PtdIns(4,5)P_2 conferred cold tolerance and increased levels of PtdIns(4,5)P_2 and Ins(1,4,5)P_3 (Stolz *et al.*, 1998). There are 15 putative inositol 5-phosphatases in *Arabidopsis* (Berdy *et al.*, 2001). One of these, At5PTase1, exhibits activity against Ins(1,4,5)P_3 and Ins(1,3,4,5)P_4 but not PtdIns(4,5)P_2 (Berdy *et al.*, 2001). Transgenic plants expressing At5PTase1 had a delay in ABA induction of a cold responsive gene (*KIN*), indicating a potential role for Ins(1,4,5)P_3 and/or Ins(1,3,4,5)P_4 in cold responses (Xiong *et al.*, 2002; Burnette *et al.*, 2003).

Cold stress also induces Ca^{2+} influx and Ca^{2+}-dependent kinase activities (Knight *et al.*, 1996; Wood *et al.*, 2000; Knight and Knight, 2001; Viswanathan and Zhu, 2002), with the rate of cooling and final temperature having the greatest impact on the Ca^{2+} response (Plieth *et al.*, 1999; Knight and Knight, 2001; Knight, 2002; Scrase-Field and Knight, 2003).

4. CELL VOLUME REGULATION IN POLLEN TUBES AND FUNCTIONAL SPECIFICITY OF PHOSPHOINOSITIDES DURING CELLULAR HOMEOSTASIS AND OSMOTIC STRESS

Flowering plants have non-motile sperm that must be delivered to the female gametophyte buried within the flower tissues to affect fertilization. The vehicle that has evolved to perform this function is the pollen tube (for reviews, Hepler *et al.*, 2001; Johnson and Preuss, 2002; Holdaway-Clarke and Hepler, 2003). Pollen tube cell biology is adapted for survival in diverse extracellular environments that are encountered after dehiscence, including germination on a receptive stigmatic surface and penetration through the style. Growth is polarized and occurs only within a narrowly restricted zone at the apex. The growth rate is not linear but displays sinusoidal oscillations from fast to slow cycles. This complex biology is achieved because the pollen tube is a dynamical system composed of interconnected biochemical and biophysical networks that can be rapidly modulated in order to attain optimum growth and function. Recent work has shown that these networks are integrated by multiple signaling pathways that process divergent input information into a cohesive cellular response. Thus, the pollen tube is an excellent model system for studies of phosphoinositide signaling in plants.

4.1. Signals that regulate pollen tube growth

A number of signaling pathways regulating pollen tube growth have been identified and include ion fluxes (Ca^{2+}, H^+, K^+, Cl^-) (for review, Holdaway-Clarke and Hepler, 2003), Rho-GTPase homologues Rac/Rop (Fu and Yang, 2001; Chen *et al.*, 2003), receptor kinases (Estruch *et al.*, 1994; Moutinho et al, 1998; Kim *et al.*, 2002), and phosphoinositides (Franklin-Tong, 1999; Kost *et al.*, 1999; Zonia *et al.*, 2002; Zonia and Munnik, 2004).

Pollen tube growth correlates with a tip-localized $[Ca^{2+}]_{cyt}$ gradient that is absent from non-growing cells. There is also influx of extracellular Ca^{2+} at the pollen tube tip. Both the intracellular $[Ca^{2+}]_{cyt}$ gradient and influx of extracellular Ca^{2+} oscillate with periods that correlate with growth oscillation periods. Initial work showed that the peak $[Ca^{2+}]_{cyt}$ gradient oscillated with the same phase as growth rate oscillations (Holdaway-Clarke *et al.*, 1997), but subsequent work showed that it is phase-delayed by ca. 35° (based on 360° cycle) with respect to growth in lily pollen tubes (Messerli *et al.*, 2000). Peak influx of extracellular Ca^{2+} lags growth rate oscillation peaks by ca. 135° in lily pollen tubes (Holdaway-Clarke *et al.*, 1997; Messerli *et al.*, 1999, 2000). Thus, while Ca^{2+} is a crucial component of signaling networks mediating pollen tube growth, other signals must also be important (see also Section 4.2.). K^+ and H^+ ions influx at the pollen tube tip but are phase-delayed by ca. 100° with respect to growth rate oscillation peaks (Messerli *et al.*, 1999).

Several studies have demonstrated the presence of PLC activity in pollen tubes (Helsper *et al.*, 1986, 1987; Franklin-Tong *et al.*, 1996). In addition, microinjection and photolysis of caged Ins(1,4,5)P_3 induces increases in $[Ca^{2+}]_{cyt}$ (Franklin-Tong *et al.*, 1996; Malhó, 1998). However, the time-lag for $[Ca^{2+}]_{cyt}$ increases were typically slow compared to the time required for photolysis of caged Ins(1,4,5)P_3 (300 s *versus* 10–20 s), and the increased levels were then sustained for a further 400 s (Franklin-Tong *et al.*, 1996). This suggests that other factors must be recruited to mediate Ins(1,4,5)P_3-induced $[Ca^{2+}]_{cyt}$ release, or it may be possible that Ins(1,4,5)P_3 is phosphorylated to higher-order inositol polyphosphate isomers that actively mobilize $[Ca^{2+}]_{cyt}$ from intracellular stores (as discussed in Section 2.2.5.). The mobilized $[Ca^{2+}]_{cyt}$ induced by photolysis of caged Ins(1,4,5)P_3 had the characteristics of a slow wave that initiated in a distal region near the nucleus and then propagated toward the apex (Franklin-Tong et al, 1996). It is also important to note that Ins(1,4,5)P_3 ultimately acts as a negative regulator of pollen tube growth (Franklin-Tong et al, 1996; Malhó, 1998; L. Zonia, personal observations).

The Rho-family GTPase Rac was shown to localize to the plasma membrane at the pollen tube apex (Kost *et al.*, 1999). PtdIns(4,5)P_2 and PIPK are known regulators of Rho GTPases in mammalian cells. Kost *et al.* (1999) used pollen tube extracts and showed that a PIPK activity interacted specifically with the pollen tube Rac and produced primarily PtdIns(4,5)P_2 from mixed phosphoinositide substrates. Finally, transient expression of a GFP construct with the PH domain of human PLCδ_1 (which binds specifically to PtdIns(4,5)P_2) accumulated at the apical plasma membrane. This work suggested that Rac, PIPK and PtdIns(4,5)P_2 all localize to the pollen tube apical region and function co-ordinately in the regulation of growth (Kost *et al.*, 1999).

4.2. Homeostatic signaling networks regulating pollen tube growth converge with those regulating cell volume during osmotic stress responses

Pollen tubes are exquisitely sensitive to changes in the extracellular osmotic potential. Upon hypoosmotic or hyperosmotic shifts, water flows relatively freely into and out of the pollen tube apical region (spanning from the apex to ca. 50 μm distal to the apex) and the cell volume swells or shrinks in response (Zonia *et al.*, 2002; Zonia and Munnik, 2004). Several signals that regulate homeostatic pollen tube growth are differentially stimulated or attenuated during these cell volume changes (Zonia and Munnik, 2004). An illustration showing the signals that are induced or attenuated in response to osmotically-induced cell volume increase or decrease is presented in Figure 2 and discussed in detail below.

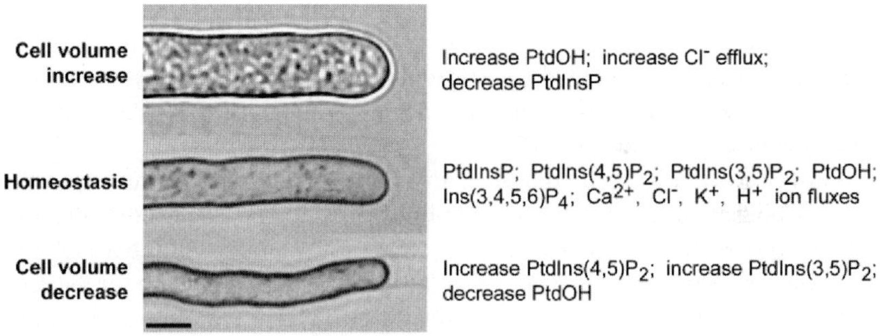

Figure 2. Phosphoinositide and phospholipid signals involved in homeostatic regulation of pollen tube growth are functionally differentiated during the response to osmotically induced cell volume changes. Imaging was performed to maximize visualization of the plasma membrane (the dark boundary around the perimeter of the pollen tube) and minimize intracellular details. The signals present during cell swelling, normal growth or cell shrinking are given at the right and discussed in detail in the text. Bar = 10 μm.

4.2.1. Homeostatic signals include Ca^{2+} and Cl^- flux, Ins(3,4,5,6)P_4, PtdOH, PtdInsP, PtdIns(4,5)P_2 and PtdIns(3,5)P_2

During a study of the affects of a number of phosphoinositides on pollen tube growth, Ins(3,4,5,6)P_4 was identified as a negative regulator (L. Zonia, personal observations). It inhibits growth and induces rapid cell volume increases. The effects of Ins(3,4,5,6)P_4 were specific and were not mimicked by either Ins(1,3,4,5)P_4 or Ins(1,3,4,5,6)P_5 (Zonia *et al.*, 2002). In mammalian cells, Ins(3,4,5,6)P_4 inhibits a Ca^{2+}-activated Cl^- channel that is involved in secretion (Vajanaphanich *et al.*, 1994; Ismailov *et al.*, 1996; Ho *et al.*, 1997; Barrett *et al.*, 1998; Nilius *et al.*, 1998; Xie *et al.*, 1998; Yang *et al.*, 1999; Carew *et al.*, 2000; Ho *et al.*, 2000, 2001, 2002; Ho and Shears, 2002; Renstrom *et al.*, 2002). Therefore, a study was initiated to explore Cl^- channels in *Nicotiana tabacum* pollen tubes, by mapping Cl^- flux along the cell surface using an ion-specific vibrating probe (Zonia *et al.*, 2001, 2002). This work showed that oscillations of Cl^- efflux occur at the apex, with peaks that are locked in phase with peaks in growth rate oscillation cycles (0° phase-shift), indicating an important link between Cl^- efflux and growth. A steady influx of Cl^- occurs along the pollen tube starting at ca. 15 μm distal to the tip. Ins(3,4,5,6)P_4 was shown to disrupt Cl^- efflux at the apex (Zonia *et al.*, 2002). Inhibition of Cl^- efflux with either Ins(3,4,5,6)P_4 or Cl^- channel blockers inhibited pollen tube growth and induced increases in the apical cell volume measured in the region spanning from the apex to 50 μm distal to the apex. Conversely, hypoosmotic shifts of the extracellular osmotic potential increased the frequency of Cl^- efflux at the apex. Extensive research has documented a role for Cl^- efflux in the regulation of cell volume and hydrostatic pressure in both plant and mammalian cells (Ward *et al.*, 1995; Teodoro *et al.*, 1998; Shabala *et al.*, 2000; Van der Wijk *et al.*, 2000; Bali *et al.*, 2001; Leonhardt *et al.*, 2001; Wondergem *et al.*, 2001). The results from pollen tubes indicate a convergence of signaling networks regulating polarized growth with those regulating cell volume, and indicate that Cl^- efflux and its negative regulator Ins(3,4,5,6)P_4 are important signals controlling these networks.

Further work identified additional signals involved in pollen tube growth. Several phosphoinositide and phospholipid signals were detected, including PtdOH, lyso-PtdOH, DGPP, PtdInsP, PtdIns(4,5)P_2, and PtdIns(3,5)P_2 (Zonia and Munnik, 2004). This indicates that these signals are necessary for the regulation of cellular homeostasis and growth. Furthermore, these signals were differentially induced or attenuated during cellular responses to hypoosmotic *versus* hyperosmotic stress, indicating that they are functionally specific. This work significantly extends previous studies of phosphoinositide and phospholipid signals involved in pollen tube growth (Helsper *et al.*, 1986, 1987; Franklin-Tong *et al.*, 1996; Kost *et al.*, 1999; Potocky *et al.*, 2003).

4.2.2. PtdOH is induced and PtdInsP is attenuated during cell swelling

Hypoosmotic stress induces increases in PLD activity and PtdOH levels and decreases the levels of PtdInsP (Zonia and Munnik, 2004). PtdOH levels begin accumulating only after a significant increase in cell volume occurs. In contrast, increased Cl^- efflux at the pollen tube apex is induced with lower cell volume increases. In mammalian cells, activation of the PLD pathway can potentiate Cl^- secretion (Vajanaphanich *et al.*, 1993; Oprins *et al.*, 2001, 2002). In *Arabidopsis* suspension cells, ABA-dependent stimulation of PLD activity was necessary for the activation of plasma membrane anion channels (Hallouin *et al.*, 2002). These results indicate considerable cross-talk between PLD pathways and anion channels in the plasma membrane of both mammalian and plant cells. Thus, it may be possible that adjustment to low levels of hypoosmosis in pollen tubes is mediated by increased Cl^- efflux while adjustment to more severe hypoosmosis and cell swelling is mediated by PtdOH acting synergistically with Cl^- efflux. Hypoosmotic stress has also has been shown to increase PtdOH levels within 2 min in the green algae *Dunaliella salina* (Einspahr *et al.*, 1988), and in suspension cells of tobacco, alfalfa, tomato and *Arabidopsis* (T. Munnik, unpublished observations). Putative targets of PtdOH include ion channels, protein kinases (CDPK, PDK1, MAPK, Raf 1-kinase), lipid kinases (type I PI4P5K), protein phosphatases (ABI1, PP1, SHP1), small G-proteins (ARF), and other proteins involved in vesicle trafficking and cytoskeletal organization (English, 1996; Munnik, 2001; Ktistakis *et al.*, 2003; Anthony *et al.*, 2004; Testerink *et al.*, 2004; Zhang *et al.*, 2004; Zhao and Wang, 2004; Testerink and Munnik, 2005).

4.2.3. PtdIns(4,5)P_2 and PtdIns(3,5)P_2 are induced and PtdOH is attenuated during cell shrinking

Hyperosmotic stress increases the levels of PtdIns(4,5)P_2 and PtdIns(3,5)P_2 and decreases PLD activity and PtdOH (Zonia and Munnik, 2004). This shows the functional specificity of signaling responses to hypoosmotic versus hyperosmotic stress. Increases in the levels of PtdIns(3,5)P_2 and/or PtdIns(4,5)P_2 in response to hyperosmotic stress have been reported for other plant tissues and for green algae (Einspahr *et al.*, 1988; Meijer *et al.*, 1999; Heilmann *et al.*, 1999, 2001; Pical *et al.*, 1999; DeWald *et al.*, 2001). Possible targets for PtdIns(3,5)P_2 include vesicle trafficking, tonoplast turnover and vacuolar integrity (Yamomoto *et al.*, 1995; Gary *et al.*, 1998; Wurmser *et al.*, 1999; Odorizzi *et al.*, 2000; Simonsen *et al.*, 2001; Shaw *et al.*, 2003; Whitley *et al.*, 2003; Rudge *et al.*, 2004; Dove *et al.*, 2004). Possible targets for PtdIns(4,5)P_2 include ion channels, proteins involved in vesicle trafficking, cytoskeletal organization and cell wall synthesis, in addition to acting as a precursor for soluble inositol polyphosphate pathways (Janmey, 1994; Martin, 1998; Janmey *et al.*, 1999; Hilgemann *et al.*, 2001; Martin, 2001; Takenawa and Itoh, 2001; McLaughlin *et al.*, 2002; Suetsugu and Takenawa, 2003; Drøbak *et al.*, 2004; Zhong and Ye, 2004; Williams *et al.*, 2005; Zhong *et al.*, 2004, 2005).

Hyperosmotic stress was previously shown to decrease PtdOH in *Dunaliella salina* (Einspahr *et al.*, 1988). However, it causes increased levels of PtdOH in other plant tissues and species (Frank *et al.*, 2000; Munnik *et al.*, 2000; Katagiri *et al.*, 2001; Meijer *et al.*, 2001a, 2002; Munnik, 2001; Munnik and Meijer, 2001; Arisz *et al.*, 2003), and activates PLD in seedlings, suspension cells and *Chlamydomonas* (for review, Meijer and Munnik, 2003). The phytohormone ABA induces guard cell shrinking and stomatal closure, and causes an increase in PLD activity and PtdOH (Jacob *et al.*, 1999). The combined evidence indicates that PtdOH signaling may be functionally differentiated depending on whether it results from activation of the PLC-DGK pathway or the PLD pathway. It is also likely that the 2 pathways are differentiated by spatial and/or developmental constraints in addition to tissue-specific expression of different phospholipase isoenzymes (for example, the *Arabidopsis* genome predicts 9 putative PLCs and 12 putative PLDs). In pollen tubes, most of the PtdOH observed during normal growth and in response to hypoosmotic stress appears to be generated by the PLD pathway (Zonia and Munnik, 2004). In contrast, activation of the PLC-DGK pathway appears to generate much of the PtdOH that accumulates in response to hyperosmotic stress in other plant tissues (Munnik *et al.*, 2000; Munnik, 2001; Meijer *et al.*, 2002; Arisz *et al.*, 2003).

4.2.4. Towards a new paradigm of plant cell stress signaling

In summary, recent work shows that phosphoinositide and phospholipid signals that are necessary for cellular homeostasis and pollen tube growth are differentially stimulated or attenuated by hypoosmotic versus hyperosmotic stress and the resultant cell volume increase or decrease (Figure 2). This indicates that signaling networks regulating cellular homeostasis and growth converge with those regulating cell volume. Ongoing research into this problem consistently yields evidence in support of this theory of pollen tube growth (L. Zonia, personal observations). It will be important to determine if signaling networks regulating cellular homeostasis converge with those regulating stress responses in other plant tissues, and to characterize their functional specificity.

5. CONCLUSIONS AND PERSPECTIVES

This review has evaluated the current understanding of plant phosphoinositide signaling in unstimulated cells and during stress responses. It is clear that signaling networks are important for normal cellular function and during plant stress responses, but to date only a subset of the potential pathways have been investigated. Even so, a number of phosphoinositides and phospholipid signals have been identified as mediators of both homeostatic and recovery/survival strategies implemented by plant cells. Dissection of signaling networks in other organisms continuously yields new data that can be utilized to push back the boundary of what is known about plant signaling networks. However, it is important to consider that plant signals and cellular responses may differ from those in other organisms because of the plant's unique biology.

New tools for research on cellular signaling are continuously being developed and are providing significant new details about the subcellular localization and dynamics of interactions between signals and their targets. Novel approaches to localize phosphoinositide signals in cells include the use of fluorescently labeled or caged phospholipids, or the use of antibodies or probes specific for PtdIns(4,5)P_2 and PtdIns(3,4,5)P_3 (Prestwich *et al.*, 2002; Watt *et al.*, 2002, 2004; Downes *et al.*, 2003; Goedhart and Gadella, 2004). The most exciting results are arguably emerging from studies employing GFP constructs fused with protein domains that bind specific phosphoinositide and phospholipid isomers (eg. FYVE, PH, PX, ENTH/ANTH, Tubby) that are important for target recognition and specificity (Cullen *et al.*, 2001; Lam *et al.*, 2002; Lemmon, 2003; Van Leeuwen *et al.*, 2004). There are a large number of genes in the *Arabidopsis* genome that contain one or more of these protein domain modules (Van Leeuwen *et al.*, 2004). Such proteins represent potential networks to relay information from the plasma membrane throughout the cell. The construction of GFP chimeras enables functional analysis and visualization of the interactions between lipid signals and their protein targets in real-time in living cells. This technology has already been used to localize Rac-GTPases and PtdIns(4,5)P_2 in the pollen tube apical region (Kost *et al.*, 1999), to track PtdIns3P function in membrane trafficking in *Arabidopsis* protoplasts (Kim *et al.*, 2001), to demonstrate the involvement of PtdIns3P and PtdIns4P in stomatal function (Jung *et al.*, 2002), to monitor PLD-induced microtubule rearrangements in tobacco BY-2 cells (Dhonukshe *et al.*, 2003), to demonstrate nuclear localization of the *Arabidopsis* inositol polyphosphate 6/3-kinase in tobacco BY2 protoplasts (Xia *et al.*, 2003), and to localize the AGC2–1 protein kinase in *Arabidopsis* root hairs (Anthony *et al.*, 2004). It is expected that future approaches such as these will continue to reveal stunning new data about phosphoinositide signaling in plants.

REFERENCES

Allen, G.J., Chu, S.P., Schumacher, K., Shimazaki, C.T., Vafaedos, D., Kemper, A., Hawke, S.D., Tallman, G., Tsien, R.Y., Harper, J.F., Chory, J. and Schroeder, J.I., 2000, Alteration of stimuli-specific guard cell calcium oscillations and stomatal closing in *Arabidopsis det3* mutant. *Science* **289:** 2338–2342.

Allen, G.J., Chu, S.P., Harrington, C.L., Schumacher, K., Hoffmann, T., Tang, Y.Y., Grill, E. and Schroeder, J.I., 2001, A defined range of guard cell calcium oscillation parameters encodes stomatal movements. *Nature* **411:** 1053–1057.

Anthony, R.G., Henriques, R., Helfer, A., Mészáros, T., Rios, G., Testerink, C., Munnik, T., Deák, M., Koncz, C. and Bögre, L., 2004, A protein kinase target of a PDK1 signalling pathway is involved in root hair growth in *Arabidopsis*. *EMBO J* **23:** 572–581.

Arisz, S.A., Valianpour, F., Van Gennip, A.H. and Munnik, T., 2003, Substrate preference of stress-activated phospholipase D in *Chlamydomonas* and its contribution to PA formation. *Plant J.* **34:** 595–604.

Bali, M.Z., Lipecka, J., Edelman, A. and Fritsch, J., 2001, Regulation of ClC-2 chloride channels in T84 cells by TGF-alpha. *Am. J. Physiol. Cell Physiol.* **280:** C1588-C1598.

Barrett, K.E., Smitham, J., Traynor-Kaplan, A. and Uribe, J.M., 1998, Inhibition of Ca^{2+}-dependent Cl^{-} secretion in T84 cells: Membrane target(s) of inhibition is agonist specific. *Am. J. Physiol. Cell Physiol.* **274:** C958-C965.

Berdy, S.E., Kudla, J., Gruissem, W. and Gillaspy, G.E., 2001, Molecular characterization of At5PTase1, an inositol phosphatase capable of terminating inositol trisphosphate signaling. *Plant Physiol.* **126:** 801–810.

Blatt, M.R., 2000, Cellular signaling and volume control in stomatal movements in plants. *Annu. Rev. Cell Dev. Biol.* **16:** 221–241.

Blume, B., Nürnberger, T., Nass, N. and Scheel, D., 2000, Receptor-mediated increase in cytoplasmic free calcium required for activation of pathogen defense in parsley. *Plant Cell* **12:** 1425–1440.

Blumwald, E., Aharon, G.S. and Lam, B.C-H., 1998, Early signal transduction pathways in plant-pathogen interactions. *Trends Plant Sci.* **3:** 342–346.

Bominaar, A.A., Van Dijken, P., Draijer, R. and Van Haastert, P.J.M., 1991, Developmental regulation of the inositol 1,4,5-trisphosphatases in *Dictyostelium discoideum*. *Differentiation* **46:** 1–5.

Burnette, R.N., Gunesekera, B.M. and Gillaspy, G.E., 2003, An *Arabidopsis* inositol 5-phosphatase gain-of-function alters abscisic acid signaling. *Plant Physiol.* **132:** 1011–1019.

Caffrey, J.J., Darden, T., Wenk, M.R. and Shears, S.B., 2001, Expanding coincident signaling by PTEN through its inositol 1,3,4,5,6-pentakisphosphate 3-phosphatase activity. *FEBS Lett.* **499:** 6–10.

Carew, M.A., Yang, X.N., Schultz, C. and Shears, S.B., 2000, myo-Inositol 3,4,5,6-tetrakisphosphate inhibits an apical calcium-activated chloride conductance in polarized monolayers of a cystic fibrosis cell line. *J. Biol. Chem.*, **275:** 26906–26913.

Carroll, K., Gomez, C. and Shapiro, L., 2004, TUBBY proteins: the plot thickens. *Nat. Rev. Mol. Cell Biol.* **5:** 55–63.

Cessna, S.G., Kim, J. and Taylor, A.T.S., 2003, Cytosolic Ca^{2+} pulses and protein kinase activation in the signal transduction pathways leading to the plant oxidative burst. *J. Plant Biol.* **46:** 215–222.

Chen, C.Y-H., Cheung, A.Y. and Wu, H-M., 2003, Actin-depolymerizing factor mediates Rac/Rop GTPase-regulated pollen tube growth. *Plant Cell* **15:** 237–249.

Cho, M.H., Shears, S.B. and Boss, W.F., 1993, Changes in phosphatidylinositol metabolism in response to hyperosmotic stress in *Daucus carota* L. cells grown in suspension culture. *Plant Physiol.* **103:** 637–647.

Cipres, A., Carrasco, S., Merino, E., Diaz, E., Krishna, U.M., Falck, J.R., Martinez, C. and Merida, I., 2003, Regulation of diacylglycerol kinase alpha by phosphoinositide 3-kinase lipid products. *J. Biol. Chem.* **278:** 35629–35635.

Cockcroft, S., 2001, Signalling roles of mammalian phospholipase D1 and D2. *Cell. Mol. Life Sci.* **58:** 1674–1687.

Cox, L.J., Larman, M.G., Saunders, C.M., Hashimoto, K., Swann, K. and Lai, F.A., 2002, Sperm phospholipase Cζ from humans and cynomoigus monkeys triggers Ca^{2+} oscillations, activation and development of mouse oocytes. *Reprod.* **124:** 611–623.

Craxton, A., Ali, N. and Shears, S.B., 1995, Comparison of the activities of a multiple inositol polyphosphate phosphatase obtained from several sources: a search for heterogeneity in this enzyme. *Biochem. J.* **305:** 491–498.

Cullen, P.J., Cozier, G.E., Banting, G. and Mellor, H., 2001, Modular phosphoinositide-binding domains – their role in signalling and membrane trafficking. *Curr. Biol.* **11:** R882-R893.

Delledonne, M., Murgia, I., Ederle, D., Sbicego, P.F., Biondani, A., Polverari, A. and Lamb, C., 2002, Reactive oxygen intermediates modulate nitric oxide signaling in the plant hypersensitive disease-resistance response. *Plant Physiol. Biochem.* **40:** 605–610.

Den Hartog, M., Musgrave, A. and Munnik, T., 2001, Nod factor-induced phosphatidic acid and diacylglycerol pyrophosphatase formation: a role for phospholipase C and D in root hair formation. *Plant J.* **25:** 55–66.

Den Hartog, M., Verhoef, N. and Munnik, T., 2003, Nod factor and elicitors activate different phospholipid signaling pathways in suspension-cultured alfalfa cells. *Plant Physiol.* **132:** 311–317.

De Jong, C.F., Laxalt, A.M., Bargmann, B.O.R., de Wit, P.J.G.M., Joosten, M.H.A.J. and Munnik, T., 2004, Phosphatidic acid accumulation is an early response in the *Cf-4/Avr4* interaction. *Plant J.* **39:** 1–12.

DeWald, D.B., Torabinejad, J., Jones, C.A., Shope, J.C., Cangelosi, A.R., Thompson, J.E., Prestwich, G.D. and Hama, H., 2001, Rapid accumulation of phosphatidylinositol 4,5-*bis*phosphate and inositol 1,4,5-*tris*phosphate correlates with calcium mobilization in salt-stressed *Arabidopsis*. *Plant Physiol.* **126:** 759–769.

Dhonukshe, P., Laxalt, A.M., Goedhart, J., Gadella, T.W.J. and Munnik, T., 2003, Phospholipase D activation correlates with microtubule reorganization in living plant cells. *Plant Cell* **15:** 2666–2679.

Dove, S.K., Piper, R.C., McEwen, R.K., Yu, J.W., King, M.C., Hughes, D.C., Thuring, J., Holmes, A.B., Cooke, F.T., Michell, R.H., Parker, P.J. and Lemmon, M.A., 2004, Svp1p defines a family of phosphatidylinositol 3,5-bisphosphate effectors. *EMBO J.* **23:** 1922–1933.

Downes, C.P., Gray, A., Watt, S.A. and Lucocq, J.M., 2003, Advances in procedures for the detection and localization of inositol phospholipid signals in cells, tissues, and enzyme assays. *Meth. Enz.* **366:** 64–84.

Drayer, A.L., Van der Kaay, J., Mayr, G.W. and Van Haastert, P.J.M., 1994, Role of phospholipase C in *Dictyostelium*: formation of inositol 1,4,5-trisphosphate and normal development in cells lacking phospholipase C activity. *EMBO J.* **13:** 1601–1609.

Drøbak, B.K. and Watkins, P.A.C., 2000, Inositol(1,4,5)*tris*phosphate production in plant cells: an early response to salinity and hyperosmotic stress. *FEBS Lett.* **481:** 240–244.

Drøbak, B.K., Franklin-Tong, V.E. and Staiger, C.J., 2004, The role of the actin cytoskeleton in plant cell signaling. *New Phytol.* **163:** 13–30.

Drøbak, B.K., Watkins, P.A.C., Valenta, R., Dove, S.K., Lloyd, C.W. and Staiger, C.J., 1994, Inhibition of plant plasma-membrane phosphoinositide phospholipase-C by the actin-binding protein, profilin. *Plant J.* **6:** 389–400.

Duman, J.G., Lee, E., Lee, G.Y., Singh, G. and Forte, J.G., 2004, Membrane fusion correlates with surface charge in exocytic vesicles. *Biochem.* **43:** 7924–7939.

Durner, J., Wendehenne, D. and Klessig, D.F., 1998, Defense gene induction in tobacco by nitric oxide, cyclic GMP, and cyclic ADP-ribose. *Proc. Natl. Acad. Sci. USA* **95:** 10328–10333.

Einspahr, K.J., Peeler, T.C. and Thompson, G.A.Jr., 1988, Rapid changes in polyphosphoinositide metabolism associated with the response of *Dunaliella salina* to hypoosmotic shock. *J. Biol. Chem.* **263:** 5775–5779.

Elge, S., Brearley, C., Xia, H-J., Kehr, J., Xue, H-W. and Mueller-Roeber, B., 2001, An *Arabidopsis* inositol phospholipid kinase strongly expressed in procambial cells: Synthesis of PtdIns(4,5)P_2 and PtdIns(3,4,5)P_3 in insect cells by 5-phosphorylation of precursors. *Plant J.* **26:** 561–571.

English, D., 1996, Phosphatidic acid: a lipid messenger involved in intracellular and extracellular signaling. *Cell Signal* **8:** 341–347.

Ercetin, M.E. and Gillaspy, G.E., 2004, Molecular characterization of an *Arabidopsis* gene encoding a phospholipid-specific inositol polyphosphate 5-phosphatase. *Plant Physiol.* **135:** 938–946.

Estruch, J.J., Kadwell, S., Merlin, E. and Crossland, L., 1994, Cloning and characterization of a maize pollen-specific calcium-dependent calmodulin-independent protein kinase. *Proc. Natl. Acad. Sci. USA* **91:** 8837–8841.

Foster, D.A. and Xu, L.Z., 2003, Phospholipase D in cell proliferation and cancer. *Mol. Cancer Res.* **1:** 789–800.

Frank, W., Munnik, T., Kerkmann, K., Salamini, F. and Bartels, D., 2000, Water deficit triggers phospholipase D activity in the resurrection plant *Craterostigma plantagineum*. *Plant Cell* **12:** 111–123.

Franklin-Tong, V.E., 1999, Signaling and the modulation of pollen tube growth. *Plant Cell* **11:** 727–738.

Franklin-Tong, V.E., Drøbak, B., Allan, A.C., Watkins, P.A.C. and Trewavas, A.J., 1996, Growth of pollen tubes of *Papaver rhoeas* is regulated by a slow-moving calcium wave propagated by inositol 1,4,5-trisphosphate. *Plant Cell* **8:** 1305–1321.

Freyberg, Z., Siddhanta, A. and Shields, D., 2003, 'Slip, sliding away': phospholipase D and the Golgi apparatus. *Trends Cell Biol.* **13:** 540–546.

Fu, Y. and Yang, Z., 2001, Rop GTPase: a master switch of cell polarity development in plants. *Trends Plant Sci.* **6:** 545–547.

Gary, J.D., Wurmser, A.E., Bonangelino, C.J., Weisoman, L.S. and Emr, S.D., 1998, Fab1p is essential for PtdIns(3)P 5-kinase activity and the maintenance of vacuolar size and membrane homeostasis. *J. Cell Biol.* **143:** 65–79.

Gawer, M., Norberg, P., Chervin, D., Guern, N., Yaniv, Z., Mazliak, P. and Kader, J.C., 1999, Phosphoinositides and stress-induced changes in lipid metabolism of tobacco cells. *Plant Sci.* **141:** 117–127.

Gilroy, S., Read, N.D. and Trewavas, A.J., 1990, Elevation of cytoplasmic calcium by caged calcium or caged inositol trisphosphate initiates stomatal closure. *Nature* **346:** 769–771.

Goedhart, J. and Gadella, T.W.J.Jr., 2004, Photolysis of caged phosphatidic acid induces flagellar excision in *Chlamydomonas*. *Biochem.* **43:** 4263–4271.

Gupta, R., Ting, J.T.L., Sokolov, L.N., Johnson, S.A. and Luan, S., 2002, A tumor suppressor homologue, AtPTEN1, is essential for pollen development in *Arabidopsis*. *Plant Cell* **14:** 2495–2507.

Hallouin, M., Ghellis, T., Brault, M., Bardat, F., Cornel, D., Miginiac, E., Rona, J.P., Sotta, B. and Jeannette, E., 2002, Plasmalemma abscisic acid perception leads to RAB18 expression via phopholipase D activation in *Arabidopsis* suspension cells. *Plant Physiol.* **130:** 265–272.

Heilmann, I., Perera, I.Y., Gross, W. and Boss, W.F., 1999, Changes in phosphoinositide metabolism with days in culture affect signal transduction pathways in *Galderia sulphuraria*. *Plant Physiol.* **119:** 1331–1339.

Heilmann, I., Perera, I.Y., Gross, W. and Boss, W.F., 2001, Plasma membrane phosphatidylinositol 4,5-*bis*phosphate levels decrease with time in culture. *Plant Physiol.* **126:** 1507–1518.

Helsper, J.P.F.G., de Groot, P.F.M., Linskens, H.F. and Jackson, J.F., 1986, Phosphatidylinositol phospholipase C activity in pollen of *Lilium longiflorum*. *Phytochem.* **25:** 2053–2055.

Helsper, J.P.F.G., Heemskerk, J.W.M. and Veerkamp, J.H., 1987, Cytosolic and particulate phosphatidylinositol phospholipase C activity in pollen of *Lilium longiflorum*. *Physiol. Plant.* **71:** 120–126.

Hepler, P.K., Vidali, L. and Cheung, A.Y., 2001, Polarized cell growth in higher plants. *Annu. Rev. Cell Dev. Biol.* **17:** 159–187.

Hetherington, A.M. and Drøbak, B.K., 1992, Inositol-containing lipids in higher-plants. *Prog. Lipid Res.* **31:** 53–63.

Hilgemann, D.W., Feng, S. and Nasuhoglu, C., 2001, The complex and intriguing lives of PIP_2 with ion channels and transporters. *Science STKE* **111:** 1–8.

Ho, M.W., Kaetzel, M.A., Armstrong, D.L. and Shears, S.B., 2001, Regulation of a human chloride channel, a paradigm for integrating input from calcium, type II calmodulin-dependent protein kinase, and inositol 3,4,5,6-tetrakisphosphate. *J. Biol. Chem.* **276:** 18673–18680.

Ho, M.W.Y., Carew, M.A., Yang, X. and Shears, S.B., 2000, Regulation of chloride channel conductance by Ins(3,4,5,6)P_4, a phosphoinositide-initiated signaling pathway that acts downstream of Ins(1,4,5)P_3. In: Biology of Phosphoinositides, Cockcroft, S. ed., Oxford University Press, New York, pp. 298–319.

Ho, M.W.Y. and Shears, S.B., 2002, Regulation of calcium-activated chloride channels by inositol 3,4,5,6-tetrakisphosphate. *Curr. Topics Membr.* **53:** 345–363.

Ho, M.W.Y., Shears, S.B., Bruzik, K.S., Duszyk, M. and French, A.S., 1997, Ins(3,4,5,6)P_4 specifically inhibits a receptor-mediated Ca^{2+}-dependent Cl^- current in CFPAC-1 cells. *Am. J. Physiol. Cell Physiol.* **272:** C1160-C1168.

Ho, M.W.Y., Yang, X.N., Carew, M.A., Zhang, T., Hua, L., Kwon, Y.U., Chung, S.K., Adelt, S., Vogel, G., Riley, A.M., Potter, B.V.L. and Shears, S.B., 2002, Regulation of Ins(3,4,5,6)P_4 signaling by a reversible kinase/phosphatase. *Curr. Biol.* **12:** 477–482.

Holdaway-Clarke, T., Feijo, J.A., Hackett, G.R., Kunkel, J.G. and Hepler, P.K., 1997, Pollen tube growth and the intracellular cytosolic calcium gradient oscillate in phase while extracellular calcium influx is delayed. *Plant Cell* **9:** 1999–2010.

Holdaway-Clarke, T.L. and Hepler, P.K., 2003, Control of pollen tube growth: role of ion gradients and fluxes. *New Phytol.* **159:** 539–563.

Hunt, L., Mills, L.N., Pical, C., Leckie, C.P., Aitken, F.L., Kopka, J., Mueller-Roeber, B., McAinsh, M.R., Hetherington, A.M. and Gray, J.E., 2003, Phospholipase C is required for the control of stomatal aperture by ABA. *Plant J.* **34:** 47–55.

Hunt, L., Otterhag, L., Lee, J.C., Lasheen, T., Hunt, J., Seki, M., Shinozaki, K., Sommarin, M., Gilmour, D.J., Pical, C. and Gray, J.E., 2004, Gene-specific expression and calcium activation of *Arabidopsis thaliana* phospholipase C isoforms. *New Phytol.* **162:** 643–654.

Irvine, R.F. and Schell, M.J., 2001, Back in the water: the return of the inositol phosphates. *Nature Rev. Mol. Cell Biol.* **2:** 327–338.

Ismailov, I.I., Fuller, C.M., Berdiev, B.K., Shlyonsky, V.G., Benos, D.J. and Barrett, K.E., 1996, A biologic function for an "orphan" messenger: D-*myo*-inositol 3,4,5,6-tetrakisphosphate selectively blocks epithelial calcium-activated chloride channels. *Proc. Natl. Acad. Sci. USA* **93:** 10505–10509.

Itoh, T. and Takenawa, T., 2004, Regulation of endocytosis by phosphatidylinositol 4,5-bisphosphate and ENTH proteins. *Curr. Top. Microbiol. Immunol.* **282:** 31–47.

Ives, E.B., Nichols, J., Wente, S.R. and York, J.D., 2000, Biochemical and functional characterization of inositol 1,3,4,5,6-pentakisphosphate 2-kinases. *J. Biol. Chem.* **275:** 36575–36583.

Jacob, T., Ritchie, S., Assmann, S.M. and Gilroy, S., 1999, Abscisic acid signal transduction in guard cells is mediated by phospholipase D activity. *Proc. Natl. Acad. Sci. USA* **96:** 12192–12197.

Janmey, P.A., 1994, Phosphoinositides and calcium as regulators of cellular actin assembly and disassembly. *Annu. Rev. Physiol.* **56:** 169–191.

Janmey, P.A., Xian, W. and Flanagan, L.A., 1999, Controlling cytoskeleton structure by phosphoinositide-protein interactions: phosphoinositide binding protein domains and effects of lipid packing. *Chem. Phys. Lipids* **101:** 93–107.

Johnson, M.A. and Preuss, D., 2002, Plotting a course: multiple signals guide pollen tubes to their targets. *Dev. Cell* **2:** 273–281.

Jung, H.W., Kim, W. and Hwang, B.K., 2003, Three pathogen-inducible genes encoding lipid transfer protein from pepper are differentially activated by pathogens, abiotic and environmental stresses. *Plant Cell Environ.* **26:** 915–928.

Jung, J-Y., Kim, Y-W., Kwak, J.M., Hwang, J-U., Young, J., Schroeder, J.I., Hwang, I. and Lee, Y., 2002, Phosphatidylinositol 3- and 4- phosphate are required for normal stomatal movements. *Plant Cell* **14:** 2399–2412.

Kamada, Y. and Muto, S., 1994, Stimulation by fungal elicitor of inositol phospholipid turnover in tobacco suspension culture cells. *Plant Cell Physiol.* **35:** 397–404.

Kanoh, H., Yamada, K. and Sakane, F., 2002, Diacylglycerol kinases: emerging downstream regulators in cell signalling systems. *J. Biol. Chem.* **131:** 629–633.

Katagiri, T., Takahashi, S. and Shinozaki, K., 2001, Involvement of a novel *Arabidopsis* phospholipase D, AtPLD delta, in dehydration-inducible accumulation of phosphatidic acid in stress signaling. *Plant J.* **26:** 595–605.

Kasparovsky, T., Blein, J.P. and Mikes, V., 2004, Ergosterol elicits oxidative burst in tobacco cells via phospholipase A_2 and protein kinase C signal pathway. *Plant Physiol. Biochem.* **42:** 429–435.

Kiegle, E., Moore, C.A., Haseloff, J., Tester, M.A. and Knight, M.R., 2000, Cell-type-specific calcium responses to drought, salt and cold in the *Arabidopsis* root. *Plant J.* **23:** 267–278.

Kim, D.H., Eu, Y-J., Yoo, C.M., Kim, Y-W., Pih, K.T., Jin, J.B., Kim, S.J., Stenmark, H. and Hwang, I., 2001, Trafficking of phosphatidylinositol 3-phosphate from the *trans*-Golgi network to the lumen of the central vacuole in plant cells. *Plant Cell* **13:** 287–301.

Kim, H.U., Cotter, R., Johnson, S., Senda, M., Dodds, P., Kulikauskas, R., Tang, W., Ezcurra, I., Herzmark, P. and McCormick, S., 2002, New pollen-specific receptor kinases identified in tomato, maize, and *Arabidopsis*: the tomato kinases show overlapping but distinct localization patterns on pollen tubes. *Plant Mol. Biol.* **50:** 1–16.

Knight, M.R., 2002, Signal transduction leading to low-temperature tolerance in *Arabidopsis thaliana*. *Phil. Trans. Roy. Soc. Lond. B* **357:** 871–874.

Knight, H. and Knight, M.R., 2001, Abiotic stress signalling pathways: specificity amd cross-talk. *Trends Plant Sci.* **6:** 262–267.

Knight, H., Trewavas, A.J. and Knight, M.R., 1996, Cold calcium signaling in *Arabidopsis* involves two cellular pools and a change in calcium signature after acclimation. *Plant Cell* **8:** 489–503.

Knight, H., Trewavas, A.J. and Knight, M.R., 1997, Calcium signaling in *Arabidopsis thaliana* responding to drought and salinity. *Plant J.* **12:** 1067–1078.

Kost, B., Lemichez, E., Spielhofer, P., Hong, Y., Tolias, K., Carpenter, C. and Chua, N-H., 1999, Rac homologues and compartmentalized phosphatidylinositol 4,5,-bisphosphate act in a common pathway to regulate polar pollen tube growth. *J. Cell Biol.* **145:** 317–330.

Kovar, D.R., Drøbak, B.K. and Staiger, C.J., 2000, Maize profilin isoforms are functionally distinct. *Plant Cell* **12:** 583–598.

Kroj, T., Rudd, J.J., Nürnberger, T., Gäbler, Y., Lee, J. and Scheel, D., 2003, Mitogen-activated protein kinases play an essential role in oxidative burst-independent expression of pathogenesis-related genes in parsley. *J. Biol. Chem.* **278:** 2256–2264.

Ktistakis, N.T., Delon, C., Manifava, M., Wood, E., Ganley, I. and Sugars, J.M., 2003, Phospholipase D1 and potential targets of its hydrolysis product, phosphatidic acid. *Biochem. Soc. Trans.* **31:** 94–97.

Kurokawa, M., Sato, K., Rissore, R.A., 2004, Mammalian fertilization: from sperm factor to phospholipase Cζ. *Biol. Cell* **96:** 37–45.

Lam, B.C-H. and Blumwald, E., 2002, Domains as functional building blocks of plant proteins. *Trends Plant Sci.* **7:** 544–549.

Larman, M.G., Saunders, C.M., Carroll, J., Lai, F.A. and Swann, K., 2004, Cell cycle-dependent Ca^{2+} oscillations in mouse embryos are regulated by nuclear targeting of PLCζ. *J. Cell Sci.* **117:** 2513–2521.

Laxalt, A.M. and Munnik, T., 2002, Phospholipid signalling in plant defence. *Curr. Op. Plant Biol.* **5:** 332–338.

Laxalt, A.M., Ter Riet, B., Verdonk, J.C., Parigi, L., Tameling, W.I.L., Vossen, J., Haring, M., Musgrave, A. and Munnik, T., 2001, Characterization of five tomato phospholipase D cDNAs: rapid and specific expression of *LePLDβ1* on elicitation with xylanase. *Plant J.* **26:** 237–247.

Lee, S., Hirt, H. and Lee, Y., 2001, Phosphatidic acid activates a wound-activated MAPK in *Glycine max*. *Plant J.* **26:** 479–486.

Lee, S., Park, J. and Lee, Y., 2003, Phosphatidic acid induces actin polymerization by activating protein kinases in soybean cells. *Mol. Cell* **15:** 313–319.

Lee, Y., Choi, Y.B., Suh, S., Lee, J., Assmann, S.M., Joe, C.O., Kelleher, J.F. and Crain, R.C., 1996, Abscisic acid-induced phosphoinositide turnover in guard cell protoplasts of *Vicia faba*. *Plant Physiol.* **110:** 987–996.

Lemtiri-Chlieh, F., MacRobbie, E.A.C. and Brearley, C.A., 2000, Inositol hexakisphosphate is a physiological signal regulating the K^+-inward rectifying conductance in guard cells. *Proc. Natl. Acad. Sci. USA* **97:** 8687–8692.

Lemtiri-Chlieh, F., MacRobbie, E.A.C., Webb, A.A.R., Manison, N.F., Brownlee, C., Skepper, J.N., Chen, J., Prestwich, G.D. and Brearley, C.A., 2003, Inositol hexakisphosphate mobilizes an endomembrane store of calcium in guard cells. *Proc. Natl. Acad. Sci. USA* **100:** 10091–10095.

Lemmon, M.A., 2003, Phosphoinositide recognition domains. *Traffic* **4:** 201–213.

Leonhardt, N., Bazin, I., Richaud, P., Marin, E., Vavasseur, A. and Forestier, C., 2001, Antibodies to the CFTR modulate the turgor pressure of guard cell protoplasts via slow anion channels. *FEBS Lett.* **494:** 15–18.

Li, W., Li, M., Zhang, W., Welti, R. and Wang, X., 2004, The plasma membrane-bound phospholipase Dδ enhances freezing tolerance in *Arabidopsis thaliana. Nature Biotech.* **22:** 427–433.

Lin, W.H., Rui, Y.E., Hui, M.A., Xu, Z.H., Xui, H.W., 2004, DNA chip-based expression profile analysis indicates involvement of the phosphatidylinositol signaling pathway in multiple plant responses to hormone and abiotic treatments. *Cell Research* **14:** 34–45.

Luan, S., Ting, J. and Gupta, R., 2001, Protein tyrosine phosphatases in higher plants. *New Phytol.* **151:** 155–164.

Luo, B., Regier, D.S., Prescott, S.M. and Topham, M.K., 2004, Diacylglycerol kinases. *Cellular Signalling* **16:** 983–989.

Luo, H.B.R., Huang, Y.E., Chen, J.M.C., Saiardi, A., Iijima, M, Ye, K.Q., Huang, Y.F., Nagata, E., Devreotes, P. and Snyder, S.H., 2003, Inositol pyrophosphates mediate chemotaxis in *Dictyostelium* via pleckstrin homology domain-PtdIns(3,4,5)P_3 interactions. *Cell* **114:** 559–572.

Malhó, R., 1998, Role of 1,4,5-inositol trisphosphate-induced Ca^{2+} release in pollen tube orientation. Sex. *Plant Reprod.* **11:** 231–235.

Maehama, T. and Dixon, J.E., 1998, The tumor suppressor, PTEN/MMAC1, dephosphorylates the lipid second messenger, phosphatidylinositol 3,4,5-trisphosphate. *J. Biol. Chem.* **273:** 13375–13378.

Martin, T.F.J., 1998, Phosphoinositide lipids as signaling molecules: Common themes for signal transduction, cytoskeletal regulation, and membrane trafficking. *Annu. Rev. Cell Dev. Biol.* **14:** 231–264.

Martin, T.F.J., 2001, PI(4,5)P_2 regulation of surface membrane traffic. *Curr. Op. Cell Biol.* **13:** 493–499.

McDermott, M., Wakelam, M.J.O. and Morris, A.J., 2004, Phospholipase D. *Biochem. Cell Biol.* **82:** 225–253.

McLaughlin, S., Wang, J., Gambhir, A. and Murray, D., 2002, PIP_2 and proteins: Interactions, organization and information flow. *Annu. Rev. Biophys. Biomol. Struct.* **31:** 151–175.

Meijer, H.J.G., Arisz, S.A., Van Himbergen, J.A.J., Musgrave, A. and Munnik, T., 2001a, Hyperosmotic stress rapidly generates lyso-phosphatidic acid in *Chlamydomonas. Plant J.* **25:** 541–548.

Meijer, H.J.G., Berrie, C.P., Iurisci, C., Divecha, N., Musgrave, A. and Munnik, T., 2001b, Identification of a new polyphosphoinositide in plants, phosphatidylinositol 5-monophosphate (PtdIns5P), and its accumulation upon osmotic stress. *Biochem. J.* **360:** 491–498.

Meijer, H.J.G., Divecha, N., Van den Ende, H., Musgrave, A. and Munnik, T., 1999, Hyperosmotic stress induces rapid synthesis of phosphatidyl-D-inositol 3,5-bisphosphate in plant cells. *Planta* **208:** 294–298.

Meijer, H.J.G., and Munnik, T., 2003, Phospholipid-based signaling in plants. *Annu. Rev. Plant Biol.* **54:** 265–306.

Meijer, H.J.G., ter Riet, B., Van Himbergen, J.A.J., Musgrave, A. and Munnik, T., 2002, KCl activates phospholipase D at two different concentration ranges: distinguishing between hyperosmotic stress and membrane depolarization. *Plant J.* **31:** 51–59.

Menniti, F.S., Oliver, K.G., Putney, J.W. and Shears, S.B., 1993, Inositol phosphates and cell signaling – new views of $InsP_5$ and $InsP_6$. *Trends Biochem. Sci.* **18:** 53–56.

Messerli, M.A., Creton, R., Jaffe, L.F. and Robinson, K.R., 2000, Periodic increases in elongation rate precede increases in cytosolic Ca^{2+} during pollen tube growth. *Dev. Biol.* **222:** 84–98.

Messerli, M.A., Danuser, G. and Robinson, K.R., 1999, Pulsatile influxes of H^+, K^+ and Ca^{2+} lag growth pulses of *Lilium longiflorum. J. Cell Sci.* **112:** 1497–1509.

Miège, C. and Maréchal, É., 1999, 1,2-*sn*-Diacylglycerol in plant cells: Product, substrate and regulator. *Plant Physiol. Biochem.* **37:** 795–808.

Mikami, K., Katagiri, T., Iuchi, S., Yamaguchi-Shinozaki, K. and Shinozaki, K., 1998, A gene encoding phosphatidylinositol-4-phosphate 5-kinase is induced by water stress and abscisic acid in *Arabidopsis thaliana. Plant J.* **15:** 563–568.

Mills, L.N., Hunt, L., Leckie, C.P., Aitken, F.L., Wentworth, M., McAinsh, M.R., Gray, J.E. and Hetherington, A.M., 2004, The effects of manipulating phospholipase C on guard cell ABA-signaling. *J. Exp. Bot.* **55:** 199–204.

Mitra, P., Zhang, Y., Rameh, L.E., Ivshina, M.P., McCollum, D., Nunnari, J.J., Hendricks, G.M., Kerr, M.L., Field, S.J., Cantley, L.C. and Ross, A.H., 2004, A novel phosphatidylinositol(3,4,5)P_3 pathway in fission yeast. *J. Cell Biol.* **166:** 205–211.

Monks, D.E., Aghoram, K., Courtney, P.D., DeWald, D.B. and Dewey, R.E., 2001, Hyperosmotic stress induces the rapid phosphorylation of a soybean phosphatidylinositol transfer protein homolog through activation of the protein kinases SPK1 and SPK2. *Plant Cell* **13:** 1205–1219.

Moutinho, A., Trewavas, A.J. and Malhó, R., 1998, Relocation of a Ca^{2+}-dependent protein kinase activity during pollen tube reorientation. *Plant Cell* **10:** 1499–1509.

Müller-Roeber, B. and Pical, C., 2002, Inositol phospholipid metabolism in *Arabidopsis*. Characterized and putative isoforms of inositol phospholipid kinase and phosphoinositide-specific phospholipase C. *Plant Physiol.* **130:** 22–46.

Munnik, T., 2001, Phosphatidic acid: an emerging plant lipid second messenger. *Trends Plant Sci.* **6:** 227–233.

Munnik, T., De Vrije, T., Irvine, R.F. and Musgrave, A., 1996, Identification of diacylglycerol pyrophosphate as a novel metabolic product of phosphatidic acid during G-protein activation in plants. *J. Biol. Chem.* **271:** 15708–15715.

Munnik, T., Arisz, S.A., De Vrije, T. and Musgrave, A., 1995, G protein activation stimulates phospholipase D signaling in plants. *Plant Cell* **7:** 2197–2210.

Munnik, T., Irvine, R.F. and Musgrave, A., 1994, Rapid turnover of phosphatidylinositol 3-phosphate in the green alga *Chlamydomonas eugametos*: signs of a phosphatidylinositide 3-kinase signaling pathway in lower plants? *Biochem. J.* **298:** 269–273.

Munnik, T., Irvine, R.F. and Musgrave, A., 1998a, Phospholipid signalling in plants. *Biochim. Biophys. Acta* **1398:** 222–272.

Munnik, T., Van Himbergen, J.A.J., Ter Riet, B., Braun, F.-J., Irvine, F.F., Van den Ende, H. and Musgrave, A., 1998b, Detailed analysis of the turnover of polyphosphoinositides and phosphatidic acid upon activation of phospholipases C and D in *Chlamydomonas* cells treated with non-permeabilizing concentrations of mastoparan. *Planta* **207:** 133–145.

Munnik, T., Ligterink, W., Meskiene, I., Calderini, O., Beyerly, J., Musgrave, A. and Hirt, H., 1999, Distinct osmo-sensing protein kinase pathways are involved in signaling moderate and severe hyper-osmotic stress. *Plant J.* **20:** 381–388.

Munnik, T. and Meijer, H.J.G., 2001, Osmotic stress activates distinct lipid and MAPK signalling pathways in plants. *FEBS Lett.* **498:** 172–178.

Munnik, T., Meijer, H.J.G., Ter Riet, B., Hirt, H., Frank, W., Bartels, D. and Musgrave, A., 2000, Hyperosmotic stress stimulates phospholipase D activity and elevates the levels of phosphatidic acid and diacylglycerol pyrophosphate. *Plant J.* **22:** 147–154.

Munnik, T., Musgrave, A. and De Vrije, T., 1994, Rapid turnover of polyphosphoinositides in carnation flower petals. *Planta* **193:** 89–98.

Nakanishi, H., de los Santos, P. and Neiman, A.M., 2004, Positive and negative regulation of a SNARE protein by control of intracellular localization. *Mol. Biol. Cell* **15:** 1802–1815.

Nilius, B., Prenen, J., Voets, T., Eggermont, J., Bruzik, K.S., Shears, S.B. and Droogmans, G., 1998, Inhibition by inositoltetrakisphosphates of calcium- and volume-activated Cl^- currents in macrovascular endothelial cells. *Pfleugers Arch. Eur. J. Physiol.* **435:** 637–644.

Nimchuk, Z., Eulgem, T., Holt, B.E. and Dangl, J.L., 2003, Recognition and response in the plant immune system. *Annu. Rev. Genet.* **37:** 579–609.

Nürnberger, T. and Scheel, D., 2001, Signal transmission in the plant immune response. *Trends Plant Sci.* **6:** 372–379.

Odom, A.R., Stahlberg, A., Wente, S.R. and York, J.D., 2000, A role for nuclear inositol 1,4,5-trisphosphate kinase in transcriptional control. *Science* **287:** 2026–2029.

Odorizzi, G., Babst, M. and Emr, S.D., 2000, Phosphoinositide signaling and the regulation of membrane trafficking in yeast. *Trends Biochem. Sci.* **25:** 229–235.

Oprins, J.C., Van der Burg, C., Meijer, H.P., Munnik, T. and Groot, J.A., 2001, PLD pathway involved in carbachol-induced Cl^- secretion: possible role of TNF-α. *Am. J. Physiol. Cell Physiol.* **280:** C789-C795.

Oprins, J.C., Van der Burg, C., Meijer, H.P., Munnik, T. and Groot, J.A., 2002, Tumor necrosis factor alpha potentiates ion secretion induced by histamine in a human intestinal epithelial cell line and in mouse colon: involvement of the phospholipase D pathway. *Gut* **50:** 314–321.

Overmyer, K., Brosché, M. and Kangasjärvi, J., 2003, Reactive oxygen species and hormonal control of cell death. *Trends Plant Sci.* **8:** 335–342.

Park, C-J., Shin, R., Park, J.M., Lee, G-J., You, J-S. and Paek, K-H., 2002, Induction of pepper cDNA encoding a lipid transfer protein during the resistance response to tobacco mosaic virus. *Plant Mol. Biol.* **48:** 243–254.

Park, J., Gu, Y., Lee, Y., Yang, Z. and Lee, Y., 2004, Phosphatidic acid induces leaf cell death in *Arabidopsis* by activating the Rho-related small G protein GTPase-mediated pathway of reactive oxygen species generation. *Plant Physiol.* **134:** 129–136.

Parsons, R., 2004, Human cancer, PTEN and the PI-3 kinase pathway. *Sem. Cell Dev. Biol.* **15:** 171–176.

Pearce, R.S., 1999, Molecular analysis of acclimation to cold. *Plant Growth Regul.* **29:** 47–76.

Perera, N.M., Michell, R.H. and Dove, S.K., 2004, Hypo-osmotic stress activates Plc1p-dependent phosphatidylinositol 4,5-bisphosphate hydrolysis and inositol hexakisphosphate accumulation in yeast. *J. Biol. Chem.* **279:** 5216–5226.

Pical, C., Westergren, T., Dove, S.K., Larsson, C. and Sommarin, M., 1999, Salinity and hyperosmotic stress induce rapid increases in phosphatidylinositol 4,5-bisphosphate, diacylglycerol pyrophosphate, and phosphatidylcholine in *Arabidopsis thaliana* cells. *J. Biol. Chem.* **274:** 38232–38240.

Plieth, C., Hansen, U-P., Knight, H. and Knight, M.R., 1999, Temperature sensing by plants: the primary characteristics of signal perception and calcium response. *Plant J.* **18:** 491–497.

Potocky, M., Elias, M., Profotova, B., Novotna, Z., Valentova, O. and Zarksy, V., 2003, Phosphatidic acid produced by phospholipase D is required for tobacco pollen tube growth. *Planta* **217:** 122–130.

Prestwich, G.D., Chen, R., Feng, L., Ozaki, S., Ferguson, C.G., Drees, B.E., Neklason, D.A., Mostert, M.J., Porter-Gill, P.A., Kang, V., Shope, J.C., Neilsen, P.O. and DeWald, D.B., 2002, In situ detection of phospholipid and phosphoinositide metabolism. *Advan. Enz. Regul.* **42:** 19–38.

Renstrom, E., Ivarsson, R. and Shears, S.B., 2002, Inositol 3,4,5,6-tetrakisphosphate inhibits insulin granule acidification and fusogenic potential. *J. Biol. Chem.* **277:** 26717–26720.

Rhee, S.G., 2001, Regulation of phospholinositide-specific phospholipase C. *Annu. Rev. Biochem.* **70:** 281–312.

Ritchie, S. and Gilroy, S., 1998, Abscisic acid signal transduction in the barley aleurone is mediated by phospholipase D activity. *Proc. Natl. Acad. Sci. USA* **95:** 2697–2702.

Ritchie, S. and Gilroy, S., 2000, Abscisic acid stimulation of phospholipase D in the barley aleurone is G-protein-mediated and localized to the plasma membrane. *Plant Physiol.* **124:** 693–702.

Rojo, E., Solano, R. and Sánchez-Serrano, J.J., 2003, Interactions between signaling compounds. *J. Plant Growth Regul.* **22:** 82–98.

Roth, M.G., 2004, Phosphoinositides in constitutive membrane traffic. *Physiol. Rev.* **84:** 699–730.

Rudge, S.A., Anderson, D.M. and Emr, S.D., 2004, Vacuole size control: regulation of PtdIns(3,5)P_2 levels by the vacuole-associated Vac14-Fig4 complex, a PtdIns(3,5)P_2-specific phosphatase. *Mol. Biol. Cell* **15:** 24–36.

Ruelland, E., Cantrel, C., Gawer, M., Kader, J.C. and Zachowski, A., 2002, Activation of phospholipases C and D is an early response to a cold exposure in *Arabidopsis* suspension cells. *Plant Physiol.* **130:** 999–1007.

Ryu, S.B., 2004, Phospholipid-derived signaling mediated by phospholipase A in plants. *Trends Plant Sci.* **9:** 229–235.

Saiardi, A., Sciambi, C., McCaffery, J., Wendland, B. and Snyder, S.H., 2002, Inositol pyrophosphates regulate endocytic trafficking. *Proc. Natl. Acad. Sci. USA* **99:** 14206–14211.

Sanchez, J-P. and Chua, N-H., 2001, *Arabidopsis* PLC1 is required for secondary responses to abscisic acid signals. *Plant Cell* **13:** 1143–1154.

Sang, Y., Cui, D. and Wang, X., 2001a, Phospholipase D and phosphatidic acid-mediated generation of superoxide in *Arabidopsis*. *Plant Physiol.* **126:** 1449–1458.

Sang, Y., Zheng, S.Q., Li, W.Q., Huang, B.R., Wang, X., 2001b, Regulation of plant water loss by manipulating the expression of phospholipase Dα. *Plant J.* **28:** 135–144.

Saunders, C.M., Larman, M.G., Parrington, J., Cox, L.J., Royse, J., Blayney, L.M., Swann, K. and Lai, F.A., 2002, PLCζ: a sperm-specific trigger of Ca^{2+} oscillations in eggs and embryo development. *Dev.* **129:** 3533–3544.

Scherer, G.F.E., Paul, R.U., Holk, A. and Marinec, J., 2002, Down-regulation by elicitors of phosphatidylcholine-hydrolyzing phospholipase C and up-regulation of phospholipase A_2 in plant cells. *Biochem. Biophys. Res. Comm.* **293:** 766–770.

Schroeder, J.I., Allen, G.J., Hugouvieux, V., Kwak, J.M. and Waner, D., 2001, Guard cell signal transduction. *Annu. Rev. Plant Physiol. Plant Mol. Biol.* **52:** 627–658.

Scrase-Field, S.A.M.G. and Knight M.R., 2003, Calcium: just a chemical switch? *Curr. Op. Plant Biol.* **6:** 500–506.

Shabala, S., Babourina, O. and Newman, I., 2000, Ion-specific mechanisms of osmoregulation in bean mesophyll cells. *J. Exp. Bot.* **51:** 1243–1253.

Shaw, J.D., Hama, H., Sohrabi, F., DeWald, D.B. and Wendland, B., 2003, PtdIns(3,5)P_2 is required for delivery of endocytic cargo into the multivesicular body. *Traffic* **4:** 479–490.

Shears, S.B., 1996, Inositol pentakis- and hexakisphosphate metabolism adds versatility to the actions of inositol polyphosphates: novel effects on ion channels and protein traffic. In: *myo*-Inositol phosphates, phosphoinositides, and signal transduction. Subcell. Biochem., Vol. 26, Plenum Press, N.Y., Chapt. 7, pp. 187–226.

Shears, S.B., 2004, How versatile are inositol polyphosphate kinases? *Biochem. J.* **377:** 265–280.

Shen, Y.J., Xu, L.Z. and Foster, D.A., 2001, Role for phospholipase D in receptor-mediated endocytosis. *Mol. Cell Biol.* **21:** 595–602.

Shigaki, T. and Bhattacharyya, M.A., 2000, Decreased inositol 1,4,5-trisphosphate content in pathogen-challenged soybean cells. *Mol. Plant-Microbe Interactions* **13:** 563–567.

Simonsen, A., Wurmser, A.E., Emr, S.D. and Stenmark, H., 2001, The role of phosphoinositides in membrane transport. *Curr. Op. Cell Biol.* **13:** 485–492.

Simpson, L. and Parsons, R., 2001, PTEN: life as a tumor suppressor. *Exp. Cell Res.* **264:** 29–41.

Song, F. and Goodman, R.M., 2002, Molecular cloning and characterization of a rice phosphoinositide-specific phospholipase C gene, *OsPI-PLC1*, that is activated in systemic acquired resistance. *Physiol. Mol. Plant Pathol.* **61:** 31–40.

Staxén, I., Pical, C., Montgomery, L.T., Gray, J.E., Hetherington, A.M. and McAinsh, M.R., 1999, Abscisic acid induces oscillations in guard-cell cytosolic free calcium that involve phosphoinositide-specific phospholipase C. *Proc. Natl. Acad. Sci. USA* **96:** 1779–1784.

Stevenson, J.M., Perera, I.Y., Heilmann, I., Persson, S. and Boss, W.F., 2000, Inositol signaling and plant growth. *Trends Plant Sci.* **5:** 252–258.

Stevenson-Paulik, J., Odom, A.R. and York, J.D., 2002, Molecular and biochemical characterization of two plant inositol polyphosphate 6-/3-/5-kinases. *J. Biol. Chem.* **277:** 42711–42718.

Stolz, L.E., Kuo, W.J., Longchamps, J., Sekhon, M. and York, J.D., 1998, *INP51*, a yeast inositol polyphosphate 5-phosphatase required for phosphatidylinositol 4,5-bisphosphate homeostasis and whose absence confers a cold-resistant phenotype. *J. Biol. Chem.* **273:** 11852–11861.

Suetsugu, S. and Takenawa, T., 2003, Regulation of cortical actin networks in cell migration. *Intl. Rev. Cytol.* **229:** 245–286.

Sulis, M.L. and Parsons, R., 2003, PTEN: from pathology to biology. *Trends Cell Biol.* **13:** 478–483.

Swann, K., Larman, M.G., Saunders, C.M. and Lai, F.A., 2004, The cytosolic sperm factor that triggers Ca^{2+} oscillations and egg activation in mammals is a novel phospholipase C: PLC zeta. *Reprod.* **127:** 431–439.

Takahashi, S., Katagiri, T., Hirayama, T., Yamaguchi-Shinozaki, K. and Shinozaki, K., 2001, Hyperosmotic stress induces a rapid and transient increase in inositol 1,4,5-trisphosphate independent of abscisic acid in *Arabidopsis* cell culture. *Plant Cell Physiol.* **42:** 214–222.

Takenawa, T. and Itoh, T., 2001, Phosphoinositides, key molecules for regulation of actin cytoskeletal organization and membrane traffic from the plasma membrane. *Biochim. Biophys. Acta-Mol. Cell Biol. Lipids* **1533:** 190–206.

Taylor, A.T.S., Kim, J. and Low, P.S., 2001, Involvement of mitogen-activated protein kinase activation in the signal transduction pathways of the soya bean oxidative burst. *Biochem. J.* **355:** 795–803.

Teodoro, A.E., Zingarelli, L. and Lado, P., 1998, Early changes in Cl^- efflux and H^+ extrusion induced by osmotic stress in *Arabidopsis thaliana* cells. *Physiol. Plant.* **102:** 29–37.

Testerink, C. and Munnik, T., 2004, Plant responses to stress: phosphatidic acid as a second messenger. In: Encyclopedia of Plant and Crop Science, (Goodman, R.M., ed.). Marcel Dekker, Inc., N.Y., pp. 995–998.

Testerink, C. and Munnik, T., 2005, Phosphatidic acid: a multifunctional stress signaling lipid in plants. *Trends Plant Sci.* **10:** 368–375.

Testerink, C., Dekker, H.L., Lim, Z.Y., Johns, M.K., Holmes, A.B., Koster, C.G., Ktistakis, N.T. and Munnik, T., 2004, Isolation and identification of phosphatidic acid targets from plants. *Plant J.* **39:** 527–536.

Thomma, B.P.H.J., Penninckx, I.A.M.A., Broekaert, W.F. and Cammue, B.P.A., 2001a, The complexity of disease signaling in *Arabidopsis*. *Curr. Op. Immunol.* **13:** 63–68.

Thomma, B.P.H.J., Tierens, K.F.M., Penninckx, I.A.M.A., Mauch-Mani, B., Broekaert, W.F. and Cammue, B.P.A., 2001b, Different micro-organisms differentially induce *Arabidopsis* disease response pathways. *Plant Physiol. Biochem.* **39:** 673–680.

Toyoda, K., Shiraishi, T., Yamada, T., Ichinose, Y. and Oku, H., 1993, Rapid changes in polyphosphoinositide metabolism in pea in response to fungal signals. *Plant Cell Physiol.* **34:** 729–735.

Vajanaphanich, M., Kachintorn, U., Barrett, K.E., Cohn, J.A., Dharmasathaphorn, K. and Traynor-Kaplan, A., 1993, Phosphatidic acid modulates Cl^- secretion in T84 cells: varying effects depending on mode of stimulation. *Am. J. Physiol. Cell Physiol.* **264:** C1210-C1218.

Vajanaphanich, M., Schultz, C., Rudolf, M.T., Wasserman, M., Enyedi, P., Craxton, A., Shears, S.B., Tsien, R.Y., Barrett, K.E. and Traynor-Kaplan, A., 1994, Long-term uncoupling of chloride secretion from intracellular calcium levels by Ins(3,4,5,6)P_4. *Nature* **371:** 711–714.

Van der Luit, A.H., Piatti, T., Van Doorn, A., Musgrave, A., Felix, G., Boller, T. and Munnik, T., 2000, Elicitation of suspension-cultured tomato cells triggers the formation of phosphatidic acid and diacylglycerol pyrophosphate. *Plant Physiol.* **123:** 1507–1515.

Van der Wijk, T., Tomassen, S.F.B., de Jong, H.R. and Tilly B.C., 2000, Signaling mechanisms involved in volume regulation of intestinal epithelial cells. *Cell Physiol. Biochem.* **10:** 289–296.

Van Dijken, P., Bergsma, J.C. and Van Haastert, P.J.M., 1997, Phospholipase C-independent inositol 1,4,5-trisphosphate formation in *Dictyostelium* cells. Activation of a plasma-membrane-bound phosphatase by receptor-stimulated Ca^{2+} influx. *Eur. J. Biochem.* **244:** 113–119.

Van Dijken, P., de Haas, J.R., Craxton, A., Erneux, C., Shears, S.B. and Van Haastert, P.J.M., 1995, A novel phospholipase C-independent pathway of inositol 1,4,5-trisphosphate formation in *Dictyostelium* and rat liver. *J. Biol. Chem.* **270:** 29724–29731.

Van Haastert, P.J.M. and Van Dijken, P., 1997, Biochemistry and genetics of inositol phosphate metabolism in *Dictyostelium*. *FEBS Lett.* **410:** 39–43.

Van Leeuwen, W., Ökrész, L., Bögre, L.. and Munnik, T., 2004, Learning the lipid language of plant signalling. *Trends Plant Sci.* **8:** 378–384.

Viswanathan, C. and Zhu, J.K., 2002, Molecular genetic analysis of cold-regulated gene transcription. *Phil. Trans. Roy. Soc. Lond. B* **357:** 877–886.

Wang, C.X., Zien, C.A., Afitlhile, M., Welti, R., Hildebrand, D.F. and Wang, X., 2000, Involvement of phospholipase D in wound-induced accumulation of jasmonic acid in *Arabidopsis*. *Plant Cell* **12:** 2237–2246.

Wang, X., 2000, Multiple forms of phospholipase D in plants: the gene family, catalytic and regulatory properties, and cellular functions. *Prog. Lipid Res.* **39:** 109–149.

Wang, X., 2001, Plant phospholipases. *Annu. Rev. Plant Physiol. Plant Mol. Biol.* **52:** 211–231.

Wang, X., 2004, Lipid signaling. *Curr. Op. Plant Biol.* **7:** 329–336.

Wang, X., Wang, C.X., Sang, Y., Qin, C.B. and Welti, R., 2002, Networking of phospholipases in plant signal transduction. *Physiol. Plant.* **115:** 331–335.

Ward, J.M., Pei, Z.-M. and Schroeder, J.I., 1995, Roles of ion channels in initiation of signal transduction in higher plants. *Plant Cell* **7:** 833–844.

Watt, S.A., Kimber, W.A., Fleming, I.N., Leslie, N.R., Downes, C.P. and Lucocq, J.M., 2004, Detection of novel intracellular agonist responsive pools of phosphatidylinositol 3,4-bisphosphate using the TAPP1 pleckstrin homology domain in immunoelectron microspray. *Biochem. J.* **377:** 653–663.

Watt, S.A., Kular, G., Fleming, I.N., Downes, C.P. and Lucocq, J.M., 2002, Subcellular localization of phosphatidylinositol 4,5-bisphosphate using the pleckstrin homology domain of phospholipase Cδ1. *Biochem. J.* **363:** 657–666.

Wehner, F., Olsen, H., Tinel, H., Kinne-Saffran, E. and Kinne, R.K.H., 2003, Cell volume regulation: osmolytes, osmolyte transport, and signal transduction. *Rev. Physiol. Biochem. Pharmacol.* **148:** 1–80.

Welti, R., Li, W.Q., Li, M.Y., Sang, Y.M., Biesiada, H., Zhou, H.E., Rajashekar, C.B., Williams, T.D. and Wang, X., 2002, Profiling membrane lipids in plant stress responses – role of phospholipase Dα in freezing-induced lipid changes in *Arabidopsis*. *J. Biol. Chem.* **277:** 31994–32002.

Welters, P., Takegawa, K., Emr, S.D. and Chrispeels, M.J., 1994, AtVPS34, a phosphatidylinositol 3-kinase of *Arabidopsis thaliana*, is an essential protein with homology to a calcium-dependent lipid-binding domain. *Proc. Natl. Acad. Sci. USA* **91:** 11398–11402.

Wendehenne, D., Pugin, A., Klessig, D.F. and Durner, J., 2001, Nitric oxide: comparative synthesis and signaling in animal and plant cells. *Trends Plant Sci.* **6:** 177–183.

Westergren, T., Dove, S.K., Sommarin, M. and Pical, C., 2001, AtPIP5K1, an *Arabidopsis thaliana* phosphatidylinositol phosphate kinase, synthesizes PtdIns(3,4)P_2 and PtdIns(4,5)P_2 in vitro and is inhibited by phosphorylation. *Biochem. J.* **359:** 583–589.

Whitley, P., Reaves, B.J., Hashimoto, M., Riley, A.M., Potter, B.V.L. and Holman, G.D., 2003, Identification of mammalian Vps24p as an effector of phosphatidylinositol 3,5-bisphosphate-dependent endosome compartmentalization. *J. Biol. Chem.* **278:** 38786–38795.

Williams, M.E., Torabinejad, J., Cohick, E., Parker, K., Drake, E.J., Thompson, J.E., Hortter, M. and DeWald, D.B., 2005, Mutations in the *Arabidopsis* phosphoinositide phosphatase gene SAC9 lead to overaccumulation of PtdIns(4,5)P_2 and constitutive expression of the stress-response pathway. *Plant Physiol.* **138:** 686–700.

Wondergem, N., Gong, W., Monen, S.H., Dooley, S.N., Gonce, J.L., Conner, T.D., Houser, M., Ecay, T.W. and Ferslew, K.E., 2001, Blocking swelling-activated chloride current inhibits mouse liver cell proliferation. *J. Physiol. Lond.* **532:** 661–672.

Wood, N.T., Allan, A.C., Haley, H., Viry-Moussaïd, M. and Trewavas, A.J., 2000, The characterization of differential calcium signalling in tobacco guard cells. *Plant J.* **24:** 335–344.

Wurmser, A.E., Gary, J.D. and Emr, S.D., 1999, Phosphoinositide 3-kinases and their FYVE domain-containing effectors as regulators of vacuolar/lysosomal membrane trafficking pathways. *J. Biol. Chem.* **274:** 9129–9132.

Xia, H-J., Brearley, C., Elge, S., Kaplan, B., Fromm, H. and Mueller-Roeber, B., 2003, *Arabidopsis* inositol polyphosphate 6-/3-kinase is a nuclear protein that complements a yeast mutant lacking a functional ArgR-Mcm1 transcription complex. *Plant Cell* **15:** 449–463.

Xie, W.W., Solomons, K.R.H., Freeman, S., Kaetzel, M.A., Bruzik, K.S., Nelson, D.J. and Shears, S.B., 1998, Regulation of Ca^{2+}-dependent Cl^- conductance in a human colonic epithelial cell line (T84): Cross-talk between Ins(3,4,5,6)P_4 and protein phosphatases. *J. Physiol. Lond.* **510:** 661–673.

Xing, T., Higgins, V.J. and Blumwald, E., 1997, Identification of G proteins mediating fungal elicitor-induced dephosphorylation of host plasma membrane H^+-ATPase. *J. Exp. Bot.* **48:** 229–237.

Xiong, L., Lee, B.H., Ishitani, M., Lee, H., Zhang, C.Q. and Zhu, J.K., 2001, The FIERY1 encoding an inositol polyphosphate 1- phosphatase is a negative regulator of abscisic acid and stress signaling in *Arabidopsis*. *Genes Dev.* **15:** 1971–1984.

Xiong, L., Schumaker, K.S. and Zhu, J-K., 2002, Cell signaling during cold, drought, and salt stress. *Plant Cell* **14:** S165-S183.

Yamamoto, A., DeWald, D.B., Boronenkov, I.V., Anderson, R.A., Emr, S.D. and Koshland, D., 1995, Novel PI(4)P 5-kinase homologue, Fab1p, essential for normal vacuole function and morphology in yeast. *Mol. Biol. Cell* **6:** 525–539.

Yang, X.N., Rudolf, M., Carew, R.A., Yoshida, M., Nerreter, V., Riley, A.M., Chung, S.K., Bruzik, K.S., Potter, B.V.L., Schultz, C. and Shears, S.B., 1999, Inositol 1,3,4-trisphosphate acts in vivo as a specific regulator of cellular signaling by insositol 3,4,5,6-tetrakisphosphate. *J. Biol. Chem.* **274:** 18973–18980.

Yin, H.L. and Janmey, P.A., 2003, Phosphoinositide regulation of the actin cytoskeleton. *Annu. Rev. Physiol.* **65:** 761–789.

York, J.D., Guo, S., Odom, A.R., Spiegelberg, B.D. and Stolz, L.E., 2001, An expanded view of inositol signaling. *Advan. Enz. Regul.* **41:** 57–71.

York, J.D., Odom, A.R., Murphy, R., Ives, E.B. and Wente, S.R., 1999, A phospholipase C-dependent inositol polyphosphate kinase pathway required for efficient messenger RNA export. *Science* **285:** 96–100.

Zabela, M.D., Fernandez-Delmond, I., Niittyla, T., Sanchez, P. and Grant, M., 2002, Differential expression of genes encoding *Arabidopsis* phospholipases after challenge with virulent or avirulent *Pseudomonas* isolates. *Mol. Plant-Microbe Interactions* **15:** 808–816.

Zhang, W.H., Qin, C.B., Zhao, J. and Wang, X., 2004, Phospholipase Dα1-derived phosphatidic acid interacts with ABI1 phosphatase 2C and regulates abscisic acid signaling. *Proc. Natl. Acad. Sci. USA* **101:** 9508–9513.

Zhang, W.H., Wang, C.X., Qin, C.B., Wood, T., Olafsdottir, G., Welti, R. and Wang, X.M., 2003, The oleate-stimulated phospholipase D, PLDδ, and phosphatidic acid decrease H_2O_2-induced cell death in *Arabidopsis*. *Plant Cell* **15:** 2285–2295.

Zhao, J. and Wang, X., 2004, *Arabidopsis* phospholipase Dα1 interacts with the heterotrimeric G-protein α-subunit through a motif analogous to the DRY motif in G-protein-coupled receptors. *J. Biol. Chem.* **279:** 1794–1800.

Zhong, R. and Ye, Z-H., 2004, Molecular and biochemical characterization of three WD-repeat-domain containing inositol polyphosphate 5-phosphatases in *Arabidopsis thaliana*. *Plant Cell Physiol.* **45:** 1720–1728.

Zhong, R., Burk, D.H., Morrison, W.H. and Ye, Z-H., 2004, FRAGILE FIBER3, an *Arabidopsis* gene encoding a Type II inositol polyphosphate 5-phosphatase, is required for secondary wall synthesis and actin organization in fiber cells. *Plant Cell* **16:** 3242–3259.

Zhong, R., Burk, D.H., Nairn, C.J., Wood-Jones, A., Morrison, W.H. and Ye, Z-H., 2005, Mutation of SAC1, an *Arabidopsis* SAC domain phosphoinositide phosphatase, causes alterations in cell morphogenesis, cell wall synthesis, and actin organization. *Plant Cell* **17:** 1449–1466.

Zimmermann, S., Nürnberger, T., Frachisse, J.M., Wirtz, W., Guern, J., Hedrich, R. and Scheel, D., 1997, Receptor-mediated activation of a plant Ca^{2+}-permeable ion channel involved in pathogen defense. *Proc. Natl. Acad. Sci. USA* **94:** 2751–2755.

Zonia, L., Cordeiro, S. and Feijo, J.A., 2001, Ion dynamics and hydrodynamics in the regulation of pollen tube growth. *Sex. Plant Reprod.* **14:** 111–116.

Zonia, L., Cordeiro, S., Tupy, J. and Feijo, J.A., 2002, Oscillatory chloride efflux at the pollen tube apex has a role in growth and cell volume regulation and is targeted by inositol 3,4,5,6-tetrakisphosphate. *Plant Cell* **14:** 2233–2249.

Zonia, L. and Munnik, T., 2004, Osmotically-induced cell swelling versus cell shrinking elicits specific changes in phospholipid signals in tobacco pollen tubes. *Plant Physiol.* **134:** 813–823.

Chapter 10

Inositols and Their Metabolites in Abiotic and Biotic Stress Responses

Teruaki Taji[1,2], Seiji Takahashi[3], and Kazuo Shinozaki[1]
[1] *Laboratory of Plant Molecular Biology, RIKEN Tsukuba Institute, 3-1-1 Koyadai, Tsukuba, Ibaraki 305-0074, Japan*
[2] *Tokyo University of Agriculture, 1-1-1 Sakuragaoka, Setagaya, Tokyo 156-8502, Japan*
[3] *Institute of Multidisciplinary Research for Advanced Materials, Tohoku University, Katahira 2-1-1, Aoba, Sendai 980-8577, Japan*

Inositols are found ubiquitously in the biological kingdom, and their metabolites play important roles in stress responses, membrane biosynthesis, growth regulation, and many other processes. In this chapter, we describe the role of inositol and its derivative metabolites in abiotic and biotic stress responses. Inositol and its metabolites function as both osmolyte and secondary messengers under these stresses. The accumulation of osmolytes during osmotic stress is a ubiquitous biochemical mechanism found in all organisms from bacteria, fungi, and algae to vascular plants and animals. The accumulated osmolytes include glycerol, *myo*-inositol, betain, taurine, proline, trehalose, and raffinose. Plants accumulate many kinds of inositol-derivative metabolites during abiotic stresses, such as drought, low temperature, and high-salinity stresses; in contrast most animals accumulate only *myo*-inositol.

In animal systems, it has been well documented that a variety of phosphoinositides and inositol phosphates function as secondary messengers in various signaling processes. Phosphoinositide-specific Phospholipase C (PI-PLC) digests phosphatidylinositol 4,5-bisphosphate (PtdIns(4,5)P_2) to generate two secondary messengers, inositol 1,4,5-trisphosphate (Ins(1,4,5)P_3) and diacylglycerol (DG). (Ins(1,4,5)P_3) induces the release of Ca^{2+} into cytoplasm, which in turn causes various cellular responses. In plants, similar systems function in response to abiotic stress, such as drought, cold, and high-salinity stresses.

A. Lahiri Majumder and B. B. Biswas (eds.), Biology of Inositols and Phosphoinositides, 239–264
© 2006 *Springer. Printed in the Netherlands.*

1. INOSITOLS AND THEIR METABOLITES AS OSMOPROTECTANTS IN RESPONSE TO ABIOTIC AND BIOTIC STRESSES

The accumulation of osmolytes in response to external osmotic change is probably universal and leads to osmotic adjustment as the major element in accomplishing tolerance. Advantages of these organic osmolytes are (1) a compatibility with macromolecular structure and function at high or variable (or both) osmolyte concentrations and (2) little effect on various proteins to function in concentrated intracellular solutions. Furthermore, osmolytes are thought to function not only in osmotic adjustment but also in the protection of cells and macromolecules, such as maintaining membrane integrity, preventing protein denaturation, protection against oxidative damage by scavenging free radicals, and by lowering the Tm value of the nucleic acid (Crowe *et al.*, 1987; Incharoensakdi *et al.*, 1986; Nomura *et al.*, 1995, 1998; Rajendrakumar *et al.*, 1997; Schobert and Tschesche, 1978; Shen *et al.*, 1997; Smirnoff and Cumbes, 1989). Actually, the main function of osmolytes may be to stabilize proteins, protein complexes, and membranes during environmental stress, because the amount of most reported osmolytes are not sufficient for osmotic adjustment. Therefore, we call such metabolites osmoprotectants.

1.1 Osmolyte strategy in animal cells

Animal cells precisely regulate their size by adjusting the cytoplasmic concentration of osmotically active substances. All nucleated cells subjected to sustained high osmolarity have developed means to preferentially accumulate compatible osmolytes, such as *myo*-inositol, sorbitol, betaine, α-glycerophosphorylcholine, and taurine, which increase intracellular osmolality, restore cell volume, and provide a buffer against osmotic stress (Burg *et al.*, 1997; Haussinger, 1996; Kwon and Handler, 1995). While some organic osmolytes are synthesized within cells, most are imported by specialized transporters to mediate their accumulation from the external environment (Garcia-Perez and Burg, 1991; Kwon and Handler, 1995; Yancey *et al.*, 1982). Four osmolyte transporters have been identified in mammals: Na^+/*myo*-inositol transporter (SMIT), betaine/γ-aminobutric acid transporter (BGT1), taurine/β-amino acid transporter (TAUT), and the system A amino acid transporter (Garcia-Perez and Burg, 1991; Kwon and Handler, 1995).

In mammals, the increased uptake of the osmolyte *myo*-inositol in response to hyperosmolarity has been described in renal medulla cells, lens epithelia, astrocytes, endothelial cells, and Kupffer cells (Paredes *et al.*, 1992; Nakanishi *et al.*, 1988; Warskulat *et al.*, 1997; Wiese *et al.*, 1996; Zhou *et al.*, 1994). The hyperosmolarity-induced *myo*-inositol accumulation inside these cells is the result of both an increase in the V_{max} of the SMIT and the increased expression of its gene (Kwon *et al.*, 1992; Nakanishi *et al.*, 1989). Conversely, hypo-osmotic exposure diminished *myo*-inositol uptake when compared with normo-osmotic incubations. The hyperosmolarity-induced *SMIT* mRNA increase was counteracted by the exogenous application of added *myo*-inositol or betaine (Warskulat *et al.*, 1997). Following promoter analysis of the cotransporter gene, a tonicity-responsive enhancer (TonE) was identified. TonE (also known as ORE), whose putative consensus sequence is TGGAAANN (C/T) N (C/T), regulates genes for SMIT, BGT1, and aldose reductase (Ferraris *et al.*, 1996; Ko *et al.*, 1997; Miyakawa *et al.*, 1998; Rim *et al.*, 1998). Furthermore, TonE binding protein (TonEBP/ OREBP/NFAT5), a transcriptional factor that stimulates transcription through its binding to TonE sequences via a Rel-like DNA binding domain was isolated (Ko *et al.*, 2000; Lopez-Rodriguez *et al.*, 2004; Miyakawa *et al.*, 1999). The TonEBP gene is ubiquitously expressed in various human tissues under normal conditions. Western blot and immunohistochemical analyzes of cells cultured in hypertonic medium revealed that exposure to hypertonicity elicits the activation of TonEBP, which is the result of an increase in the amount of TonEBP and its translocation to the nucleus (Miyakawa *et al.*, 1999). Lopez-Rodriguez generated NAFT5 (TonEBP/OREBP) null mice. The homozygous NFAT5 null allele resulted in midembryonic lethality with incomplete penetrance. Surviving mutant mice showed progressive growth retardation and impaired activation of osmoprotective genes, including *AR*, *SMIT*, *BGT1*, and *TAUT*. *AR* is controlled by a TonE, so that cells lacking NFAT5 do not express *AR* mRNA. These results demonstrate a central role for NAFT5/TonEBP/OREBP as a tonicity-responsive transcription factor (Lopez-Rodriguez *et al.*, 2004).

Similarities exist between the intracellular signaling pathways that regulate the responses to osmotic stress in yeast and those found in animals. In both the cases, a member of the MAP kinase family, $p38^{MAPK}$/Hog1, is a key regulatory protein. In yeast cells exposed to hyperosmolarity, a mitogen-activated protein kinase, Hog1, is activated (Brewster *et al.*, 1993; Han *et al.*, 1994). Activated Hog1 upregulates the expression of glycerol-3-phosphatate dehydrogenase and glycerol-3-phosphatase isoform 2 (Posas *et al.*, 2000; Rep *et al.*, 2000), which are essential for the accumulation of glycerol, the principal osmolyte in yeast (Ohmiya *et al.*, 1995; Pahlman *et al.*, 2001). Yeast *hog1* mutant cannot survive in hyperosmorality, but the induction of p38, the mammalian homolog of Hog1, can complement the phenotype (Han *et al.*, 1994). Treatment with pyridinyl imidazole, SB203580, a specific inhibitor of p38, inhibits the hypertonic induction of *SMIT*, *BGT1*, and *AR* gene expression through ORE/TonE as well as *myo*-inositol and betaine uptake (Denkert *et al.*, 1998; Ko *et al.*, 2002; Sheikh-Hamad *et al.*, 1998). In addition to the activation of p38, osmotic stress induces an increase in protein tyrosine phosphorylation, which is mediated in part by Fyn kinase, an ubiquitously expressed member of the Src family of protein tyrosine kinases (Kapus *et al.*, 1999; Szaszi *et al.*, 1997; Thomas and Brugge, 1997). Hypertonic activation of the ORE/TonE reporter was partially blocked by pharmacological inhibition of Fyn (Ko *et al.*, 2002). Importantly, inhibiting p38 in Fyn-deficient cells almost completely suppressed the induction of ORE/TonE reporter activity, indicating that p38 and Fyn are the major signaling pathways for the hypertonic activation of *OREBP/TonEBP* (Ko *et al.*, 2002).

It is known that high-salinity stress damages DNA, and that DNA damage activates ataxia telangiectasia mutated (ATM) kinase. *TonEBP/OREBP* includes consensus sites for ATM kinase phosphorylation. Recently, it was found that high-salinity activated ATM. High urea and radiation also activates ATM, but they do not increase *TonEBP/OREBP* transcriptional activity like high NaCl does (Irarrazabal *et al.*, 2004). Wortmannin, an ATM inhibitor, reduces NaCl-induced *TonEBP/OREBP* transcriptional activation and *BGT1* mRNA increase. Overexpression of wild type *TonEBP/OREBP* increases ORE/TonE reporter activity much more than does overexpression of *TonEBP/OREBP* mutagenized the phosphorylation sites. These results suggest that the signaling via ATM is necessary for full activation of *TonEBP/OREBP* by high salinity (Irarrazabal *et al.*, 2004).

1.2 Osmolyte strategy in plant cells

Plant growth is greatly affected by environmental abiotic stresses, such as drought, high salinity, and low temperature. Plants respond and adapt in order to survive during stressful conditions. Among these abiotic stresses, drought or water deficit is the most severe limiting factor against plant growth and crop production. Water-deficit stress induces various biochemical and physiological responses in plants. It is known that higher plants have sophisticated mechanisms for adaptation against drought and salt stress such as stomatal closure, accumulation of salt into the vacuole, whilst leaves become succulent, and cuticularized leaves to protect their cells from these stresses. Various osmoprotectants and antioxidants accumulate during water stress in plant cells.

The unicellular eukaryotic *Saccharomyces cerevisiae* is an ideal model for studying the cellular and molecular mechanisms of osmotic stress tolerance in higher plants. Yeast cells accumulate compatible osmolytes, mainly glycerol and some trehalose during osmotic stress. However, it is known that plant species accumulate a greater number of osmolytes compared to yeasts and animals. The accumulated osmolytes in plants include amino acids, their derivatives (proline, glycine betaine, beta-alanine betaine, and proline betaine), tertiary amines, sulfonium compounds (choline *o*-sulfate, and dimethylsulfoniopropionate), sugars, and polyols (glycerol, mannitol, sorbitol, trehalose, fructans, and inositols) (Bohnert and Jensen, 1996a,b; Hanson *et al.*, 1994; McCue and Hanson, 1990; Trossat *et al.*, 1996). This suggests that plants have adapted many environments. We can find plants anywhere under various climatic conditions differing in temperature, light quantity, nutrient quantity, and availability of water. Plants growing in extreme conditions are adapted to these severe conditions with changes to their growth and development. This adaptation is essential for plants to survive in any environment because plants cannot escape unfavorable conditions due to their sessile growth habit. Thus, plants accumulate a number of osmolytes whilst adapting to their surrounding environments during the process of evolution.

Recently, glycine, betaine, mannitol, ononitol, trehalose, fructan proline, and raffinose have been shown to be important for the improvement of stress tolerance in plants by the manipulation of genes encoding key enzymes of osmolyte synthesis or degradation pathways (Hayashi *et al.*, 1997; Holmström *et al.*, 1996; Kavi Kishor *et al.*, 1995; Nanjo *et al.*, 1999a,b; Nuccio *et al.*, 1998; Pilon-Smits *et al.*, 1995; Sakamoto *et al.*, 1998; Sheveleva *et al.*, 1998; Taji *et al.*, 2002; Takabe *et al.*, 1998; Tarczynski *et al.*, 1993; Thomas *et al.*, 1995).

In the next section, we describe the osmolyte strategy plants adopt against drought, high salinity, and cold stresses, especially the role of *myo*-inositol derivatives including *myo*-inositol-1-phosphate, D-ononitol, D-pinitol, galactinol, and raffinose.

1.2.1 Osmolyte strategy using inositol and inositol derivatives including *myo*-inositol-1-phosphate, D-ononitol, and D-pinitol in plant

The pathway from glucose 6-phosphate to *myo*-inositol 1-phosphate and *myo*-inositol is essential for the synthesis of various metabolites. The first step for the synthesis of *myo*-inositol 1-phosphate is catalyzed by *myo*-inositol 1-phosphate synthase (MIPS) through an internal oxidoreduction reaction involving NAD^+. Free inositol is generated by dephosphorylation of the MIPS product by a specific Mg^{2+}-dependent inositol-1-phosphate phosphatase. This mechanism is conserved among all *myo*-inositol producing organisms (Majumder *et al.*, 1997). The gene coding for cytosolic MIPS, named *INO1*, was first identified in the yeast, *S. cerevisiae* (Klig and Henry, 1984; Majumder *et al.*, 1981), and to date, over 60 genes homologous to INO1 have been identified from various organisms (Majumder *et al.*, 2003).

In some plant species inositol provides for the production and accumulation of the osmolytes D-ononitol and D-pinitol. In the halophyte *Mesembryanthemum crystallinum* (common ice plant), which has a remarkable tolerance against drought, high salinity, and cold stress, inositol is methylated to D-ononitol and subsequently epimerized to D-pinitol. This plant accumulates a large amount of these inositol derivatives during the above stresses (Adams *et al.*, 1992; Paul and Cockburn, 1989; Vernon *et al.*, 1993). The gene encoding *myo*-inositol *O*-methyl-transferase, *IMT1*, is transcriptionally induced by osmotic stresses, while neither transcripts nor enzyme activity are detectable in normal growth conditions in ice plants (Rammesmayer *et al.*, 1995; Vernon and Bohnert, 1992). IMT1 uses S-adenosylmethionine (SAM) as a methyl donor and produces D-ononitol and S-adenosylhomocysteine (SAH). Following the ononitol synthesis, pinitol is synthesized by ononitol epimerase, OEP1 (Ishitani *et al.*, 1996).

The *Inps1* gene, which encodes a *myo*-inositol-1-phosphate synthase was first isolated from an ice plant. *Inps1* transcripts can be detected under normal growth conditions. Following high-salinity stress *Inps1* is upregulated as well as the *Imt1* gene, and a large amount of free inositol and D-pinitol accumulate during salinity stress (Ishitani *et al.*, 1996). In contrast, *Arabidopsis thaliana*, a typical glycophyte model plant, does not show upregulation of *Inps1* or any accumulation of inositol and methylated inositols including D-pinitol and D-ononitol (Ishitani *et al.*, 1996). This result demonstrates a fundamental difference between halophytes and glycophytes in transcriptional and metabolic regulation that is causally related to salinity tolerance. By the immunocytological analysis of INPS and IMT, INPS is present in all cells, but IMT is repressed during normal growth conditions. After salinity stress, the amount of INPS was enhanced in leaves but repressed in roots, while IMT was induced in all cell types (Nelson *et al.*, 1998). Regulation of the amount of enzyme is controlled by transcriptional level of the *Inps* and *Imt*. The regulation of the *Imt* promoter is mediated by an element very close to the TATA box, but the precise *cis*-element has not been identified (Nelson *et al.*, 1998). Inositol and ononitol constitute major phloem carbohydrates that seem to have two functions in roots. Inositol is essential for root growth, and the amounts of inositol and ononitol are correlated with the amount of sodium uptake to the leaves through the xylem (Nelson *et al.*, 1998).

To generate transgenic plants with a capacity for higher polyol production, IMT1 cDNA from ice plant has been transferred into *Nicotiana tabacum* cultivar SR1. Methylated inositols including ononitol and pinitol are usually absent in tobacco. The transgenic tobacco plants show normal phenotypes under normal growth conditions. When the transgenic plants were exposed to salinity or drought stresses, a large amount of D-ononitol accumulates and the plants showed enhanced tolerance to drought and salt stress compared with control tobacco plants (Sheveleva *et al.*, 1997). After these stress treatments, the transgenic leaves were slower to lose turgor and their photosynthetic rates were less affected than in wild type plants (Sheveleva *et al.*, 1997). High accumulation of D-ononitol was thought to occur because the levels of *myo*-inositol increase during high salinity or drought stresses, thus providing additional substrate for IMT1. Recently, a novel salt-tolerant MIPS, *myo*-inositol-1-phosphate synthase, has been co-isolated from *Porteresia coarctata* (Roxb.) Tateoka, a halophytic wild rice (Majee *et al.*, 2004). Purified bacterially expressed PINO1 protein shows salt tolerance compared with the protein expressed by the homologous gene, *RINO1* from *Oryza sativa* L. There is a stretch of 37 amino acids difference between *PINO1* and *RINO1*. Interestingly, the mutagenized PINO1 protein where the extra 37 amino acids were deleted lost its salt tolerance, suggesting that the structure of PINO1 protein is stable against salinity stress (Majee *et al.*, 2004). To assess the salt tolerance effect of the *PINO1* gene in planta, *RINO1*, *PINO1*, and the mutagenized *PINO1* genes were transformed to tobacco plants to induce constitutive high expression. *RINO1* overexpressed plants produced ~1.5- to 3-fold higher inositol in comparison with the control plants, plants transformed with the empty vector, or the mutagenized *PINO1* plants. *PINO1* overexpressed plants showed a two- to sevenfold increase in their levels of inositol under high-salinity stress compared with the *RINO1*, the mutagenized *PINO1* or the vector control plants under the same conditions. Moreover, *PINO1* overexpressed plants were able to grow in 200–300 mM NaCl, a salt concentration that inhibited the growth of vector control and *RINO1* overexpressed plants markedly (Majee *et al.*, 2004).

1.2.2 Osmolyte strategy using *myo*-inositol as a substrate including galactinol and raffinose in plants

Recently, it has been suggested that raffinose family oligosaccharides (RFO), such as raffinose and stachyose, are also able to function as osmolytes. Galactinol synthase (GolS) catalyzes the first committed step in the biosynthesis of RFO (Figure 1), and galactinol is used as a substrate in rafffinose and stachyose synthesis. Therefore, GolS is believed to be a key enzyme in RFO metabolism. Galactinol is synthesized from *myo*-inositol.

RFO accumulate during seed development and are thought to play a role in seed desiccation tolerance. For example, RFO accumulate at the late stage of maturation and desiccation in soybean seeds (Castillo *et al.*, 1990; Saravitz *et al.*, 1987). In maize, raffinose accumulates during the seed desiccation process and is thought to function in stress tolerance, whereas sucrose accumulates independently of desiccation tolerance. Desiccation tolerance of seeds is not achieved in the absence of raffinose accumulation (Brenac *et al.*, 1997). These results suggest that the ratio of sucrose to RFO is critical for desiccation tolerance of seeds rather than the total amount of sugars. Although young excised soybean seeds are not tolerant to desiccation, slow dehydration induces stress tolerance, which is strongly correlated with a significant increase in stachyose content (Blackman *et al.*, 1992).

We examined the functions of RFO in *A. thaliana* plants under drought, salt, and cold stress conditions based on the analyzes of function and expression of genes involved in RFO biosynthesis. Sugar analysis showed that drought-, high salinity- and cold- treated Arabidopsis plants accumulate a large amount of raffinose and galactinol, but not stachyose. Raffinose and galactinol were not detected in unstressed plants. This suggests that raffinose and galactinol are involved in the tolerance to drought, high salinity, and cold stresses. We identified three stress-responsive *GolS* genes (*AtGolS1, AtGolS2,* and *AtGolS3*) among all of eight Arabidopsis *GolS* genes in Arabidopsis genome. *AtGolS1* and *AtGolS2* were induced by drought and high-salinity stresses but not by cold stress. By contrast, *AtGolS3* was induced by cold stress but not by drought or salt stress (Taji *et al.*, 2002). Overexpression of *AtGolS2* in transgenic Arabidopsis caused an increase in endogenous galactinol and raffinose, plants also showed reduced transpiration from leaves thereby improving drought tolerance (Taji *et al.*, 2002). These results show that stress-inducible GolS plays a key role in the accumulation of galactinol and raffinose under abiotic stress conditions,

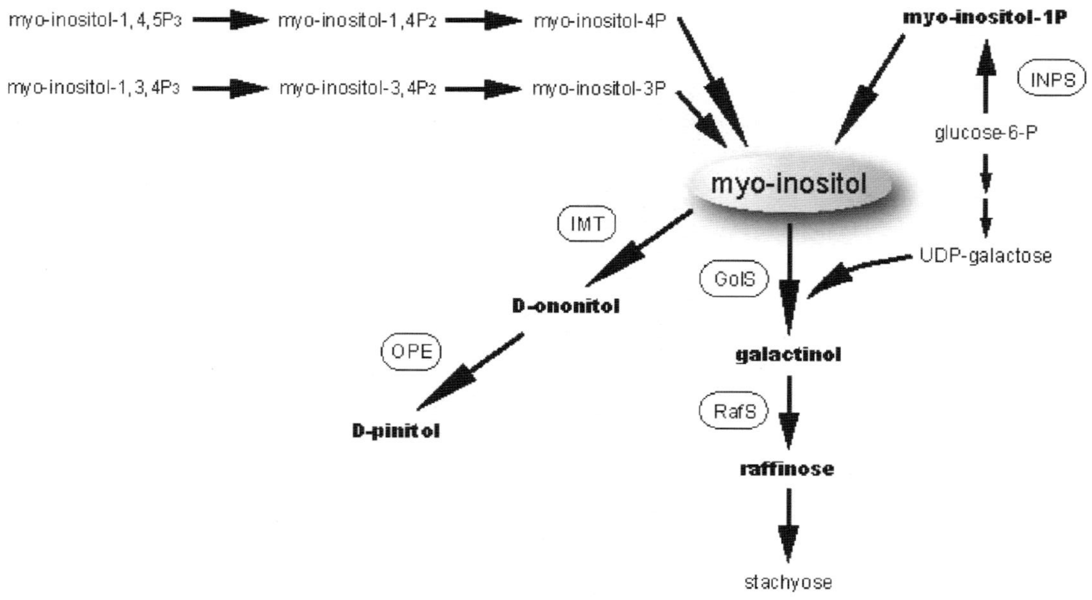

Figure 1. Metabolic pathway for the synthesis of *myo*-inositol and it is derivatives in plants.

and that galactinol and raffinose may function as osmoprotectants in drought stress tolerance of plants.

It is well known that DRE binding proteins *DREB1/CBF* and *DREB2*, which bind to dehydration-responsive element, function as transcription factors in cold, drought, and high-salinity stress responses. Expression of the *DREB1A/CBF3* gene was induced by cold stress but not by drought and high-salinity stresses, as was the *AtGolS3* gene. Overexpression of transcription factors, *DREB1A/CBF3* and *DREB1B/CBF1*, improved the tolerance to drought, high salinity, and cold stresses markedly (Jaglo-Ottosen *et al.*, 1998; Kasuga *et al.*, 1999; Liu *et al.*, 1998), and induced overexpression of many stress-inducible target genes under normal growth conditions (Gilmour *et al.*, 2000; Kasuga *et al.*, 1999; Maruyama *et al.*, 2004; Seki *et al.*, 2001). Recently, it was revealed that *DREB1A/CBF3* overexpression plants accumulated more proline and sugars, especially raffinose compared to wild type plants (Gilmour *et al.*, 2000). Expression analysis of *AtGolS* genes in *DREB1A/CBF3* overexpression plants revealed that only *AtGolS3* mRNA was significantly observed under normal growth conditions (Taji *et al.*, 2002). In the *AtGolS3* promoter, there are two DRE (TACCGACAT) and two DRE-like A/GCCGAC core motifs (Yamaguchi-Shinozaki and Shinozaki, 1994). These results suggest that *AtGolS3* is a target gene of DREB1A but the other *AtGolS* genes are not.

Heat shock factors (HSFs) are transcriptional regulators of the heat shock response. The major targets of HSFs are the genes encoding heat shock proteins (HSPs), which are known to have a protective function against cytotoxic effects. Recently, *AtGolS1* was identified as a novel HSF-dependent heat shock gene. *AtGolS1* shows heat inducibility in wild type, but shows constitutive high expression in *HSF3* overexpressor plants (Panikulangara *et al.*, 2004). Raffinose contents in wild type plants increase upon heat stress but not in *AtGolS1* mutant plants. Furthermore, the level of raffinose is enhanced and stachyose is also present in *HSF3* overexpressor plants under normal temperature, though raffinose and especially the stachyose content are hardly detectable at this temperature in wild type plants (Panikulangara *et al.*, 2004). Thus, RFO contribute not only to cold, drought, and high-salinity stress tolerances, but also to heat stress tolerance in plants.

Raffinose degradation proceeds by the action of α-galactosidase, which catalyzes the hydrolysis of the α-1,6 linkage of raffinose oligosaccharides. To investigate the function of α-galactosidase in RFO metabolism and to elucidate the role of raffinose in freezing stress tolerance in plants, the expression of α-galactosidase was modified in transgenic petunia (Pennycooke *et al.*, 2003). The tomato α-galactosidase gene under the control of a constitutive promoter was introduced into petunia in the sense and antisense orientations, so that α-galactosidase transcripts were reduced in antisense lines and increased in sense lines, respectively. Raffinose content of the antisense plants increased 12- to 22-fold compared to wild type under normal growth conditions, and increased 22- to 53-fold after cold acclimation. Downregulating α-galactosidase in petunia increased the freezing tolerance due to the over accumulation of raffinose, whereas overexpression of α-galactosidase gene caused a decrease in endogenous raffinose and impaired freezing tolerance (Pennycooke *et al.*, 2003). These results confirm a role for α-galactosidase in the inhibition of RFO metabolism.

2. INOSITOL PHOSPHATES AS SIGNALING MOLECULES IN RESPONSE TO ABIOTIC STRESS

To survive even after the serious damages caused by abiotic stress, land plants respond and adapt to the stress by means of physiological, developmental, and biochemical changes. After the perception of abiotic stress, the stimuli are transferred to the cytoplasm and nuclei, causing a dramatic change in the expression of many genes via complex signal transduction cascades mediated by protein phosphorylation/dephosphorylation, regulation of phytohormone levels, and generation of secondary messengers, such as cytosolic free Ca^{2+} and reactive oxygen species (ROS). Compounds derived from membrane components, such as phospholipids, also play important roles as secondary messengers in transmembrane signaling. The phosphoinositide-signaling pathway is known to be one of the most pathway in conveying extracellular stimuli into specific short- and long-term intracellular responses in animal systems (Berridge, 1993). The strongest evidence of a phosphoinositide-signal transduction system in plant cells has been provided by the identification of genes that show homology to animal phosphoinositide-metabolizing enzymes (reviewed in Mueller-Roeber and Pical, 2002; Munnik *et al.*, 1998; Wang, 2004) and phosphoinositides/inositol phosphates themselves (Brearley and Hanke, 1995; Irvine *et al.*, 1989; Munnik *et al.*, 1994; Parmar and Brearley, 1995). A number of reports have indicated that phosphoinositide- and inositol phosphate-turnover in plant cells is activated by various external stimuli. Moreover, genes encoding phosphoinositides- or inositol phosphate-metabolic enzyme have been shown to be upregulated by abiotic stress. In recent reports, forward- and reverse-genetics have provided the direct evidences that phosphoinositide/inositol phosphate turnover plays important roles in response to abiotic stress in plants.

2.1 Metabolic regulation of phosphoinositides/inositol phosphates in response to abiotic stress

2.1.1 Increase of inositol 1,4,5-trisphosphate levels in response to abiotic stress

In animal systems, it is well documented that phosphoinositides and inositol phosphates participate in various signal transduction processes. One of the most studied signaling events involved in phosphoinositide turnover is the generation of two secondary messengers, Ins(1,4,5)P_3, and DG, from hydrolysis of PtdIns(4,5)P_2 catalyzed by PI-PLC. In animal cells, activation of PI-PLC is one of the earliest key events in the regulation of various cell functions by more than 100 extracellular signaling molecules (reviewed in Rebecchi and Pentyala, 2000; Rhee and Bae, 1997). Ins(1,4,5)P_3 and DG mediate the release of intracellular Ca^{2+} from Ins(1,4,5)P_3-receptor-coupling Ca^{2+} channels and the activation of protein kinase C, respectively (Berridge, 1993; Majerus *et al.*, 1990; Nishizuka, 1992). Although no direct evidence has been presented for the existence of animal type Ins(1,4,5)P_3 receptors in plants, production of Ins(1,4,5)P_3 in response to extracellular stimuli has been studied as an initial response that triggers intracellular signal transduction cascades.

In plant systems, a number of reports have indicated that the level of intracellular Ins(1,4,5)P_3 increases in response to a variety of extracellular stimuli, such as light, pathogen elicitors, cell fusion, H_2O_2, and ethanol (reviewed in Chapman, 1998; Drøbak, 1992; Munnik *et al.*, 1998; Stevenson *et al.*, 2000). In gravistimulated maize pulvini and oat shoot pulvini, both transient and long-term Ins(1,4,5)P_3 increase prior to gravitropic curvature have been shown, suggesting a role for Ins(1,4,5)P_3 as a secondary messenger in the establishment of tissue polarity caused by gravistimulus (Perera *et al.*, 1999, 2001; Stevenson *et al.*, 2000). Franklin-Tong *et al.* (1996) demonstrated the regulatory effect of Ins(1,4,5)P_3 on *Papaver rhoeas* pollen tube growth during self-incompatibility response. The release of Ins(1,4,5)P_3 in pollen tubes causes an increase in the level of cytosolic free Ca^{2+} and inhibition of pollen tube growth, which is blocked partially by PI-PLC inhibitors. These results indicate that Ins(1,4,5)P_3-triggered signaling is also involved in the Ca^{2+} gradient and the regulation of cell growth. Recently, blue-light-induced transient elevation of cytosolic Ca^{2+}, involving a blue-light receptor, phot2, has been shown to be mediated by the activation of PI-PLC in Arabidopsis, suggesting that the increase of Ins(1,4,5)P_3 via PI-PLC activation in response to blue-light may induce the release of Ca^{2+} from internal Ca^{2+} stores (Harada *et al.*, 2003).

Plants have complex signaling networks to respond and adapt to abiotic stresses (Shinozaki *et al.*, 2000, 2003; Xiong *et al.*, 2002). Signal transduction pathways mediated by Ins(1,4,5)P_3 production have been proposed to participate in abiotic stress responses. Winter oil seed rape leaf discs exposed to freezing temperature (-3°C to -7°C) were shown to induce transient formation of Ins(1,4,5)P_3 30 min after treatment (Smolenska-Sym and Kacperska, 1996). Transient- and long-term increase of Ins(1,4,5)P_3 were also shown in Arabidopsis suspension cells submitted to a cold exposure at 0°C (Ruelland *et al.*, 2002). These findings suggest a role of Ins(1,4,5)P_3 as a second messenger in the cold or freezing response in plants. Hyperosmotic, dehydration, or high-salinity stress has been noted to increase the level of Ins(1,4,5)P_3 from a number of reports about phosphoinositide signaling in various plants, such as red beet slices (0.2 M mannitol; Srivastava *et al.*, 1989), carnation flowers (dehydration; Drory *et al.*, 1992), winter oil seed rape (30% PEG; Smolenska-Sym and

Kacperska, 1996), and suspension-cultured carrot cells (mannitol, some salts; Drøbak and Watkins, 2000). Although these reports suggest the involvement of Ins(1,4,5)P_3 in response to osmotic/water-stresses, it is not clear whether the Ins(1,4,5)P_3 production is caused by the activation of PI-PLC or other changes in phosphoinositide turnover. In order to elucidate the contribution to the Ins(1,4,5)P_3 production in plant cells, inhibitors of PI-PLC, neomycin and U73122, has been applied. In Arabidopsis seedlings grown in liquid media, the level of both PtdIns(4,5)P_2 and Ins(1,4,5)P_3 was shown to increase rapidly in a nearly logarithmic fashion for 30 min by the treatment with 0.25 M sodium chloride, 0.25 M potassium chloride, and 1 M sorbitol, and these responses are blocked by U73122 (DeWald *et al.*, 2001). Changes in Ins(1,4,5)P_3 level in response to hyperosmotic shock or salinity were also observed in *A. thaliana* T87 cultured cells (Takahashi *et al.*, 2001). In this report, a hyperosmotic shock, caused by mannitol, NaCl, or dehydration, induced a rapid and transient increase of Ins(1,4,5)P_3 in T87 cells within a few seconds, which was inhibited by neomycin and U73122. A rapid increase in PtdIns(4,5)P_2 in response to the hyperosmotic shock also occurred, but its time course of increase was much slower than that of Ins(1,4,5)P_3. These findings indicate that the transient-Ins(1,4,5)P_3 production is not due to an increase in the level of PtdIns(4,5)P_2 but due to the activation of PI-PLC in response to hyperosmotic stress. Moreover, PI-PLC inhibitors also inhibit hyperosmotic stress responsive expression of some dehydration-inducible genes, such as *rd29A* (*lti78/cor78*) (Yamaguchi-Shinozaki and Shinozaki, 1994; Yamaguchi-Shinozaki *et al.*, 1992), and *rd17* (*cor47*) (Gilmour *et al.*, 1992; Iwasaki *et al.*, 1997), which are controlled by a *cis*-acting element termed the dehydration-responsive element (DRE)/C-repeat (CRT) in their promoters (Baker *et al.*, 1994; Yamaguchi-Shinozaki and Shinozaki, 1994). The different pattern of the Ins(1,4,5)P_3 increase in response to hyperosmotic stress in Arabidopsis between two reports described above may be due to the difference between single cell culture and intact seedlings or the existence of some Ins(1,4,5)P_3-mediated signaling pathways in different tissues. Taken together, these results suggest the involvement of PI-PLC and Ins(1,4,5)P_3 in a hyperosmotic stress signal transduction pathway in higher plants.

In animal system, intracellular Ca^{2+} release mediated by Ins(1,4,5)P_3 production leads directly or indirectly to changes in the activity of a wide range of effector enzymes. While Ca^{2+} is also accepted as an important secondary messenger, and many Ca^{2+}-regulated enzymes involved in a range of processes have been identified in plants (reviewed in Sanders *et al.*, 1999; Scrase-Field and Knight, 2003), the existence of Ins(1,4,5)P_3-coupling Ca^{2+} channels in plants (and yeast) remains to be elucidated. However, the rapid and transient increase of Ins(1,4,5)P_3 occurred within a few seconds in Arabidopsis T87 cells (Takahashi *et al.*, 2001) strongly reminds us of a transient increase in cytosolic free Ca^{2+} of Arabidopsis seedlings within a few seconds in response to osmotic stress (Knight *et al.*, 1997, 1998). Extracellular Ca^{2+} channel blockers and PI-PLC inhibitors, neomycin and U73122, are able to partially inhibit the mannitol-induced Ca^{2+} elevation, suggesting that cytoplasmic Ca^{2+} increase in response to hyperosmotic stress is due to both influx of extracellular Ca^{2+} and release of Ca^{2+} from intracellular stores mediated by phosphoinositide signaling (Knight *et al.*, 1997). DeWald *et al.* (2001) have also shown that U73122 effectively blocked not only the Ins(1,4,5)P_3 accumulation but also the increase of intracellular Ca^{2+} in Arabidopsis root tips in high-salinity conditions. In guard cells surrounding stomatal pores, Ins(1,4,5)P_3 inhibited the plasma membrane inward K^+ channel, and release from microinjected "caged" Ins(1,4,5)P_3 not only increases cytosolic free Ca^{2+} but also triggers stomatal closure (Blatt *et al.*, 1990; Gilroy *et al.*, 1990). These reports also suggest an important role of a sequential signal transduction pathway from the PI-PLC activation to the Ca^{2+} release in cellular responses to hyperosmotic stress in higher plants. However, actual intracellular targets of Ins(1,4,5)P_3 in plants still remain to be elucidated.

2.1.2 Involvement of inositol 1,4,5-trisphosphate in ABA signaling

In plant systems Ins(1,4,5)P_3 formation is also induced by the treatment of phytohormones, such as cytokinins (Grabowski *et al.*, 1991), gibberellins (Murthy *et al.*, 1989), and auxins (Ettlinger and Lehle, 1988; Grabowski *et al.*, 1991). Evidences that Ins(1,4,5)P_3 is involved in abscisic acid (ABA) signal transduction cascades in various plants have also been reported. ABA mediates many important cellular processes, such as adaptation to environmental stresses including drought, salinity, cold, and wounding, the regulation of seed dormancy, seed maturation, and vegetative growth during plant development. Endogenous ABA level increases significantly in many plants under water-deficit conditions, resulting in the induction of tolerance to stress by physiological responses and the induction of a series of genes (Finkelstein *et al.*, 2002; Himmelbach *et al.*, 2003). ABA triggers the closing of stomata, which are composed of a pair of guard cells, to optimize the loss of water vapor (Hetherington, 2001; Schroeder *et al.*, 2001). In the analysis of ABA signal transduction cascades, guard cells are a well-developed system for understanding how components interact within signaling cascades from the perception of ABA to ABA-induced stomatal closure. A rapid- and transient-Ins(1,4,5)P_3 spike in response to ABA treatment was observed in guard cells of *Vicia faba* (Lee *et al.*, 1996). In this report, Ins(1,4,5)P_3 levels increased by 90% relative to controls after 10 s of 10 μM ABA treatment and then returned to the control level after 1 min. Microinjection of Ins(1,4,5)P_3 into guard cells has been shown to induce stomatal closure, and Ins(1,4,5)P_3-induced Ca^{2+} increase in guard cells have also been correlated with stomatal closure (Gilroy *et al.*, 1990). Involvement of PI-PLC activation in ABA-induced stomatal closure has been shown by biochemical analysis using PI-PLC inhibitors. The PI-PLC inhibitor U73122 inhibits both ABA-induced increase of intracellular Ca^{2+}

levels and stomatal closure (MacRobbie, 2000; Staxen *et al.*, 1999). The increase of Ins(1,4,5)P_3 in leaves or guard cells by exogenous ABA treatment suggests the potential role of Ins(1,4,5)P_3 as a secondary messenger in ABA-induced stomatal closure under dehydration conditions.

2.1.3 Analysis of enzymes that regulate the Ins(1,4,5)P_3 level

In the phosphoinositide-turnover system, PI-PLC functions as a key enzyme in the production of secondary messengers, Ins(1,4,5)P_3 and DG (Figure 2). PI-PLCs identified from mammals have been classified into three subfamilies based on their domain organization. They are designated as PLCβ, PLCγ and PLCδ (reviewed in Rebecchi and Pentyala, 2000). Recently, new isozymes of PLCs, termed PLCε (Lopez *et al.*, 2001; Song *et al.*, 2001) and PLCζ, (Saunders *et al.*, 2002) were identified. PI-PLC genes, which show homology to mammalian PI-PLC, have been identified from several plant species (Mueller-Roeber and Pical, 2002). All PI-PLCs isolated from plants have similar domain structure, which is the most simple among PI-PLCs from all species. Previously, plant PI-PLCs were regarded as PLCδ-related family, although plant PI-PLC lacks pleckstrin-homology (PH) domain in the N-terminal region, which functions in recognition of target membrane phospholipids (Essen *et al.*, 1996). However, identification of a novel sperm-specific PLCζ, which also lacks the PH domain, revealed that the domain structure of plant PI-PLC is more similar to PLCζ. In the Arabidopsis genome, there are nine genes that show homology to PI-PLC, termed AtPLC1 to AtPLC9 (Hirayama *et al.*, 1995, 1997; Mueller-Roeber and Pical, 2002; Yamamoto *et al.*, 1995). Among them, the expression of AtPLC1, AtPLC4, and AtPLC5 is increased by several abiotic stresses, such as drought, cold, and high salinity (Hirayama *et al.*, 1995; Hunt *et al.*, 2004). The expression of *Vr-PLC3* identified from *Vigna radiata* L. is also reported to be induced under abiotic stress conditions, such as drought and high salinity (Kim *et al.*, 2004). The expression of potato (*Solanum tuberosum*) PI-PLC, *StPLC1*, is reduced by wounding and wilting, while that of *StPLC2* is induced by these stresses (Kopka *et al.*, 1998). These reports suggest the involvement of PI-PLC in abiotic stress responses in plants. Direct evidences of the involvement in plants have been shown by reverse-genetics approaches to investigate the function of the PI-PLC gene. Downregulation of *AtPLC1* gene in transgenic Arabidopsis seedlings expressing the antisense *AtPLC1* transgene reduced ABA-induced Ins(1,4,5)P_3 production 60 min after the treatment with ABA, accompanying the suppression of some physiological responses to ABA, such as inhibition of seed germination and seedling growth, and ABA-responsive gene expression (Sanchez and Chua, 2001). The requirement of PI-PLC in ABA signaling has also been shown in ABA-dependent inhibition of light-activated stomatal opening in *N. tobacum* (Hunt *et al.*, 2003). These reports indicate that the increase of Ins(1,4,5)P_3 mediated by PI-PLC activation plays an important role in ABA signaling. However, the mechanism that regulates

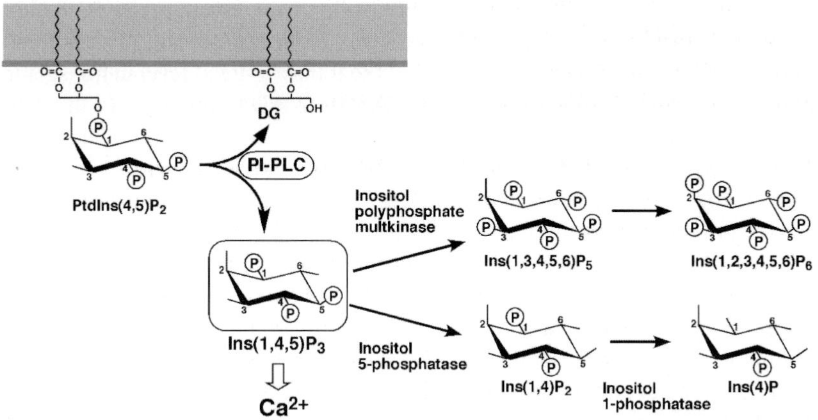

Figure 2. Metabolic pathway of Ins(1,4,5)P_3 in plants.

PI-PLC activity in plants is unknown. In animal systems, activation mechanism of the PLCζ subfamily, which has the most similar domain structure to plants' PI-PLCs, also remains to be elucidated. The tetradecapeptide mastoparan which is known to be a potent G-protein activator, has been shown to increase the formation of Ins(1,4,5)P_3 in carrot suspension cultures, soybean, and green algae *Chlamydomonas* (reviewed in Munnik *et al.*, 1998), indicating the involvement of heterotrimeric G-proteins in the regulation of PI-PLC in plants. However, this effect may also result from pore formation in plasma membranes by mastoparan (Suh *et al.*, 1998). In the analysis of PI-PLC *in vitro*, most of enzymes require Ca^{2+} to show full activity (Hirayama *et al.*, 1995; Hunt *et al.*, 2004). This fact raises a hypothesis that PI-PLC activity in plants is regulated by changes in cytosolic Ca^{2+} levels, which are caused by extracellular influx or release from intracellular stores mediated by potential secondary messengers, such as H_2O_2 (McAinsh *et al.*, 1996), cyclic ADP ribose (Leckie *et al.*, 1998; Wu *et al.*, 1997), nicotinic acid adenine dinucleotide phosphate (Navazio *et al.*, 2000), sphingosine 1-phosphate (Ng *et al.*, 2001), and inositol hexakisphosphate ($InsP_6$; Lemtri-Chlieh *et al.*, 2003).

To function as a secondary messenger, intracellular levels of Ins(1,4,5)P_3 must be tightly regulated at both the levels of production and degradation. Increased Ins(1,4,5)P_3 is terminated by phosphorylation or dephosphorylation of inositol polyphosphate (Figure 2). In animal systems, Ins(1,4,5)P_3 is metabolized by Ins(1,4,5)P_3 3-kinases, which are apparently restricted to the animal kingdom, inositol polyphosphate multikinases, or inositol polyphosphate 5-phosphatases (reviewed in Irvine and Schell, 2001; Shears, 2004). In contrast to the 3-kinase family, inositol polyphosphate multikinase is widely distributed in nature. In Arabidopsis, two inositol polyphosphate kinases, designated AtIpk2α and AtIpk2β, have been identified and shown to have inositol polyphosphate 6-/3-/5-kinase activities (Stevenson-Paulik *et al.*, 2002; Xia *et al.*, 2003), although no direct evidence of the involvement in the regulation of abiotic stress-induced Ins(1,4,5)P_3 level has been shown. By contrast, manipulation of inositol 5-phosphatase at the transcriptional level in transgenic plants indicates the importance of Ins(1,4,5)P_3 in ABA signaling. In the Arabidopsis genome, 15 genes showing high homology to inositol polyphosphate 5-phosphatase have been found (Berdy *et al.*, 2001). Among them, overexpression of *AtIP5PII/At5PTase2* and *At5PTase1* in Arabidopsis seedlings have been shown to reduce not only ABA-induced Ins(1,4,5)P_3 generation but also the physiological effects of ABA, such as inhibition of seed germination and seedling growth, and the expression of ABA-inducible genes (Burnette *et al.*, 2003; Sanchez and Chua, 2001). The involvement of another inositol polyphosphate phosphatase in abiotic stress responses has been revealed in a genetic screen of Arabidopsis mutants that exhibited an enhanced induction of the stress-responsive gene *RD29A* under cold, drought, salt, and ABA treatment (Xiong *et al.*, 2001). One of the mutants, termed *fiery1* (*fry1*), shows not only superinduction of ABA- and stress-responsive genes but also enhanced tolerance to freezing, drought, and salt stress. The *FRY1* gene encodes a bifunctional enzyme, SAL1, with 3′(2′),5′-bisphosphate nucleotidase and inositol polyphosphate 1-phosphatase activities, which was initially identified as a gene that confers salt tolerance when expressed in yeast (Quintero *et al.*, 1996). The SAL1 protein has been shown to dephosphorylate Ins(1,4)P_2 and Ins(1,3,4)P_3, which is the typical activity of inositol polyphosphate 1-phosphatases. Although the *fry1-1* mutant exhibits higher levels of Ins(1,4,5)P_3 than wild type seedlings with/without ABA treatment, direct evidence that FRY1/SAL1 shows 1-phosphatase activity against Ins(1,4,5)P_3 remains to be shown. Because the Ins(1,4,5)P_3 signal in plants is considered to be terminated mainly by the activity of inositol polyphosphate 5-phosphatase or inositol polyphosphate multikinase, sustained accumulation of Ins(1,4,5)P_3 in *fry1-1* mutant plants may result from down-regulation of the 5-phosphatese activity against Ins(1,4,5)P_3, which is caused by unusual accumulation of Ins(1,4)P_2, a potential substrate of FRY1/SAL1. Taken together, these results correspond with the data from transgenic plants with repressed levels of the PI-PLC gene, indicating the requirement of Ins(1,4,5)P_3 in ABA signaling and abiotic stress tolerance.

2.2 Inositol 1,2,3,4,5,6-hexakisphosphate as a signaling mediator

Inositol 1,2,3,4,5,6-hexakisphosphate ($InsP_6$), known as phytic acid, was the first inositol phosphate discovered and the most abundant inositol phosphate in nature. A significant amount of $InsP_6$ accumulates in plant seeds and other storage tissues as the principal storage mechanism of phosphorus (Raboy, 2003). Although $InsP_6$ was previously considered to be found only in plants and the erythrocytes of a few animals, recent reports have demonstrated that $InsP_6$ is ubiquitous in all eukaryotic species and typically the most abundant inositol phosphate in cells (Sasakawa *et al.*, 1995). $InsP_6$ and its derivatives are reported to function in mRNA export, DNA repair, and DNA recombination in endocytosis and vesicular trafficking (Hanakahi and West, 2002; Luo *et al.*, 2002; Saiardi *et al.*, 2002), and as antioxidants (Graf *et al.*, 1987). $InsP_6$ is also reported to function as anticancer compound when it is dosed with inositol (Vucenik and Shamsuddin, 2003). The involvement of $InsP_6$ in environmental stress responses has been reported from the analysis of *Schizosaccharomyces pombe*, showing that the level of $InsP_6$ increases more than threefold in response to hyperosmotic shock within a few minutes (Ongusaha *et al.*, 1998). Although no direct evidence of a correlation between $InsP_6$ regulation and environmental stress responses has been reported in plants, recent studies indicate that $InsP_6$ is involved in ABA signal transduction in guard cells, suggesting role for $InsP_6$ in osmotic stress response.

In plants, ABA level increases in response to dehydration, which affects the activity of ion channels in guard cells, such as Ca^{2+}-independent activation of the plasmalemma outward K^+-channel and Ca^{2+}-dependent inactivation of the plasmalemma inward K^+-channel. The long-term efflux of both K^+ and counter anions from guard cells contributes to the loss of guard cell turgor, resulting in stomatal closing (Schroeder *et al.*, 2001). Lemtiri-Chlieh *et al.* (2000) has shown that the level of $InsP_6$ in guard cells of *S. tuberosum* rapidly increased more than fivefold with the treatment of ABA. In this system, $InsP_6$ was also shown to mimic the effect of ABA, inhibiting the plasmalemma inward K^+-channel in guard cells of *S. tuberosum* and *V. faba* in a Ca^{2+}-dependent manner. Moreover, $InsP_6$ is approximately 100-fold more potent than Ins(1,4,5)P_3 for the inactivation of inward rectifying K^+ current in this system. Release of $InsP_6$ in guard cell protoplasts of *V. faba* has been shown to induce transient increase of Ca^{2+} (Lemtiri-Chlieh *et al.*, 2003). The fact that $InsP_6$ did not affect Ca^{2+} permeable channels in the plasma membrane and that $InsP_6$ could inhibit the plasmalemma inward K^+-channel without external Ca^{2+} indicates that the $InsP_6$-induced transient Ca^{2+} increase in guard cells does not result from influx of external Ca^{2+} but release of Ca^{2+} from internal stores. Taken together, $InsP_6$ functions as a signaling mediator in ABA signal transduction in guard cells, triggering a Ca^{2+} release that results in the inhibition of inward K^+-channels. Other potent signaling mediators, Ins(1,4,5)P_3 (Blatt *et al.*, 1990; Gilroy *et al.*, 1990), cyclic adenosine diphosphoribose (cADPR) (Leckie *et al.*, 1998), and sphingosine 1-phosphate (Ng *et al.*, 2001) have been shown to trigger the cytosolic Ca^{2+} increase when these compounds are introduced into guard cells directly. $InsP_6$ induces a rapid increase in Ca^{2+} reaching the maximum level within a few seconds (Lemtiri-Chlieh *et al.*, 2003). In contrast to this, Ins(1,4,5)P_3 induced relatively slow increase in cytosolic Ca^{2+} which was sustained for 5 min (Gilroy *et al.*, 1990). The differences between these inositol polyphosphates and the fact that $InsP_6$ is more potent than Ins(1,4,5)P_3 for inactivation of inward rectifying K^+ current (Lemtiri-Chlieh *et al.*, 2000) suggests that Ins(1,4,5)P_3 applied into the guard cells functions as metabolic precursor for increasing the level of $InsP_6$ in guard cells.

Although the presence of $InsP_6$ (phytic acid) in plants has been known for over a century, elucidation of the detailed biosynthetic pathway of $InsP_6$ has progressed only recently. In higher plants, two major $InsP_6$ biosynthetic pathways have been reported; the inositol phosphate intermediate pathway, which proceeds from Ins(3)P to $InsP_6$ with sequential phosphorylation and the phosphatidyl inositol phosphate intermediate pathway which proceeds via Ins(1,4,5)P_3 generated by the action of PI-PLC (Irvine and Schell, 2001; Raboy, 2003) (Figure 3). Evidence of the former pathway in plants has been shown from the analysis of duckweed, *Spirodela polyrhiza* L. (Brearley and Hanke, 1996). This pathway is similar to that found in *Dictyostelium*, which was the first elucidated $InsP_6$ synthetic pathway, except for two intermediates, Ins(3,4)P_2 and Ins(3,4,5,6)P_4 (Stephens and Irvine, 1990). The maize low-phytic acid mutant *lpa2*, which accumulates intermediates of inositol phosphates, has been reported to have a defect in the inositol phosphate kinase gene, *ZmIpk*, which shows homology to the Arabidopsis Ins(1,3,4)P_3 5/6-kinase gene (Shi *et al.*, 2003). The levels of seed phytic acid in this mutant were reduced by 30%, indicating that ZmIpk is one of the kinases responsible for $InsP_6$ biosynthesis in maize seed and that other redundant pathway also functions in $InsP_6$ biosynthesis. In the inositol phosphate intermediate pathway, inositol 3-kinase or Ins(3)P kinase is essential for the first phosphorylation step. However, genes encoding such kinds of enzyme remain to be identified.

The latter, PI-PLC-dependent $InsP_6$ biosynthetic pathway, has been well studied in yeast. In studies using *S. pombe*, osmotic stress enhanced the production of $InsP_6$, which is mainly mediated by the pathway from Ins(1,4,5)P_3 with stepwise phosphorylation (Ongusaha *et al.*, 1998). Attempts to identify factors that function in mRNA export in *S. cerevisiae* revealed that one PI-PLC gene, *PLC1*, and two potential inositol phosphate kinase genes play essential roles in mRNA export (Odom *et al.*, 2000; York *et al.*, 1999). In these studies, a yeast strain deficient in *PLC1* was shown to be unable to produce detectable amounts of $InsP_6$, indicating that $InsP_6$ is biosynthesized mainly from Ins(1,4,5)P_3 generated from PtdIns(4,5)P_2, and two inositol phosphate kinases, termed Ipk1 and Ipk2, were shown to have enzyme activities of Ins(1,3,4,

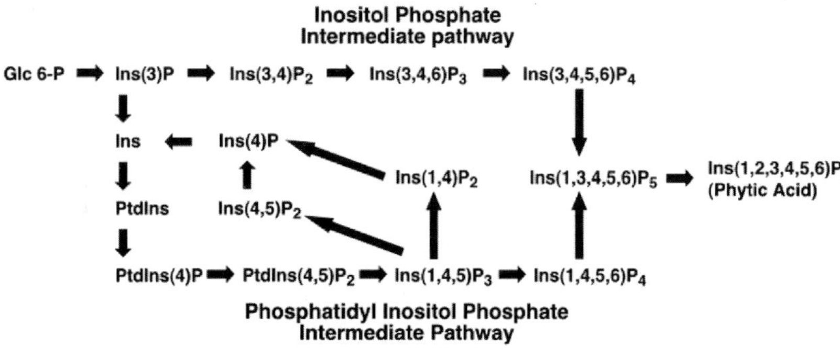

Figure 3. A part of the metabolic pathway from glucose 6-phosphate to $InsP_6$ in plants.

5,6)P_5 2-kinase (York *et al.*, 1999) and $InsP_3$/$InsP_4$ 6-/3-kinase (Odom *et al.*, 2000; Saiardi *et al.*, 2000), respectively. In a recent study, two inositol phosphate kinases, designated AtIpk2α and AtIpk2β, that show homology with Ipk2 in Arabidopsis have been isolated and analysis indicated that these kinases have 6-/3-kinase activities that phosphorylate Ins(1,4,5)P_3 to Ins(1,3,4,5,6)P_5 *via* Ins (1,4,5,6)P_4 intermediate and 5-kinase activity that phosphorylate Ins(1,3,4,6)P_4 and Ins(1,2,3,4,6)P_5 *in vitro* (Stevenson-Paulik *et al.*, 2002; Xia *et al.*, 2003). These results suggest an Ins(1,4,5)P_3-mediated $InsP_6$ biosynthesis pathway exists in Arabidopsis similar to the pathway identified in yeast. However, no direct evidence exists to show this pathway and the involvement of AtIpk2α and AtIpk2β in this pathway. Detailed elucidation of PI-PLC-dependent $InsP_6$ pathways and signaling target of $InsP_6$ in yeast and plants may explain one of reasons why the counterpart of Ins(1,4,5)P_3-receptor-coupling Ca^{2+} channel cannot find in yeasts and plants nevertheless Ins(1,4,5)P_3 acts as a potent signaling mediator in plants.

3. CONCLUSION AND PERSPECTIVES

Inositol and its derivatives play important roles in various abiotic stress tolerances in both animals and plants as osmolytes and as secondary messengers. The mechanism for the accumulation of these osmolytes in the cytosol of plants suffering osmotic stress is different from that found in animals. In animals, *myo*-inositol is imported from the external environment to the cytosol by a SMIT, whereas plants synthesize these osmolytes in the cytosol directly, and upregulate the genes encoding these metabolic enzymes in response to osmotic stress. Especially, plants accumulate many kinds of inositol derivatives during abiotic stresses, such as drought, low temperature and high-salinity stresses. Galactinol and RFO are found to accumulate in response to various abiotic stresses in both glycopytes and halophytes, whereas ononitol and pinitol are found only in halophytes. Recently, the halophytic plant species, *Thellungiella halophila* (salt cress), was reported to grow in high-salinity coastal areas in eastern China. It is closely related to *A. thaliana* and it has a gene composition with >90% nucleotide identity with that of Arabidopsis, nevertheless it shows extreme salt tolerance and can grow in the presence of 500 mM NaCl, conditions that inhibit the growth of Arabidopsis severely (Bressan *et al.*, 2001; Zhu, 2001). Microarray analysis revealed that only a few genes were induced by salt stress in *T. halophila* in contrast to Arabidopsis. Notably a large number of known abiotic- and biotic-stress inducible genes were expressed in *T. halophila* at high levels even in the absence of stress. The genes induced by salt stress in *T. halophila* included the *myo*-inositol-1-phosphate synthase (INPS) gene (Taji *et al.*, 2004). We are currently conducting metabolome analysis to show different metabolic profiles between Arabidopsis and *T. halophila*. It will be interesting to discover what kind of inositol derivatives *T. halophila* uses as osmoprotectants.

Furthermore, phosphoinositide/inositol phosphate turnover plays an important role in the production of secondary messengers during abiotic stress in plants. Within a few seconds, hyperosmotic shock induced a rapid and transient increase of Ins(1,4,5)P_3 in Arabidopsis cultured cells. This fact strongly suggests the PI turnover system functions near an osmosensor; however, the molecule that functions as the osmosensor has not been identified, except in yeast and bacteria. In the near future, we hope to determine which molecule can activate the PI turnover system in response to hyperosmotic stress in plants.

ACKNOWLEDGMENT

We thank Dr. Kieran John David Lee for critical reading of this manuscript.

REFERENCES

Baker, S.S., Wilhelm, K.S., and Thomashow, M.F., 1994, The 5′-region of *Arabidopsis thaliana cor15a* has *cis*-acting elements that confer cold-, drought- and ABA-regulated gene expression. *Plant Mol. Biol.* **24:** 701–713.

Berdy, S.E., Kudla, J., Gruissem, W., and Gillaspy, G.E., 2001, Molecular characterization of At5PTase1, an inositol phosphatase capable of terminating inositol trisphosphate signaling. *Plant Physiol.* **126:** 801–810.

Berridge, M.J., 1993, Inositol trisphosphate and calcium signaling. *Nature* **361:** 315–325.

Blatt, M.R., Thiel, G., and Trentham, D.R., 1990, Reversible inactivation of K^+ channels of *Vicia* stomatal guard cells following the photolysis of caged inositol 1,4,5-trisphosphate. *Nature* **346:** 766–769.

Brearley, C.A., and Hanke, D.E., 1995, Evidence for substrate-cycling of 3-, 3,4-, 4-, and 4,5-phosphorylated phosphatidylinositols in plants. *Biochem. J.* **311:** 1001–1007.

Brearley, C.A., and Hanke, D.E., 1996, Inositol phosphates in the duckweed *Spirodela polyrhiza* L. *Biochem. J.* **314:** 215–225.

Brewster, J.L., de Valoir, T., Dwyer, N.D., Winter, E., and Gustin, M.C., 1993, An osmosensing signal transduction pathway in yeast. *Science* **259**(5102): 1760–1763.

Burg, M.B., Kwon, E.D., and Kultz, D., 1997, Regulation of gene expression by hypertonicity. *Annu. Rev. Physiol.* **59:** 437–455.

Burnette, R.N., Gunesekera, B.M., and Gillaspy, G.E., 2003, An Arabidopsis inositol 5-phosphatase gain-of-function alters abscisic acid signaling. *Plant Physiol.* **132:** 1011–1019.

Chapman, K.D., 1998, Phospholipase activity during plant growth and development and in response to environmental stress. *Trends Plant Sci.* **3:** 419–426.

Chauhan, S. *et al.*, 2000, Na+/*myo*-inositol symporters and Na+/H+-antiport in *Mesembryanthemum crystallinum*. *Plant J.* **24**(4): 511–522.

Dahl, S.C., Handler, J.S., and Kwon, H.M., 2001, Hypertonicity-induced phosphorylation and nuclear localization of the transcription factor TonEBP. *Am. J. Physiol. Cell Physiol.* **280**(2): C248–C253.

Denkert, C., Warskulat, U., Hensel, F., and Haussinger, D., 1998, Osmolyte strategy in human monocytes and macrophages: Involvement of p38MAPK in hyperosmotic induction of betaine and *myo*inositol transporters. *Arch. Biochem. Biophys.* **354**(1): 172–180.

DeWald, D.B., Torabinejad, J., Jones, C.A., Shope, J.C., Cangelosi, A.R., Thompson, J.E., Prestwich, G.D., and Hama, H., 2001, Rapid accumulation of phosphatidylinositol 4,5-bisphosphate and inositol 1,4,5-trisphosphate correlates with calcium mobilization in salt-stressed *Arabidopsis*. *Plant Physiol.* **126:** 759–769.

Drøbak, B.K., 1992, The plant phosphoinositide system. *Biochem. J.* **288:** 697–712.

Drøbak, B.K., and Watkins, P.A., 2000, Inositol(1,4,5)trisphosphate production in plant cells: An early response to salinity and hyperosmotic stress. *FEBS Lett.* **481:** 240–244.

Drory, A., Borochov, A., and Mayak, S., 1992, Transient water stress and phospholipid turnover in carnation flowers. *J. Plant Physiol.* **140:** 116–120.

Ettlinger, C., and Lehle, L., 1988, Auxin induces rapid changes in phosphatidylinositol metabolites. *Nature* **331:** 176–178.

Essen, L.O., Perisic, O., Cheung, R., Katan, M., and Williams, R.L., 1996, Crystal structure of a mammalian phosphoinositide-specific phospholipase Cδ. *Nature* **380:** 595–602.

Franklin-Tong, V.E., Drøbak, B.K., Allan, A.C., Watkins, P.A.C., and Trewavas, A.J., 1996, Growth of pollen tubes of *Papaver rhoeas* is regulated by a slow-moving calcium wave propagated by inositol 1,4,5-trisphosphate. *Plant Cell* **8:** 1305–1321.

Finkelstein, R.R., Gampala, S.S., and Rock, C.D., 2002, Abscisic acid signaling in seeds and seedlings. *Plant Cell*, 15–45.

Finkelstein, R.R., Gampala, S.S., and Rock, C.D., 2002, Abscisic acid signaling in seeds and seedlings. *Plant Cell* **14:** S15–S45.

Garcia-Perez, A., and Burg, M.B., 1991, Role of organic osmolytes in adaptation of renal cells to high osmolality. *J. Membr. Biol.* **119**(1): 1–13.

Gilmour, S.J., Artus, N.N., and Thomashow, M.F., 1992, cDNA sequence analysis and expression of two cold-regulated genes of *Arabidopsis thaliana*. *Plant Mol. Biol.* **18:** 13–21.

Gilroy, S., Read, N.D., and Trewavas, A.J., 1990, Elevation of cytoplasmic calcium by caged calcium or caged inositol triphosphate initiates stomatal closure. *Nature* **346:** 769–771.

Grabowski, L., Heim, S., and Wagner, K.G., 1991, Rapid changes in the enzyme activities and metabolites of the phosphatidylinositol-cycle upon induction by growth substrates of auxin-starved suspension cultured *Catharanthus roseus* cells. *Plant Sci.* **75:** 33–38.

Graf, E., Empson, K.L., and Eaton, J.W., 1987, Phytic acid. A natural antioxidant. *J. Biol. Chem.* **262:** 11647–11650.

Han, J., Lee, J.D., Bibbs, L., and Ulevitch, R.J., 1994, A MAP kinase targeted by endotoxin and hyperosmolarity in mammalian cells. *Science* **265**(5173): 808–811.

Hanakahi, L.A., and West, S.C., 2002, Specific interaction of IP_6 with human Ku70/80, the DNA-binding subunit of DNA-PK. *EMBO J.* **21:** 2038–2044.

Harada, A., Sakai, T., and Okada, K., 2003, Phot1 and phot2 mediate blue light-induced transient increases in cytosolic Ca^{2+} differently in *Arabidopsis* leaves. *Proc. Natl. Acad. Sci. USA* **100:** 8583–8588.

Haussinger, D., 1996, The role of cellular hydration in the regulation of cell function. *Biochem. J.* **313**(Pt 3): 697–710.

Hetherington, A.M., 2001, Guard cell signaling. *Cell* **107:** 711–714.

Himmelbach, A., Yang, Y., and Grill, E., 2003, Relay and control of abscisic acid signaling. *Curr. Opin. Plant Biol.* **6:** 470–479.

Hirayama, T., Mitsukawa, N., Shibata, D., and Shinozaki, K., 1997, AtPLC2, a gene encoding phosphoinositide-specific phospholipase C, is constitutively expressed in vegetative and floral tissues in *Arabidopsis thaliana*. *Plant Mol. Biol.* **34:** 175–180.

Hirayama, T., Ohto, C., Mizoguchi, T., and Shinozaki, K., 1995, A gene encoding a phosphatidylinositol-specific phospholipase C is induced by dehydration and salt stress in *Arabidopsis thaliana*. *Proc. Natl. Acad. Sci. U.S.A.* **92:** 3903–3907.

Hong, S.W., Lee, U., and Vierling, E., 2003, Arabidopsis hot mutants define multiple functions required for acclimation to high temperatures. *Plant Physiol.* **132**(2): 757–767.

Hong, S.W., and Vierling, E., 2000, Mutants of *Arabidopsis thaliana* defective in the acquisition of tolerance to high temperature stress. *Proc. Natl. Acad. Sci. U.S.A.* **97**(8): 4392–4397.

Hong, S.W., and Vierling, E., 2001, Hsp101 is necessary for heat tolerance but dispensable for development and germination in the absence of stress. *Plant J.* **27**(1): 25–35.

Hunt, L., Mills, L.N., Pical, C., Leckie, C.P., Aitken, F.L., Kopka, J., Mueller-Roeber, B., McAinsh, M.R., Hetherington, A.M., and Gray, J.E., 2003, Phospholipase C is required for the control of stomatal aperture by ABA. *Plant J.* **34:** 47–55.

Hunt, L., Otterhag, L., Lee, J.C., Lasheen, T., Hunt, J., Seki, M., Shinozaki, K., Sommarin, M., Gilmour, D.J., Pical, C., and Gray, J.E., 2004, Gene-specific expression and calcium activation of *Arabidopsis thaliana* phospholipase C isoforms. *New Phytol.* **162:** 643–654.

Irarrazabal, C.E., Liu, J.C., Burg, M.B., and Ferraris, J.D., 2004, ATM, a DNA damage-inducible kinase, contributes to activation by high NaCl of the transcription factor TonEBP/OREBP. *Proc. Natl. Acad. Sci. U.S.A.* **101**(23): 8809–8814.

Irvine, R.F., Letcher, A.J., Lander,D.J., Drøbak, B.K., Dawson, A.P., and Musgrave, A., 1989, Phosphatidylinositol 4,5-bisphosphate and phosphatidylinositol 4-phosphate in plant tissues. *Plant Physiol.* **89:** 888–892.

Irvine, R.F., and Schell, M.J., 2001, Back in the water: The return of the inositol phosphates. *Nat. Rev. Mol. Cell Biol.* **2:** 327–338.

Ishitani, M., Majumder, A.L., Bornhouser, A., Michalowski, C.B., Jensen, R.G., and Bohnert, H.J., 1996, Coodinate transcriptional induction of *myo*-inositol metabolism during environmental stress. *Plant J.* **9**(4): 537–548.

Iwasaki, T., Kiyosue, T., Yamaguchi-Shinozaki, K., and Shinozaki, K., 1997, The dehydration-inducible RD17 (Cor47) gene and its promoter region in *Arabidopsis thaliana* (accession no. AB004872) (PGR 97-156). *Plant Physiol.* **115:** 1287.

Jiang, Z., Chung, S.K., Zhou, C., Cammarata, P.R., and Chung, S.S., 2000, Overexpression of Na(+)-dependent *myo*-inositol transporter gene in mouse lens led to congenital cataract. *Invest. Ophthalmol. Vis. Sci.* **41**(6): 1467–1472.

Kapus, A., Szaszi, K., Sun, J., Rizoli, S., and Rotstein, O.D., 1999, Cell shrinkage regulates Src kinases and induces tyrosine phosphorylation of cortactin, independent of the osmotic regulation of Na+/H+ exchangers. *J. Biol. Chem.* **274**(12): 8093–8102.

Kempf, B., and Bremer, E., 1998, Uptake and synthesis of compatible solutes as microbial stress responses to high-osmolality environments. *Arch. Microbiol.* **170**(5): 319–330.

Kim, Y.J., Kim, J.E., Lee, J.H., Lee, M.H., Jung, H.W., Bahk, Y.Y., Hwang, B.K., Hwang, I., and Kim, W.T., 2004, The Vr-PLC3 gene encodes a putative plasma membrane-localized phosphoinositide-specific phospholipase C whose expression is induced by abiotic stress in mung bean (*Vigna radiata* L.). *FEBS Lett.* **556:** 127–136.

Knight, H., Brandt, S., and Knight, M.R., 1998, A history of stress alters drought calcium signalling pathways in *Arabidopsis*. *Plant J.* **16:** 681–687.

Knight, H., Trewavas, A.J., and Knight, M.R., 1997, Calcium signalling in *Arabidopsis thaliana* responding to drought and salinity. *Plant J.* **12:** 1067–1078.

Ko, B.C., Ruepp, B., Bohren, K.M., Gabbay, K.H., and Chung, S.S., 1997, Identification and characterization of multiple osmotic response sequences in the human aldose reductase gene. *J. Biol. Chem.* **272**(26): 16431–16437.

Ko, B.C. *et al.*, 2002, Fyn and p38 signaling are both required for maximal hypertonic activation of the osmotic response element-binding protein/tonicity-responsive enhancer-binding protein (OREBP/TonEBP). *J. Biol. Chem.* **277**(48): 46085–46092.

Kopka, J., Pical, C., Gray, J.E., and Muller-Roeber, B., 1998, Molecular and enzymatic characterization of three phosphoinositide-specific phospholipase C isoforms from potato. *Plant Physiol.* **116:** 239–250.

Kwon, H.M., and Handler, J.S., 1995, Cell volume regulated transporters of compatible osmolytes. *Curr. Opin. Cell Biol.* **7**(4): 465–471.

Kwon, O.S., Park, J., and Churchich, J.E., 1992, Brain 4-aminobutyrate aminotransferase. Isolation and sequence of a cDNA encoding the enzyme. *J. Biol. Chem.* **267**(11): 7215–7216.

Leckie, C.P., McAinsh, M.R., Allen, G.J., Sanders, D., and Hetherington, A.M., 1998, Abscisic acid-induced stomatal closure mediated by cyclic ADP-ribose. *Proc. Natl. Acad. Sci. U.S.A.* **95:** 15837–15842.

Lee, Y., Choi, Y.B., Suh, S., Lee, J., Assmann, S.M., Joe, C.O., Kelleher, J.F., and Crain, R.C., 1996, Abscisic acid-induced phosphoinositide turnover in guard cell protoplasts of *Vicia faba*. *Plant Physiol.* **110:** 987–996.

Lemtiri-Chlieh, F., MacRobbie, E.A., and Brearley, C.A., 2000, Inositol hexakisphosphate is a physiological signal regulating the K^+-inward rectifying conductance in guard cells. *Proc. Natl. Acad. Sci. U.S.A.* **97:** 8687–8692.

Lemtiri-Chlieh, F., MacRobbie, E.A., Webb, A.A., Manison, N.F., Brownlee, C., Skepper, J.N., Chen, J., Prestwich, G.D., and Brearley, C.A., 2003, Inositol hexakisphosphate mobilizes an endomembrane store of calcium in guard cells. *Proc. Natl. Acad. Sci. U.S.A.* **100:** 10091– 10095.

Lopez, I., Mak, E.C., Ding, J., Hamm, H.E., and Lomasney, J.W., 2001, A novel bifunctional phospholipase C that is regulated by $G(_{12}$ and stimulates the Ras/mitogen-activated protein kinase pathway. *J. Biol. Chem.* **276:** 2758–2765.

Lopez-Rodriguez, C. *et al.*, 2004, Loss of NFAT5 results in renal atrophy and lack of tonicity-responsive gene expression. *Proc. Natl. Acad. Sci. U.S.A.* **101**(8): 2392–2397.

Luo, H.R., Saiardi, A., Yu, H., Nagata, E., Ye, K., and Snyder. S.H., 2002, Inositol pyrophosphates are required for DNA hyperrecombination in protein kinase c1 mutant yeast. *Biochemistry* **41:** 2509–2515.

MacRobbie, E.A., 2000, ABA activates multiple Ca^{2+} fluxes in stomatal guard cells, triggering vacuolar K^+ (Rb^+) release. *Proc. Natl. Acad. Sci. U.S.A.* **97:** 12361–12368.

Majee, M. *et al.*, 2004, A novel salt-tolerant L-*myo*-inositol-1-phosphate synthase from *Porteresia coarctata* (Roxb.) Tateoka, a halophytic wild rice: Molecular cloning, bacterial overexpression, characterization, and functional introgression into tobacco-conferring salt tolerance phenotype. *J. Biol. Chem.* **279**(27): 28539–28552.

Majerus, P.W., Ross, T.S., Cunningham, T.W., Caldwell, K.K., Jefferson, A.B., and Bansal, V.S., 1990, Recent insights in phosphatidylinositol signaling. *Cell* **63:** 459–465.

Majumder, A.L., Chatterjee, A., Ghosh Dastidar, K., and Majee, M., 2003, Diversification and evolution of L-*myo*-inositol 1-phosphate synthase. *FEBS Lett.* **553**(1–2): 3–10.

Maruyama, K. *et al.*, 2004. Identification of cold-inducible downstream genes of the Arabidopsis DREB1A/CBF3 transcriptional factor using two microarray systems. *Plant J.* **38**(6): 982–993.

McAinsh, M.R., Clayton, H., Mansfield, T.A., and Hetherington, A.M., 1996, Changes in Stomatal Behavior and Guard Cell Cytosolic Free Calcium in Response to Oxidative Stress. *Plant Physiol.* **111:** 1031–1042.

Miyakawa, H., Woo, S.K., Dahl, S.C., Handler, J.S., and Kwon, H.M., 1999, Tonicity-responsive enhancer binding protein, a rel-like protein that stimulates transcription in response to hypertonicity. *Proc. Natl. Acad. Sci. U.S.A.* **96**(5): 2538–2542.

Mueller-Roeber, B., and Pical, C., 2002, Inositol phospholipid metabolism in Arabidopsis. Characterized and putative isoforms of inositol phospholipid kinase and phosphoinositide-specific phospholipase C. *Plant Physiol.* **130:** 22–46.

Munnik, T., Irvine, R.F., and Musgrave, A., 1998, Phospholipid signalling in plants. *Biochim. Biophys. Acta* **1389:** 222–272.

Munnik, T., Musgrave, A., and de Vrije, T., 1994, Rapid turnover of polyphosphoinositides in carnation flower petals. *Planta* **93:** 89–98.

Murthy, P.P.N., Renders, J.M., and Keranen, L.M., 1989, Phosphoinisitides in barley aleurone layers and gibberellic acid-induced changes in metabolism. *Plant Physiol.* **91:** 1266–1269.

Nakanishi, T., Balaban, R.S., and Burg, M.B., 1988, Survey of osmolytes in renal cell lines. *Am. J. Physiol.* **255**(2 Pt 1): C181–C191.

Nakanishi, T., Turner, R.J., and Burg, M.B., 1989, Osmoregulatory changes in *myo*-inositol transport by renal cells. *Proc. Natl. Acad. Sci. USA* **86**(15): 6002–6006.

Navazio, L., Bewell, M.A., Siddiqua, A., Dickinson, G.D., Galione, A., and Sanders, D., 2000, Calcium release from the endoplasmic reticulum of higher plants elicited by the NADP metabolite nicotinic acid adenine dinucleotide phosphate. *Proc. Natl. Acad. Sci. USA.* **97:** 8693–8698.

Nelson, D.E., Rammesmayer, G., and Bohnert, H.J., 1998, Regulation of cell-specific inositol metabolism and transport in plant salinity tolerance. *Plant Cell* **10**(5): 753–764.

Ng, C.K., Carr, K., McAinsh, M.R., Powell, B., and Hetherington, A.M., 2001, Drought-induced guard cell signal transduction involves sphingosine-1-phosphate. *Nature* **410:** 596–599.

Nishizuka, Y., 1992, Intracellular signaling by hydrolysis of phospholipids and activation of protein kinase C. *Science* **258:** 607–614.

Odom, A.R., Stahlberg, A., Wente, S.R., and York, J.D., 2000, A role for nuclear inositol 1,4,5-trisphosphate kinase in transcriptional control. *Science* **287:** 2026–2029.

Ohmiya, R., Yamada, H., Nakashima, K., Aiba, H., and Mizuno, T., 1995, Osmoregulation of fission yeast: Cloning of two distinct genes encoding glycerol-3-phosphate dehydrogenase, one of which is responsible for osmotolerance for growth. *Mol. Microbiol.* **18**(5): 963–973.

Ongusaha, P.P., Hughes, P.J., Davey, J., and Michell. R.H., 1998, Inositol hexakisphosphate in *Schizosaccharomyces pombe*: Synthesis from Ins(1,4,5)P_3 and osmotic regulation. *Biochem. J.* **335:** 671–679.

Pahlman, A.K., Granath, K., Ansell, R., Hohmann, S., and Adler, L., 2001, The yeast glycerol 3-phosphatases Gpp1p and Gpp2p are required for glycerol biosynthesis and differentially involved in the cellular responses to osmotic, anaerobic, and oxidative stress. *J. Biol. Chem.* **276**(5): 3555–3563.

Panikulangara, T.J., Eggers-Schumacher, G., Wunderlich, M., Stransky, H., and Schoffl, F., 2004, Galactinol synthase1. A novel heat shock factor target gene responsible for heat-induced synthesis of raffinose family oligosaccharides in Arabidopsis. *Plant Physiol.* **136**(2): 3148–3158.

Paredes, A., McManus, M., Kwon, H.M., and Strange, K., 1992, Osmoregulation of Na(+)-inositol cotransporter activity and mRNA levels in brain glial cells. *Am. J. Physiol.* **263**(6 Pt 1): C1282–C1288.

Parmar, P.N., and Brearley, C.A., 1995, Metabolism of 3- and 4- phosphorylated phosphatidylinositols in stomatal guard cells of *Commelina communis* L. *Plant J.* **8:** 425–433.

Pennycooke, J.C., Jones, M.L., and Stushnoff, C., 2003, Down-regulating alpha-galactosidase enhances freezing tolerance in transgenic petunia. *Plant Physiol.* **133**(2): 901–909.

Perera, I.Y., Heilmann, I., and Boss, W.F., 1999, Transient and sustained increases in inositol 1,4,5-trisphosphate precede the differential growth response in gravistimulated maize pulvini. *Proc. Natl. Acad. Sci. U.S.A.* **96:** 5838–5843.

Perera, I.Y., Heilmann, I., Chang, S.C., Boss, W.F., and Kaufman, P.B., 2001, A role for inositol 1,4,5-trisphosphate in gravitropic signaling and the retention of cold-perceived gravistimulation of oat shoot pulvini. *Plant Physiol.* **125:** 1499–1507.

Posas, F. *et al.*, 2000. The transcriptional response of yeast to saline stress. *J. Biol. Chem.* **275**(23): 17249–17255.

Quintero, F.J., Garciadeblas, B., and Rodriguez-Navarro, A., 1996 The SAL1 gene of Arabidopsis, encoding an enzyme with 3′(2′),5′-bisphosphate nucleotidase and inositol polyphosphate 1-phosphatase activities, increases salt tolerance in yeast. *Plant Cell* **8:** 529–537.

Raboy, V., 2003, *myo*-Inositol-1,2,3,4,5,6-hexakisphosphate. *Phytochemistry* **64:** 1033–1043.

Rebecchi, M., and Pentyala, S.N., 2000, Structure, function, and control of phosphoinositide-specific phospholipase C. *Physiol. Rev.* **80:** 1291–1335.

Rep, M., Krantz, M., Thevelein, J.M., and Hohmann, S., 2000, The transcriptional response of *Saccharomyces cerevisiae* to osmotic shock. Hot1p and Msn2p/Msn4p are required for the induction of subsets of high osmolarity glycerol pathway-dependent genes. *J. Biol. Chem.* **275**(12): 8290–8300.

Rhee, S.G., and Bae, Y.S., 1997, Regulation of phosphoinositide-specific phospholipase C isozymes. *J. Biol. Chem.* **272:** 15045–15048.

Rim, J.S. *et al.*, 1998. Transcription of the sodium/*myo*-inositol cotransporter gene is regulated by multiple tonicity-responsive enhancers spread over 50 kilobase pairs in the 5(-flanking region. *J. Biol. Chem.* **273**(32): 20615–20621.

Ross, H.A., McRae, D., and Davies, H.V., 1996, Sucrolytic enzyme activities in cotyledons of the faba bean (developmental changes and purification of alkaline invertase). *Plant Physiol.* **111**(1): 329–338.

Ruelland, E., Cantrel, C., Gawer, M., Kader, J.C., and Zachowski, A., 2002, Activation of phospholipases C and D is an early response to a cold exposure in Arabidopsis suspension cells. *Plant Physiol.* **130:** 999–1007.

Saiardi, A., Caffrey, J.J., Snyder, S.H., and Shears, S.B., 2000, The inositol hexakisphosphate kinase family. Catalytic flexibility and function in yeast vacuole biogenesis. *J. Biol. Chem.* **275:** 24686–24692.

Saiardi, A., Sciambi, C., McCaffery, J.M., Wendland, B., and Snyder, S.H., 2002, Inositol pyrophosphates regulate endocytic trafficking. *Proc. Natl. Acad. Sci. U.S.A.* **99:** 14206–14211.

Sanchez, J.P., and Chua, N.H., 2001, Arabidopsis PLC1 is required for secondary responses to abscisic acid signals. *Plant Cell* **13:** 1143–1154.

Sanders, D., Brownlee, C., and Harper, J.F., 1999, Communicating with calcium. *Plant Cell* **11:** 691–706.

Sasakawa, N., Sharif, M., and Hanley, M.R., 1995, Metabolism and biological activities of inositol pentakisphosphate and inositol hexakisphosphate. *Biochem. Pharmacol.* **50:** 137–146.

Saunders, C.M., Larman, M.G., Parrington, J., Cox, L.J., Royse, J., Blayney, L.M., Swann, K., and Lai, F.A., 2002, PLC zeta: A sperm-specific trigger of Ca^{2+} oscillations in eggs and embryo development. *Development* **129:** 3533–3544.

Schroeder, J.I., Kwak, J.M., and Allen, G.J., 2001, Guard cell abscisic acid signalling and engineering drought hardiness in plants. *Nature* **410:** 327–330.

Scrase-Field, S.A., and Knight, M.R., 2003, Calcium: just a chemical switch? *Curr. Opin. Plant Biol.* **6:** 500–506.

Shears, S.B., 2004, How versatile are inositol phosphate kinases? *Biochem. J.* **377:** 265–280.

Sheikh-Hamad, D. *et al.*, 1998, p38 Kinase activity is essential for osmotic induction of mRNAs for HSP70 and transporter for organic solute betaine in Madin-Darby canine kidney cells. *J. Biol. Chem.* **273**(3): 1832–1837.

Sheveleva, E., Chmara, W., Bohnert, H.J., and Jensen, R.G., 1997, Increased salt and drought tolerance by D-ononitol production in transgenic *Nicotiana tabacum* L. *Plant Physiol.* **115**(3): 1211–1219.

Shi, J., Wang, H., Wu, Y., Hazebroek, J., Meeley, R.B., and Ertl, D.S., 2003, The maize low-phytic acid mutant *lpa2* is caused by mutation in an inositol phosphate kinase gene. *Plant Physiol.* **131:** 507–515.

Shinozaki, K., and Yamaguchi-Shinozaki, K., 2000, Molecular responses to dehydration and low temperature: Differences and cross-talk between two stress signaling pathways. *Curr. Opin. Plant Biol.* **3:** 217–223.

Shinozaki, K., Yamaguchi-Shinozaki, K., and Seki, M., 2003, Regulatory network of gene expression in the drought and cold stress responses. *Curr. Opin. Plant Biol.* **6:** 410–417.

Smolenska-Sym, G., and Kacperska, A., 1996, Inositol 1,4,5-trisphosphate formation in leaves of water oilseed rape plants in response to freezing, tissue water potential and abscisic acid. *Physiol. Plant* **96:** 692–698.

Song, C., Hu, C.D., Masago, M., Kariyai, K., Yamawaki-Kataoka, Y., Shibatohge, M., Wu, D., Satoh, T., and Kataoka, T., 2001, Regulation of a novel human phospholipase C, PLC epsilon, through membrane targeting by Ras. *J. Biol. Chem.* **276:** 2752–2757.

Srivastava, A., Pines, M., and Jacoby, B., 1989, Enhanced potassium uptake and phosphatidylinositol-phosphate turnover by hypertonic mannitol shock. *Physiol. Plant* **77:** 320–325.

Staxen, I., Pical, C., Montgomery, L.T., Gray, J.E., Hetherington, A.M., and McAinsh, M.R., 1999, Abscisic acid induces oscillations in guard-cell cytosolic free calcium that involve phosphoinositide-specific phospholipase C. *Proc. Natl. Acad. Sci. U.S.A.* **96:** 1779–1784.

Steeves, C.L. *et al.*, 2003, The glycine neurotransmitter transporter GLYT1 is an organic osmolyte transporter regulating cell volume in cleavage-stage embryos. *Proc. Natl. Acad. Sci. U.S.A.* **100**(24): 13982–13987.

Stephens, L.R., and Irvine, R.F., 1990, Stepwise phosphorylation of *myo*-inositol leading to *myo*-inositol hexakisphosphate in *Dictyostelium*. *Nature* **346:** 580–583.

Stevenson, J.M., Perera, I.Y., Heilmann, I., Persson, S., and Boss, W.F., 2000, Inositol signaling and plant growth. *Trends Plant Sci.* **5:** 252–258.

Stevenson-Paulik, J., Odom, A.R., and York, J.D., 2002, Molecular and biochemical characterization of two plant inositol polyphosphate 6-/3-/5-kinases. *J. Biol. Chem.* **277:** 42711–42718.

Suh, B.C., Lee, I.S., Chae, H.D., Han, S., and Kim, K.T., 1998 Characterization of Mas-7-induced pore formation in SK-N-BE(2)C human neuroblastoma cells. *Mol. Cells* **8:** 162–168.

Szaszi, K., Buday, L., and Kapus, A., 1997, Shrinkage-induced protein tyrosine phosphorylation in Chinese hamster ovary cells. *J. Biol. Chem.* **272**(26): 16670–16678.

Taji, T. *et al.*, 2002, Important roles of drought- and cold-inducible genes for galactinol synthase in stress tolerance in *Arabidopsis thaliana*. *Plant J.* **29**(4): 417–426.

Taji, T. *et al.*, 2004, Comparative genomics in salt tolerance between Arabidopsis and aRabidopsis-related halophyte salt cress using Arabidopsis microarray. *Plant Physiol.* **135**(3): 1697–1709.

Takahashi, S., Katagiri, T., Hirayama, T., Yamaguchi-Shinozaki, K., and Shinozaki, K., 2001, Hyperosmotic stress induces a rapid and transient increase in inositol 1,4,5-trisphosphate independent of abscisic acid in Arabidopsis cell culture. *Plant Cell Physiol.* **42:** 214–222.

Thomas, S.M., and Brugge, J.S., 1997, Cellular functions regulated by Src family kinases. *Annu. Rev. Cell Dev. Biol.* **13:** 513–609.

Uchida, S., Yamauchi, A., Preston, A.S., Kwon, H.M., and Handler, J.S., 1993, Medium tonicity regulates expression of the Na(+)- and Cl(−)-dependent betaine transporter in Madin-Darby canine kidney cells by increasing transcription of the transporter gene. *J. Clin. Invest.* **91**(4): 1604–1607.

Vucenik, I., and Shamsuddin, A.M., 2003, Cancer inhibition by inositol hexaphosphate (IP_6) and inositol: From laboratory to clinic. *J. Nutr.* **133:** 3778S–3784S.

Wang, X., 2004, Lipid signaling. *Curr. Opin. Plant Biol.* **7:** 329–336.

Warskulat, U., Weik, C., and Haussinger, D., 1997, *myo*-Inositol is an osmolyte in rat liver macrophages (Kupffer cells) but not in RAW 264.7 mouse macrophages. *Biochem. J.* **326** (Pt 1): 289–295.

Wiese, T.J. *et al.*, 1996, Osmotic regulation of Na-*myo*-inositol cotransporter mRNA level and activity in endothelial and neural cells. *Am. J. Physiol.* **270**(4 Pt 1): C990–C997.

Wu, Y., Kuzma, J., Marechal, E., Graeff, R., Lee, H.C., Foster, R., and Chua, N.H., 1997, Abscisic acid signaling through cyclic ADP-ribose in plants. *Science* **278:** 2126–2130.

Xia, H.J., Brearley, C., Elge, S., Kaplan, B., Fromm, H., and Mueller-Roeber, B., 2003, Arabidopsis inositol polyphosphate 6-/3-kinase is a nuclear protein that complements a yeast mutant lacking a functional ArgR-Mcm1 transcription complex. *Plant Cell* **15:** 449–463.

Xiong, L., Schumaker, K.S., and Zhu, J.K., 2002, Cell signaling during cold, drought, and salt stress. *Plant Cell* **2002:** 165–183.

Xiong, L., Lee, B.h., Ishitani, M., Lee, H., Zhang, C., and Zhu, J.K., 2001, *FIERY1* encoding an inositol polyphosphate 1-phosphatase is a negative regulator of abscisic acid and stress signaling in Arabidopsis. *Genes Dev.* **15:** 1971–1984.

Yamaguchi-Shinozaki, K., Koizumi, M., Urao, S., and Shinozaki, K., 1992, Molecular cloning and characterization of 9 cDNAs for genes that are responsive to desiccation in *Arabidopsis thaliana*: Sequence analysis of one cDNA clone that encodes a putative transmembrane channel protein. *Plant Cell Physiol.* **33:** 217–224.

Yamaguchi-Shinozaki, K., and Shinozaki, K., 1994, A novel *cis*-acting element in an *Arabidopsis* gene is involved in responsiveness to drought, low-temperature, or high-salt stress. *Plant Cell* **6:** 251–264.

Yamamoto, Y.T., Conkling, M.A., Sussex, I.M., and Irish, V.F., 1995, An Arabidopsis cDNA related to animal phosphoinositide-specific phospholipase C genes. *Plant Physiol.* **107:** 1029–1030.

Yamauchi, A., Uchida, S., Preston, A.S., Kwon, H.M., and Handler, J.S., 1993, Hypertonicity stimulates transcription of gene for Na(+)-*myo*-inositol cotransporter in MDCK cells. *Am. J. Physiol.* **264**(1 Pt 2): F20–F23.

Yancey, P.H., Clark, M.E., Hand, S.C., Bowlus, R.D., and Somero, G.N., 1982, Living with water stress: Evolution of osmolyte systems. *Science* **217**(4566): 1214–1222.

York, J.D., Odom, A.R., Murphy, R., Ives, E.B., and Wente, S.R., 1999, A phospholipase C-dependent inositol polyphosphate kinase pathway required for efficient messenger RNA export. *Science* **285:** 96–100.

Zhang, Z., Ferraris, J.D., Brooks, H.L., Brisc, I., and Burg, M.B., 2003, Expression of osmotic stress-related genes in tissues of normal and hyposmotic rats. *Am. J. Physiol. Renal. Physiol.* **285**(4): F688–F693.

Zhou, Y., Wang, W., Ren, B., and Shou, T., 1994, Receptive field properties of cat retinal ganglion cells during short-term IOP elevation. *Invest. Ophthalmol. Vis. Sci.* **35**(6): 2758–2764.

Chapter 11

Inositol Phosphates and Phosphoinositides in Health and Disease

Yihui Shi*, Abed N. Azab*, Morgan N. Thompson, and Miriam L. Greenberg
Department of Biological Sciences, Wayne State University, Detroit, MI 48202, USA

1. INTRODUCTION

Inositol is an essential molecule found ubiquitously in biological systems. Phosphorylation of the cyclic inositol ring produces two related families of molecules, inositol phosphates and phosphinositides. Research into the roles of inositol and its derivatives has been hampered by the complex and multitudinous interactions of these molecules in multiple cellular pathways. However, the potential rewards of such studies are immense, as inositol phosphates and phosphoinositides (PIs) play a role in numerous human diseases. This review addresses current knowledge of the role of inositol phosphates and PIs in human health and disease.

Part one of this review focuses on the role of inositol phosphates in cellular signaling pathways. Specifically, we focus on inositol 1,4,5-triphosphate ($InsP_3$) and inositol hexaphosphate ($InsP_6$) because these are the most highly studied inositol phosphates in relation to human disease. $InsP_3$ plays an essential role as a secondary messenger in the $InsP_3/Ca^{2+}$ signal transduction pathway, which is responsible for modulating the activity of numerous cellular processes. Perturbation of this pathway has been implicated in a variety of disorders including bipolar affective disorder, Alzheimer's disease (AD), Parkinson's disease, and malignant hyperthermia (MH). $InsP_6$ may be the most abundant inositol phosphate and is found ubiquitously in mammalian cells. Recently, $InsP_6$ has been identified as a potential antineoplastic therapy due to

*Contributed equally

A. Lahiri Majumder and B. B. Biswas (eds.), Biology of Inositols and Phosphoinositides, 263–290
© 2006 *Springer. Printed in the Netherlands.*

its antioxidant properties. As a result of the important functions of these molecules, increasing interest in examining the roles of other inositol phosphates has led to identification of novel functions. We briefly discuss the significance of recent findings regarding a variety of other inositol phosphates that provide promising avenues for future research.

In part two, we review the role of PIs in human disease. Although PIs are not abundant in biological systems, they have displayed numerous important functions in multiple signal transduction pathways. A number of human diseases are characterized by dysfunctional PI pathways, including cancer, type 2 diabetes, Lowe syndrome, myotubular myopathy, and Charcot-Marie-Tooth disease.

2. INOSITOL PHOSPHATES

The existence of inositol phosphates has been known for over 80 years (Posternak, 1919). Inositol, a six-carbon cyclitol found ubiquitously in all biological systems (Bachhawat and Mande, 1999; Chen *et al.*, 2000; Majumder *et al.*, 2003), exists in eight possible isomeric forms (*myo, chiro, scyllo, neo, cis, epi, allo, and muco*), of which *myo* is physiologically the most common and important stereoisomer. Phosphorylation of the inositol ring at one or more positions generates numerous PIs and inositol phosphates. The study of and interest in inositol phosphates is complicated by three major factors (Irvine and Schell, 2001): (1) there are many of them (63 possible isomers for inositol monophosphates alone), a potential that can be expanded further by attaching more than one phosphate on the same position, as in inositol pyrophosphates; (2) the multiplicity of metabolic pathways makes it hard to understand how their levels are regulated in cells; and (3) inositol phosphates are suspected to play a role in multiple signaling pathways (especially due to their involvement in Ca^{2+} metabolism), and, therefore, it is difficult to "bring them together" in one place.

Although the cellular roles of these molecules are not fully understood, inositol phosphates have been shown to convey signals for a variety of hormones, growth factors, and neurotransmitters (Berridge, 1993; Berridge and Irvine, 1989). As mentioned, cells contain a large array of inositol phosphates, some of which respond to receptor stimulation, providing the basis for multiple and complex responses. Among the inositol phosphates, we focus on $InsP_3$ and $InsP_6$, the most widely studied in relation to human health and disease.

2.1 Inositol 1,4,5-triphosphate ($InsP_3$) – A major role in neurological disorders

Streb and coworkers (1983) discovered that $InsP_3$ is a Ca^{2+}-mobilizing second messenger. Since then, a huge body of data has accumulated regarding its pivotal roles in the regulation of multiple cellular pathways, and its possible association with multiple illnesses, mainly neurological disorders.

The demonstration that $InsP_3$ causes release of Ca^{2+} from intracellular stores laid the foundation for the current understanding of the function of the phosphatidylinositol turnover pathway (Streb *et al.*, 1983). Briefly, the pathway involves receptor-mediated activation of phospholipase C (PLC), which cleaves phosphatidylinositol(4,5)biphosphate[$PtdIns(4,5)P_2$] to produce 1,2-diacylglycerol (DAG) and $Ins(1,4,5)P_3$ (here referred to as $InsP_3$) (Gould *et al.*, 2004a; Irvine and Schell, 2001; Irvine *et al.*, 1984; Streb *et al.*, 1983; Woodcock, 1997). DAG activates various isomers of protein kinase C (PKC), and $InsP_3$ initiates rises in Ca^{2+} (Berridge, 1987; Streb *et al.*, 1983). $InsP_3$ amplifies its cellular effects (Ca^{2+}-mobilization) by activating $InsP_3$ receptors (IP_3Rs) (Streb *et al.*, 1983). The IP_3R has a tetrameric structure similar to other Ca^{2+} channels (Patterson *et al.*, 2004). The wide distribution of these receptors in cells likely reflects the multiplicity of functions of $InsP_3$. Initially, these receptors were identified in the endoplasmic reticulum, but they also reside in the Golgi apparatus, plasma membrane, nucleoplasmic reticulum, and other cellular organelles (Patterson *et al.*, 2004).

The $InsP_3/Ca^{2+}$ signal transduction pathway modulates the activity of a multitude of intracellular events. Numerous receptors (mostly G-protein coupled receptors, particularly $G_{q/11}$) are associated with PLC-induced $InsP_3$ release. For example, in the central nervous system, the excitatory neurotransmitters glutamate and aspartate, and also M_1 and M_3 muscarinic, β_1-adrenergic, 5-HT_2 serotonergic, H_1 histaminic, and vasopressin V_1 receptors, among others, are all known to increase $InsP_3/Ca^{2+}$ release (Bloom, 2001).

Calcium is a very important signaling molecule within cells (Berridge, 1993). Many cellular functions are directly or indirectly regulated by free cytosolic Ca^{2+} ($[Ca^{2+}]_i$). The Ca^{2+} ions needed to control the activity of the cell can be supplied to the cytosol from the extracellular space or from intracellular stores (mainly from endoplasmic and sarcoplasmic reticulum, but also from mitochondria and Golgi apparatus). Interestingly, $InsP_3/Ca^{2+}$ release is known to take place in the nucleus of various cell types, and IP_3Rs have been found in the inner nuclear membrane (Martelli *et al.*, 2004). $InsP_3$ regulation of nuclear Ca^{2+} levels is important for several processes that take place in the nucleus, such as protein transport across the nuclear envelope and regulation of gene expression, among others (Martelli *et al.*, 2004).

The *time*, *space*, and *amplitude* of the fluctuating changes in ($[Ca^{2+}]_i$) concentration are very important and strictly regulated because cells extract specific information from these three parameters. Because Ca^{2+} is such an important signaling molecule, mutations causing drastic functional alteration in ($[Ca^{2+}]_i$) homeostasis are most likely not compatible with life (Lorenzon and Beam, 2000). Mutations or abnormalities in the proteins involved in ($[Ca^{2+}]_i$) regulation, which may cause only trivial alterations in the function of the protein *in vitro*, often lead to a diverse array of diseases (Missiaen *et al.*, 2000). For example, alterations in the function of proteins related to ($[Ca^{2+}]_i$) regulation are associated with AD, skeletal muscle pathology, heart disease, MH, visual disturbances, and skin diseases, among others (Missiaen *et al.*, 2000).

The role of $InsP_3$ in the regulation of Ca^{2+} release is not universal for all cell types and signal transduction pathways. The effects of $InsP_3$ on Ca^{2+} mobilization from intracellular stores primarily occur in non-excitable cells, in which appropriate receptor stimulation causes rapid release of $InsP_3$. However, it should be kept in mind that this pattern is not observed in some excitable cells, *e.g.*, cardiomyocytes, smooth muscle, and skeletal muscle cells, in which voltage-regulated channels are major contributors to Ca^{2+} control mechanisms and release (Woodcock, 1997). Nevertheless, it is an oversimplification to say that $InsP_3$ does not play a role in these excitable cells (cardiomyocytes for instance). IP_3Rs are present ubiquitously, and their presence opens up the possibility that these receptors have other functions (Kijima *et al.*, 1993). In addition, evidence suggests that $InsP_3$ is not always the primary inositol phosphate released, is not always formed from PtdIns(4,5)P_2, and is often present in unstimulated cells at concentrations sufficient to activate/saturate its receptors (Woodcock, 1997). This further complicates the understanding of the regulation and cellular functions of this pivotal second messenger.

Perturbation of the $InsP_3/Ca^{2+}$ signaling pathway leads to a variety of disorders. However, our focus in this section is on the role of $InsP_3/Ca^{2+}$ signaling in the pathophysiology of neurological disorders, mainly bipolar affective disorder and AD.

2.1.1 Bipolar disorder (BAD)

Bipolar disorder (BAD, manic-depressive illness) is a severe and chronic illness, which is a major public health problem, in any given year affecting approximately 1–3% of the US population (Narrow *et al.*, 2002). In the World Health Organization Global Burden of Disease study, BAD ranked sixth among all medical disorders in years of life lost to death or disability worldwide, and is projected to have a greater impact in the future (Murray and Lopez, 1996).

Although a number of mood-stabilizing drugs are commonly used in the treatment of BAD, lithium and the anticonvulsants valproate (VPA) and carbamazepine are the only drugs for which long-term efficacy has been established, and are therefore used for maintenance treatment of BAD (Belmaker, 2004). However, these agents are far from the perfect medications, and they are ineffective and not well tolerated by a significant portion of patients.

The mechanism of action of mood-stabilizing drugs in the treatment of BAD is not fully understood. The inositol-depletion hypothesis has been suggested to explain the mechanism of action of lithium (Berridge and Irvine, 1989; Berridge *et al.*, 1982). This hypothesis postulates that the therapeutic effects of lithium are due to uncompetitive inhibition of inositol monophosphatase (IMPase), which leads to depletion of *myo*-inositol in brain cells, and consequently, to dampening of PI signaling. The action of IMPase is the final step in inositol synthesis. Inositol polyphosphate 1-phosphatase (IPPase) removes phosphate from Ins (1,4)-biphosphate. Both enzymes appear to be critical for the maintenance of *myo*-inositol levels and continuation of PI-mediated signaling. Therefore, direct inhibition of IMPase by lithium (Hallcher and Sherman, 1980; Naccarato *et al.*, 1974) and IPPase (Inhorn and Majerus, 1988) could potentially lead to inositol depletion. Indeed, lithium has consistently been shown to decrease free inositol levels in human brain sections and in brains of rodents (Atack, 2000). Lithium treatment also decreases inositol levels in human subjects (Moore *et al.*, 1999). Furthermore, lithium and VPA were found to normalize the altered PI cycle in BAD patients (Silverstone *et al.*, 2002).

Controversy regarding the inositol depletion hypothesis has centered around a number of arguments. Lithium produces therapeutic effects only after chronic administration, whereas direct inhibition of IMPase by lithium is rapid (Atack, 2000; Pollack *et al.*, 1994). As there are currently no specific IMPase inhibitors available for clinical use, it is difficult to directly test the inositol depletion hypothesis in BAD patients (Atack, 2000). Recent evidence points toward other signaling molecules and signal transduction pathways as targets for the mood-stabilizing effects of anti-bipolar drugs (Gould *et al.*, 2004a; Jope, 2003). Furthermore, accumulating data suggest that severe mood disorders are associated with impairments of structural plasticity and cellular resilience, and that BAD patients may suffer from a reduction in CNS volume (Gould *et al.*, 2004b). Gould and coworkers have suggested that the fact that currently used mood stabilizers take weeks to produce their therapeutic effects may implicate changes in gene expression, protein function, and more importantly, general neural plasticity (Gould *et al.*, 2004a). The anti-bipolar effect of lithium is now attributed, at least in part, to its neuroprotective effect, achieved mainly by inhibiting the activity of glycogen synthase kinase-3 (GSK-3) (Gould *et al.*, 2004a,b). GSK-3 is the only kinase known to be inhibited by lithium at nearly therapeutic concentrations ($\sim$1 mM) (Jope, 2003). Lithium inhibits GSK-3 activity in two ways: (i) by *directly* inhibiting catalytic activity, and (ii) by *indirectly* increasing phosphorylation that inhibits activity (Jope, 2003). Since GSK-3 plays a major role in multiple cellular pathways (Gould *et al.*, 2004a,b), inhibition of its activity by lithium may have far-reaching effects. Consistent with these observations, GSK-3 inhibitors were found to possess mood-stabilizing effects in the forced swim test in mice and rats (Gould *et al.*, 2004c; Kaidanovich-Beilin *et al.*, 2004). The neuroprotective effects of lithium are also attributed to its ability to increase the levels of the neuroprotective protein bcl-2 (Manji *et al.*, 1999). Similar to lithium, VPA was also found to elicit neuroprotective effects through the inhibition of GSK-3 and increase of bcl-2 levels (Chen *et al.*, 1999a,b).

The inositol depletion hypothesis was revived by the observations that VPA, like lithium, also decreases intracellular inositol levels (Vaden *et al.*, 2001), and that anti-bipolar drugs increase the growth cone area of sensory neurons in an inositol-dependent manner (Williams *et al.*, 2002). VPA decreases intracellular inositol in yeast, and this decrease is accompanied by a significant derepression of *INO1*, the gene encoding 1-D-*myo*-inositol-3-phosphate (MIP) synthase, which is derepressed in response to inositol limitation (Vaden *et al.*, 2001). VPA decreases intracellular inositol levels by indirectly inhibiting MIP synthase activity *in vivo* (Ju *et al.*, 2004). MIP synthase catalyzes the rate-limiting step in inositol synthesis, the conversion of D-glucose-6-phosphate (G-6-P) to MIP (Loewus *et al.*, 1980). VPA was shown to decrease MIP synthase activity (by 50%) in crude homogenate of human postmortem prefrontal cortex (Shaltiel *et al.*, in press). The inositol depletion hypothesis was further strengthened by the observation that three mood-stabilizing drugs – lithium, VPA, and carbamazepine – inhibit the contraction of sensory neuron growth cones and increase growth cone area in an inositol-dependent manner, providing evidence correlating inositol depletion and neuronal function (Williams *et al.*, 2002).

As mentioned above, anti-bipolar drugs cause a rapid decrease in intracellular inositol levels, but their therapeutic effects are apparent only after chronic treatment (Pollack *et al.*, 1994). While this may seem to argue against the inositol depletion hypothesis, it is important to view inositol depletion in light of the pivotal role of inositol as a major metabolic sensor for regulating major cellular pathways, including protein secretion, the unfolded protein response pathway, and the glucose response pathway (Carman and Henry, 1999; Ju, 2004). Therefore, the acute depletion of inositol by anti-bipolar drugs may have far-reaching cellular responses that contribute to their therapeutic effects.

Taken together, the data reviewed suggest that inositol may play a role in the pathophysiological mechanisms underlying BAD, and that the inositol depletion hypothesis may help to explain the mood-stabilizing effects of anti-bipolar drugs.

2.1.2 Alzheimer's disease (AD)

Alzheimer's disease (AD) is the leading cause of dementia in the elderly population; in the United States alone, almost 4 million patients suffer from the disease (Brookmeyer *et al.*, 1998). Each year, approximately 360,000 new cases of AD are identified, and by the year 2050, more than 14 million persons in the United States alone may likely suffer from this neurodegenerative disease (Brookmeyer *et al.*, 1998). The pathophysiological mechanisms underlying AD are unclear. However, one of the most characteristic neuropathological lesions in the brains of patients with AD are amyloid plaques that are composed primarily of a peptide known as amyloid-β (Aβ) (Selkoe, 1991). The formation and aggregation of Aβ plaques are known to cause accumulation of microglia and reactive astrocytes, leading to acute phase/inflammatory response, which may damage neurons and exacerbate the pathological processes of the disease (Aisen, 1997; McGeer and McGeer, 1995; Selkoe, 1991). It has also been suggested that cerebrovascular disease may contribute to the severity of the disease (Riekse *et al.*, 2004).

Interestingly, disruption of the $InsP_3/Ca^{2+}$ signaling pathway is implicated in AD and other neurogenerative disorders (LaFerla, 2002; Mattson *et al.*, 2000a). For example, it was shown that IP_3R levels were significantly decreased in the cerebellum, superior temporal, and superior frontal cortices of AD patients, as compared to matched control brains (Garlind *et al.*, 1995). Similarly, in a postmortem analysis of eight AD patients, IP_3R protein levels were decreased significantly in the temporal and frontal cortices, compared to matched control cases (Haug *et al.*, 1996). These degenerative changes may be responsible, at least in part, for the dysregulation of calcium homeostasis seen in AD. Moreover, abnormal function of IP_3R1 was observed in animal models of AD (Mattson *et al.*, 2000b). $InsP_3$-induced Ca^{2+} release is known to serve numerous signaling functions in neurons, including modulation of membrane excitability (Yamamoto *et al.*, 2002), synaptic plasticity (Fujii *et al.*, 2000), and gene expression (Mellstrom and Naranjo, 2001). Recent attention has focused on presenilin 1 (PS1), an endoplasmic-reticulum-localized protein which modulates $InsP_3$-induced Ca^{2+} release, and is required for the proteolysis of amyloid precursor protein (Selkoe, 2001). Mutant forms of the *PS1* gene have been shown to contribute to the majority of early-onset AD cases (Selkoe, 2001). Recently, it was found that dysregulation of $InsP_3/Ca^{2+}$ signaling, which was induced by a *PS1* mutation, has enhanced neuronal Ca^{2+} liberation and signaling, and altered membrane excitability (Stutzmann *et al.*, 2004). These data indicate that perturbation of the $InsP_3/Ca^{2+}$ signaling pathway may contribute to the pathology of AD.

2.1.3 Other neurological disorders

IP_3R1 is predominant in cerebellar Purkinje cells and is also present in other neural and peripheral tissues (Matsumoto *et al.*, 1996). Importantly, the binding sites for $InsP_3$ correlated with IP_3R levels were significantly decreased in multiple regions of postmortem brain tissue from Parkinson's disease patients, compared with those of age-matched controls (Kitamura *et al.*, 1989). Similar results were observed in postmortem brain regions of patients with Huntington's disease, in which the binding sites for $InsP_3$ were significantly reduced (Warsh *et al.*, 1991). Consistent with these findings, it was found that IP_3R1 is related to the dysregulation of calcium signaling in a mouse model of Huntington's disease (Tang *et al.*, 2003). Abnormal IP_3R1 function in humans is associated with neurological abnormalities, the most prominent of which is ataxia (Zecevic *et al.*, 1999). Disruption of the IP_3R1 gene in mice resulted in a severe reduction of $InsP_3$-induced Ca^{2+} release in brain tissue (Matsumoto *et al.*, 1996). Very few embryos carrying the disruption survive to birth, pointing to a significant role for IP_3R1 during embryonic development. Those animals that did survive to birth exhibited severe neurological alteration, including ataxia and epilepsy. Accordingly, Street and coworkers have also observed that abnormal function of IP_3R1 is associated with epilepsy in mice (Street *et al.*, 1997).

2.1.4 Other pathologies

Alteration in $InsP_3/Ca^{2+}$ signaling is one of the suggested mechanisms for MH in humans, a disorder characterized by uncontrolled, severely elevated body temperature and muscle contractions (Wappler *et al.*, 1997). Interestingly, this study has shown that $InsP_3$ levels were significantly increased in the skeletal muscles but not in the plasma of MH patients (Wappler *et al.*, 1997). The underlying mechanism of $InsP_3$-induced MH is not clear. However, in a swine model of MH (induced by halothane challenge), it was suggested that an increase in $InsP_3$ activity may cause a drastic increase in Ca^{2+} concentration, leading to metabolic changes resulting in MH (Tonner *et al.*, 1995). Moreover, altered $InsP_3/Ca^{2+}$ signaling was observed during myocardial ischemia/reperfusion cycles in rats (Mouton *et al.*, 1991). Adrenergic stimulation led to ischemia that was accompanied by a significant increase in $InsP_3$ levels in the myocardial tissue. Interestingly, an increase in $InsP_3$ levels was also observed during the reperfusion phase (Mouton *et al.*, 1991).

Calcium overload appears to be involved in ischemia/reperfusion cycles. However, due to the differences in the ischemia-inducing protocols, the exact role of Ca^{2+} overload is unclear. Mouton and coworkers have concluded that calcium overload (due to the increased $InsP_3$ activity) is involved in the pathophysiology of myocardial ischemia/reperfusion cycles. In contrast to this study, Woodcock and coworkers have shown that during global myocardial ischemia, the levels of $InsP_3$ were significantly decreased in rat ventricles, whereas levels were increased during reperfusion (Woodcock *et al.*, 1997). These authors have also shown that increased Ca^{2+} concentrations during ischemia (induced by coronary artery ligation) may further enhance $InsP_3$ release, and this may account for the Ca^{2+} overload seen in the ischemic myocardial tissue (Woodcock *et al.*, 1996). This Ca^{2+} overload was suggested as a major contributor to the severe cardiac arrhythmias seen during the ischemia/reperfusion cycles.

2.2 Inositol hexaphosphate ($InsP_6$) – A potential antineoplastic therapy

$InsP_6$, also known as phytic acid, was the first inositol phosphate discovered (Posternak, 1919). Initially, $InsP_6$ was thought to exist only in plants (mostly legumes) and in the erythrocytes of a few animals. However, it is now known to be ubiquitous in mammalian (and probably all eukaryotic) cells (Heslop *et al.*, 1985), and may actually be the most abundant inositol phosphate.

Chemically, $InsP_6$ is a highly charged molecule. It interacts with positively charged groups on proteins and with low molecular weight cations, and in doing so, can compete with other molecules that would bind these proteins and cations *in vivo*.

The earliest (and probably the most important) proposed function for $InsP_6$ in plants is that of a phosphate store for seeds (Posternak, 1919). It is now known that $InsP_6$ exhibits strong antioxidant properties (Hawkins *et al.*, 1993) and can cause complete inhibition of Fe^{3+}-catalyzed hydroxyl-radical formation. Functions proposed more recently for $InsP_6$ include protein phosphatase inhibition (Larsson *et al.*, 1997) and activation of PKC (Efanov *et al.*, 1997). $InsP_6$ was also found in the nucleus, and York and coworkers observed that $InsP_6$ may modulate mRNA transport out of the nucleus (York *et al.*, 1999).

One of the most striking findings regarding $InsP_6$ is that it may confer anticancer properties (Fox and Eberl, 2002; Vucenik and Shamsuddin, 2003). Numerous studies suggest that wheat bran is antineoplastic, especially for colon cancer (Fox and Eberl, 2002). Further research has indicated that the potential active ingredient for this effect is $InsP_6$. It has been extensively studied *in vitro* and in *in vivo* animal models of cancer, and in the majority of them it exhibited anticancer properties (Fox and Eberl, 2002; Vucenik and Shamsuddin, 2003). Interestingly, the most consistent and potent anticancer results were obtained from the combination of $InsP_6$ with inositol (which itself has a moderate anticancer effect) (Vucenik and Shamsuddin, 2003). For example, $InsP_6$ was found to inhibit neoplastic growth in many types of cancer, including breast, colon, liver, lung, prostate, rhabdomyosarcoma, skin, and others (Fox and Eberl, 2002; Vucenik and Shamsuddin, 2003). $InsP_6$ was effective (as an antineoplastic) in a dose-dependent manner given either before or after carcinogen (such as 1,2-dimethylhydrazine and azoxymethane) administration to rats and mice with colon cancer. Importantly, $InsP_6$ was found to act synergistically with standard chemotherapeutic agents, and therefore, significantly augmented the anticancer effect of the treatment (Tantivejkul *et al.*, 2003). One of the most important studies to elucidate the anticancer properties of $InsP_6$ was the pilot clinical trial performed in patients with advanced colorectal cancer (Druzijanic *et al.*, 2002). The results demonstrated enhanced antitumor activity of the standard treatment when $InsP_6$ (plus inositol) was added, without compromising the patients' quality of life.

The molecular mechanisms underlying the anticancer effect of $InsP_6$ are not understood. Proposed mechanisms of action for $InsP_6$ are (i) increase in natural killer cell activity and enhanced host immunity, (ii) alteration in signal transduction, (iii) stimulation of genes toward greater cell differentiation, (iv) reduction in cell proliferation, and (v) antioxidant activity (Fox and Eberl, 2002).

Exogenously administered $InsP_6$ is rapidly taken up by cells and dephosphorylated to other inositol phosphates, *e.g.*, $InsP_4$ and $InsP_5$, which interfere with signal transduction pathways and cell cycle arrest (Vucenik and

Shamsuddin, 2003). One of the advantages attributed to $InsP_6$ as a potential anticancer agent is that it has only a minimal effect on normal cells (Deliliers *et al.*, 2002; Vucenik and Shamsuddin, 2003). Other beneficial health effects were attributed to $InsP_6$ treatment *in vivo*, including inhibition of kidney stone formation and reduction in the risk of developing cardiovascular diseases (via hypocholesterolemic and platelets antiaggregation effects) (Vucenik and Shamsuddin, 2003). The inhibitory effect on kidney stone formation may be attributed to potent inhibition of calcium salt crystallization (Grases *et al.*, 2004). Because $InsP_6$ is abundant in the normal diet, is efficiently absorbed from the gastrointestinal system, and is considered safe, $InsP_6$ holds promise as a new strategy for the prevention and treatment of cancer. Not coincidently, $InsP_6$ is now being promoted extensively in health food stores as a "natural" anticancer compound. Importantly, the antineoplastic effects of $InsP_6$ were accompanied by serious side effects, such as chelation of multivalent cations and an increase in bladder and renal papillomas.

In summary, *in vivo* and *in vitro* studies indicate that $InsP_6$ may have significant potential as an effective agent for the prevention and treatment of cancer. To the best of our knowledge, no published controlled clinical trials have examined the anticancer effect of $InsP_6$.

2.3 Other inositol phosphates

The significance of $InsP_3$ as an important second messenger in multiple cellular pathways has led to the examination of the functions of other inositol phosphates.

2.3.1 Inositol 1,4-biphosphate ($InsP_2$)

The termination of $InsP_3$ intracellular signaling is achieved via the activity of a 5-phosphatase that converts $InsP_3$ to $InsP_2$. $InsP_2$ is a substrate for IPPase that removes a phosphate to form Ins(1)P_1. The latter is dephosphorylated by IMPase to give free *myo*-inositol. To the best of our knowledge, no established physiological function has been attributed to Ins(1)P_1 to date. On the other hand, $InsP_2$ was found to stimulate DNA polymerase-α (Sylvia *et al.*, 1988) and may play a role in the control of DNA replication (York *et al.*, 1994). Furthermore, in the heart, $InsP_2$ appears to be the primary inositol phosphate released (more than $InsP_3$) following "normal" adrenergic stimulation. $InsP_2$ levels were significantly reduced (70–90%) during global myocardial ischemia in rats (Woodcock *et al.*, 1997). This study has demonstrated that $InsP_3$ content is also reduced significantly during the ischemic phase but predominates under conditions of postischemic reperfusion. It is possible that the release of $InsP_2$ substitutes for $InsP_3$ release because in some experimental models, enhanced $InsP_3$ signaling (and Ca^{2+} accumulation) is proarrhythmic in the heart (Jacobsen *et al.*, 1996).

2.3.2 Inositol-tetrakisphosphates

Inositol-tetrakisphosphates are inositol phosphates containing four phosphates. Of these, inositol-1,3,4,5-tetrakisphosphate (Ins(1,3,4,5)P_4), along with its immediate catabolic product Ins(1,3,4)P_3, were among the first wave of inositol phosphates to be discovered (Irvine *et al.*, 1984). One of the most important functions of Ins(1,3,4,5)P_4 is that it can protect Ins(1,4,5)P_3 against hydrolysis, probably by competing for the hydrolytic enzyme inositol 5-phosphatase (Connolly *et al.*, 1987). Ins(1,3,4,5)P_4 is hydrolyzed by the same 5-phosphatase that hydrolyzes Ins(1,4,5)P_3, but the enzyme has a 10-fold higher affinity for Ins(1,3,4,5)P_4. In addition, Ins(1,3,4,5)P_4 can act in synergy with Ins(1,4,5)P_3 to mobilize Ca^{2+} and to activate its entry to cells (Morris *et al.*, 1987; Smith *et al.*, 2000). Furthermore, in endothelial cells (Luckhoff and Clapham, 1992) and in neurons (Tsubokawa *et al.*, 1996), Ins(1,3,4,5)P_4 can *independently* activate Ca^{2+} channels in the plasma membrane.

Another member of the inositol-tetrakisphosphates is Ins(3,4,5,6)P_4. In epithelial cells, Ins(3,4,5,6)P_4 seems to be a physiologically important inhibitor of Ca^{2+}-regulated chloride channels (CLCA), and is now considered to be a regulator of chloride secretion (Vajanaphanich *et al.*, 1994). Similar findings were observed by Ismailov and coworkers, in which the effect of Ins(3,4,5,6)P_4 on CLCA was found to be *bi-phasic*, namely, an initial phase of activation, followed by a prominent inhibition phase (Ismailov *et al.*, 1996). These effects were not observed under treatment with other tetrakisphosphate isomers. The significance of this effect is particularly important for patients with cystic fibrosis (Ismailov *et al.*, 1996) in whom the epithelial cyclic AMP-regulated chloride channel [or cystic fibrosis transmembrane conductance regulator, (CFTR)] is compromised, and CLCA is probably the most important remaining functional channel (Rudolf *et al.*, 2003). It is worth noting that the majority of cystic fibrosis patients have a mutation in the gene encoding the CFTR (Modiano *et al.*, 2004), resulting in defective hydration of mucosal membranes. Interestingly, the activity of CLCA was found to be enhanced in CF patients, a phenomenon that was suggested as a compensatory mechanism for the defect in CFTR activity (Leung *et al.*, 1995). Therefore, restoring transepithelial chloride secretion by augmenting the activity of CLCA may have beneficial effects in the treatment of CF. This hypothesis was tested in a human colonic epithelial cell line by blocking the inhibitory effect of Ins(3,4,5,6)P_4 [using Ins(3,4,5,6)P_4 derivatives that can bind to but not inhibit the CLCA] on CLCA, which resulted in enhanced chloride secretion (Rudolf *et al.*, 2003). These findings may serve as a basis for a new direction in the treatment of CF.

2.3.3 Inositol-1,3,4,5,6-pentakisphosphate

Ins(1,3,4,5,6)P_5 is the InsP_5 isomer that predominates in mammalian cells (Stephens *et al.*, 1991). In most cells, receptor-stimulated release of

Ins(1,4,5)P_3 is accompanied by hydrolysis of Ins(1,3,4,5,6)P_5 and the generation of Ins(3,4,5,6)P_4 (Ye *et al.*, 1995). Interestingly, InsP_5 acts as a potent antagonist of the IP_3R by competitive receptor binding (Lu *et al.*, 1996). This inhibitory effect may have physiological significance because it dampens calcium mobilization. One of the most documented physiological functions for InsP_5 is regulation of the hemoglobin–O_2 interaction, as it was found that InsP_5 alters the affinity of hemoglobin for O_2 (Liang *et al.*, 2001; Riera *et al.*, 1991). A role for nuclear InsP_5 (and InsP_4) production in regulating gene expression was observed in yeast (York *et al.*, 2001). The significance of these observations is still not well understood and now seems to be more complex than originally thought.

2.3.4 Inositol pyrophosphates

A fascinating discovery was made that an inositol ring containing six phosphates can accommodate additional phosphates (Menniti *et al.*, 1993; Stephens *et al.*, 1991). The simplest versions of inositol pyrophosphates are InsP_7 and InsP_8 [diphosphoinositol pentakisphosphate (InsP_5PP) and bis (diphospho) inositol tetrakisphosphate (Ins$P_4(PP)_2$), respectively]. In InsP_7, the pyrophosphate occurs at the 1 position, whereas in InsP_8, the pyrophosphates occur at the 1 and 2, or 1 and 4 positions (Stephens *et al.*, 1993). In mammalian cells, these metabolites turn over very rapidly and their most likely function is as an ATP-generating system (Voglmaier *et al.*, 1996), as they were found to supply energy for vesicle formation and/or transport systems. These results suggest that pyrophosphates constitute an energy reservoir in mammalian cells, analogous to InsP_6 in plants. Additional pyrophosphates may yet be identified on different positions of the ring, further expanding the family of inositol phosphates. The physiological functions of these inositol phosphates remain to be ascertained.

3. PHOSPHOINOSITIDES

Phosphoinositides (PIs) are low in abundance but have been shown to play important roles in many signal transduction pathways. Various PIs have been found to interact with high-affinity PI-binding proteins and affect the activity and/or localization of these proteins (Toker, 2002). These specific PI-binding proteins further recruit and regulate specific signaling proteins that mediate many physiological processes, including cell growth, proliferation, apoptosis, insulin action, cytoskeletal assembly, and vesicle trafficking. The spatiotemporal control of signal transduction requires precise regulation of PI generation and turnover by PI-metabolizing enzymes, including specific kinases and phosphatases.

There are five free hydroxyl groups on the inositol ring of PIs. To date, three of these, D3, D4, and D5 are known to be phosphorylated *in vivo* (Toker,

2002). Eight PIs have been identified in the cell, including PI, PI(3)P, PI(4)P, PI(5)P, PI(3,4)P_2, PI(4,5)P_2, PI(3,5)P_2, and PI(3,4,5)P_3 (Figure 1). The intracellular levels of the PIs are strictly regulated by PI-metabolizing enzymes. In this part of the review, we focus on diseases caused by perturbation of PI metabolism.

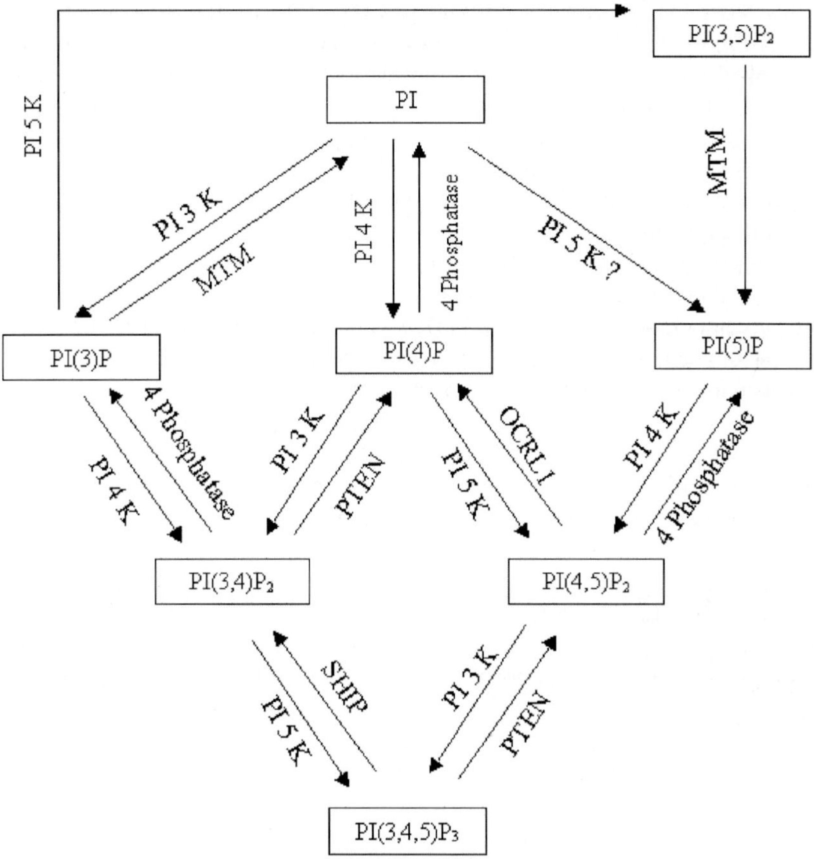

Figure 1. Synthesis of phosphoinositides (Martin, 1998; Pendaries *et al.*, 2003).

3.1 PI3K/AKT pathway in cancer

PI3 kinase (PI3K) was first brought to the attention of cancer researchers in the mid-1980s, when it was discovered that the viral oncoprotein SRC and polyomavirus middle T-antigen could induce PI3K activity (Whitman *et al.*, 1985). PI3Ks are classified into three groups according to their structures and specific substrates. Activated PI3Ks specifically phosphorylate PI, PI(4)P, and PI(4,5)P_2 to generate PI(3)P, PI(3,4)P_2, and PI(3,4,5)P_3. PI3Ks are heterodimeric enzymes composed of a catalytic subunit, p110, and a regulatory subunit, p85 (Cantley, 2002). The class I PI3Ks are activated by *r*eceptor *t*yrosine *k*inases (RTK), and catalyze the conversion of PI(4,5)P_2 to PI(3,4,5)P_3. PI(3,4,5)P_3 is a quantitatively minor PI, normally undetectable and transiently increased when PI3K is activated by various agonist-mediated stimuli. Acting as a second messenger, PI(3,4,5)P_3 recruits pleckstrin homology (PH) domain-containing proteins, such as the cellular homolog of Akt retroviral oncogene protein serine–threonine kinase (AKT), also called protein kinase B (PKB), and PI-dependent kinase 1 (PDK1) to the plasma membrane. AKT is subsequently phosphorylated at Thr308 by PDK1 and at Ser473 by a putative PDK2. Activated AKT phosphorylates multiple downstream proteins on serine and threonine residues (Vivanco and Sawyers, 2002). Through phosphorylation of these targets, AKT carries out its important role in the regulation of many aspects of cellular physiology, including glucose metabolism, cell proliferation, cell growth, and survival (Figure 2). Constitutive activation of PI3K/AKT has been implicated in various human cancers. Increased expression of the gene encoding the catalytic subunit (p110) of PI3K was found in ovarian, breast, and colon cancer. In addition, a gain of function mutation in the regulatory subunit p85 has been identified in ovarian and colon cancer (Vivanco and Sawyers, 2002).

PTEN (*p*hosphatase and *ten*sin homolog deleted on chromosome *ten*)/*MMAC* (*m*utated in *m*ultiple *a*dvanced *c*ancers)/*TEP-1* (*T*GF-β-regulated and *e*pithelial cell-enriched *p*hosphatase) was originally identified in 1997 by two different groups as a tumor suppressor gene on chromosome 10q23, using traditional positional-cloning strategies (Li *et al.*, 1997; Steck *et al.*, 1997). Mutations in the *PTEN* gene were associated with breast cancer, glioblastomas, prostate, endometrial, renal and small cell lung carcinoma, melanoma, and meningioma. Subsequent studies confirmed that the *PTEN* gene is defective in a large number of human cancers (Cantley and Neel, 1999). In addition, germ-line mutations in *PTEN* result in three rare, dominant, inherited diseases: Cowden disease, Lhermitte-Duclos disease, and Bannayan-Zonana syndrome (Liaw *et al.*, 1997; Marsh *et al.*, 1998; Nelen *et al.*, 1997). These disorders are characterized by multiple hamartomas and increased risk of developing cancers. These genetic data suggest that PTEN is important for normal cell growth and that *PTEN* dysfunction contributes to carcinogenesis.

Although analysis of the *PTEN* sequence suggested that it is a dual-specificity phosphatase, it is difficult to identify the biologically relevant targets of PTEN. In 1998, Maehama and Dixon showed that PTEN is actually a PI phosphatase, rather than a protein phosphatase (Maehama and Dixon, 1998). PTEN could dephosphorylate the D3 position of PI(3,4,5)P_3 both *in vitro* and *in vivo*, leading to decreased PI(3,4,5)P_3 levels. This finding led to a model in which PTEN is a negative regulator of the PI3K/AKT pathway. The deletion or inactivation of *PTEN* results in constitutive PI3K/AKT activation and aberrant cell growth.

Aside from PTEN, SH-2 domain-containing *i*nositol 5' *p*hosphatases SHIP1 and SHIP2 also regulate PI(3,4,5)P_3 levels by removing the D5 phosphate.

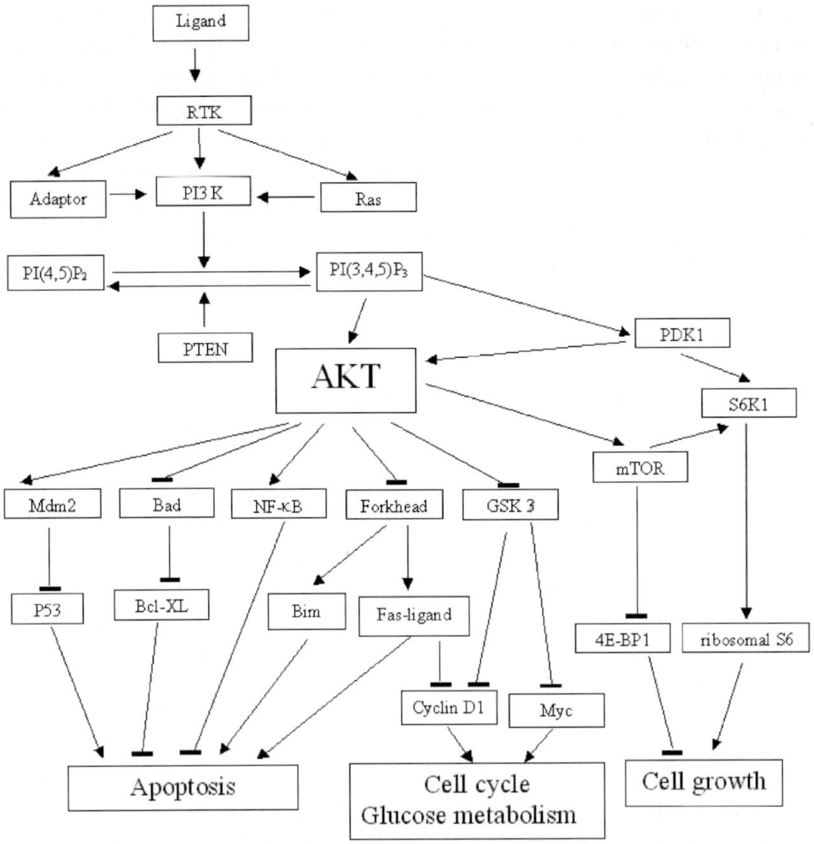

Figure 2. The PI3 Kinase (PI3K)/AKT pathway (Luo *et al.*, 2003). Growth factor *r*eceptor *t*yrosine *k*inases (RTKs) receive ligand stimulation and recruit class I PI3K through direct interaction, adaptor molecules, or Ras. At the membrane, PI3K phosphorylates PI (4,5) P2 to PI(3,4,5)P_3. PTEN removes the D3 phosphate from PI (3,4,5) P3, thus regulating the intracellular PI(3,4,5)P_3 level. PI (3,4,5) P3 recruits the cellular homolog of Akt retroviral oncogene protein serine–threonine kinase (AKT) and *p*hosphoinositide-*d*ependent *k*inase 1 (PDK1) to the membrane. AKT is subsequently phosphorylated by PDK1 and a putative PDK2. Activated AKT regulates a number of downstream targets. AKT promotes cell survival by inhibition of apoptosis, regulates cell cycle, and stimulates glucose metabolism. By activating the *m*ammalian *t*arget *o*f *r*apamycin (mTOR) pathway, AKT also stimulates protein synthesis and boosts cell growth.

SHIP1 is expressed exclusively in hematopoietic cells (Geier *et al.*, 1997; Liu *et al.*, 1998). Homozygous disruption of *SHIP1* in mice leads to a myeloproliferative syndrome characterized by a dramatic increase in the number of granulocyte–macrophage progenitor cells in the marrow and spleen (Helgason *et al.*, 1998). Based on the *SHIP1* gene knockout mouse phenotype and the hydrolysis of PI(3,4,5)P_3, SHIP1 protein is likely a negative regulator of the PI3K/AKT pathway. Consistent with this hypothesis, *SHIP1* gene expression and protein half-life were decreased in primary neoplastic cells from patients with chronic myelogenous leukemia (Sattler *et al.*, 1999). Expression of constitutively active SHIP1 protein in the leukemic Jurkat cell line decreases the PI(3,4,5)P_3 level and the activity of AKT (Freeburn *et al.*, 2002). Recently, a mutation in the human *SHIP1* gene has been found in acute myeloid leukemia (Luo *et al.*, 2003). The mutation reduced the catalytic activity of SHIP1 and led to enhanced PI3K/AKT activation following IL-3 stimulation. Overall, these data suggest that SHIP1 plays an important role in controlling the PI(3,4,5)P_3 level and downstream PI3K/AKT pathway activity.

3.2 Type 2 diabetes

3.2.1 PI3K/AKT pathway in type2 diabetes

One of the downstream targets of the PI3K/AKT pathway is glucose metabolism (Figure 2). AKT phosphorylates and regulates GSK-3, which is critical in insulin-mediated glucose metabolism (Shepherd *et al.*, 1998). Overexpression of AKT or the PI3K catalytic subunit p110 stimulates insulin-mediated glucose metabolism (Katagiri *et al.*, 1996; Ueki *et al.*, 1998), which is blocked upon loss of function of the PI3K regulatory subunit p85 (Calera *et al.*, 1998; Sharma *et al.*, 1998). Insulin stimulates glucose uptake in adipose and muscle tissue by translocating the glucose transporter (GLUT4) from intracellular sites to the cell surface. Deficiency in GLUT4 translocation and glucose uptake in response to insulin stimulation is an important cause of type 2 diabetes (Shulman, 2000). GLUT4 translocation has been shown to be regulated by the PI3K/AKT pathway (Baumann *et al.*, 2000; Kanzaki and Pessin, 2003; Saltiel and Kahn, 2001). Insulin-stimulated translocation of GLUT4 to the cell surface and glucose uptake into cells is blocked by the PI3K inhibitors wortmannin and LY294002 (Kanai *et al.*, 1993; Okada *et al.*, 1994).

As discussed above, the activity of the PI3K/AKT pathway is also regulated by PI(3,4,5)P_3 D3 phosphatase (PTEN) and D5 phosphatase (SHIP1, SHIP2). Overexpression of PTEN reduces insulin-induced PI3K/AKT pathway activity, GLUT4 translocation, and glucose uptake into cells (Nakashima *et al.*, 2000; Ono *et al.*, 2001). Microinjection of an anti-PTEN antibody increases insulin-stimulated translocation of GLUT4 to the cell surface and glucose uptake into cells (Nakashima *et al.*, 2000). These results suggest that PTEN reduces insulin sensitivity, which is increased upon inhibition of PTEN.

SHIP1 is expressed only in hematopoietic cells, while SHIP2 is expressed ubiquitously. SHIP2 dephosphorylates PI(3,4,5)P_3 to PI(3,4)P_2 and attenuates insulin-stimulated PI3K/AKT activity. Overexpression of SHIP2 protein inhibits insulin-stimulated PI3K/AKT activity and leads to GSK-3 inactivation followed by reduced glycogen synthetase activity (Blero *et al.*, 2001). Genetic deletion of *SHIP2* in mice resulted in increased PI3K/AKT activity and insulin sensitivity *in vivo* (Clement *et al.*, 2001). Interestingly, mutations in SHIP2 that result in increased SHIP2 activity have been associated with type 2 diabetes in both mice and humans (Marion *et al.*, 2002). Altogether, these results suggest that SHIP2 also reduces insulin signaling, and inhibitors of SHIP2 may be candidates for the treatment of type 2 diabetes.

3.2.2 PI5P in insulin signaling

As mentioned above, alteration of GLUT4 translocation contributes to the development of insulin resistance and type 2 diabetes. Aside from the PI3K/AKT dependent pathway, GLUT4 translocation is also mediated by a PI3K/AKT independent pathway (Bachhawat and Mande, 1999). Recently, it has been suggested that PI5P is a novel intermediate for insulin signaling and GLUT4 translocation (Sbrissa *et al.*, 2004). PI5P is the most recently identified member of the PI family and its physiological function is not clear. After insulin stimulation, a transient increase in PI5P was observed in both CHO-T and 3T3-L1 adipocytes. The PI3K inhibitor wortmannin could not block this PI5P increase. In 3T3-L1 adipocytes, microinjected PI5P, but not other PIs, mimicked the effect of insulin in translocating GLUT4 to the cell surface. Taken together, these results suggest that PI5P mediates insulin signaling and GLUT4 translocation. However, there is currently no direct evidence for altered levels of PI5P in type 2 diabetes patients.

3.3 Lowe syndrome

Lowe syndrome is a rare X-linked disorder characterized by severe congenital cataracts, mental retardation, and renal tubular dysfunction. It is also called OCRL syndrome (*o*culo-*c*erebro-*r*enal syndrome of *L*owe) because of the three major organ systems affected. The gene responsible for Lowe syndrome, *OCRL1*, has been cloned and different mutations have been identified (Addis *et al.*, 2004; Attree *et al.*, 1992; Lin *et al.*, 1997; Monnier *et al.*, 2000). *OCRL1* encodes a 105 kDa PI(4,5)P_2 5-phosphatase (Zhang *et al.*, 1995). The OCRL1 protein is expressed in almost all tissues examined, except hematopoietic cells (Janne *et al.*, 1998; Olivos-Glander *et al.*, 1995). It was originally reported that OCRL1 protein was localized in the Golgi apparatus in fibroblasts (Olivos-Glander *et al.*, 1995; Suchy *et al.*, 1995). Cells deficient in the *OCRL1* gene accumulate PI(4,5)P_2, consistent with loss of function of the phosphatase activity (Zhang *et al.*, 1998). In cells from Lowe syndrome patients, high levels of extracellular lysosomal enzymes were found, which may contribute to the pathogenesis of the disease (Olivos-Glander *et al.*, 1995; Suchy *et al.*, 1995; Ungewickell and Majerus, 1999; Zhang *et al.*, 1998). A subsequent study showed that OCRL1 was located in the trans-Golgi network in fibroblasts and two kidney epithelial cell lines (Dressman *et al.*, 2000; Suchy *et al.*, 1995). Recent evidence also indicates that OCRL1 was also localized in endosomes (Ungewickell *et al.*, 2004). The localization of OCRL1 in organelles involved in trafficking suggests that defective OCRL1 may lead to abnormal vesicle trafficking.

Abnormalities in the actin cytoskeleton, including decreased actin stress fibers and altered response to depolymerizing agents were also observed in Lowe patients (Suchy and Nussbaum, 2002). Two actin-binding proteins, gelsolin and alpha-actinin, which are regulated by both PIP_2 and Ca^{2+} were found to be abnormally distributed. Actin polymerization plays a key role in cell – cell contacts, which may be defective in Lowe syndrome.

3.4 Diseases associated with the myotubularin family

The myotubularin (MTM) family was originally identified as a member of the *p*hospho*t*yrosine *p*hosphatases (PTP)/*d*ual-*s*pecificity *p*hosphatases (DSPs), which are able to dephosphorylate phosphotyrosine and phosphoserine/ threonine residues (Laporte *et al.*, 1996). MTMs contain the conserved sequence H*C*SDGWD*R*TXE, which fits with the CX_5R signature of the PTP/DSPs. Recent studies demonstrate that the preferred substrates of MTMs are PI 3-phosphatases (Blondeau *et al.*, 2000; Taylor *et al.*, 2000). The MTM family in humans consists of 14 members, including MTM1 and MTM-related (MTMR) proteins 1–13 (Laporte *et al.*, 1998; Maehama *et al.*, 2001; Wishart *et al.*, 2001). Among these, *MTM1* was found to be mutated in X-linked myotubular myopathy (XLMTM). *MTMR*2 and *MTMR*13 are responsible for Charcot-Marie-Tooth disease type 4B1 (CMT4B1) and Charcot-Marie-Tooth disease type 4B2 (CMT4B2), respectively.

3.4.1 Myotubular myopathy (XLMTM)

Myotubular myopathy (XLMTM) is an X-linked severe recessive disorder characterized by muscle weakness and hypotonia that affects 1 out of 50,000 newborn males. The histological analysis of patients' muscles shows small rounded fibers with central nuclei. This is a characteristic of fetal myotubes, whereas mature muscle fibers have peripherally located nuclei (Sarnat, 1990). *MTM*1, the gene responsible for XLMTM and located on chromosome Xq28, was cloned in 1996 (Laporte *et al.*, 1996). Mutations in *MTM*1 have been reported in more than 300 patients (de Gouyon *et al.*, 1997; Herman *et al.*, 2002; Laporte *et al.*, 2000; Nishino *et al.*, 1998; Tanner *et al.*, 1999). *MTM*1 codes for myotubularin and is expressed ubiquitously. Although MTM1 shows a dual-specificity protein phosphatase activity *in vitro*, the preferred substrate of MTM1 *in vivo* is PI(3)P (Blondeau *et al.*, 2000; Taylor *et al.*, 2000). Thus, MTM1 may regulate PI(3)P levels by directly degrading PI(3)P to PI. A recent study found that MTM1 could dephosphorylate both $PI(3,5)P_2$ and PI(3)P. The GRAM (*G*lucosyltransferase, *R*ab-like GTPase *A*ctivator and *M*yotubularins) domain of MTM1 binds to $PI(3,5)P_2$ with high affinity *in vivo*. This GRAM-$PI(3,5)P_2$ interaction is essential for translocation of MTM1 to the late endosomal compartment in response to growth factor stimulation (Tsujita *et al.*, 2004). These findings suggest that MTM1 may also regulate $PI(3,5)P_2$ levels, and may function in late endosomal trafficking *in vivo*. *MTM1* knockout mice are viable and show normal muscle differentiation at birth, but a progressive myopathy appears a few weeks after birth (Buj-Bello *et al.*, 2002). This indicates that MTM1 is essential for skeletal muscle maintenance but not for myogenesis in mice. Studies in knock-out mice will help to elucidate the function and the regulatory mechanisms of MTM1 and increase understanding of the pathology of XLMTM.

3.4.2 Charcot-Marie-Tooth (CMT) disease

Charcot-Marie-Tooth (CMT) disease is a group of disorders characterized by chronic motor and sensory polyneuropathy. CMT4B1 is an autosomal recessive neuropathy with focally folded myelin sheaths and demyelination. Mutations in *MTMR*2 on chromosome 11q22 cause CMT4B1 (Bolino *et al.*, 2000). In CMT4B1 patients, 3' phosphatase activity is dramatically reduced. PI(3)P is a substrate of both MTM1 and MTMR2. But the preferred substrate of MTMR2 is $PI(3,5)P_2$ (Berger *et al.*, 2002). Misregulation of $PI(3,5)P_2$ in neurons may affect neural membrane recycling and membrane trafficking. The different substrate preferences of MTMR2 and MTM1 may account for the different pathologies of the diseases associated with mutations of these two genes.

Mutations in the *MTMR*13 gene cause CMT4B2 (Azzedine *et al.*, 2003). *MTMR*13 is located on chromosome 11p15 and encodes an inactive phosphatase (Senderek *et al.*, 2003). How this inactive phosphatase causes disease remains uncertain. However, it is possible that MTMR13 acts as an adaptor or directly regulates MTMR2 activity due to the similar phenotype of CMT4B2 and CMT4B1(Pendaries *et al.*, 2003).

4. SUMMARY

In the past two decades, considerable progress has been made toward understanding inositol phosphates and PI metabolism. However, there is still much to learn. The present challenge is to understand how inositol phosphates and PIs are compartmentalized, identify new targets of inositol phosphates and PIs, and elucidate the mechanisms underlying spatial and temporal regulation of the enzymes that metabolize inositol phosphates and PIs. Answers to these questions will help clarify the mechanisms of the diseases associated with these molecules and identify new possibilities for drug design.

ACKNOWLEDGMENTS

The Greenberg laboratory is supported by grants HL62263 and MH56220 from the National Institutes of Health.

REFERENCES

Addis, M., Loi, M., Lepiani, C., Cau, M., and Melis, M.A., 2004, OCRL mutation analysis in Italian patients with Lowe syndrome. *Hum. Mutat.* **23:** 524–525.

Aisen, P.S., 1997, Inflammation and Alzheimer's disease: Mechanisms and therapeutic strategies. *Gerontology* **43:** 143–149.

Atack, J.R., 2000, Lithium, phosphatidylinositol signaling, and bipolar disorder. In: Manji, H.K., Bowden, C.L., and Belmaker, R.H. (eds.), Bipolar Medications: Mechanism of Action. American Psychiatric Press, Inc., Washington, DC.

Attree, O., Olivos, I.M., Okabe, I., Bailey, L.C., Nelson, D.L., Lewis, R.A., McInnes, R.R., and Nussbaum, R.L., 1992, The Lowe's oculocerebrorenal syndrome gene encodes a protein highly homologous to inositol polyphosphate-5-phosphatase. *Nature* **358:** 239–242.

Azzedine, H., Bolino, A., Taieb, T., Birouk, N., Di Duca, M., Bouhouche, A., Benamou, S., Mrabet, A., Hammadouche, T., Chkili, T., Gouider, R., Ravazzolo, R., Brice, A., Laporte, J., and LeGuern, E., 2003, Mutations in MTMR13, a new pseudophosphatase homologue of MTMR2 and Sbf1, in two families with an autosomal recessive demyelinating form of Charcot-Marie-Tooth disease associated with early-onset glaucoma. *Am. J. Hum. Genet.* **72:** 1141–1153.

Bachhawat, N., and Mande, S.C., 1999, Identification of the INO1 gene of Mycobacterium tuberculosis H37Rv reveals a novel class of inositol-1-phosphate synthase enzyme. *J. Mol. Biol.* **291:** 531–536.

Baumann, C.A., Ribon, V., Kanzaki, M., Thurmond, D.C., Mora, S., Shigematsu, S., Bickel, P.E., Pessin, J.E., and Saltiel, A.R., 2000, CAP defines a second signalling pathway required for insulin-stimulated glucose transport. *Nature* **407:** 202–207.

Belmaker, R.H., 2004, Bipolar disorder. *N. Engl. J. Med.* **351:** 476–486.

Berger, P., Bonneick, S., Willi, S., Wymann, M., and Suter, U., 2002, Loss of phosphatase activity in myotubularin-related protein 2 is associated with Charcot-Marie-Tooth disease type 4B1. *Hum. Mol. Genet.* **11:** 1569–1579.

Berridge, M.J., 1987, Inositol trisphosphate and diacylglycerol: Two interacting second messengers. *Annu. Rev. Biochem.* **56:** 159–193.

Berridge, M.J., 1993, Inositol trisphosphate and calcium signalling. *Nature* **361:** 315–325.

Berridge, M.J., Downes, C.P., and Hanley, M.R., 1982, Lithium amplifies agonist-dependent phosphatidylinositol responses in brain and salivary glands. *Biochem. J.* **206:** 587–595.

Berridge, M.J., and Irvine, R.F., 1989, Inositol phosphates and cell signalling. *Nature* **341:** 197–205.

Blero, D., De Smedt, F., Pesesse, X., Paternotte, N., Moreau, C., Payrastre, B., and Erneux, C., 2001, The SH2 domain containing inositol 5-phosphatase SHIP2 controls phosphatidylinositol 3,4,5-trisphosphate levels in CHO-IR cells stimulated by insulin. *Biochem. Biophys. Res. Commun.* **282:** 839–843.

Blondeau, F., Laporte, J., Bodin, S., Superti-Furga, G., Payrastre, B., and Mandel, J.L., 2000, Myotubularin, a phosphatase deficient in myotubular myopathy, acts on phosphatidylinositol 3-kinase and phosphatidylinositol 3-phosphate pathway. *Hum. Mol. Genet.* **9:** 2223–2229.

Bloom, F.E., 2001, Neurotransmission and the central nervous system. In: Hradman, J.G., Limbird, L.E., and Gilman, A.G. (eds.), Goodman and Gilman's The Pharmacological Basis of Therapeutics, 10th ed., Section III, Chapter 12. The McGraw-Hill companies.

Bolino, A., Muglia, M., Conforti, F.L., LeGuern, E., Salih, M.A., Georgiou, D.M., Christodoulou, K., Hausmanowa-Petrusewicz, I., Mandich, P., Schenone, A., Gambardella, A., Bono, F., Quattrone, A., Devoto, M., and Monaco, A.P., 2000, Charcot-Marie-Tooth type 4B is caused by mutations in the gene encoding myotubularin-related protein-2. *Nat. Genet.* **25:** 17–19.

Brookmeyer, R., Gray, S., and Kawas, C., 1998, Projections of Alzheimer's disease in the United States and the public health impact of delaying disease onset. *Am. J. Public Health* **88:** 1337–1342.

Buj-Bello, A., Laugel, V., Messaddeq, N., Zahreddine, H., Laporte, J., Pellissier, J.F., and Mandel, J.L., 2002, The lipid phosphatase myotubularin is essential for skeletal muscle maintenance but not for myogenesis in mice. *Proc. Natl. Acad. Sci. U. S. A.* **99:** 15060–15065.

Calera, M.R., Martinez, C., Liu, H., Jack, A.K., Birnbaum, M.J., and Pilch, P.F., 1998, Insulin increases the association of Akt-2 with Glut4-containing vesicles. *J. Biol. Chem.* **273:** 7201–7204.

Cantley, L.C., 2002, The phosphoinositide 3-kinase pathway. *Science* **296:** 1655–1657.

Cantley, L.C., and Neel, B.G., 1999, New insights into tumor suppression: PTEN suppresses tumor formation by restraining the phosphoinositide 3-kinase/AKT pathway. *Proc. Natl. Acad. Sci. U. S. A.* **96:** 4240–4245.

Carman, G.M., and Henry, S.A., 1999, Phospholipid biosynthesis in the yeast Saccharomyces cerevisiae and interrelationship with other metabolic processes. *Prog. Lipid Res.* **38:** 361–399.

Chen, G., Huang, L.D., Jiang, Y.M., and Manji, H.K., 1999a, The mood-stabilizing agent valproate inhibits the activity of glycogen synthase kinase-3. *J. Neurochem.* **72:** 1327–1330.

Chen, G., Zeng, W.Z., Yuan, P.X., Huang, L.D., Jiang, Y.M., Zhao, Z.H., and Manji, H.K., 1999b, The mood-stabilizing agents lithium and valproate robustly increase the levels of the neuroprotective protein bcl-2 in the CNS. *J. Neurochem.* **72:** 879–882.

Chen, L., Zhou, C., Yang, H., and Roberts, M.F., 2000, Inositol-1-phosphate synthase from Archaeoglobus fulgidus is a class II aldolase. *Biochemistry* **39:** 12415–12423.

Clement, S., Krause, U., Desmedt, F., Tanti, J.F., Behrends, J., Pesesse, X., Sasaki, T., Penninger, J., Doherty, M., Malaisse, W., Dumont, J.E., Le Marchand-Brustel, Y., Erneux, C., Hue, L., and Schurmans, S., 2001, The lipid phosphatase SHIP2 controls insulin sensitivity. *Nature* **409:** 92–97.

Connolly, T.M., Bansal, V.S., Bross, T.E., Irvine, R.F., and Majerus, P.W., 1987, The metabolism of tris- and tetraphosphates of inositol by 5-phosphomonoesterase and 3-kinase enzymes. *J. Biol. Chem.* **262:** 2146–2149.

de Gouyon, B.M., Zhao, W., Laporte, J., Mandel, J.L., Metzenberg, A., and Herman, G.E., 1997, Characterization of mutations in the myotubularin gene in twenty six patients with X-linked myotubular myopathy. *Hum. Mol. Genet.* **6:** 1499–1504.

Deliliers, G.L., Servida, F., Fracchiolla, N.S., Ricci, C., Borsotti, C., Colombo, G., and Soligo, D., 2002, Effect of inositol hexaphosphate (IP(6)) on human normal and leukaemic haematopoietic cells. *Br. J. Haematol.* **117:** 577–587.

Dressman, M.A., Olivos-Glander, I.M., Nussbaum, R.L., and Suchy, S.F., 2000, Ocrl1, a PtdIns(4,5)P(2) 5-phosphatase, is localized to the trans-Golgi network of fibroblasts and epithelial cells. *J. Histochem. Cytochem.* **48:** 179–190.

Druzijanic, N., Juricic, J., Perko, Z., and Kraljevic, D., 2002, IP-6 & inosito: Adjuvant chemotherapy of colon cancer. A pilot clinical trial. *Rev. Oncologia* **4:** 171.

Efanov, A.M., Zaitsev, S.V., and Berggren, P.O., 1997, Inositol hexakisphosphate stimulates non-Ca2+-mediated and primes Ca2+-mediated exocytosis of insulin by activation of protein kinase C. *Proc. Natl. Acad. Sci. U. S. A.* **94:** 4435–4439.

Fox, C.H., and Eberl, M., 2002, Phytic acid (IP6), novel broad spectrum anti-neoplastic agent: A systematic review. *Complement. Ther. Med.* **10:** 229–234.

Freeburn, R.W., Wright, K.L., Burgess, S.J., Astoul, E., Cantrell, D.A., and Ward, S.G., 2002, Evidence that SHIP-1 contributes to phosphatidylinositol 3,4,5-trisphosphate metabolism in T lymphocytes and can regulate novel phosphoinositide 3-kinase effectors. *J. Immunol.* **169:** 5441–5450.

Fujii, S., Matsumoto, M., Igarashi, K., Kato, H., and Mikoshiba, K., 2000, Synaptic plasticity in hippocampal CA1 neurons of mice lacking type 1 inositol-1,4,5-trisphosphate receptors. *Learn. Mem.* **7:** 312–320.

Garlind, A., Cowburn, R.F., Forsell, C., Ravid, R., Winblad, B., and Fowler, C.J., 1995, Diminished [3H]inositol(1,4,5)P3 but not [3H]inositol(1,3,4,5)P4 binding in Alzheimer's disease brain. *Brain Res.* **681:** 160–166.

Geier, S.J., Algate, P.A., Carlberg, K., Flowers, D., Friedman, C., Trask, B., and Rohrschneider, L.R., 1997, The human SHIP gene is differentially expressed in cell lineages of the bone marrow and blood. *Blood* **89:** 1876–1885.

Gould, T.D., Einat, H., Bhat, R., and Manji, H.K., 2004c, AR-A014418, a selective GSK-3 inhibitor, produces antidepressant-like effects in the forced swim test. *Int. J. Neuropsychopharmacol.* 1–4.

Gould, T.D., Quiroz, J.A., Singh, J., Zarate, C.A., and Manji, H.K., 2004a, Emerging experimental therapeutics for bipolar disorder: Insights from the molecular and cellular actions of current mood stabilizers. *Mol. Psychiatry* **9:** 734–755.

Gould, T.D., Zarate, C.A., and Manji, H.K., 2004b, Glycogen synthase kinase-3: A target for novel bipolar disorder treatments. *J. Clin. Psychiatry.* **65:** 10–21.

Grases, F., Perello, J., Prieto, R.M., Simonet, B.M., and Torres, J.J., 2004, Dietary *myo*-inositol hexaphosphate prevents dystrophic calcifications in soft tissues: A pilot study in Wistar rats. *Life Sci.* **75:** 11–19.

Hallcher, L.M., and Sherman, W.R., 1980, The effects of lithium ion and other agents on the activity of *myo*-inositol-1-phosphatase from bovine brain. *J. Biol. Chem.* **255:** 10896–10901.

Haug, L.S., Ostvold, A.C., Cowburn, R.F., Garlind, A., Winblad, B., Bogdanovich, N., and Walaas, S.I., 1996, Decreased inositol (1,4,5)-trisphosphate receptor levels in Alzheimer's disease cerebral cortex: Selectivity of changes and possible correlation to pathological severity. *Neurodegeneration* **5:** 169–176.

Hawkins, P.T., Poyner, D.R., Jackson, T.R., Letcher, A.J., Lander, D.A., and Irvine, R.F., 1993, Inhibition of iron-catalysed hydroxyl radical formation by inositol polyphosphates: A possible physiological function for *myo*-inositol hexakisphosphate. *Biochem. J.* **294**(Pt 3): 929–934.

Helgason, C.D., Damen, J.E., Rosten, P., Grewal, R., Sorensen, P., Chappel, S.M., Borowski, A., Jirik, F., Krystal, G., and Humphries, R.K., 1998, Targeted disruption of SHIP leads to hemopoietic perturbations, lung pathology, and a shortened life span. *Genes Dev.* **12:** 1610–1620.

Herman, G.E., Kopacz, K., Zhao, W., Mills, P.L., Metzenberg, A., and Das, S., 2002, Characterization of mutations in fifty North American patients with X-linked myotubular myopathy. *Hum. Mutat.* **19:** 114–121.

Heslop, J.P., Irvine, R.F., Tashjian, A.H., Jr., and Berridge, M.J., 1985, Inositol tetrakis- and pentakisphosphates in GH4 cells. *J. Exp. Biol.* **119:** 395–401.

Inhorn, R.C., and Majerus, P.W., 1988, Properties of inositol polyphosphate 1-phosphatase. *J. Biol. Chem.* **263:** 14559–14565.

Irvine, R.F., Letcher, A.J., Lander, D.J., and Downes, C.P., 1984, Inositol trisphosphates in carbachol-stimulated rat parotid glands. *Biochem. J.* **223:** 237–243.

Irvine, R.F., and Schell, M.J., 2001, Back in the water: The return of the inositol phosphates. *Nat. Rev. Mol. Cell Biol.* **2:** 327–338.

Ismailov, II, Fuller, C.M., Berdiev, B.K., Shlyonsky, V.G., Benos, D.J., and Barrett, K.E., 1996, A biologic function for an "orphan" messenger: D-*myo*-inositol 3,4,5,6-tetrakisphosphate selectively blocks epithelial calcium-activated chloride channels. *Proc. Natl. Acad. Sci. U. S. A.* **93:** 10505–10509.

Jacobsen, A.N., Du, X.J., Lambert, K.A., Dart, A.M., and Woodcock, E.A., 1996, Arrhythmogenic action of thrombin during myocardial reperfusion via release of inositol 1,4,5-triphosphate. *Circulation* **93:** 23–26.

Janne, P.A., Suchy, S.F., Bernard, D., MacDonald, M., Crawley, J., Grinberg, A., Wynshaw-Boris, A., Westphal, H., and Nussbaum, R.L., 1998, Functional overlap between murine Inpp5b and Ocrl1 may explain why deficiency of the murine ortholog for OCRL1 does not cause Lowe syndrome in mice. *J. Clin. Invest.* **101:** 2042–2053.

Jope, R.S., 2003, Lithium and GSK-3: One inhibitor, two inhibitory actions, multiple outcomes. *Trends Pharmacol. Sci.* **24:** 441–443.

Ju, S., Greenberg, M.L., in press, 1D-*myo*-inositol 3-phosphate synthase: Conversion, regulation, and putative target of mood stabilizers. *Clin. Neurosci. Res.*

Ju, S., Shaltiel, G., Shamir, A., Agam, G., and Greenberg, M.L., 2004, Human 1-D-*myo*-inositol-3-phosphate synthase is functional in yeast. *J. Biol. Chem.* **279:** 21759–21765.

Kaidanovich-Beilin, O., Milman, A., Weizman, A., Pick, C.G., and Eldar-Finkelman, H., 2004, Rapid antidepressive-like activity of specific glycogen synthase kinase-3 inhibitor and its effect on beta-catenin in mouse hippocampus. *Biol. Psychiatry* **55:** 781–784.

Kanai, F., Ito, K., Todaka, M., Hayashi, H., Kamohara, S., Ishii, K., Okada, T., Hazeki, O., Ui, M., and Ebina, Y., 1993, Insulin-stimulated GLUT4 translocation is relevant to the phosphorylation of IRS-1 and the activity of PI3-kinase. *Biochem. Biophys. Res. Commun.* **195:** 762–768.

Kanzaki, M., and Pessin, J.E., 2003, Insulin signaling: GLUT4 vesicles exit via the exocyst. *Curr. Biol.* **13:** R574–R576.

Katagiri, H., Asano, T., Ishihara, H., Inukai, K., Shibasaki, Y., Kikuchi, M., Yazaki, Y., and Oka, Y., 1996, Overexpression of catalytic subunit p110alpha of phosphatidylinositol 3-kinase increases glucose transport activity with translocation of glucose transporters in 3T3-L1 adipocytes. *J. Biol. Chem.* **271:** 16987–16990.

Kijima, Y., Saito, A., Jetton, T.L., Magnuson, M.A., and Fleischer, S., 1993, Different intracellular localization of inositol 1,4,5-trisphosphate and ryanodine receptors in cardiomyocytes. *J. Biol. Chem.* **268:** 3499–3506.

Kitamura, N., Hashimoto, T., Nishino, N., and Tanaka, C., 1989, Inositol 1,4,5-trisphosphate binding sites in the brain: Regional distribution, characterization, and alterations in brains of patients with Parkinson's disease. *J. Mol. Neurosci.* **1:** 181–187.

LaFerla, F.M., 2002, Calcium dyshomeostasis and intracellular signalling in Alzheimer's disease. *Nat. Rev. Neurosci.* **3:** 862–872.

Laporte, J., Biancalana, V., Tanner, S.M., Kress, W., Schneider, V., Wallgren-Pettersson, C., Herger, F., Buj-Bello, A., Blondeau, F., Liechti-Gallati, S., and Mandel, J.L., 2000, MTM1 mutations in X-linked myotubular myopathy. *Hum. Mutat.* **15:** 393–409.

Laporte, J., Blondeau, F., Buj-Bello, A., Tentler, D., Kretz, C., Dahl, N., and Mandel, J.L., 1998, Characterization of the myotubularin dual specificity phosphatase gene family from yeast to human. *Hum. Mol. Genet.* **7:** 1703–1712.

Laporte, J., Hu, L.J., Kretz, C., Mandel, J.L., Kioschis, P., Coy, J.F., Klauck, S.M., Poustka, A., and Dahl, N., 1996, A gene mutated in X-linked myotubular myopathy defines a new putative tyrosine phosphatase family conserved in yeast. *Nat. Genet.* **13:** 175–182.

Larsson, O., Barker, C.J., Sj-oholm, A., Carlqvist, H., Michell, R.H., Bertorello, A., Nilsson, T., Honkanen, R.E., Mayr, G.W., Zwiller, J., and Berggren, P.O., 1997, Inhibition of phosphatases and increased Ca2+ channel activity by inositol hexakisphosphate. *Science* **278:** 471–474.

Leung, A.Y., Wong, P.Y., Gabriel, S.E., Yankaskas, J.R., and Boucher, R.C., 1995, cAMP- but not Ca(2+)-regulated Cl- conductance in the oviduct is defective in mouse model of cystic fibrosis. *Am. J. Physiol.* **268:** C708–C712.

Li, J., Yen, C., Liaw, D., Podsypanina, K., Bose, S., Wang, S.I., Puc, J., Miliaresis, C., Rodgers, L., McCombie, R., Bigner, S.H., Giovanella, B.C., Ittmann, M., Tycko, B., Hibshoosh, H., Wigler, M.H., and Parsons, R., 1997, PTEN, a putative protein tyrosine phosphatase gene mutated in human brain, breast, and prostate cancer. *Science* **275:** 1943–1947.

Liang, Y., Hua, Z., Liang, X., Xu, Q., and Lu, G., 2001, The crystal structure of bar-headed goose hemoglobin in deoxy form: The allosteric mechanism of a hemoglobin species with high oxygen affinity. *J. Mol. Biol.* **313:** 123–137.

Liaw, D., Marsh, D.J., Li, J., Dahia, P.L., Wang, S.I., Zheng, Z., Bose, S., Call, K.M., Tsou, H.C., Peacocke, M., Eng, C., and Parsons, R., 1997, Germline mutations of the PTEN gene in Cowden disease, an inherited breast and thyroid cancer syndrome. *Nat. Genet.* **16:** 64–67.

Lin, T., Orrison, B.M., Leahey, A.M., Suchy, S.F., Bernard, D.J., Lewis, R.A., and Nussbaum, R.L., 1997, Spectrum of mutations in the OCRL1 gene in the Lowe oculocerebrorenal syndrome. *Am. J. Hum. Genet.* **60:** 1384–1388.

Liu, Q., Shalaby, F., Jones, J., Bouchard, D., and Dumont, D.J., 1998, The SH2-containing inositol polyphosphate 5-phosphatase, ship, is expressed during hematopoiesis and spermatogenesis. *Blood* **91:** 2753–2759.

Loewus, M.W., Loewus, F.A., Brillinger, G.U., Otsuka, H., and Floss, H.G., 1980, Stereochemistry of the *myo*-inositol-1-phosphate synthase reaction. *J. Biol. Chem.* **255:** 11710–11712.

Lorenzon, N.M., and Beam, K.G., 2000, Calcium channelopathies. *Kidney Int.* **57:** 794–802.

Lu, P.J., Shieh, W.R., and Chen, C.S., 1996, Antagonistic effect of inositol pentakisphosphate on inositol triphosphate receptors. *Biochem. Biophys. Res. Commun.* **220:** 637–642.

Luckhoff, A., and Clapham, D.E., 1992, Inositol 1,3,4,5-tetrakisphosphate activates an endothelial Ca(2+)-permeable channel. *Nature* **355:** 356–358.

Luo, J., Manning, B.D., and Cantley, L.C., 2003, Targeting the PI3K-Akt pathway in human cancer: Rationale and promise. *Cancer Cell* **4:** 257–262.

Luo, J.M., Yoshida, H., Komura, S., Ohishi, N., Pan, L., Shigeno, K., Hanamura, I., Miura, K., Iida, S., Ueda, R., Naoe, T., Akao, Y., Ohno, R., and Ohnishi, K., 2003, Possible dominant-negative mutation of the SHIP gene in acute myeloid leukemia. *Leukemia* **17:** 1–8.

Maehama, T., and Dixon, J.E., 1998, The tumor suppressor, PTEN/MMAC1, dephosphorylates the lipid second messenger, phosphatidylinositol 3,4,5-trisphosphate. *J. Biol. Chem.* **273:** 13375–13378.

Maehama, T., Taylor, G.S., and Dixon, J.E., 2001, PTEN and myotubularin: Novel phosphoinositide phosphatases. *Annu. Rev. Biochem.* **70:** 247–279.

Majumder, A.L., Chatterjee, A., Ghosh Dastidar, K., and Majee, M., 2003, Diversification and evolution of L-*myo*-inositol 1-phosphate synthase. *FEBS Lett.* **553:** 3–10.

Manji, H.K., Moore, G.J., and Chen, G., 1999, Lithium at 50: Have the neuroprotective effects of this unique cation been overlooked? *Biol. Psychiatry* **46:** 929–940.

Marion, E., Kaisaki, P.J., Pouillon, V., Gueydan, C., Levy, J.C., Bodson, A., Krzentowski, G., Daubresse, J.C., Mockel, J., Behrends, J., Servais, G., Szpirer, C., Kruys, V., Gauguier, D., and Schurmans, S., 2002, The gene INPPL1, encoding the lipid phosphatase SHIP2, is a candidate for type 2 diabetes in rat and man. *Diabetes* **51:** 2012–2017.

Marsh, D.J., Coulon, V., Lunetta, K.L., Rocca-Serra, P., Dahia, P.L., Zheng, Z., Liaw, D., Caron, S., Duboue, B., Lin, A.Y., Richardson, A.L., Bonnetblanc, J.M., Bressieux, J.M., Cabarrot-Moreau, A., Chompret, A., Demange, L., Eeles, R.A., Yahanda, A.M., Fearon, E.R., Fricker, J.P., Gorlin, R.J., Hodgson, S.V., Huson, S., Lacombe, D., and Eng, C., 1998, Mutation spectrum and genotype-phenotype analyses in Cowden disease and Bannayan-Zonana syndrome, two hamartoma syndromes with germline PTEN mutation. *Hum. Mol. Genet.* **7:** 507–515.

Martelli, A.M., Manzoli, L., and Cocco, L., 2004, Nuclear inositides: Facts and perspectives. *Pharmacol. Ther.* **101:** 47–64.

Martin, T.F., 1998, Phosphoinositide lipids as signaling molecules: Common themes for signal transduction, cytoskeletal regulation, and membrane trafficking. *Annu. Rev. Cell Dev. Biol.* **14:** 231–264.

Matsumoto, M., Nakagawa, T., Inoue, T., Nagata, E., Tanaka, K., Takano, H., Minowa, O., Kuno, J., Sakakibara, S., Yamada, M., Yoneshima, H., Miyawaki, A., Fukuuchi, Y., Furuichi, T., Okano, H., Mikoshiba, K., and Noda, T., 1996, Ataxia and epileptic seizures in mice lacking type 1 inositol 1,4,5-trisphosphate receptor. *Nature* **379:** 168–171.

Mattson, M.P., LaFerla, F.M., Chan, S.L., Leissring, M.A., Shepel, P.N., and Geiger, J.D., 2000a, Calcium signaling in the ER: Its role in neuronal plasticity and neurodegenerative disorders. *Trends Neurosci.* **23:** 222–229.

Mattson, M.P., Zhu, H., Yu, J., and Kindy, M.S., 2000b, Presenilin-1 mutation increases neuronal vulnerability to focal ischemia in vivo and to hypoxia and glucose deprivation in cell culture: Involvement of perturbed calcium homeostasis. *J. Neurosci.* **20:** 1358–1364.

McGeer, P.L., and McGeer, E.G., 1995, The inflammatory response system of brain: Implications for therapy of Alzheimer and other neurodegenerative diseases. *Brain Res. Brain Res. Rev.* **21:** 195–218.

Mellstrom, B., and Naranjo, J.R., 2001, Mechanisms of Ca(2+)-dependent transcription. *Curr. Opin. Neurobiol.* **11:** 312–319.

Menniti, F.S., Miller, R.N., Putney, J.W., Jr., and Shears, S.B., 1993, Turnover of inositol polyphosphate pyrophosphates in pancreatoma cells. *J. Biol. Chem.* **268:** 3850–3856.

Missiaen, L., Robberecht, W., van den Bosch, L., Callewaert, G., Parys, J.B., Wuytack, F., Raeymaekers, L., Nilius, B., Eggermont, J., and De Smedt, H., 2000, Abnormal intracellular ca(2+)homeostasis and disease. *Cell Calcium* **28:** 1–21.

Modiano, G., Bombieri, C., Ciminelli, B.M., Belpinati, F., Giorgi, S., Georges, M.D., Scotet, V., Pompei, F., Ciccacci, C., Guittard, C., Audrezet, M.P., Begnini, A., Toepfer, M., Macek, M., Ferec, C., Claustres, M., and Pignatti, P.F., 2004, A large-scale study of the random variability of a coding sequence: A study on the CFTR gene. *Eur. J. Hum. Genet.*

Monnier, N., Satre, V., Lerouge, E., Berthoin, F., and Lunardi, J., 2000, OCRL1 mutation analysis in French Lowe syndrome patients: Implications for molecular diagnosis strategy and genetic counseling. *Hum. Mutat.* **16:** 157–165.

Moore, G.J., Bebchuk, J.M., Parrish, J.K., Faulk, M.W., Arfken, C.L., Strahl-Bevacqua, J., and Manji, H.K., 1999, Temporal dissociation between lithium-induced changes in frontal lobe *myo*-inositol and clinical response in manic-depressive illness. *Am. J. Psychiatry* **156:** 1902–1908.

Morris, A.P., Gallacher, D.V., Irvine, R.F., and Petersen, O.H., 1987, Synergism of inositol trisphosphate and tetrakisphosphate in activating Ca2+-dependent K+ channels. *Nature* **330:** 653–655.

Mouton, R., Huisamen, B., and Lochner, A., 1991, Increased myocardial inositol trisphosphate levels during alpha 1-adrenergic stimulation and reperfusion of ischaemic rat heart. *J. Mol. Cell. Cardiol.* **23:** 841–850.

Murray, C.J.L., and Lopez, A.D. (eds.), 1996, The Global Burden of Disease. Harvard University Press, Cambridge, MA.

Naccarato, W.F., Ray, R.E., and Wells, W.W., 1974, Biosynthesis of *myo*-inositol in rat mammary gland. Isolation and properties of the enzymes. *Arch. Biochem. Biophys.* **164:** 194–201.

Nakashima, N., Sharma, P.M., Imamura, T., Bookstein, R., and Olefsky, J.M., 2000, The tumor suppressor PTEN negatively regulates insulin signaling in 3T3-L1 adipocytes. *J. Biol. Chem.* **275:** 12889–12895.

Narrow, W.E., Rae, D.S., Robins, L.N., and Regier, D.A., 2002, Revised prevalence estimates of mental disorders in the United States: Using a clinical significance criterion to reconcile 2 surveys' estimates. *Arch. Gen. Psychiatry* **59:** 115–123.

Nelen, M.R., van Staveren, W.C., Peeters, E.A., Hassel, M.B., Gorlin, R.J., Hamm, H., Lindboe, C.F., Fryns, J.P., Sijmons, R.H., Woods, D.G., Mariman, E.C., Padberg, G.W., and Kremer, H., 1997, Germline mutations in the PTEN/MMAC1 gene in patients with Cowden disease. *Hum. Mol. Genet.* **6:** 1383–1387.

Nishino, I., Minami, N., Kobayashi, O., Ikezawa, M., Goto, Y., Arahata, K., and Nonaka, I., 1998, MTM1 gene mutations in Japanese patients with the severe infantile form of myotubular myopathy. *Neuromuscul. Disord.* **8:** 453–458.

Okada, T., Sakuma, L., Fukui, Y., Hazeki, O., and Ui, M., 1994, Blockage of chemotactic peptide-induced stimulation of neutrophils by wortmannin as a result of selective inhibition of phosphatidylinositol 3-kinase. *J. Biol. Chem.* **269:** 3563–3567.

Olivos-Glander, I.M., Janne, P.A., and Nussbaum, R.L., 1995, The oculocerebrorenal syndrome gene product is a 105-kD protein localized to the Golgi complex. *Am. J. Hum. Genet.* **57:** 817–823.

Ono, H., Katagiri, H., Funaki, M., Anai, M., Inukai, K., Fukushima, Y., Sakoda, H., Ogihara, T., Onishi, Y., Fujishiro, M., Kikuchi, M., Oka, Y., and Asano, T., 2001, Regulation of phosphoinositide metabolism, Akt phosphorylation, and glucose transport by PTEN (phosphatase and tensin homolog deleted on chromosome 10) in 3T3-L1 adipocytes. *Mol. Endocrinol.* **15:** 1411–1422.

Patterson, R.L., Boehning, D., and Snyder, S.H., 2004, Inositol 1,4,5-trisphosphate receptors as signal integrators. *Annu. Rev. Biochem.* **73:** 437–465.

Pendaries, C., Tronchere, H., Plantavid, M., and Payrastre, B., 2003, Phosphoinositide signaling disorders in human diseases. *FEBS Lett.* **546:** 25–31.

Pollack, S.J., Atack, J.R., Knowles, M.R., McAllister, G., Ragan, C.I., Baker, R., Fletcher, S.R., Iversen, L.L., and Broughton, H.B., 1994, Mechanism of inositol monophosphatase, the putative target of lithium therapy. *Proc. Natl. Acad. Sci. U. S. A.* **91:** 5766–5770.

Posternak, S., 1919, Sur la synthese de l' ether hexaphosphorique de l' inosite avec le principe phosphoorganique de reserve des plantes vertes. *C. R. Acad. Sci.* **169:** 138–140.

Riekse, R.G., Leverenz, J.B., McCormick, W., Bowen, J.D., Teri, L., Nochlin, D., Simpson, K., Eugenio, C., Larson, E.B., and Tsuang, D., 2004, Effect of vascular lesions on cognition in Alzheimer's disease: A community-based study. *J. Am. Geriatr. Soc.* **52:** 1442–1448.

Riera, M., Fuster, J.F., and Palacios, L., 1991, Role of erythrocyte organic phosphates in blood oxygen transport in anemic quail. *Am. J. Physiol.* **260:** R798–R803.

Rudolf, M.T., Dinkel, C., Traynor-Kaplan, A.E., and Schultz, C., 2003, Antagonists of *myo*-inositol 3,4,5,6-tetrakisphosphate allow repeated epithelial chloride secretion. *Bioorg. Med. Chem.* **11:** 3315–3329.

Saltiel, A.R., and Kahn, C.R., 2001, Insulin signalling and the regulation of glucose and lipid metabolism. *Nature* **414:** 799–806.

Sarnat, H.B., 1990, Myotubular myopathy: Arrest of morphogenesis of myofibres associated with persistence of fetal vimentin and desmin. Four cases compared with fetal and neonatal muscle. *Can. J. Neurol. Sci.* **17:** 109–123.

Sattler, M., Verma, S., Byrne, C.H., Shrikhande, G., Winkler, T., Algate, P.A., Rohrschneider, L.R., and Griffin, J.D., 1999, BCR/ABL directly inhibits expression of SHIP, an SH2-containing polyinositol-5-phosphatase involved in the regulation of hematopoiesis. *Mol. Cell. Biol.* **19:** 7473–7480.

Sbrissa, D., Ikonomov, O.C., Strakova, J., and Shisheva, A., 2004, Role for a novel signaling intermediate, phosphatidylinositol 5-phosphate, in insulin-regulated F-actin stress fiber breakdown and GLUT4 translocation. *Endocrinology* **145:** 4853–4865.

Selkoe, D.J., 1991, The molecular pathology of Alzheimer's disease. *Neuron* **6:** 487–498.

Selkoe, D.J., 2001, Alzheimer's disease: Genes, proteins, and therapy. *Physiol. Rev.* **81:** 741–766.

Senderek, J., Bergmann, C., Weber, S., Ketelsen, U.P., Schorle, H., Rudnik-Schoneborn, S., Buttner, R., Buchheim, E., and Zerres, K., 2003, Mutation of the SBF2 gene, encoding a novel member of the myotubularin family, in Charcot-Marie-Tooth neuropathy type 4B2/11p15. *Hum. Mol. Genet.* **12:** 349–356.

Shaltiel, G., Shamir, A., Shapiro, J., Ding, D., Dalton, E., Bialer, M., Harwood, J.A., Belmaker, R.H., Greenberg, M.L., and Agam, G., in press, Valproate decreases inositol biosynthesis. *Mol. Psychiatry*.

Sharma, P.M., Egawa, K., Huang, Y., Martin, J.L., Huvar, I., Boss, G.R., and Olefsky, J.M., 1998, Inhibition of phosphatidylinositol 3-kinase activity by adenovirus-mediated gene transfer and its effect on insulin action. *J. Biol. Chem.* **273:** 18528–18537.

Shepherd, P.R., Withers, D.J., and Siddle, K., 1998, Phosphoinositide 3-kinase: The key switch mechanism in insulin signalling. *Biochem. J.* **333**(Pt 3): 471–490.

Shulman, G.I., 2000, Cellular mechanisms of insulin resistance. *J. Clin. Invest.* **106:** 171–176.

Silverstone, P.H., Wu, R.H., O'Donnell, T., Ulrich, M., Asghar, S.J., and Hanstock, C.C., 2002, Chronic treatment with both lithium and sodium valproate may normalize phosphoinositol cycle activity in bipolar patients. *Hum. Psychopharmacol.* **17:** 321–327.

Smith, P.M., Harmer, A.R., Letcher, A.J., and Irvine, R.F., 2000, The effect of inositol 1,3,4,5-tetrakisphosphate on inositol trisphosphate-induced Ca2+ mobilization in freshly isolated and cultured mouse lacrimal acinar cells. *Biochem. J.* **347**(Pt 1): 77–82.

Steck, P.A., Pershouse, M.A., Jasser, S.A., Yung, W.K., Lin, H., Ligon, A.H., Langford, L.A., Baumgard, M.L., Hattier, T., Davis, T., Frye, C., Hu, R., Swedlund, B., Teng, D.H., and Tavtigian, S.V., 1997, Identification of a candidate tumour suppressor gene, MMAC1, at chromosome 10q23.3 that is mutated in multiple advanced cancers. *Nat. Genet.* **15:** 356–362.

Stephens, L., Radenberg, T., Thiel, U., Vogel, G., Khoo, K.H., Dell, A., Jackson, T.R., Hawkins, P.T., and Mayr, G.W., 1993, The detection, purification, structural characterization, and metabolism of diphosphoinositol pentakisphosphate(s) and bisdiphosphoinositol tetrakisphosphate(s). *J. Biol. Chem.* **268:** 4009–4015.

Stephens, L.R., Hawkins, P.T., Stanley, A.F., Moore, T., Poyner, D.R., Morris, P.J., Hanley, M.R., Kay, R.R., and Irvine, R.F., 1991, *myo*-Inositol pentakisphosphates. Structure, biological occurrence and phosphorylation to *myo*-inositol hexakisphosphate. *Biochem. J.* **275**(Pt 2): 485–499.

Streb, H., Irvine, R.F., Berridge, M.J., and Schulz, I., 1983, Release of Ca2+ from a nonmitochondrial intracellular store in pancreatic acinar cells by inositol-1,4,5-trisphosphate. *Nature* **306:** 67–69.

Street, V.A., Bosma, M.M., Demas, V.P., Regan, M.R., Lin, D.D., Robinson, L.C., Agnew, W.S., and Tempel, B.L., 1997, The type 1 inositol 1,4,5-trisphosphate receptor gene is altered in the opisthotonos mouse. *J. Neurosci.* **17:** 635–645.

Stutzmann, G.E., Caccamo, A., LaFerla, F.M., and Parker, I., 2004, Dysregulated IP3 signaling in cortical neurons of knock-in mice expressing an Alzheimer's-linked mutation in presenilin1 results in exaggerated Ca2+ signals and altered membrane excitability. *J. Neurosci.* **24:** 508–513.

Suchy, S.F., and Nussbaum, R.L., 2002, The deficiency of PIP2 5-phosphatase in Lowe syndrome affects actin polymerization. *Am. J. Hum. Genet.* **71:** 1420–1427.

Suchy, S.F., Olivos-Glander, I.M., and Nussabaum, R.L., 1995, Lowe syndrome, a deficiency of phosphatidylinositol 4,5-bisphosphate 5-phosphatase in the Golgi apparatus. *Hum. Mol. Genet.* **4:** 2245–2250.

Sylvia, V., Curtin, G., Norman, J., Stec, J., and Busbee, D., 1988, Activation of a low specific activity form of DNA polymerase alpha by inositol-1,4-bisphosphate. *Cell* **54:** 651–658.

Tang, T.S., Tu, H., Chan, E.Y., Maximov, A., Wang, Z., Wellington, C.L., Hayden, M.R., and Bezprozvanny, I., 2003, Huntingtin and huntingtin-associated protein 1 influence neuronal calcium signaling mediated by inositol-(1,4,5) triphosphate receptor type 1. *Neuron* **39:** 227–239.

Tanner, S.M., Schneider, V., Thomas, N.S., Clarke, A., Lazarou, L., and Liechti-Gallati, S., 1999, Characterization of 34 novel and six known MTM1 gene mutations in 47 unrelated X-linked myotubular myopathy patients. *Neuromuscul. Disord.* **9:** 41–49.

Tantivejkul, K., Vucenik, I., Eiseman, J., and Shamsuddin, A.M., 2003, Inositol hexaphosphate (IP6) enhances the anti-proliferative effects of adriamycin and tamoxifen in breast cancer. *Breast Cancer Res. Treat.* **79:** 301–312.

Taylor, G.S., Maehama, T., and Dixon, J.E., 2000, Inaugural article: Myotubularin, a protein tyrosine phosphatase mutated in myotubular myopathy, dephosphorylates the lipid second messenger, phosphatidylinositol 3-phosphate. *Proc. Natl. Acad. Sci. U. S. A.* **97:** 8910–8915.

Toker, A., 2002, Phosphoinositides and Signal Transduction.

Tonner, P.H., Scholz, J., Richter, A., Loscher, W., Steinfath, M., Wappler, F., Wlaz, P., Hadji, B., Roewer, N., and Schulte am Esch, J., 1995, Alterations of inositol polyphosphates in skeletal muscle during porcine malignant hyperthermia. *Br. J. Anaesth.* **75:** 467–471.

Tsubokawa, H., Oguro, K., Robinson, H.P., Masuzawa, T., and Kawai, N., 1996, Intracellular inositol 1,3,4,5-tetrakisphosphate enhances the calcium current in hippocampal CA1 neurones of the gerbil after ischaemia. *J. Physiol.* **497**(Pt 1): 67–78.

Tsujita, K., Itoh, T., Ijuin, T., Yamamoto, A., Shisheva, A., Laporte, J., and Takenawa, T., 2004, Myotubularin regulates the function of the late endosome through the gram domain-phosphatidylinositol 3,5-bisphosphate interaction. *J. Biol. Chem.* **279:** 13817–13824.

Ueki, K., Yamamoto-Honda, R., Kaburagi, Y., Yamauchi, T., Tobe, K., Burgering, B.M., Coffer, P.J., Komuro, I., Akanuma, Y., Yazaki, Y., and Kadowaki, T., 1998, Potential role of protein kinase B in insulin-induced glucose transport, glycogen synthesis, and protein synthesis. *J. Biol. Chem.* **273:** 5315–5322.

Ungewickell, A.J., and Majerus, P.W., 1999, Increased levels of plasma lysosomal enzymes in patients with Lowe syndrome. *Proc. Natl. Acad. Sci. U. S. A.* **96:** 13342–13344.

Ungewickell, A., Ward, M.E., Ungewickell, E., and Majerus, P.W., 2004, The inositol polyphosphate 5-phosphatase Ocrl associates with endosomes that are partially coated with clathrin. *Proc. Natl. Acad. Sci. U. S. A.* **101:** 13501–13506.

Vaden, D.L., Ding, D., Peterson, B., and Greenberg, M.L., 2001, Lithium and valproate decrease inositol mass and increase expression of the yeast INO1 and INO2 genes for inositol biosynthesis. *J. Biol. Chem.* **276:** 15466–15471.

Vajanaphanich, M., Schultz, C., Rudolf, M.T., Wasserman, M., Enyedi, P., Craxton, A., Shears, S.B., Tsien, R.Y., Barrett, K.E., and Traynor-Kaplan, A., 1994, Long-term uncoupling of chloride secretion from intracellular calcium levels by Ins(3,4,5,6)P4. *Nature* **371:** 711–714.

Vivanco, I., and Sawyers, C.L., 2002, The phosphatidylinositol 3-Kinase AKT pathway in human cancer. *Nat. Rev. Cancer* **2:** 489–501.

Voglmaier, S.M., Bembenek, M.E., Kaplin, A.I., Dorman, G., Olszewski, J.D., Prestwich, G.D., and Snyder, S.H., 1996, Purified inositol hexakisphosphate kinase is an ATP synthase: Diphosphoinositol pentakisphosphate as a high-energy phosphate donor. *Proc. Natl. Acad. Sci. U. S. A.* **93:** 4305–4310.

Vucenik, I., and Shamsuddin, A.M., 2003, Cancer inhibition by inositol hexaphosphate (IP6) and inositol: From laboratory to clinic. *J. Nutr.* **133:** 3778S–3784S.

Wappler, F., Scholz, J., Kochling, A., Steinfath, M., Krause, T., and Schulte am Esch, J., 1997, Inositol 1,4,5-trisphosphate in blood and skeletal muscle in human malignant hyperthermia. *Br. J. Anaesth.* **78:** 541–547.

Warsh, J.J., Politsky, J.M., Li, P.P., Kish, S.J., and Hornykiewicz, O., 1991, Reduced striatal [3H]inositol 1,4,5-trisphosphate binding in Huntington's disease. *J. Neurochem.* **56:** 1417–1422.

Whitman, M., Kaplan, D.R., Schaffhausen, B., Cantley, L., and Roberts, T.M., 1985, Association of phosphatidylinositol kinase activity with polyoma middle-T competent for transformation. *Nature* **315:** 239–242.

Williams, R.S., Cheng, L., Mudge, A.W., and Harwood, A.J., 2002, A common mechanism of action for three mood-stabilizing drugs. *Nature* **417:** 292–295.

Wishart, M.J., Taylor, G.S., Slama, J.T., and Dixon, J.E., 2001, PTEN and myotubularin phosphoinositide phosphatases: Bringing bioinformatics to the lab bench. *Curr. Opin. Cell Biol.* **13:** 172–181.

Woodcock, E.A., 1997, Inositol phosphates and inositol phospholipids: How big is the iceberg? *Mol. Cell. Endocrinol.* **127:** 1–10.

Woodcock, E.A., Lambert, K.A., and Du, X.J., 1996, Ins(1,4,5)P3 during myocardial ischemia and its relationship to the development of arrhythmias. *J. Mol. Cell. Cardiol.* **28:** 2129–2138.

Woodcock, E.A., Lambert, K.A., Phan, T., and Jacobsen, A.N., 1997, Inositol phosphate metabolism during myocardial ischemia. *J. Mol. Cell. Cardiol.* **29:** 449–460.

Yamamoto, K., Hashimoto, K., Nakano, M., Shimohama, S., and Kato, N., 2002, A distinct form of calcium release down-regulates membrane excitability in neocortical pyramidal cells. *Neuroscience* **109:** 665–676.

Ye, W., Ali, N., Bembenek, M.E., Shears, S.B., and Lafer, E.M., 1995, Inhibition of clathrin assembly by high affinity binding of specific inositol polyphosphates to the synapse-specific clathrin assembly protein AP-3. *J. Biol. Chem.* **270:** 1564–1568.

York, J.D., Guo, S., Odom, A.R., Spiegelberg, B.D., and Stolz, L.E., 2001, An expanded view of inositol signaling. *Adv. Enzyme Regul.* **41:** 57–71.

York, J.D., Odom, A.R., Murphy, R., Ives, E.B., and Wente, S.R., 1999, A phospholipase C-dependent inositol polyphosphate kinase pathway required for efficient messenger RNA export. *Science* **285:** 96–100.

York, J.D., Saffitz, J.E., and Majerus, P.W., 1994, Inositol polyphosphate 1-phosphatase is present in the nucleus and inhibits DNA synthesis. *J. Biol. Chem.* **269:** 19992–19999.

Zecevic, N., Milosevic, A., and Ehrlich, B.E., 1999, Calcium signaling molecules in human cerebellum at midgestation and in ataxia. *Early Hum. Dev.* **54:** 103–116.

Zhang, X., Hartz, P.A., Philip, E., Racusen, L.C., and Majerus, P.W., 1998, Cell lines from kidney proximal tubules of a patient with Lowe syndrome lack OCRL inositol polyphosphate 5-phosphatase and accumulate phosphatidylinositol 4,5-bisphosphate. *J. Biol. Chem.* **273:** 1574–1582.

Zhang, X., Jefferson, A.B., Auethavekiat, V., and Majerus, P.W., 1995, The protein deficient in Lowe syndrome is a phosphatidylinositol-4,5-bisphosphate 5-phosphatase. *Proc. Natl. Acad. Sci. U. S. A.* **92:** 4853–4856.

Chapter 12

Mammalian Inositol 3-phosphate Synthase: Its Role in the Biosynthesis of Brain Inositol and its Clinical Use as a Psychoactive Agent

Latha K. Parthasarathy, L., Ratnam S. Seelan, Carmelita Tobias, Manuel F. Casanova, and Ranga N. Parthasarathy
Molecular Neuroscience and Bioinformatics Laboratories and Autism Research Unit, Mental Health, Behavioral Science and Research Services, VA Medical Center (151), Louisville, Kentucky, 40206, USA. Departments of Psychiatry, Biochemistry and Molecular Biology, University of Louisville, Kentucky, 40202, USA.

Abbreviations used: EMSA, electrophoretic mobility shift analysis; IMPase 1, inositol monophosphatase 1; inositol synthase, *Myo*-inositol 3-phosphate synthase; NCBI, National Center for Biotechnology Information; PCR, polymerase chain reaction; Rb, retinoblastoma protein; *tss*, transcriptional start site; IP_3 , Inositol trisphosphate; DAG, diacyl glycerol; FISH, fluorescent *in situ* hybridization.

1. INTRODUCTION

myo-inositol,[1] a carbocyclic sugar that is abundant in brain and other mammalian tissues, mediates cell signal transduction in response to a variety of hormones, neurotransmitters and growth factors and participates in osmoregulation (Agranoff and Fisher, 2001; Berridge and Irvine, 1989; Fisher *et al.*, 2002; Thurston *et al.*, 1989; Parthasarathy and Eisenberg, 1986). The *myo*-inositol structure represents a unique paradigm for the illustration of chirality and optical and geometrical stereoisomerisms (Parthasarathy and Eisenberg, 1991). Among the nine possible isomers of inositol (*myo, neo, scyllo, epi, D-chiro, L-chiro, cis,*

[1] In this chapter, inositol indicates the *myo*-isomer.

A. Lahiri Majumder and B. B. Biswas (eds.), Biology of Inositols and Phosphoinositides, 293–314
© 2006 *Springer. Printed in the Netherlands.*

muco and allo-inositols) the *myo*-isomer is most abundant and biologically active molecule in nature. *myo*-inositol is a major biologically active cyclitol having a single axial hydroxyl group at carbon-2 leading to only one plane of symmetry (Parthasarathy and Eisenberg, 1986). It is the substrate for the synthesis of cell membrane inositol phospholipids linked to calcium-mobilizing receptors in mammalian brain.

myo-inositol exists in mammalian tissues in several bound and free forms: (i) as the primary cell membrane inositol phospholipid, phosphatidylinositol (PI), which by stepwise phosphorylation gives rise to mono-, bis- and tris-

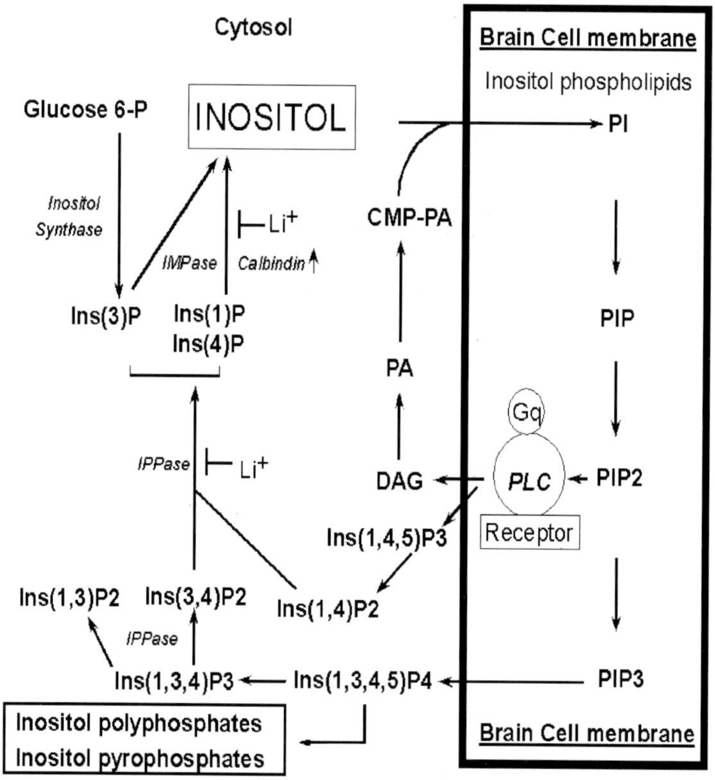

Figure 1. Schematic representation of the brain inositol signaling system. The quantities of IMPase isoenzymes and IPPase are increased by chronic lithium treatment occurring at either the gene or protein levels. Inositol in this diagram indicates the *myo*-inositol isomer. Calbindin – calcium binding protein; DAG- diacyl glycerol; Gq-GTP binding protein; IMPase 1– inositol mono phosphatase 1; IPPase- inositol polyphosphate 1-phosphatase; Ins(1)P, Ins(3)P, Ins(4)P-inositol monophosphates; Ins(1,3)P_2 – inositol 1,3-bisphosphate; Ins(1,4)*P*2 – inositol 1,4-bisphosphate; Ins(3,4)*P*2- inositol 3,4-bisphosphate; Ins (1,4,5)*P*3 – inositol 1,4,5-trisphosphate; Ins(1,3,4)*P*3 – inositol 1,3,4-trisphosphate; Li^+-lithium; PA – phosphatidic acid; PI- phosphatidyl inositol; PIP- phosphatidyl inositol 4-phosphate; PIP_2- phosphatidyl inositol 4,5-bisphosphate; PIP_3- phosphatidyl inositol 3,4,5 trisphosphate; PLC – phospholipase-C, VPA-valproate.

phospho- inositides (PIP, PIP_2 and PIP_3; Parthasarathy *et al.*, 1993); (ii) as water soluble inositol mono-, bis-, tris- and poly-phosphates; (iii) as the glycosyl-phosphatidylinositol (GPI) anchors of membrane proteins and enzymes, and (iv) as free inositol which is the precursor for the brain inositol signal system (Agranoff and Fisher, 2001; Majerus *et al.*, 1999; Parthasarathy *et al.*, 1994). In mammalian brain, the inositol signal system serves as a major pathway linking serotonergic, muscarinic, adrenergic, histaminergic, and metabotropic receptors, and cholecystokinin, tachykinin, neurotensin, and platelet activating factor receptor systems (Agranoff and Fisher, 2001; Vadnal *et al.*, 1997). Extracellular signaling occurs through a series of receptors and transducing proteins, such as GTP-binding protein (G_q), which activate membrane bound phospholipase C. Phospholipase C generates two second messengers- inositol 1,4,5-trisphosphate (IP_3) and diacylglycerol(DAG)- in the cytosol, liberating calcium ions from the endoplasmic reticulum (ER) which then stimulate protein kinase C (Figure 1). Calcium ions activate a number of enzymes and receptors. IP_3 can be further phosphorylated by different inositol phosphate kinases to form highly phosphorylated forms such as IP_7 and IP_8, which exhibit profound biological activity. Recently, Snyder's group has shown (Saiardi *et al.*, 2004) that a subset of proteins can be directly phosphorylated by IP_7 in an enzyme-free manner. These reactions reveal new avenues to control a bewildering array of biological functions especially in the brain (York and Hunter, 2004).

2. ENZYMATIC AND STRUCTURAL ASPECTS OF INOSITOL SYNTHASE

Eisenberg at NIH originally identified mammalian inositol synthase in crude rat tissue homogenates (Eisenberg, 1967). The biosynthesis of inositol in mammalian tissues takes place in two stages utilizing glucose 6-phosphate as the substrate. β-Glucose 6-phosphate is first isomerized to D-*myo*-inositol 3-monophosphate by inositol synthase (E.C. 5.5.1.4), an NAD^+ requiring enzyme (Eisenberg and Parthasarathy, 1987; Parthasarathy and Eisenberg, 1986; Mauck *et al.*, 1980; Wong and Sherman, 1980). The second stage liberates free inositol from inositol 3-monophosphate in a hydrolytic reaction mediated by inositol monophosphatase 1 (IMPase 1), a Mg^{2+} requiring lithium- sensitive enzyme (Hallcher and Sherman, 1980; Naccarato *et al.*, 1974). IMPase 1 differs from other nonspecific phosphatases such as alkaline and acid phosphatases in physical, structural and enzymatic characteristics and it cannot differentiate between the isomeric forms of inositol monophosphates (D-1 and D-3 monophosphates; Eisenberg and Parthasarathy, 1984). Basic enzyme characteristics, without the addition of any cations, were first studied in rat inositol synthase isolated from a variety of tissues (Maeda and Eisenberg, 1980; Eisenberg and Parthasarathy, 1987). Addition of any cations and addition of different cations (final concentration 1 mM) stimulated synthase activity several fold. Both NH_4^+ and K^+ monovalent ions are stimulatory and divalent cations are inhibitory to inositol synthase activity (Eisenberg and Parthasarathy, 1987; Maeda and Eisenberg, 1980; Mauck *et al.*, 1980). Lithium, a widely used mood stabilizer, sharply decreased the K_m of its activity. The chronic intake of lithium may have the ability to stimulate inositol synthase activity by decreasing its K_m thus facilitating the formation of inositol 3-phosphate from glucose 6-phosphate under conditions where there is a continued shortage of inositol, leading to inositol monophosphates accumulation. Inositol monophosphate accumulation during chronic lithium intake has been shown by Sherman *et al.* (1981a,b). The therapeutic nature of lithium may reside in its ability to decrease the K_m of inositol synthase activity, which supports the inositol depletion hypothesis proposed to explain the therapeutic effect of lithium. While some groups have recently challenged this hypothesis (Berry *et al.*, 2004; Brambilla *et al.*, 2004; Friedman *et al.*, 2004;), recent observations by others in a number of cell types and isolated neurons (Harwood, 2005) support it. Inhibitors such as 2-deoxy-glucose 6-phosphate and glucitol 6-phosphate (Eisenberg and Parthasarathy, 1987; Maeda and Eisenberg, 1980) sharply decrease the activity of inositol synthase.

3. INTERMEDIATES OF INOSITOL SYNTHASE REACTION

The overall simplified reaction of inositol biosynthesis is: Glucose 6-phosphate → inositol + inorganic phosphate (Adhikari and Majumder, 1988; Eisenberg and Parthasarathy, 1987; Jin and Geiger, 2004; Jin *et al.*, 2004; Parthasarathy and Eisenberg, 1986;). The first reaction converts glucose 6-phosphate to D-inositol 3-phosphate by synthase. Synthase action in the cyclization of glucose 6-phosphate is a rate limiting reaction (Hasegawa and Eisenberg, 1981). Biosynthetic formation of one molecule of free inositol is accompanied by the formation of one molecule of D-inositol 3-phosphate by inositol synthase (Eisenberg and Parthasarathy, 1987). Although NAD^+ is required, no net change in this coenzyme concentration is observed in the complete isomerization reaction leading from glucose 6-phosphate to inositol 3-phosphate (Parthasarathy and Eisenberg, 1986; Figure 2).

The mechanism of enzymatic isomerization of glucose 6-phosphate to D-inositol 3-monophosphate by synthase has been intensively studied for the past three decades (Eisenberg *et al.*, 1964; Loewus *et al.*, 1980; Parthasarathy and Eisenberg, 1986; Sherman *et al.*, 1981a,b). Eisenberg and coworkers studied the formation of various intermediates in the reaction using electrophoretically homogeneous rat synthase (Eisenberg and Parthasarathy, 1987). The first intermediate they postulated to occur in the reaction was 5-ketoglucose 6-phosphate which was believed to be tightly bound to synthase leading to a head-to-tail intramolecular aldol condensation forming the second intermediate, *myo*-inosose-2 1-phosphate.

NAD^+

β-Glucose 6-phosphate → [5-keto glucose 6-phosphate]* → *myo*-inosose-2 1-phosphate

Inositol
Synthase

$NADH^+$

→ D-*myo*-inositol 3-phosphate + NAD^+ → free *myo*-inositol.

Inositol
monophosphatase 1

Figure 2. The pathway of biosynthesis of *myo*-inositol to free inositol by synthase reaction in mammalian tissues. * This is a postulated intermediate presumed to occur during the course of the reaction. Crystallographic studies with yeast synthase support the presence of this intermediate.

NADH generated from the oxidative step reduces the second intermediate to D-inositol 3-monophosphate. In the early work on the elucidation of the mechanism of synthase action it was noticed that the intermediates and coenzymes were tightly bound (this was prior to the advent of crystallographic studies and recombinant DNA analysis). Eisenberg and Parthasarathy (1987) disrupted the synthase reaction by adding large amounts of sodium borotritide (200-milliCurie NaB^3H_4/enzyme reaction), which enabled them to trap the intermediates in the form of reduced epimeric sugar alcohols representing the putative intermediates. The products thus obtained were a mixture of ^{3}H labeled *myo*-inositol and its 2-epimer, *scyllo*-inositol, which were derived from the unused substrate and a putative intermediate. The reduction of this intermediate (*myo*-inosose-2, 1-phosphate) by borotritide generated *myo*-inositol. The expected products from 5-keto glucose 6-phosphate conversion are D-[1,5-^{3}H]glucitol and its 5-epimer L-[1,5-^{3}H]iditol. In these experiments although abundant label was found in hydrogen-1 of glucitol (H1), no label was found in iditol and hydrogen-5 of glucitol (Figure 3). Therefore, it was concluded that 5-ketoglucose 6-phosphate had no finite existence and was very tightly bound to synthase and that cyclization to inosose-2, 1-phosphate occurs as quickly as its formation from glucose 6-phosphate. Experiments with purified rat synthase clearly demonstrated the formation of inosose-2, 1-phosphate that was sufficiently long-lived to be isolated and that the reduction by NADH must be the rate-limiting step, similar to the observations made using kinetic isotopic experiments. Synthase essentially closes the C-6 and C-1 bonds of glucose 6-phosphate, and with other isomerization reactions that lead from glucose 6-phosphate to the formation of fructose 6-phosphate and glucose 1-phosphate, accounts for almost all the branch points of glucose metabolism in nature. Besides the closure of the carbon-carbon bond, the transfer of hydrogen is a specific mechanism in the synthase reaction and has been corroborated by recent crystallographic and chemical studies (Stein and Geiger, 2000, 2002). The hydrogen shuttle

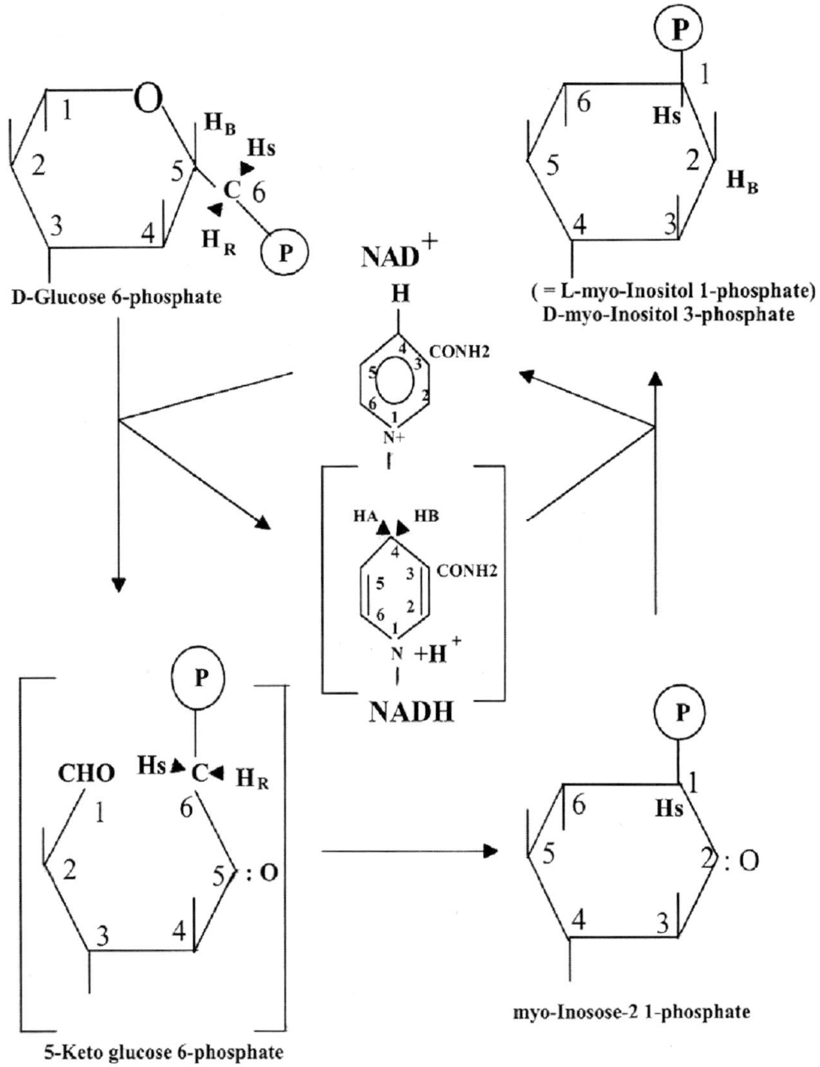

Figure 3. Isomerization of D-glucose 6-phosphate to D-*myo*-inositol 3-phosphate by inositol synthase. The hypothetical intermediate is within parenthesis.

between substrate and coenzyme leads to the oxidation at carbon-5 of glucose 6-phosphate by NAD^+ and reduction at carbon-2 of inosose-2 1-phosphate by NADH. Carbon-4 of reduced nicotinamide carries a pair of hydrogen atoms (HA and HB, see Figure 3), which can be stereoselectively, used for the synthase reaction. HB of NADH is apparently transferred while HA is not (Byun and Jenness, 1981). A similar stereo-selective approach occurs when C-6 and C-1 are linked to form D-*myo*-inositol 3-phosphate. The paired hydrogen atoms at C-6 of 5-ketoglucose have hydrogen atoms Hs and HR. Hs is retained in the cyclization reaction while HR is lost to the solution after cyclization (Loewus *et al.*, 1980). In 1995, Migaud and Frost synthesized inosose-2, 1-phosphate and observed that it not only acted as a substrate but was also a potent competitive inhibitor of inositol synthase. The crystal structures of *S. cerevisiae* (Stein and Geiger, 2000, 2002) and *M. tuberculosis* enzymes (Norman *et al.*, 2002) have been studied at 2.0–2.5 Å resolutions. Both studies have unequivocally demonstrated the presence of bound NAD^+ at the active site.

4. STRUCTURAL ASPECTS OF INOSITOL SYNTHASE

Native mammalian inositol synthase is a trimer composed of 558 amino acids per subunit with a native molecular weight of 210,000 ± 2,000 as determined by chromatography, electrophoresis and ultra-centrifugal sedimentation equilibrium methods (Eisenberg and Parthasarathy, 1987; Maeda and Eisenberg, 1980; Mauck *et al.*, 1980). The trimeric nature of mammalian inositol synthase is in sharp contrast to yeast and plant forms where it exists as a tetramer (Majumder *et al.*, 1997 and 2003). The subunit molecular weight of inositol synthase is approximately 69,000 ± 600 (Eisenberg and Parthasarathy, 1987; Maeda and Eisenberg, 1980). Amino acid sequence analysis of rat testicular inositol synthase showed that the amino terminus was blocked precluding sequence analysis of the subunits after SDS-PAGE. Only cyanogen bromide (CNBr) was able to cleave the protein to generate two peptide fragments (EPAAEILVDSPDVIF and ESLRPRPSVYIPEFIAAN).

5. GENETIC REGULATION OF INOSITOL SYNTHASE

Early studies by Henry and co-workers (Carman and Henry, 1999; Culbertson *et al.*, 1976; Chapter 6 in this volume) focused on the molecular aspects of the yeast inositol synthase gene (*INO1*), which is tightly regulated by inositol levels (Culbertson *et al.*, 1976). When yeast is grown in media containing inositol, the *INO1* gene is completely repressed while its absence completely de-represses *INO1*. Expression of *INO1* is dependent on the upstream activation sequence, UAS_{INO}, which is the binding target for the heterodimer Ino2p/Ino4p (Culbertson *et al.*, 1976; Klig and Henry, 1984; Hirsch and Henry, 1986). Binding is facilitated by limiting inositol levels, which derepresses *INO1* expression (Loewy and Henry, 1984) while repression is mediated by the transcriptional regulator Opi1p (Greenberg *et al.*, 1982; Jiranek *et al.*, 1998; Lopes and Henry, 1991). Widely used mood stabilizing agents such as lithium and valproate stimulate *INO1* gene expression, but valproate inhibits the synthesis of inositol 3-phosphate, thus

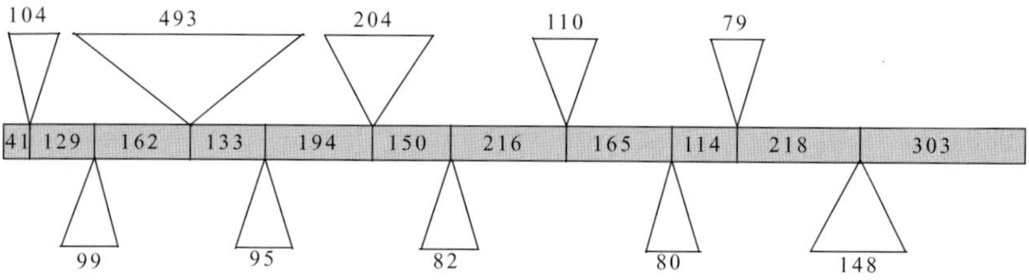

Figure 4. The gene structure of the human *ISYNA1*. The *ISYNA1* gene contains 11 exons (shaded) and 10 introns (triangles). The sizes of the exons (within boxes) and introns (above and below triangles) are indicated. The size of the first exon was determined after identification of the transcription start site (*tss*).

suggesting that inositol synthase may be a target for valproate action in yeast. Most of the information on human *ISYNA1* (the mammalian counterpart of *INO1*) is only available recently (AF220250 to AF220259; AF220530; AF314170; AF288525; AF251265; AH009098). The human gene spans 11 exons (Figure 4) and is located on chromosome 19p13.1; a related processed pseudogene is located on 4p15 (Figure 5). It is interesting that a recent genome-wide linkage analysis provides a strong support for an autism susceptible locus at 19p13.11 where *ISYNA1* resides (McCauley *et al.*, 2005). mRNA structure and characterization indicate that it is ~1.8 kb long (Figure 6). Transcriptional start site (*tss)* analysis of *ISYNA1* mRNA from liver by an Inverse 5′ PCR protocol (Zeiner and Gehring, 1994) indicates that the *tss* is 20 bp upstream from the 3′ end of the non-coding, exon 1. However, a brain cDNA deposited in the NCBI database (AF220530) is longer at the 5′-end by an additional 21 bp. The differences between the two mRNAs with respect to their *tss* may be due to tissue-specific initiation sites, typical of many house-keeping genes possessing a G/C-rich promoter region. Indeed, the proximal 500 bp of the *ISYNA1* promoter, including exon 1, is extremely G/C rich (~70%). Majumder et al., (1997, 2003) have compared mRNA sequences from diverse species and have identified conserved amino acid motifs in inositol synthase. Given the importance of this gene in inositol signaling and that perturbances in cellular inositol homeostasis can lead to neurological symptoms, a detailed analysis of its promoter has been undertaken recently (discussed below; Seelan *et al.*, 2004).

Functional expression of the human inositol synthase cDNA in yeast cells devoid of the *INO1* gene shows that it can complement inositol auxotrophy and excrete inositol (Ju *et al.*, 2004). When grown in valproate (0.6 mM), these cells show a 35 and 25% decrease in synthase activity and inositol levels, respectively. However, valproate does not directly inhibit synthase activity at this concentration (0.6 mM) implying that this mood stabilizer works probably at the translational or transcriptional levels. Inositol synthase is present in a wide variety of organisms (protozoa, fungi, plants and mammals) and has been

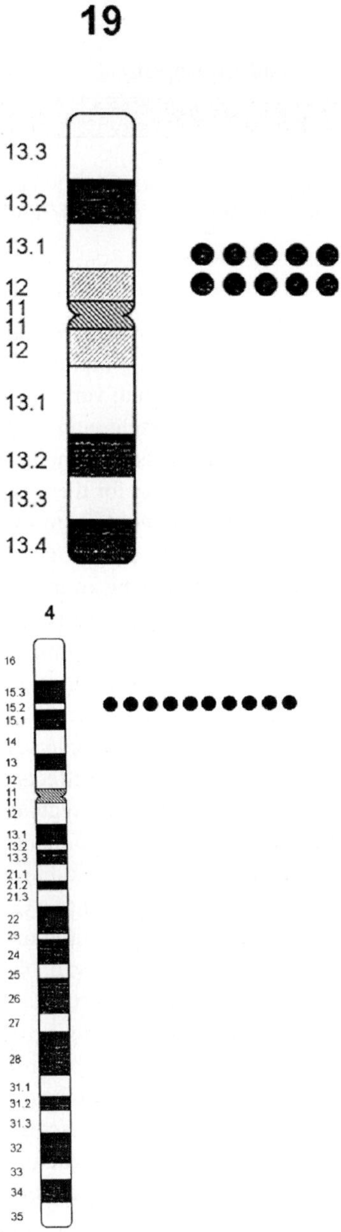

Figure 5. FISH mapping of *ISYNA1* gene. Each dot represents the double FISH signals detected on chromosome 19 and 4, after analysis of 100 metaphase chromosomal spreads. Chromosome 19p13.1 harbors the expressed gene and chromosome 4p15, the pseudogene.

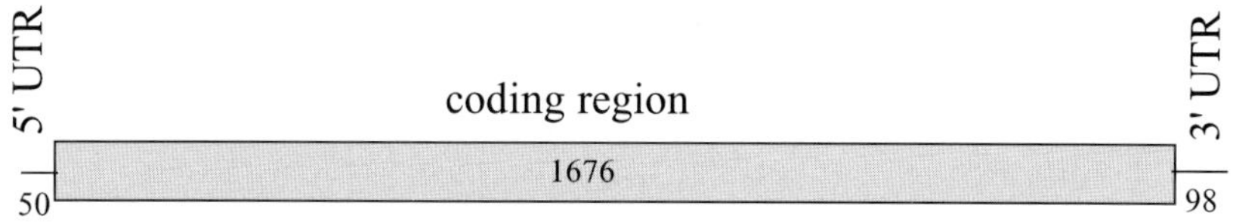

Figure 6. The cDNA structure of the human *ISYNA1* gene. The gene encodes a 1676 bp coding region with a 5′ UTR (untranslated region) of 50 bp and a 3′ UTR of 98 bp (GenBank Accession #AF220530).

purified from diverse sources (Majumder *et al.*, 1997 and 2003). Inositol synthase is expressed in all tissues and has the highest expression in testis, followed by heart, pancreas, ovary and placenta; very low expression is observed in blood leukocytes, thymus, skeletal muscle, and colon (Guan *et al.*, 2003). Paradoxically, it is only marginally expressed in the brain (Guan *et al.*, 2003), an organ where inositol synthase is a target for the therapeutic intervention as a mood stabilizer (Agam *et al.*, 2002). In contrast to *INO1* regulation in yeast, studies on *ISYNA1* show its regulation to be more complex. Rats administered therapeutic levels of lithium for 10 days show an increase in inositol synthase expression in the hippocampus, but not in the frontal cortex (Shamir *et al.*, 2003). Inositol synthase activity in rat testes is significantly decreased in long-term diabetes when compared to controls (Whiting *et al.*, 1979). Hasegawa and Eisenberg (1981) demonstrated decreased expression of inositol synthase activity in reproductive organs and liver of hypophysectomized male rats and in liver of thyroidectomized male rats, and these effects can be reversed by hormone treatments. It is also the target for estrogen regulation in the uterus of rats (Rivera-Gonzalez *et al.*, 1998). A recent study (Guan *et al.* 2003) has shown that in HepG2 (liver-derived) cell lines, glucose and lovastain increase *ISYNA1* expression, the latter suggesting the involvement of a G-protein coupled signal transduction system. When HepG2 cells are cultured in the presence of lithium (10 mM), a 50% suppression in activity is observed. No effects were observed with inositol, estrogen, thyroid hormone or insulin. The lack of an inositol effect in HepG2 is surprising since the yeast gene is highly regulated by inositol levels; also, while lithium is inhibitory to the enzyme in HepG2 cells (Guan *et al.*, 2003), it appears to stimulate expression in the rat hippocampus (Shamir *et al.*, 2003). These contrasting effects may underscore some of the complexities associated with human inositol synthase regulation and may, in part, be attributed to tissue- or species-specific gene expression responses.

6. HUMAN *ISYNA1* PROMOTER ELEMENTS

Characterization of the core promoter of *ISYNA1* (Seelan *et al.*, 2004) indicates that the promoter is up regulated by E2F1, a cell-cycle regulator. That E2F1 transactivates *ISYNA1* has been demonstrated by the following observations:

(i) the 5′-flanking region is stimulated by ectopic expression of E2F1; (ii) E2F1 induction is localized to the minimal promoter region between −261 and −34; sequences downstream of −34 do not significantly induce promoter activity; (iii) the promoter activity of the minimal promoter region shows a dose-dependent increase with exogenous expression of E2F1; (iv) the induction by E2F1 can be suppressed by the exogenous expression of Rb (retinoblastoma protein), a key negative modulator of E2F1 expression in cells (Dyson, 1998); and, (v) the presence of at least one functional E2F binding element at −117 (Seelan *et al.*, 2004), as inferred by electrophoretic mobility shift assay (EMSA) and E2F1 antibody super-shift analysis.

Manual scan of the minimal promoter (−261 to –34) identified seven possible sequences with close homology to the consensus binding motifs for E2F1 (Tao *et al.*, 1997). These are: TTCCGCC at −188, TGGCGG at −180, TCGGGC at −146, TTGGGCC at −117, TCCCGC at −90, TTCCCCG at −81, and TCGCGCG at −37 (Figure 7). Two of these sites at −81 and −90 were also identified with the program TFBind (Tsunoda and Takagi, 1999). EMSA and

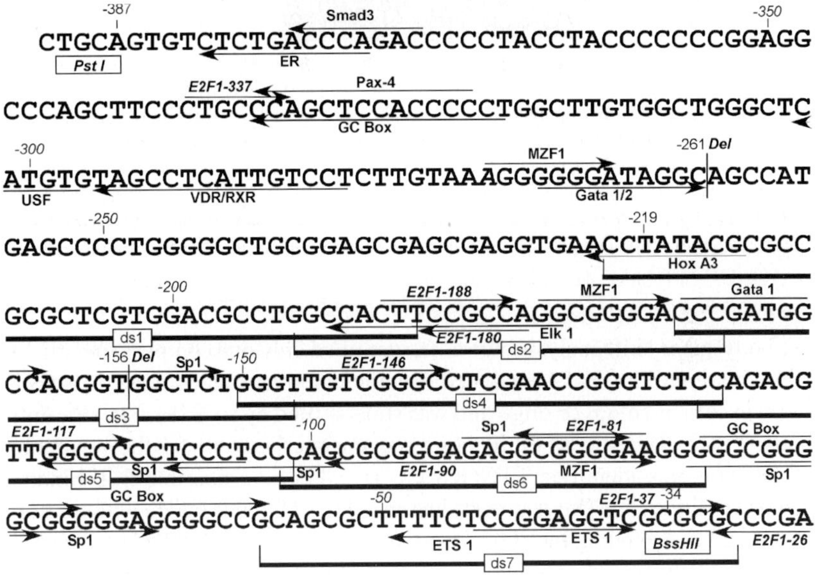

Figure 7. Features of the *ISYNA1* minimal promoter. The sequence of the minimal promoter between the *Pst* I (−387) and *BssH* II (−34) sites is shown. Potential transcription factor binding sites, obtained by various database searches are indicated by arrows and names. Putative E2F motifs are indicated with their nucleotide locations. −261Del and –156Del (vertical bars) indicate the site of truncation of the minimal promoter plasmid used for deletion analysis. ds1 – ds7 represent the oligos used for EMSA. ER, estrogen receptor; Elk 1, member of ETS oncogene family; ETS 1, E26 avian leukemia oncogene1; GATA 1, GATA binding protein 1; Hox A3, homeobox A3; MZF1, myeloid zinc finger 1; Pax-4, Paired box gene 4; Smad3, MAD homology 3; Sp1, transacting transcription factor 1; USF, upstream stimulating factor; VDR/RXR, Vitamin D receptor/Retinoid X receptor. A functional E2F element is located at –117 (Seelan *et al.*, 2004; reprinted with permission from Elsevier).

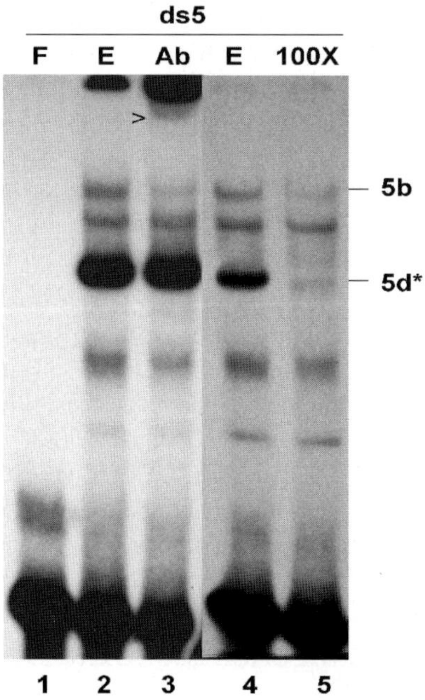

Figure 8. Analysis of E2F1 binding to ds5 of the *ISYNA1* minimal promoter by EMSA. ds5 binding was assessed with HeLa nuclear extracts alone (E), or preincubated with an antibody, KH129 (ActiveMotif, Carlsbad, CA), raised against E2F1 (Ab), or competed with a 100-fold molar excess of an unlabeled E2F oligo (100X), prior to the addition of labeled probe. F, free probe. The super-shifted band by E2F specific antibody is denoted by an arrowhead. Complex 5b is the E2F1-specific complex while complex 5d* is a spurious band (Seelan *et al.*, 2004; reprinted with permission from Elsevier).

antibody super-shift analysis, however, identified only one complex specific for E2F1 binding. This E2F1 mediated specific complex was competed partially by an excess of cold E2F oligo and was supershifted by an E2F1- specific antibody (Figure 8). Analysis of the oligo sequence responsible for the generation of this complex identified TTGGGCCC at −117 as a possible E2F binding sequence (Figure 7). This sequence has an 8/11 contiguous match with the context based E2F binding motif of Kel *et al.* (2001)–TTTSGGCSMDR. Interestingly, this E2F binding complex is also competed by an oligo containing the Sp1 motif, suggesting that Sp1 may also interact at this site (not shown). An analysis of the oligo sequence indicates that the E2F1 site overlaps an Sp1 element on the opposite strand (Figure 7). Thus the −117 element may be occupied by E2F1, Sp1, or by both simultaneously, and this may explain the partial competition observed using the E2F1 oligo in competition experiments.

Further detailed promoter analysis suggestes that E2F responsiveness is mediated by more than one E2F element in the minimal promoter. Because only one weak affinity E2F1 complex at −117 was identified experimentally, we conclude that the transactivation of the *ISYNA1* promoter by E2F1 occurs through the cooperative interaction of several low-affinity binding sites present in the minimal promoter of *ISYNA*1 (Seelan *et al.*, 2004). Some of these putative sites have been indicated in Figure 7. The significance of E2F regulation of *ISYNA1* promoter implies that this enzyme may have other important functions besides its known biosynthetic function. Alternatively, inositol biosynthesis may be intricately associated with several vital processes such as cell division, DNA synthesis or apoptosis. Interestingly, E2F1 plays a major role in the transcriptional regulation of E2F target genes in the adult testes (Wells *et al.*, 2002), an organ where inositol synthase also has the highest activity (Guan *et al.*, 2003).

We have also observed several other specific protein/DNA complexes binding to the minimal promoter of *ISYNA1*. Some of these complexes include Sp1. Sp1 recognizes the consensus sequence, GGGCGG (Kadonaga *et al.*, 1986) and also loosely binds to GC-rich or GT-rich sequences (Suske,1999). EMSA analysis of the promoter revealed that several of the DNA/protein complexes are competed by the Sp1 oligo (Seelan *et al.*, 2004). Both high affinity Sp1 binding (as inferred by the intense banding and ready competition) and weak Sp1 interactions were evident (Seelan *et al.*, 2004).

7. PRECLINICAL AND CLINICAL STUDIES WITH INOSITOL

Allison and Stewart (1971) first demonstrated that lithium (acute model) depleted rat brain inositol by 30%. These studies along with others served as the basis for the "inositol depletion hypothesis" to explain lithium's mode of action in mood disorders (Berridge *et al.*, 1989). However, chronic lithium studies with rats have shown conflicting results. Renshaw *et al.* (1986) observed an increase while Honchar *et al.*, (1989) observed no change in inositol monophosphatase 1 levels in rat cerebral cortex. Whitworth and Kendall (1989) showed an increase in free inositol levels in rat cerebral cortex with no change in the striatum. Results from chronic lithium studies are difficult to assess because they had different experimental designs. In humans, inositol enters the blood stream and CSF quickly after oral intake (Groenen *et al.*, 2003; Levine *et al.*, 1993b, 1996;). Proton magnetic resonance studies also showed that brain inositol levels initially increased after oral inositol and subsequently returned to baseline levels. NMR techniques also show that brain inositol phosphates are increased after 7 and 14 days of lithium treatment (Yildiz *et al.*, 2001). It should be noted that preliminary clinical studies on the beneficial effects of oral inositol in alleviating psychiatric symptoms have only been conducted with a limited number of patients and they have to be elaborated in a larger number of patients for the understanding of the full psychoactive effects of inositol. Some of the studies in various psychiatric conditions are summarized below:

7.1 Depression

Barkai *et al.* (1978) and Frey *et al.*, (1998) have shown that depressed patients had significantly lower levels of CSF inositol, although others could not replicate this observation (Levine *et al.*, 1995b). Placebo-controlled double-blind[2] study compared the effects of 12 gm/day of inositol versus glucose in a group of depressed patients, including both manic-depressive and major depressive disorders (Levine *et a*l., 1995b). All medications were stopped 3-7 days prior to the start of the study and patients received only inositol or the placebo, glucose. At 4 weeks, significant improvement was noted in the depressed group taking inositol versus placebo, demonstrating that inositol had a considerable anti-depressant effect. One patient experienced a hypomanic phase with placebo; no manic phases occurred with inositol. Laboratory analysis did not indicate any changes in liver, blood, or kidney functions. Based upon these studies and those of Chengappa *et al.* (2000) it can be considered that inositol can possibly be used as a psychotherapeutic agent in the treatment of depression. However, further studies with a larger number of patients are needed.

7.2 Obsessive-Compulsive Disorder (OCD)

Using 18 g/day of inositol, versus placebo, a double blind,[2] and crossover study was performed on 13 patients for six weeks. The Yale-Brown Obsessive Compulsive Scale[3] was used for analysis, and patients taking the inositol had significantly lower scores than those receiving placebo (Fux *et al.*, 1996, 1999). Since inositol appears to be effective in depression, panic disorder and OCD, it has the clinical profile of a serotonin selective reuptake inhibitor (SSRI). Higher doses were used in this study compared to the depression and panic disorder studies previously described. Only a small number of patients were used.

7.3 Panic Disorder

Benjamin *et al.* (1995) conducted a double blind crossover study[2] with a group of 21 patients suffering from panic disorder. Four weeks of inositol

[2]Double blind cross: A clinical study in which neither the patients nor the investigating physician know the intervention to which the patients have been assigned. A double-blind design is considered to provide the most reliable data from a clinical trial. In a crossover study the subject is randomized into at least two treatment groups. After an appropriate wash out period each subject is reassigned to the alternate treatment group. A crossover study must have at least 2 treatment groups.

[3]Yale-Brown Obsessive Compulsive Scale is a reliable psychological measuring method to determine the severity of illness in patients suffering with obsessive-compulsive disorder with a range of severity and types of obsessive-compulsive symptoms.

(12 gm/day) or a proper placebo was administered. This study group contained patients with panic disorder with or without agoraphobia.[4] The results demonstrated that patients with panic disorder showed statistically lower frequency and severity of panic attacks and agoraphobia when treated with inositol. The number of panic attacks per week decreased from a baseline score of 9.7 $\pm$ 15 to 6.3 $\pm$ 9 after 4 weeks in the placebo group, compared to a significant reduction from 9.7 $\pm$ 15 to 3.7 $\pm$ 4 in the inositol treated group. Statistically significant differences were found in the number of panic attacks and phobia scores. Inositol was found to be an effective psychotherapeutic agent in the treatment of panic disorder in this initial study as well as in later studies (Palatnik *et al.*, 2001). It should be noted that traditional treatments with SSRIs such as fluoxetine, tricyclic antidepressants (TCAs; *e.g.*, imipramine), monoamine oxidase inhibitors (MAOIs; phenelzine) and triazolobenzodiazepines (*e.g.*, alprazolam), exhibit significant side effects.

7.4 Schizophrenia

Levine *et al.* (1993a) conducted the first clinical study using inositol as a psychoactive agent in schozophrenia. Using 6 gms of inositol per day in a schizophrenic group, this double-blind add-on study found no significant therapeutic effects. These patients were maintained on their usual antipsychotic medication during the study period. Both the length of the study (days) and the dose of inositol (6 gm/day) were noted to be sub-optimal.

7.5 Post-Traumatic Stress Disorder (PTSD)

A double-blind crossover study was conducted on a group of patients suffering from post-traumatic stress disorder (Kaplan *et al.*, 1996). Thirteen patients were studied, after excluding any history of drug abuse or organicity. Twelve grams of inositol per day or placebo (glucose) were used, after a washout period of two weeks, with patients remaining on study drugs for four weeks for each phase of the study. No other medications were allowed during this study. The outcome scales used were the Impact of Event Scale, SCL-90, HAM-D, and HAM-A, with no statistically significant effects noted on measures of intrusion or avoidance scales specific for PTSD. In general, inositol treatment did not demonstrate any therapeutic effect in the core symptoms of PTSD.

[4]Agoraphobia is a common phobic disorder. Most of the people suffering form this disorder show the fear of being alone in public places from which the person thinks escape would be difficult or help unavailable if he or she were incapacitated.

7.6 Attention Deficit Disorder (ADD)

Inositol has been tested in attention deficit disorder with hyperactivity (ADDH). Eleven children were given either inositol (200 mg/kg) or dextrose (placebo) in a crossover, double-blind study, over an eight-week period (Levine *et al.*, 1995a). No other medications were allowed during the study. Of the eleven patients, eight had a history of responding to previous stimulant medications, one had been treatment-resistant, and two had not previously been treated. There was a trend towards worsening of the disorder with inositol versus placebo, and appeared to worsen core symptoms of this syndrome. Based upon these observations it was suggested that a low inositol diet might be of benefit in ADDH. The potential negative effects of dietary inositol in this condition require further cautious studies in children.

7.7 Autism

There has been an increasing interest in the use of serotonin reuptake inhibitors (SSRIs) in the treatment of autism (McDougle *et al.*, 1996). Serotonin binds to several types of receptors including the 5-HT2, and 5-HT1c receptors, which are linked to inositol signaling. A double-blind crossover study was completed at a dose of 200 mg/kg per day, and the results demonstrated no significant improvement on inositol therapy (Levine *et al.*, 1997).

7.8 Alzheimer's Disease

Pacheco and Jope (1996) described inositol system dysfunction in the brains of Alzheimers patients. A double-blind placebo controlled crossover study using a 6 gm/day dose of inositol (versus glucose) for 4 weeks demonstrated no significant changes in 11 Alzheimer's patients (Barak *et al.*, 1996). The Cambridge Mental Disorders of the Elderly Examination (CAMCOG) overall scores demonstrated an improving trend with inositol, but results were not statistically significant. This study is encouraging, and needs to be repeated with larger patient numbers, higher doses, and for more extended times. A several month study would be desirable. It would be interesting to find out whether inositol as a natural brain sugar could delay the progression of Alzheimer's disease, or improve cognitive functions.

7.9 Other conditions

While Nemets *et al.* (2002) showed that 12 gm of oral inositol per day did not show any beneficial effects in premenstrual dysphoric disorder, 18 gms per day of inositol was therapeutic for patients suffering from bulimia nervosa and binge eating (Gelber *et al.*, 2001). In this study, the Global Clinical Impresion, the Visual Analogue Scale and binge eating criteria were followed for analysis. This finding is similar to the serotonin selective reuptake inhibitor used often with patients.

8. CONCLUSION

myo-inositol is a simple, naturally occurring sugar present in all mammalian cells and is composed of a cyclohexane ring with a hydroxyl group bound to each of the six carbons in a precise stereospecific orientation. It is the principal isomer of the brain and its presence in millimolar amounts contributes to osmoregulation. Although, it is not directly involved in the generation of energy, as is glucose, it is uniquely involved in brain signal transduction generating many inositol mono- and polyphosphates. The dietary route is one of the three pathways to maintain inositol homeostasis, the others being the receptor mediated salvage pathway involving IMPase 1 and a *de novo* pathway involving inositol synthase. In brain, inositol is a precursor for the formation of the inositol phospholipids, which on receptor stimulation produce two-second messengers, IP_3 and DAG. Inositol pathway has been implicated in the pathogenesis of bipolar disorder, with the mood stabilizers valproate and lithium targeting inositol synthase and IMPase 1, respectively. Recent observations of the inhibitory effect of valproate on inositol synthase suggest that this biosynthetic enzyme may be a potential therapeutic target for modulating brain inositol levels. Inositol synthase and its substrate/coenzyme (glucose 6-phosphate/NAD^+) are sufficient for the biosynthesis of inositol. Mammalian inositol synthase is a homotrimer and its isomerization of glucose 6-phosphate occurs through two intermediates as enzyme bound forms (5-ketoglucose 6-phosphate and inosose-2, 1-phosphate). The genomic structure, chromosomal localization, and characterization of the minimal promoter and identification of several transcription factor-binding elements in *ISYNA1* have been determined. The gene is located at 19p13.11, a candidate locus for autism spectrum disorder. Preclinical and selective clinical studies show that inositol can be used as a psychoactive agent and inositol synthase may be a promising target for designing novel anti-depressants thereby modulating brain inositol levels. These new mood modulators will potentially decrease the huge amounts of oral inositol taken to achieve therapeutic effects in neuropsychiatric disorders.

ACKNOWLEDGEMENTS

This work was supported by the Office of Research and Development, Medical Research Service, Department of Veteran Affairs, Washington DC, USA

(R.N.P.), the Clinical Research Foundation, Louisville, KY (R.S.S., L.K.P., R.N.P.) and by the National Institute of Health, Bethesda, Maryland, USA (M.F.C).

REFERENCES

Adhikari, J., and Majumder, A.L., 1988, L-*myo*-inositol-1-phosphate synthase from mammalian brain: Partial purification and characterisation of the fetal and adult enzyme. *Indian J. Biochem. Biophys.* **25:** 408–412.

Agam, G., Shamir, A., Shaltiel, G., and Greenberg, M.L., 2002, *Myo*-inositol-1-phosphate (MIP) synthase: A possible new target for antibipolar drugs. *Bipolar Disord. Suppl.* **4:** 15–20.

Agranoff, B.W., and Fisher, S.K., 2001, Inositol, lithium and the brain. *Psychopharmacol. Bull.* **35:** 5–8.

Allison, J.H., and Stewart, M.A., 1971, Reduced brain inositol in lithium treated rats. *Nature New Biol.* **233:** 262–268.

Atack, J.R., 1996, Inositol monophosphatase, the putative therapeutic target for lithium. *Brain Res. Rev.* **22:** 183–190.

Barak, Y., Levine, J., Glasman, A., Elizur, A., and Belmaker, R.H., 1996, Inositol treatment of Alzheimer's disease: A double blind, cross-over placebo controlled trial. *Prog. Neuro-psychopharmacol. Biol. Psychiatry* **20:** 729–735.

Barkai, A., Dunner, D.L., Gross, H.A., Mayo, P., and Fieve, R.R., 1978, Reduced *myo*-inositol levels in cerebrospinal fluid from patients with affective disorder. *Biol. Psychiatry* **13:** 65–72.

Benjamin, J., Levine, J., Fux, M., Aviv, A., Levy, D., and Belmaker, R.H., 1995, Double-blind, placebo-controlled, crossover trial of inositol treatment for panic disorder. *Am. J. Psychiatry* **152:** 1084–1086.

Berridge, M.J., Downes, C.P., and Hanley, M.R., 1989, Neural and developmental actions of lithium: A unifying hypothesis. *Cell* **59:** 411–419.

Berridge, M.J., and Irvine, R.F., 1989, Inositol phosphates and cell signaling. *Nature* **341:** 197–205.

Berry, G.T., Buccafusca, R., Greer, J.J., and Eccleston, E., 2004, Phosphoinositide deficiency due to inositol depletion is not a mechanism of lithium action in brain. *Mol. Genet. Metab.* **82:** 87–92.

Brambilla, P., Stanley, J.A., Sassi, R.B., Nicoletti, M.A., Mallinger, A.G., Keshavan, M.S., and Soares, J.C., 2004, ^{1}H MRS study of dorsolateral prefrontal cortex in healthy individuals before and after lithium administration. *Neuropsychopharmacology* **29:** 1918–1924.

Byun, S.M., and Jenness, R., 1981, Stereospecificity of L-*myo*-inositol 1-phosphate synthase for nicotinamide adenine dinucleotide. *Biochemistry,* **20:** 5174–5177.

Carman, G.M., and Henry, S.A., 1999, Phospholipid biosynthesis in the yeast *Saccharomyces cerevisiae* and interrelationship with other metabolic processes. *Prog. Lipid Res.* **38:** 361–399.

Chengappa, K.N., Levine, J., Gershon, S., Mallinger, A.G., Hardan, A., Vagnucci, A., Pollock, B., Luther, J., Buttenfield, J., Verfaille, S., and Kupfer, D.J., 2000, Inositol as an add-on treatment for bipolar depression. *Bipolar Disord.* **2:** 47–55.

Culbertson, M.R., Donahue, T.F., and Henry, S.A., 1976, Control of inositol biosynthesis in *Saccharomyces cerevisiae*; inositol-phosphate synthetase mutants. *J. Bacteriol.* **126:** 243–250.

Dyson, N., 1998, The regulation of E2F by pRB-family proteins. *Gene Dev.* **12:** 2245–2262.

Eisenberg, F., Jr., 1967, D-*myo*-Inositol 1-phosphate as product of cyclization of glucose 6-phosphate and substrate for a specific phosphatase in rat testis. *J. Biol. Chem.* **242:** 1375–1382.

Eisenberg, F., Jr., Bolden, A.H., and Loewus, F.A., 1964, Inositol formation by cyclization of glucose chain in rat testis. *Biochem. Biophys. Res. Commun.* **14:** 419–424.

Eisenberg, F, Jr., and Parthasarathy, R., 1984, *Myo*-inositol 1-phosphate. In: Bergmeyer, H.U. (ed.), Methods of Enzymatic Analysis, 3rd ed., Vol. 6. Verlag Chemie, Weinheim, pp. 371–375.

Eisenberg, F. Jr., and Parthasarathy, R., 1987, Measurement of biosynthesis of *myo*-inositol from glucose 6-phosphate. *Methods Enzymol.* **14:** 127–143.

Fisher, S.K., Novak, J.E., and Agranoff, B.W., 2002, Inositol and higher inositol phosphates in neural tissues: Homeostasis, metabolism and functional significance. *J. Neurochem.* **82:** 736–754.

Frey, R., Metzler, D., Fischer, P., Heiden, A., Scharfetter, J., Moser, E., and Kasper, S. 1998, Myo-inositol in depressive and healthy subjects determined by frontal ^{1}H-magnetic resonance spectroscopy at 1.5 tesla. *J. Psychaitr. Res.,* **32:** 411–420.

Friedman, S.D., Dager, S.R., Parow, A., Hirashima, F., Demopulos, C., Stoll, A.L., Lyoo, I.K., Dunner, D.L., and Renshaw, P.F., 2004, Lithium and valproic acid treatment effects on brain chemistry in bipolar disorder. *Biol. Psychiatry* **56:** 340–348.

Fux, M., Benjamin, J., and Belmaker, R.H., 1999, Inositol versus placebo augmentation of serotonin reuptake inhibitors in the treatment of obsessive-compulsive disorder: A double-blind cross-over study. *Int. J Neuropsychopharmacol.* **2:** 193–195.

Fux, M., Levine, J., Aviv, A., and Belmaker, R.H., 1996, Inositol treatment of obsessive-compulsive disorder. *Am. J. Psychiatry* **153:** 1219–1221.

Gelber, D., Levine, J., and Belmaker, R.H., 2001, Effect of inositol on bulimia nervosa and binge eating. *Int. J. Eat. Disord.* **29:** 345–348.

Greenberg, M.L., Reiner, B., and Henry, S.A., 1982, Regulatory mutations of inositol biosynthesis in yeast: Isolation of inositol-excreting mutants. *Genetics,* **100:** 19–33.

Groenen, P.M., Merkus, H.M., Sweep, F.C., Wevers, R.A., Janssen, F.S., and Steegers-Theunissen, R.P., 2003, Kinetics of *myo*-inositol loading in women of reproductive age. *Ann. Clin. Biochem.* **40:** 79–85.

Guan, G., Dai, P., and Shechter, I., 2003, cDNA cloning and gene expression analysis of human *myo*-inositol 1-phosphate synthase. Arch. *Biochem. Biophys.* **417:** 251–259.

Hallcher, L.M., and Sherman, W.R., 1980, The effects of the lithium ion and other agents on the activity of *myo*-inositol 1-phosphatase from bovine brain. *J. Biol. Chem.* **255:** 10896–10901.

Harwood, A.J., 2005, Lithium and bipolar mood disorder: The inositol-depletion hypothesis revisited. *Mol. Psychiatry* **10:** 117–126.

Hasegawa, R., and Eisenberg, F. Jr., 1981, Selective hormonal control of myo-inositol biosynthesis in reproductive organs and liver of the male rat. *Proc. Natl. Acad. Sci. U.S.A.* **78:** 4863–4866.

Hirsch, J.P., and Henry, S.A., 1986, Expression of the *Saccharomyces cerevisiae* inositol-1-phosphate synthase (*INO1*) gene is regulated by factors that affect phospholipid synthesis. *Mol. Cell. Biol.* **6:** 3320–3328.

Honchar, M.P., Ackerman, K.E., and Sherman, W.R., 1989, Chronically administered lithium alters neither *myo*-inositol monophosphatase activity nor phosphoinositide levels in rat brain. *J. Neurochem.* **53:** 590–594.

Jin, X., Foley, K.M., and Geiger, J.H., 2004, The structure of the 1L-*myo*-inositol-1-phosphate synthase-NAD-2-deoxy-D-glucitol 6-(E)-vinylhomophosphonate complex demands a revision of the enzyme mechanism. *J. Biol. Chem.* **279:** 13889–13895.

Jin, X., and Geiger, J.H., 2004, Structures of NAD^{+}- and NADH-bound 1-L-myo-inositol 1phosphate synthase. *Acta. Crystallogr. D. Biol. Crystallogr.* **59:** 1154–1164.

Jiranek, V., Graves, J.A., and Henry, S.A., 1998, Pleiotropic effects of the *opi1* regulatory mutation of yeast: Its effects on growth and on phospholipid and inositol metabolism. *Microbiology* **144:** 2739–2748.

Ju, S., Shaltiel, G., Shamir, A., Agam, G., and Greenberg, M.L., 2004, Human 1-D-*myo*-inositol-3-phosphate synthase is functional in yeast. *J. Biol. Chem.* **279:** 21759–21765.

Kadonaga, J.T., Jones, K.A., and Tjian, R., 1986, Promoter-specific activation of RNA polymerase II transcription by Sp1. *Trends Biochem. Sci.* **11:** 20–23.

Kaplan, Z., Amir, M., Swartz, M., and Levine, J., 1996, Inositol treatment of post-traumatic stress disorder. *Anxiety* **2:** 51–52.

Kel, A.E., Kel-Margoulis, O.V., Farnham, P.J., Bartley, S.M., Wingender, E., and Zhang, M.Q., 2001, Computer-assisted identification of cell cycle-related genes: New targets for E2F transcription factors. *J. Mol. Biol.* **309:** 99–120.

Klig, L.S., and Henry, S.A., 1984, Isolation of the yeast *INO1* gene: Located on an autonomously replicating plasmid, the gene is fully regulated. *Proc. Natl. Acad. Sci. U.S.A.* **82:** 3816–3820.

Levine, J., Aviram, A., Holan, A., Ring, A., Barak, Y., and Belmaker, R.H., 1997, Inositol treatment of autism. *J. Neural Transm.* **104:** 307–310.

Levine, J., Barak, Y., Gonzales, M., Szor, H., Elizur, A., Kofman, O., and Belmaker, R.H., 1995b, Double-blind, controlled trial of inositol treatment of depression. *Am. J. Psychiatry* **152:** 792–794.

Levine, J., Kurtzman, L., Rapport, A., Zimmerman, J., Bersudsky, Y., Shapiro, J., Belmaker, R.H., and Agam, G., 1996, CSF inositol does not predict antidepressant response to inositol. *J. Neural Transm.* **103:** 1457–1462.

Levine, J., Rapaport, A., Lev, L., Bersudsky, Y., Kofman, O., Belmaker, R.H., Shapiro, J., and Agam, G., 1993b, Inositol treatment raises CSF inositol levels. *Brain Res.* **627:** 168–170.

Levine, J., Ring, A., Barak, Y., Elizur, A., and Belmaker, R.H., 1995a, Inositol may worsen attention deficit disorder with hyperactivity. *Human Psychopharmacol.* **10:** 481–484.

Levine, J., Umansky, R., Ezrielev, G., and Belmaker, R.H., 1993a, Lack of effect of inositol treatment in chronic schizophrenia. *Biol. Psychiatry* **33:** 673–675.

Loewus, M.W., Loewus, F.A., Brillinger, G.U., Otsuka, H., and Floss, H.G., 1980, Stereochemistry of the *myo*-inositol-1-phosphate synthase reaction. *J. Biol. Chem.* **255:** 11710–11712.

Loewy, B.S., and Henry, S.A., 1984, The *INO2* and *INO4* loci of *Saccharomyces cerevisiae* are pleiotropic regulatory genes. *Mol. Cell Biol.* **4:** 2479–2485.

Lopes, J.M., and Henry, S.A., 1991, Interaction of trans and cis regulatory elements in the INO1 promoter of *Saccharomyces cerevisiae. Nucleic Acids Res.* **19:** 3987–3994.

Maeda, T., and Eisenberg, F. Jr., 1980, Purification, structure, and catalytic properties of L-*myo*-inositol 1-phosphate synthase from rat testis. *J. Biol. Chem.* **255:** 8458–8464.

Majerus, P.W., Kisseleva, M.V., and Norris, F.A., 1999, The role of phosphatases in inositol signaling reactions. *J. Biol. Chem.* **274:** 10669–10672.

Majumder, A.L., Chatterjee, A., Dastidar, K.G., and Majee, M., 2003, Diversification and evolution of L-*myo*-inositol 1-phosphate synthase. *FEBS Lett.* **553:** 3–10.

Majumder, A.L., Johnson, M.D., and Henry, S.A., 1997, 1L-*myo*-inositol-1-phosphate synthase, Biochim. *Biophys. Acta,* **1348:** 245–256.

Mauck, L.A., Wong, Y.H., and Sherman, W.R., 1980, L-*myo*-Inositol-1-phosphate synthase from bovine testis: Purification to homogeneity and partial characterization. *Biochemistry* **19:** 3623–3629.

McCauley, J.L., Li, C., Jiang, L., Olson, L.M., Crockett, G., Gainer, K., Folstein, S.E., Haines, J.L., and Sutcliffe, J.S., 2005, Genome-wide and ordered-subset linkage analyses provide support for autism loci on 17q and 19p with evidence of phenotypic and interlocus genetic correlates. *BMC. Med. Genet.* **6:** 1–15.

McDougle, C.J., Naylor, S.T., Cohen, D.J., Volkmar, F.R., Heninger, G.R., and Price, L.H., 1996, A double-blind, placebo-controlled study of fluvoxamine in adults with autistic disorder. *Arch. Gen. Psychiatry,* **53:** 1001–1008.

Migaud, M.E., and Frost, J.W., 1995, Inhibition of *myo*-inositol 1-phosphate synthase reaction by a reaction coordinate intermediate. *J. Am. Chem. Soc.* **117:** 5154–5155.

Naccarato, W.F., Ray, R.E., and Wells, W.W., 1974, Biosynthesis of *myo*-inositol in rat mammary gland. Isolation and properties of the enzymes. *Arch. Biochem. Biophys.* **162:** 194–201.

Nemets, B., Talesnick, B., Belmaker, R.H., and Levine, J., 2002, *Myo*-inositol has no beneficial effect on premenstrual dysphoric disorder. *World J. Biol. Psychiatry* **3:** 147–149.

Norman, R.A., McAlister, M.S., Murray-Rust, J., Movahedzadeh, F., Stoker, N.G., and McDonald, N.Q., 2002, Crystal structure of inositol 1-phosphate synthase from *Mycobacterium tuberculosis*, a key enzyme in phosphatidylinositol synthesis. *Structure* **10:** 393–402.

Pacheco, M.A., and Jope, R.S., 1996, Phosphoinositide signaling in human brain. *Prog. Neurobiol.* **50:** 255–273.

Palatnik, A., Frolov, K., Fux, M., and Benjamin, J., 2001, Double blind, controlled, crossover trial of inositol versus fluvoxamine for the treatment of panic disorder. *J. Clin. Psycho-pharmacol.* **21:** 335–339.

Parthasarathy, R., and Eisenberg, F. Jr., 1986, The inositol phopholipids: A stereochemical view of biological activity. *Biochem. J.* **235:** 313–322.

Parthasarathy, R., and Eisenberg, F. Jr., 1991, Inositol Phosphates and Derivatives: Synthesis, Biochemistry and Therapeutic Potential. American Chemical Society, Washington, DC, pp. 1–19.

Parthasarathy, R., Parthasarathy, L., and Vadnal, R.E., 1993, Identification of phosphatidyl inositol tris-phosphate in rat brain. In: Fain, J. (ed.), Methods in Neurosciences, Vol. 18. Academic Press, N.Y., pp. 113–124.

Parthasarathy, L., Vadnal, R.E., Parthasarathy, R., and Devi, C.S.S., 1994, Biochemical and molecular properties of lithium-sensitive *myo*-inositol monophosphatase. *Life Sci.* **54:** 1127–1142.

Renshaw, P.F., Joseph, N.E., and Leigh, J.S., 1986, Chronic dietary lithium induces increased levels of *myo*-inositol 1-phosphatase activity in rat cerebral cortex homogenates. *Brain Res.* **380:** 401–404.

Rivera-Gonzalez, R., Petersen, D.N., Tkalcevic, G., Thompson, D.D., and Brown, T.A., 1998, Estrogen-induced genes in the uterus of ovariectomized rats and their regulation by droloxifene and tamoxifen. *J. Steroid Biochem. Mol. Biol.* **64:** 13–24.

Saiardi, A., Bhandari, R., Resnick, A.C., Snowman, A.M., and Snyder, S.H., 2004, Phosphorylation of proteins by inositol pyrophosphates. *Science* **306:** 2101–2105.

Seelan, R.S., Parthasarathy, L., and Parthasarathy, R., 2004, E2F1 regulation of the human *myo*-inositol 1-phosphate synthase (*ISYNA1*) gene promoter. *Arch. Biochem. Biophys.* **431:** 95–106.

Shaltiel,G, Shamir, A., Shapiro, J., Ding, D., Dalton, E., Bialer, M., Harwood, A.J., Belmaker, R.H., Greenberg, M.L. and Agam, G., 2004, Valproate decreases inositol biosynthesis. *Biol Psychiatry,* **56:** 868–874.

Shamir, A., Shaltiel, G., Greenberg, M.L., Belmaker, R.H., and Agam, G, 2003, The effect of lithium on expression of genes for inositol biosynthetic enzymes in mouse hippocampus: A comparison with the yeast model. *Brain Res. Mol. Brain Res.* **115:** 104–110 (Erratum in Brain Res. Mol. Brain Res, 2004, 123: 137).

Sherman, W.R., Leavitt, A.L., Honchar, M.P., Hallcher, L.M., Packman, P.M., and Phillips, B.E., 1981b, Evidence that lithium alters phosphoinositide metabolism: Chronic administration elevates primarily D-myo-inositol-1-phosphate in cerebral cortex of the rat. *J. Neurochem.* **36:** 1947–1951.

Sherman, W.R., Loewus, M.W., Pina, M.Z., and Wong, Y.H., 1981a, Studies on myo-inositol-1-phosphate from *Lilium longiflorum* pollen, *Neurospora crassa* and bovine testis. Further evidence that a classical aldolase step is not utilized. *Biochim. Biophys. Acta.* **660:** 299–305.

Stein, A.J., and Geiger, J.H., 2000, Structural studies of MIP synthase. *Acta. Crystallogr. D. Biol. Crystallogr.* **56:** 348–350.

Stein, A.J., and Geiger, J.H., 2002, The crystal structure and mechanism of 1-L-*myo*-inositol-1-phosphate synthase. *J. Biol. Chem.* **277:** 9484–9491.

Suske, G., 1999, The Sp-family of transcription factors. Gene 238: 291–300.

Tao, Y., Kassatly, R.F., Cress, W.D., and Horowitz, J.M., 1997, Subunit composition determines E2F DNA-binding site specificity. *Mol. Cell. Biol.* **17:** 6994–7007.

Thurston, J.H., Sherman, W.R., Hauhart, R.E., and Kloepper, R.F., 1989, *myo*-inositol: A newly identified non-nitrogenous osmoregulatory molecule in mammalian brain. *Pediatr. Res.* **26:** 482–485.

Tsunoda, T., and Takagi, T., 1999, Estimating transcription factor bindability on DNA. *Bioinformatics* **15:** 622–630.

Vadnal, R.E., Parthasarathy, L., and Parthasarathy, R., 1997, Role of inositol in psychiatric disorderss: Basic and clinical aspects. *CNS Drugs* **7:** 6–16.

Wells, J., Graveel, C.R., Bartley, S.J., Madore, S.J., and Farnham, P.J., 2002, The identification of E2F1-specific target genes. *Proc. Natl. Acad. Sci. U.S.A.* **99:** 3890–3895.
Whitworth, P., and Kendall, D.A., 1989, Effects of lithium on inositol phospholipid hydrolysis and inhibition of dopamine D1 receptor-mediated cyclic AMP formation by carbachol in rat brain slices. *J. Neurochem.* **53:** 536–541.
Whiting, P.H., Palmano, K.P., and Hawthorne, J.N., 1979, Enzymes of *myo*-inositol and inositol lipid metabolism in rats with streptozotocin-induced diabetes. *Biochem. J.* **179:** 549–553.
Wong, Y.H., and Sherman, W.R., 1980, Anomeric and other substrate specificity studies with *myo*-inositol 1-P synthase. *J. Biol. Chem.* **260:** 11083–11090.
Yildiz, A., Demopulos, C.M., Moore, C.M., Renshaw, P.F., and Sachs, G.S., 2001, Effect of lithium on phosphoinositide metabolism in human brain: A proton decoupled $^{(31)}$P magnetic resonance spectroscopy study. *Biol. Psychiatry* **50:** 3–7.
York, J.D., and Hunter, T., 2004, Signal transduction: Unexpected mediators of protein phosphorylation. *Science* **306:** 2053–2055.
Zeiner, M., and Gehring, U., 1994, Cloning of 5′ cDNA regions by inverse PCR. *Biotechniques* **17:** 1052–1054.

Chapter 13

Evolutionary Divergence of L-*myo*-Inositol 1-Phosphate Synthase: Significance of a "Core Catalytic Structure"

Krishnarup GhoshDastidar*, Aparajita Chatterjee*, Anirban Chatterjee*, and Arun Lahiri Majumder
Plant Molecular and Cellular Genetics, Bose Institute (Centenary Campus), P-1/12 C.I.T Scheme VII M, Kolkata 700054, India

"It seems that there are few limits to the number of molecules and biochemical pathways into which evolution has incorporated myo-inositol"

— Hinchliffe and Irvine, 1997.

1. INTRODUCTION

Darwin's theory of natural selection and theories derived thereof explaining the processes of natural evolution had given rise to a dogma of adaptationist arguments about most biological variations. The first major deviation from such adaptationist arguments became necessary to explain the observations made at the molecular level.

With the determination of amino acid sequences of hemoglobins and cytochrome C from many mammalian sources, sequence comparisons revealed that the number of amino acid divergences between different pair of mammals seemed to be roughly proportional to the time since they had diverged from one another, as inferred from the fossil record (Holsinger, 2004). As a possible explanation, Zuckerkandl and Pauling (1965) propounded a "molecular clock hypothesis" in which they proposed the presence of a constant rate of amino acid substitution over time.

*Contributed equally.

A. Lahiri Majumder and B. B. Biswas (eds.), Biology of Inositols and Phosphoinositides, 313–338
© 2006 *Springer. Printed in the Netherlands.*

The molecular clock hypothesis was used to propose the probable divergence date of humans and apes to approximately 5 million years ago (Sarich and Wilson, 1967). A year before this, the presence of surprising amounts of genetic variability within populations using the then novel techniques of protein electrophoresis was reported (Harris, 1966; Hubby and Lewontin, 1966). Harris (1966) had identified three loci with polymorphic alleles, and one locus with three alleles after studying only 10 loci in human populations. Evolutionary geneticists of the time were hard pressed to explain both sets of observations through any evolutionary mechanism that could produce a constant rate of substitution. Nor could natural selection be implicated to maintain so much polymorphism within populations (Holsinger, 2004). The problems in explaining the natural phenomena were too empirical in nature.

The hypothesis forwarded by Kimura and others (Kimura, 1968; King and Jukes, 1969) proposed a way to solve all such empirical problems. In their view if it is assumed that the vast majority of amino acid substitutions are selectively neutral, then substitutions will occur at approximately a constant rate (assuming that mutation rates do not vary over time) and it will be easy to maintain lots of polymorphism within populations with apparently no cost of selection.

The neutral theory asserts that alternative alleles at variable protein loci are selectively neutral irrespective of the locus being important or unimportant (Kimura, 1968). Selectively neutral allele means that the selectional force acting on different genotypes at a locus is weak enough to allow the interaction of weaker evolutionary forces like mutation, drift, mating system and migration in maintaining the pattern of variation observed (Kimura, 1968; Holsinger, 2004).

A possible outcome of neutral evolution on protein is the accumulation of changes in regions of protein sequence, not directly involved in the functional activities of the protein. Such "silent" changes, may lead to considerable differences in protein sequences across diverse phyla, even if the functionality of the proteins remain unaltered. However, differences observed in proteins with same or similar functions might occur not only due to "silent" mutations, but also due to adaptive changes for cellular localizations and diverse intracellular environment or even due to convergent evolution. Neutral theory of evolution predicts that there exists a significant correlation between protein dispensability and evolutionary rate (Kimura and Ohta, 1974) and dispensable (non essential or unimportant) proteins tend to evolve faster than the indispensable (essential or important) ones (Wilson *et al.*, 1977; Zhang and He, 2005). However, it has been found that evolutionary rate of protein and their fitness effect are correlated only when fitness value is <0.5 (Hirsh and Fraser, 2001). Besides this correlation has been found insignificant in case of singleton genes (Yang *et al.*, 2003), covariation with gene expression (Pal *et al.*, 2003) and after gene expression level is controlled for (Rocha and Danchin, 2004). This has also been found to be insignificant in prokaryotes. However, using larger data set and improved method significant correlation can be obtained (Hirsh and Fraser, 2003).

All such changes need to consider the possibility of maintaining a certain "core structure" to keep the functionality of the protein unaltered. The preservation of such "core structure" will only highlight the evolutionary significance of selectional forces acting on a few amino acids of the protein, while allowing the effects of other forces to modify the rest of the amino acid sequence, as proposed by the neutral theory. In short, proteins with variable degrees of sequence difference, yet a remarkable degree of functional similarity across phyla will be examples of the neutral theory as a paradigm complementary to natural selection.

Such a hypothesis is well documented in the case of the inositol biosynthetic machinery .The statement quoted at the beginning of this chapter exemplifies the ubiquity of inositol along the evolutionary diverse organisms. This has been amply confirmed by the vast majority of the role played by inositol and its various derivatives in the biology of the prokaryotic, archaeal and eukaryotic systems. The earlier 12 chapters of this volume dealt with such enormous importance of the cyclitol in the biological kingdom. Interestingly enough, throughout the biological kingdom, inositol is synthesized by the one and only enzyme protein L-myo-inositol 1-phosphate synthase (EC 5.5.1.4; hitherto referred to as MIPS). When we look at the evolutionary time frame, which originates from a predicted cenancestor through cyanobacteria, eubacteria and archaea to ultimately higher eukaryotes like plant and human, it seems quite a long period since MIPS started evolving with continuous divergence (Figure 1). This in turn also highlights not only the indispensability of MIPS but also of the whole inositol biosynthetic pathway in a given organism. Presence of this protein throughout the spectrum of life has also made the stand of MIPS as an important protein very rigid. It is apparent now that the antiquity of MIPS catalytic activity is achieved through the conservation of a sequence of amino acids, which maintain the "core catalytic domain" of the enzyme protein isolated from evolutionary diverse organisms despite variations in the nucleotide sequence of the individual gene(s). Probably MIPS is one of the important examples which, being an indispensable component of a highly conserved biochemical pathway, evolved over such a long evolutionary time scale accumulating most of its amino acid substitution apparently in the functionally less important regions albeit keeping a core catalytic region unperturbed across diverse phyla. Although this effect seems to decline with evolutionary time, a scope of detailed statistical analysis is ahead which may lead to quantitative conclusion regarding protein dispensability and evolutionary rate. The present chapter attempts to explain such a phenomenon in the light of knowledge acquired from analysis of the various Ino1 gene(s) and crystal structure of the enzyme from different sources

2. L-MYO-INOSITOL 1- PHOSPHATE SYNTHASE: TRACING THE GENE/PROTEIN ALONG THE EVOLUTIONARY LINEAGE

L-myo-inositol 1-phosphate synthase (MIPS) is the first enzyme for the synthesis of L-myo-inositol 1-phosphate from D-glucose 6-phosphate (G6P), which is

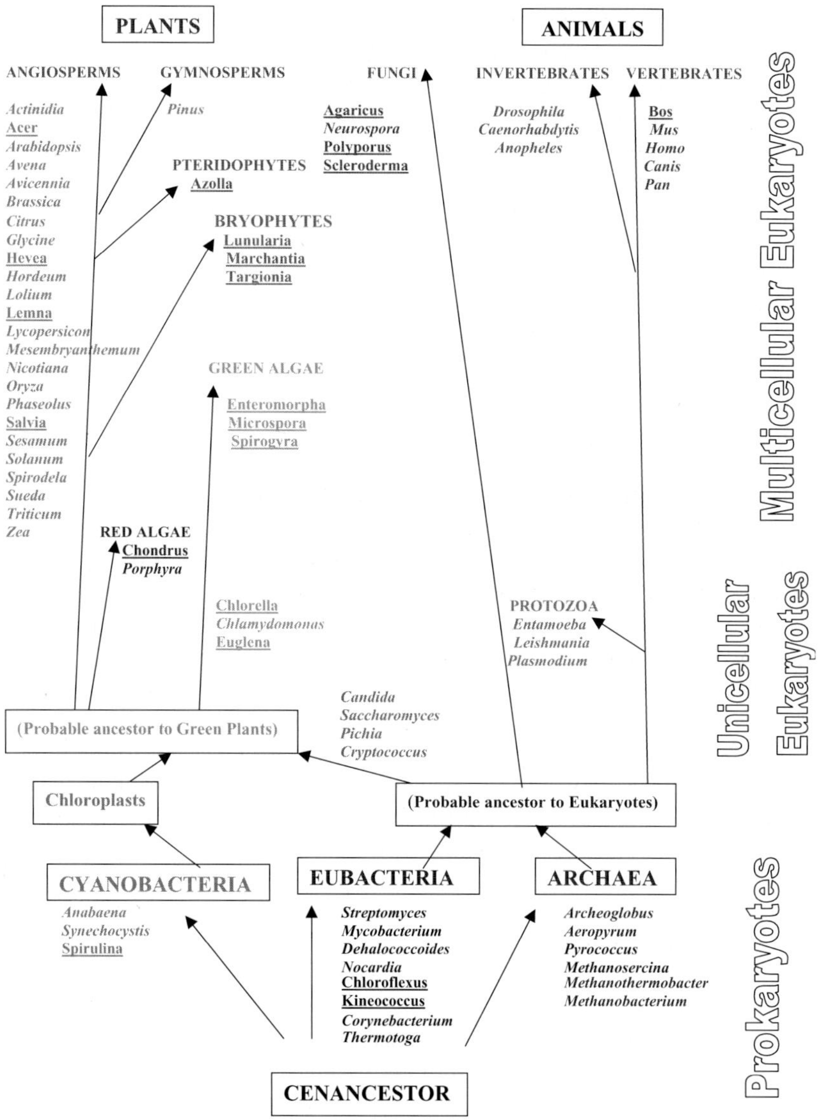

Figure 1. Distribution of MIPS across the spectrum of life showing possible gene lineages. The names in normal underlined font indicate organisms with a report of MIPS protein only, while names in *italics* indicate organisms for which a gene encoding MIPS has been identified (complete or partial). (Adapted from Majumder *et al.*, 2003)

converted into free inositol via dephosphorylation of the MIPS reaction product by a specific Mg^{2+}-dependent inositol 1-phosphate phosphatase (EC 3.1.3.25). The structural gene coding for MIPS, termed Ino1, was first identified from yeast, *Saccharomyces cerevisiae* (Donahue and Henry, 1981; Majumder, 1981), cloned (Klig and Henry, 1984) and sequenced (Johnson and Henry, 1989). Subsequently, Ino1 homologues have been identified from a number of higher plants, fungi and protozoan parasites (Majumder *et al.*, 1997, 2003).

A survey of the sources of MIPS identified till date shows that MIPS enzyme activity is distributed in evolutionary diverse phyla from eubacteria, archaebacteria, cyanobacteria, algae, fungi, higher plants and animal (Figure 1). Between the years of 2000 and 2004, a large number of MIPS coding sequences (Ino1 homologues) have been reported from prokayotic sources. Current study reveals the presence of MIPS coding genes in several new prokaryotic organisms and some of the protein products had been subjected to heterologous overexpression and characterization. The first report of a probable eubacterial MIPS was from the bacterium *Streptomyces coelicolor* (Pittner *et al.*, 1979). However, the first cloning and experimental evidence of a eubacterial MIPS was from the *Mycobacterium tuberculosis* by Bachhawat and Mande (1999). The mycobacterial MIPS gene was identified through bioinformatic comparisons, cloning and genetic complementation in an Ino1-disrupted inositol auxotrophic yeast strain *FY250*. Subsequently, the same gene-product has been bacterially expressed, purified and crystallized (Norman *et al.*, 2002).

The first MIPS to be reported from an archaebacterial source was from *Archaeoglobus fulgidus* (Chen *et al.*, 2000). The *Archaeoglobus* Ino1 homologue has been bacterially overexpressed, characterized and the activity has been studied by P^{32} NMR analysis (Chen *et al.*, 2000). The report showed that the MIPS from *Archaeoglobus* is more active in an order of magnitude than other MIPS enzymes, and was operative at extremely high temperatures (above 60 °C) having an optimal enzyme activity at around 85 °C. Subsequently, several other archaebacterial sources have been found where from MIPS homologues have been reported, like *Pyrococcus, Aeropyrum, Methanothermobacter, Methanosercina* etc. But none of these MIPS coding genes have been bacterially overexpressed and characterized.

In line with the large number of genome sequencing projects, especially for the microbes, several more putative MIPS sequences have been reported , like those from *Bacillus, Magnetospirillum, Novosphingobium* etc. However, the bacterial expression and functional characterization of these MIPS proteins are still awaited. Recently one eubacterial MIPS from *Thermotoga maritima* has been cloned overexpressed and crystallized (Table 1). Further, a novel MIPS has been reported from the cyanobacterium, *Synechocystis sp* PCC6803, encoded by the hitherto unassigned, hypothetical ORF sll1722 (Chatterjee *et al.*, 2004). MIPS proteins have also been reported from other phylogenetically diverse organisms such as red algae, fungi, several bryophytes and gymnosperms (Figure 1).

Table 1. Details of available crystal structures of MIPS, till date, from five different organisms

System	Enzyme	Resolution in Å	Ion bound	Small molecule bound	Subunit organization	PDB ID	References
S. cerevisiae	MIPS	2.4		NAD^+	Homotetramer (Dimer of two asymmetric dimeric units)	1JKF	*J.Biol.Chem.* 277 pp. 9484 (2002)
		2.2	NH_4^+	NADH and 2deoxy-glucitol-6 phosphate		1JKI	
		2.6	—	—		1P1F	*Acta Crystallogr., Sect. D* 59 pp. 1154 (2003)
		1.95		NAD^+		1P1H	*Acta Crystallogr., Sect. D* 59 pp. 1154 (2003)
		2.4		NAD^+		1P1I	*Acta Crystallogr., Sect. D* 59pp. 1154 (2003)
	MIPS	1.7	PO_4^{3-}	NADH, Propane 1,2,3 triol and glycerol		1P1J	
		2.1	EDTA	NADH		1P1K	
		2.65		NAD^+ and selenomethionine		1LA2	*J. Struct. Funct. Genom.* 2 pp. 129 (2002)
M. tuberculosis	MIPS	1.95	Zn^+	NAD^+	Homotetramer (Dimer of two asymmetric dimeric units)	1GRO	*Structure Fold.Des.* 10 pp. 393 (2002)
			$C_2H_6O_2$				

	SeMetMIPS (Hypothetical 40.1 Kda Protein Rv0046C)	2.6	As				
A. fulgidus dsm 4304	MIPS	1.9	K^+	NAD^+	Homotetramer (Dimer of two asymmetric dimeric units)	1U1I	*Biochemistry* 44 pp. 213 (2005)
T. maritima	MIPS (Tm1419)	1.7	Mg^{2+} Cl^-	PO_4^{3-} $2H_2O$ NAD^+	Homotetramer	1VJP	Joint Center for Structural Genomics (JCSG)
C. elegans	MIPS	2.3	Cl^- I^- K^+ $P_2O_7^{2-}$	NAD^+	Homotetramer	1VKO	Joint Center for Structural Genomics (JCSG)

Several fungi such as *Candida albicans*, *Pichia pastoris*, *Neurospora crassa* have been reported to code for a MIPS gene highly homologous to the *Saccharomyces* MIPS. Interestingly, all the fungal MIPS protein sequences have the unique feature of having an extended N-terminal tail of around 28 amino acids, which is absent in all other MIPS sequences. So far all the five MIPS crystallized and that of *Drosophila* (Park and Kim, 2004) have been found to be tetrameric. Among the eukaryotic MIPS sequences, the fungal MIPSs are also tetrameric in nature.

The most numerous of all eukaryotic MIPS identified till date are those from plant sources. The first MIPS gene to be cloned from a plant source was *Spirodela polyrrhiza* (Smart and Fleming, 1993). These workers showed that tur1, a cDNA, which is upregulated early by ABA-induction in *Spirodela*, is a gene coding for MIPS. The tur1 gene was found to have high sequence similarity to the INO1 gene (coding for MIPS) from yeast. Subsequent analysis showed that the protein encoded by tur1 was indeed a MIPS. Since then, different plant MIPS-coding genes have been identified, like *Citrus* (Abu-abied and Holland, 1993), *Arabidopsis* (Johnson, 1994; Johnson and Sussex, 1995), *Phaseolus* (Johnson and Wang, 1996), *Mesembryanthemum* (Ishitani *et al.*, 1996), *Zea* and *Hordeum* (Keller *et al.*, 1998), *Oryza* (Yoshida *et al.*, 1999), *Brassica* (Larson and Raboy, 1999) etc. (Table 2). MIPS has been reported from a number of lower plant groups, (Dagupta *et al.*,1984) as well as higher plant sources such as rice cell cultures (Funkhouser and Loewus, 1975), *Vigna* (Majumder and Biswas, 1973), rice grains (Hayakawa and Kurasaw, 1976), and both cytosolic and chloroplastic fraction of pea leaves (Imhoff and Bourdu, 1973), *Euglena*, *Vigna* (Adhikari *et al.*, 1987) and *Oryza* (Ray Chaudhuri *et al.*, 1997; Hait *et al.*, 2002). MIPS coding genes have also been reported from several plant sources which are either important economically like *Glycine max* (Iqbal *et al.*, 2002), *Sesamum* (Chun *et al.*, 2003) or halophytes like *Porteresia* (GenBank accession number AAP74579, Majee *et al.*, 2004) or the resurrection plant *Xerophyta* (GenBank accession number AY323824, Majee *et al.*, 2005). Recently presence of a MIPS has been reported from the aquatic fern, *Azolla filiculoides* (Benaroya *et al.*, 2004).

3. MIPS PROTEIN SEQUENCE COMPARISONS

Till date over 70 Ino1 genes have been reported from evolutionarily diverse organisms, both prokaryotic and eukaryotic (Table 2). When different representative MIPS amino acid sequences (available at http://www.ncbi.nlm.nih.gov/entrez) are compared against each other with a multiple alignment tool like MULTALIN (http://prodes.toulouse.inra.fr/multalin/multalin.html) (Corpet, 1988), a clear difference between the prokaryotic and the eukaryotic sequences become evident (data not shown). As discussed earlier (Chun *et al.*, 2003; Majumder *et al.*, 1997), stretches of amino acid residues like GWGGNNG, LWTANTERY, NGSPQNTFVPGL, and SYNHLGNNDG are conserved in MIPS proteins of all eukaryotes.

Table 2. Diversity of MIPS across the spectrum of life, with key features of the gene/enzyme (names of organisms arranged alphabetically)

Organisms	Accession number		Gene and protein size		Remarks
	Protein	Nucleotide	Gene (kb)	Protein (kDa) (subunit/holoenzyme)	
1. *Actinidia arguta*	AAF97409	AY005128	~1.5	~58/–	(partial)
2. *Aeropyrum pernix*	F72632	APE1517	~1.1	~40/–	(putative)
3. *Anopheles gambii*	EAA00329	AAAB01008986	~1.5	~60/–	(putative)
4. *Arabidopsis thaliana*	T50021	AY065415	~1.5	~60/~180	(native/exp)
5. *Archaeoglobus fulgidus*	AAB89456	AE000979	~1.2	~45/–	(exp)
6. *Aspergillus nidulans*	XP_411762	EAA61811		~50/–	(putative)
7. *Aster tripolium*	BAC57963	AB090886.1			(partial)
8. *Avena sativa*	BAB40956	AB059557	~1.5	~60/–	(putative)
9. *Avicennia marina*	AAK21969	AY028259			(partial)
10. *Bacillus cereus*	AAP09606	AE017006			(putative)
11. *Bacteroides thetaiotaomicron*	AAO76633	AE016932.1			(partial)
12. *Branchiostoma belcheri*	AAL02140	AY043320	~0.5		(partial)
13. *Brassica napus*	AAB06756	U66307	~1.5	~60/~180	(putative)
14. *Caenorhabditis elegans*	T18569	NM_064098	~1.6	~60/~180	(putative)
15. *Candida albicans*	S45452	L22737	~1.6	~65/~240	(exp)
16. *Canis familiaris*	XP_533872	XM_533872		~60/–	(putative)
17. *Chlamydomonas reinhardtii Chlamy db*	20021010.7198.1		~1.5	~60/–	(putative)
18. *Citrus paradisi*	CAA83565	Z32632	~1.5	~60/~180	(putative)
19. *Corynebacterium glutamicum*	BAC00390	AP005283	~1.2	~45/–	(putative)
20. *Debaryomyces hansenii*	CAG90267	CR382139		~60/–	(putative)
21. *Dehalococcoides ethenogenes*	YP_181702	NC_002936		~40/–	(putative)
22. *Drosophila melanogaster*	AAD02819	AF071104	~1.6	~60/–	(exp)

(continued)

Table 2. (continued)

Organisms	Accession number		Gene and protein size		Remarks
	Protein	Nucleotide	Gene (kb)	Protein (kDa) (subunit/holoenzyme)	
23. *Entamoeba histolytica*	CAA72135	Y11270	~1.5	~60/~180	(exp)
24. *Giardia lamblia*	EAA38884.1	AACB01000086			(putative)
25. *Glycine max*	AAK49896	AF293970	~1.5	~60/~180	(native/exp)
26. *Homo sapiens*	AAF26444	AF220530	~1.7	~70/–	(native)
27. *Hordeum vulgare*	T04399	AF056325	~1.5	~60/~180	(native/exp)
28. *Leishmania amazonensis*	AAB51376	U91965	~1.6	~65/–	(exp)
29. *Leishmania major*	CAB94019	AL358652	~1.6	~65/–	(putative)
30. *Leishmania mexicana*	CAC69873	AJ344544	~1.6	~65/–	(partial)
31. *Lolium perenne*	AAN52772	AY154382	~1.5	~60/–	(putative)
32. *Lycopersicon esculentum*	AAG14461	AF293460_1	~0.5		(partial)
33. *Magnetospirillum magnetotacticum*	ZP_00048843	NZ_AAAP01001385			(partial)
34. *Mesembryanthemum crystallinum*	AAB03687	U32511	~1.5	~60/~180	(native/exp)
35. *Methanosarcina acetivorans*	AAM03529	AE010664	~1.1	~40/–	(putative)
36. *Methanosarcina mazei*	AAM31066	AE013370	~1.1	~40/–	(putative)
37. *Methanothermobacter thermoautotrophicus*	NP_276233	NC_000916	~1.1	~40/–	(putative)
38. *Mus musculus*	AAF90201	AF288525	~1.6	~65/~200	(exp)
39. *Mycobacterium leprae*	AAC43244	U00015	~1.1	~40/~160	(putative)
40. *Mycobacterium tuberculosis*	P71703	NP_334460	~1.1	~40/~160	(crystal)
41. *Mycobacterium bovis*	NP_853716	BX248334			(putative)
42. *Nicotiana paniculata*	BAA84084	AB032073	~1.5	~60/~180	(putative)
43. *Nicotiana tabacum*	BAA95788	AB009881	~1.5	~60/~180	(putative)

44. *Neurospora crassa*	CAD70896	BX294019			(putative)
45. *Nocardia farcinica*	YP_121768	NC_006361		~40/–	(putative)
46. *Novosphingobium aromaticivorans*	ZP_00096038	NZ_AAAV01000170			(partial)
47. *Oryza sativa*	BAA25729	AB012107	~1.5	~60/~180	(native/exp)
48. *Phaseolus vulgaris*	T10964	U38920.1	~1.5	~60/~180	(exp)
49. *Pichia pastoris*	AAC33791	AF078915	~1.6	~65/~240	(exp)
50. *Pinus taeda Pine Genomic Seq Contig*	7989		~1.5	~60	(putative)
51. *Plasmodium falciparum*	CAD51482	AL929352	~1.8	~70/–	(putative)
52. *Plasmodium yoelli*	EAA15800	AABL01001197	~1.8	~70/–	(putative)
53. *Porteresia coarctata*	AAP74579	AF412340	~1.5	~60/~180	(native/exp)
54. *Pyrobaculum aerophilum*	AAL63705	AE009838	~1.1	~40/~160	(putative)
55. *Pyrococcus abyssi*	NP_126250	AJ248284	~1.1	~40/–	(putative)
56. *Pyrococcus furiosus*	AAL81740	AE010261	~1.1	~40/–	(putative)
57. *Pyrococcus horikoshii*	B75175	PAB1989	~1.1	~40/–	(putative)
58. *Saccharomyces cerevisiae*	A30902	L23520	~1.6	~65/~230	(native/exp/cryst)
59. *Sesamum indicum*	AAG01148	AF284065	~1.5	~60/~180	(native/exp)
60. *Solanum tuberosum*	AAK26439	AF357837_1	~1.5	~60/~180	(partial)
61. *Spirodela polyrrhiza*	P42803	Z11693	~1.5	~60/~180	(native/exp)
62. *Streptomyces coelicolor*	CAB38887	AL939115	~1.1	~40	(putative)
63. *Suaeda maritima*	AAL28131	AF433879	~1.5	~60/~180	(putative)
64. *Synechocystis PCC6803*	BAA17443	BA000022	~1.3	~50/~200	(exp)
65. *Sulfolobus solfataricus*	AAK41169	AE006710.1			(partial)
66. *Thermotoga maritima*	CAC21207	AJ401010.1	~1.1	~40/–	(putative)
67. *Thermotoga neapolitana*	CAC21211	AJ401014	~1.1	~40/–	(partial)
68. *Triticum aestivum*	AAD26332	AF120146	~1.5	~60/~180	(putative)
69. *Xenopus laevis*	AAH44073.1	BC044073			(putative)
70. *Xerophyta viscosa*	AY323824		~1.5	~60/~180	(exp)
71. *Yarrowia lipolytica*	CAG82716	CR382128			(partial)
72. *Zea mays*	AAC15756	AF056326	~1.5	~60/~180	(putative)

Among plants, larger stretches of amino acid residues are conserved throughout the length of the protein showing much higher degree of preservation of sequence identity irrespective of whether the plant is a monocot, dicot, green alga, or a gymnosperm. This certainly indicates a monophyletic origin of the higher plant MIPS. In case of fungi, such as *S. cerevisiae*, *Pichia pastoris* and *Candida albicans*, there is an extra amino acid stretch at the N-terminal end, unique to this group. This sequence is highly conserved among the fungi and is probably due to a later addition in the early fungal ancestral sequence after its divergence from the main eukaryotic stock. Since the fungal MIPS has a homotetrameric association, as compared to the presumed homotrimeric association of the other eukaryotic proteins, a relationship between the additional N-terminal tail and the homotetrameric structure may be conjectured. In case of animals, the MIPS sequence has an extra C-terminal sequence, the significance of which is not clear as yet.

From the phylogenetic tree derived by both Majumder *et al.* (2003) and Bachhawat and Mande (2000), it is quite obvious that MIPS genes form a closely related tight cluster with relation to the eukaryotic family and have little or weak sequence similarity to the prokaryotic MIPS genes. Further, the monophyletic origin of the eukaryotic MIPS genes is evident whereas the same is certainly not true for the prokaryotic MIPS genes. The prokaryotic MIPS genes show much less sequence similarity among each other and do not form a tight cluster during any phylogenetic tree generation (data not presented).

Amino acid sequence alignment of prokaryotic MIPS sequences highlights the lack of sequence similarity. The *Archaeoglobus* MIPS (Chen *et al*., 2000) has more sequence similarity to the eukaryotic MIPS than the other known prokaryotic ones. The distribution of the MIPS sequences in the prokaryotic cluster however, show that some other archaeal sequences (*e.g.*, *Aeropyrum, Pyrococcus*) share closer homology with eubacterial sequences (*e.g.*, *Thermotoga maritima* and *T. horikoshii*) than with the other archaeal sequences like *Archaeoglobus* or the *Methanosercina* and *Methanothermobacter*. The eubacterial MIPS sequences of *Mycobacterium* and *Streptomyces*, share closer homology with each other than with other eubacterium like *Thermotoga*. It seems likely that the eukaryotic MIPS progenitor had diverged early from the prokaryotic stock. Since then the prokaryotic MIPS have undergone profound changes in amino acid sequence.

4. STRUCTURAL INSIGHTS INTO MIPS PROTEINS FROM DIVERSE SOURCES

Characteristics of the crystal structures for MIPS proteins from five different organisms viz. *S cerevisiae* (Stein and Geiger, 2002, Jin and Geiger, 2003, Jin *et al.*, 2004), *M.tuberculosis* (Normal *et al.*, 2002), *A.fulgidus* (Stieglitz *et al.*, 2005), *T.maritima,* and *C. elegans* as available till date are presented in Table 1. Such structural analyses for two MIPS proteins, viz *Saccharomyces* and *Mycobacterium* have shed much light on MIPS evolution. The *Saccharomyces*

MIPS crystal structure is characterized by a homotetrameric association, with a 222 symmetry where two monomers are related by a non crystallographic twofold axis in an asymmetric unit and two such molecules are related by a crystallographic two-fold axis at one end. The holoenzyme seems to have three well defined domains, where the N and C terminal ends are a part of the central domain which is involved in subunit interactions, an NAD binding domain containing a modified Rossman fold and a catalytic domain which contains the active site amino acids and residues that occur at the tetramerization interface (Jin *et al.*, 2004; Majumder *et al.*, 2003; Stein and Geiger, 2002).

Mycobacterial MIPS structure shows two well-defined domains (D1 and D2) linked through two hinged regions. The D1 domain can be presumed to be containing two parts, the D1a domain and the D1b domain. The D1a domain is characterized by the presence of a modified Rossman fold (with the GXGXXG motif), whereas the D1b contains the C-terminal end. D2 domain contains the residues involved in tetramerization interface (Norman *et al.*, 2002). It has been conjectured that the D1 domain of the prokaryotic interface might have evolved to more complex structures through selection giving rise to the two well defined domains for NAD binding (probably from D1a) and the central domain (from D1b) in eukaryotes (Majumder *et al.*, 2003). The archaeal MIPS forms an overall homotetrameric structure with a pair of asymmetric dimeric unit like other MIPS but monomers are organized in two distinct domains like mycobacterial MIPS and unlike the yeast MIPS. The first domain (the dinucleotide binding domain) has a gap that leads to the formation of a loosely defined Rossman motif in archaeal MIPS whereas the mycobacterial MIPS has an insertion here. A striking observation in archaeal MIPS is the presence of a solvent accessible region, which is followed by the Rossman motif. This solvent accessible region at the N terminus is also present in the yeast MIPS but poorly conserved among other bacterial MIPSs. This solvent accessible loop contains highly charged residues, which sometimes contribute to the solubility of MIPS (*e.g.*, *Pyrococcus abyssii*, Vieille and Zeikus, 2001). Recently a novel salt tolerant MIPS, isolated from *Porteresia coarctata*, has also been found to contain an extra 37 amino acid stretch rich in charged residues like Arg and Asp which can be considered as an insertion over the Rossman fold in the yeast MIPS structure. Our recent findings suggested a coiled structure of this region of 37 amino acid stretch is likely to be responsible for the salt-tolerance property of the enzyme. The stretch seems to be exposed and accessible to the solvent. This highly charged solvent exposed coiled region may be responsible for shielding the salt ions contributing to the high solubility of the enzyme at high ionic strength thus providing a possible explanation for the salt tolerance of the enzyme-protein under *in vitro* condition. A large deletion from the N terminus of the archaeal MIPS has made it closer to mycobacterial MIPS with respect to interaction among the subunit. In yeast MIPS, interactions among the dimers are stronger than those among the tetramers which is just the reverse in case of mycobacterial and archaeal MIPS (Stieglitz *et al.*, 2005).

Comparison of the two structures revealed striking similarities in the alignment of motif, presence of functional as well as structural domains in the two structures. Details of such comparison can be found in Majumder *et al.* (2003). The most striking of these was the absence of electron structure for a stretch of amino acids in the two crystals (26 amino acids, between 241 and 267 for *Mycobacterium* MIPS and 58 amino acids, between 351 and 409 for *Saccharomyces* MIPS). Both of these two regions showing a lack of defined structure happen to be from the tetramerization domains of the respective enzymes. However, when the *Saccharomyces* MIPS protein was crystallized in presence of glucitol-6 phosphate, a substrate analogue of the enzyme, nucleation around these 58 amino acids resulted in an observed electron density in their diffraction pattern. The results exemplify the "induced fit" model hypothesized earlier by Koshland (1958) whereby it is presumed that the substrate is catalyzing the proper folding of the enzyme at the tetramerization interface and thereby initiating the tetramerization of the enzyme followed by activation of the catalytic activity. It seems possible that the same holds true for the *Mycobacterium* MIPS and the unorganized (electron-density less) region would be similarly nucleated by a substrate. It has been hypothesized that such a mechanism of "induced fit" might have evolved quite early in evolution of MIPS and remained largely unchanged through selectional pressure (Majumder et al., 2003). However, in case of archaeal MIPS the corresponding region (263–285) is less unorganized but fully folded producing a large cavity in the interior of protein for the substrate that is lined with sets of positively charged, negatively charged and hydrophobic amino acid residues. In all known structures this is a common feature and probably due to the nonneutralization and charge shielding effects this region has been found to be totally unorganized in case of structures except the archaeal MIPS in absence of the substrate. But in this case probably three factors are responsible for having the region fully folded even in absence of the substrate. These are: (i) high positive field density due to the presence of greater number of positively charged residues which are almost neutralized by the negatively charged residues; (ii) a crystalographically identified putative K^1, directly ligated to the NAD^1, which takes care of excess negative charges; and (iii) inorganic phosphate molecule, which has been found to occupy the active site cavity at the anion binding site, taking care of excess positive charges. An overall surface electrostatic field calculation revealed that archaeal MIPS active site has more negative field potential and is much more rigid in terms of its solubility, activity, and stability compared to other known MIPS till date (Stieglitz *et al.*, 2005). In a more recent publication, while elucidating the reaction mechanism of *S. cerevisiae* MIPS, the authors have put forward strong arguments to support the characteristics for two of the prokaryotic MIPS from *Mycobacterium* and *Archaeoglobus* through their models of *Saccharomyces* MIPS and its reaction mechanism (Jin *et al.*, 2004).

Mechanistically both archaeal MIPS and yeast MIPS(s) are almost similar. Recent detailed mutational study on archaeal MIPS revealed the involvement of same amino acid residues those implicated in the yeast structure and belong to the core catalytic region (Neelon *et al.*, 2005; Majumder *et al.*, 2003). In the first step of the newly proposed model for yeast MIPS reaction mechanism, D-glucose 6-phosphate is oxidized at C5 by NAD^+ involving a direct hydride transfer from C5 of D-glucose 6-phosphate to C4 of the nicotinamide moiety of NAD^+. During this process a proton is transferred from the C5 hydroxyl group of D-glucose 6-phosphate to the Lys-369 terminal nitrogen atom (archaeal Lys-274), which can be accepted by the structurally adjacent residue of Asp-320 (archaeal Asp-225), in a proton-shuffling system. In the next step of enolization, the pro-R hydrogen of C6 is proposed to be eliminated, where phosphate monoester or Lys-489 (archaeal Lys-367) may act as the enolization base. The developing negative charge on the enolate oxygen is stabilized by the Lys-369 (archaeal Lys-274 and Lys-367) (Neelon *et al.*, 2005). Reduction by NADH is the third and the last step, where, the hydride that was transferred in the first step returns to the C5 of the intermediate myo-2-inosose 1-phosphate from C4 of the nicotinamide. In return, a proton could then be transferred to the C5 ketone oxygen from Asp-320, via Lys-369, using the same proton-shuffling system (Jin *et al.*, 2004). One of the most important facts about the active site of archaeal MIPS is that this active site has been found to be lined with several Glutamic acid residues at one side of the cavity (Stieglitz *et al.*, 2005). Divergence in the mechanism between the archaeal and eukaryotic MIPS occurs with respect to the involvement of Lewis acid in the actual catalysis. Involvement of Lewis acid is a common feature of all MIPSs. Although in case of eukaryotic MIPS, electron density that corresponds to a divalent metal ion had been mapped near the NAD^+, under *in vitro* condition only NH_4^+ has been found to activate the enzyme. Its actual role in the catalysis has also been implicated in the yeast structure (Jin *et al.*, 2004). On the other hand, in case of archaeal MIPS, involvement of two metal ions have been implicated; (i) a putative K^+ ("the structural metal ion"), crystalographically identified beside the NAD^+ which serves for the proper positioning of nicotinamide ring, and (ii) a divalent metal ion like Zn/Mg ("catalytic metal ion") that stabilizes the developing negative charge on the oxygen atom during aldol cyclization (Stieglitz *et al.*, 2005). The same role of a divalent Zn has also been implicated in mycobacterial MIPS.

5. "CORE CATALYTIC STRUCTURE" AND ITS EVOLUTIONARY IMPLICATIONS

From the phylogenetic analyses done from this laboratory, eukaryotic MIPS seems to have evolved from one common stock, probably from the fusion of an archaebacterial and a eubacterial MIPS gene (Majumder *et al.*, 2003), as postulated for the origin of eukaryotic genomes (Kleiger and Eisenberg, 2002). Since its origin and early adaptations, the eukaryotic MIPS sequence has maintained striking conservation across the different eukaryotic groups as evidenced from sequence comparisons (Majumder *et al.*, 2003). Further analysis have shown common structural motifs or domains found in different MIPS sequences from different organisms (Table 3). While some of the motifs are shared by a number of organisms, some motifs are characteristically unique for a few of them. Examples of such motifs are diaminopimelate dehydrogenase in *Mycobacterium*, ABC transporter ATPase domain and alpha-beta hydrolase in *Entamoeba*, aldolase DAHP synthetase in *Saccharomyces* and TPP binding domain in *Synechocystis* (Table 3).

Analysis of the MIPS protein sequences across the evolutionary diverse organisms revealed interesting features. Some of the important amino acid residues identified in the active site of both *S. cerevisiae* and archaeal MIPS (namely, *S. cerevisiae* Q325, L352, N354, L360, K369, I400, I402, K412, D438, and K489, *archaeal* D225, K274, K278, K306, K367, N255, L257, and D332) (Neelon *et al.*, 2005) are highly conserved in all the eukaryotic as well as prokaryotic MIPS enzymes. This has also been documented in the recently isolated salt tolerant MIPS from *Porteresia*, PINO1, where deletion of a stretch of 20 amino acid belonging to the core region (342ndN to 361stK) containing the two important residues, N342 (N354 in yeast MIPS) and K357 (K369 in yeast MIPS) resulted in a catalytically inactive enzyme (Majee *et al.*, 2004). However, sequence comparison indicated several distinct residues for archaeal, bacterial, or eukaryotic MIPS (*e.g.*, D259 in archaea has been replaced by N354 in yeast and N233 in *Mycobacterium*, respectively). These amino acids can be considered as part of a "eukaryotic core structure" which has been conserved through selectional forces, even though the rest of the protein sequence has changed along time, probably due to non-selectional evolutionary forces like mutation, drift etc. (Majumder *et al.*, 2003; Figure 2).

6. SIGNATURE SEQUENCES OF MIPS PROTEINS

An interesting fact emerges through the different sequence alignment studies of eukaryotic MIPS from different systems. Almost all eukaryotic MIPS sequences can be clustered in one domain in a phylogenetic tree with respect to the presence of only some highly conserved stretches, apart from Rossmann fold motif (GXGGXXG), which can be considered as signature sequences of MIPS. These highly conserved clusters are – FXGWDISXXN, VVLWTANTERY, XGDDFKSGQTXX, XSYNHLGNNDGX, SKSNVVDD, and INGSPQNTXVPGXXXLAXX. Out of these five highly conserved clusters the last three belong to the proposed "core catalytic region" and share many important key amino acid residues involved in the MIPS catalysis. However, this "core catalytic region" itself maintains more than 70% sequence homology among the eukaryotes and all through share almost all probable active site amino acid residues. When compared with the

Table 3. Structural motifs/domains commonly found in MIPS sequences of different organisms

Motif/Domain name	Eubacterial MIPS	Archaeal MIPS	Fungal MIPS	Animal MIPS	Plant MIPS	Cyano MIPS
Glycosyltransferases (*e.g.*, Beta-glycanases)	*Mycobacterium tuberculosis*	*Archaeoglobus fulgidus*	*Saccharomyces cerevisiae*		*Arabidopsis thaliana*	*Synechocystis sp* (sll1722)
Methyltransferase (*e.g.*, Cobalt precorrin-4 methyltransferase)				*Entamoeba histolytica*	*Arabidopsis thaliana*	*Synechocystis sp* (sll1722)
P-loop containing nucleotide binding fold (*e.g.*, G proteins)			*Saccharomyces cerevisiae*	*Entamoeba histolytica*		
NAD(P)-binding Rossmann-fold (*e.g.*, formate/glycerate dehydrogenase)	*Mycobacterium tuberculosis*	*Archaeoglobus fulgidus*		*Entamoeba histolytica*		*Synechocystis sp* (sll1722)
NAD(P)-linked oxidoreductase (*e.g.*, Aldo-keto reductases–NADP)	*Mycobacterium tuberculosis*	*Archaeoglobus fulgidus*		*Entamoeba histolytica*		*Synechocystis sp* (sll1722)
Enolase	*Mycobacterium tuberculosis*			*Entamoeba histolytica*		
Other transferases (*e.g.*, PLP-dependent transferases.)			*Saccharomyces cerevisiae*			*Synechocystis sp* (sll1722)
Diaminopimelate dehydrogenase	*Mycobacterium tuberculosis*					
ABC transporter ATPase domain				*Entamoeba histolytica*		
TPP binding domain						*Synechocystis sp* (sll1981)
Aldolase DAHP synthetase			*Saccharomyces cerevisiae*			
Alpha-beta Hydrolase				*Entamoeba histolytica*		

```
           195 #                                        #   *  *              *   *      264
Mycobtuber  GDDIKSQVGA TITHRVLAKL FEDRGVQLDR TMQLNVGGNM DFLNMLERER LESKKISKTQ AVTSNLKREF
Pyroc-abys  GDD--GATGA TPLTADILGH LAQRNRHVLD IAQFNIGGNT DFLALTDKER NKSKEYTKSS VVEDILGYDA
Thermomari  GDD--GATGA TPFTADVLSH LAQRNRYVKD VAQFNIGGNM DFLALTDDGK NKSKEFTKSS IVKDILGYDA
Streptomyc  GSD--GKTGE TLVKSVLAPM FARRALRVRS WSGTNLLGGG DGATLADPER VVSKNASKGL VLEAEL----
Archaeoglo  GND--GKTGE TLVKTTLAPM FAYRNMEVVG WMSYNILGDY DGKVLSARDN KESKVLSKDK VLEKML----
Entamoebah  GDD--FKTGQ TKIKSVLADF LVSSGLKLQS IVSYNHLGNN DGKNLSSPSQ FRSKEISKSN VVSDVVKSNN
Arabidopst  GDD--FKSGQ TKMKSVLVDF LVGAGIKPTS IVSYNHLGNN DGMNLSAPQT FRSKGDSKSN VVDDMVASNG
Mesembcrys  GDD--FKSGQ TKMKSVLVDF LVGAGIKPTS IVSYNHLGNN DGMNLSAPQT FRSKEISKSN VVDDMVASNG
SesamumInd  GDD--FKSGQ TKMKSVLVDF LVGAGIKPTS IVSYNHLGNN DGMNLSAPQT FRSKEISKSN VVDDMVASNG
GlycineMax  GDD--FKSGQ TKMKSVLVDF LVGAGIKPTS IVSYNHLGNN DGMNLSAPQT FRSKEISKSN VVDDMVNSNA
Oryzasativ  GDD--FKSGQ TKMKSVLVDF LVGAGIKPTS IVSYNHLGNN DGMNLSAPQT FRSKEISKSN VVDDMVSSNA
Drosophila  GDD--FKSGQ TKIKSVLVDF LVGAGIKPVS IASYNHLGNN DGKNLSAPQQ FRSKEISKSN VVDDMVASNR
HomoSapien  GDD--FKSGQ TKVKSVLVDF LIGSGLKTMS IVSYNHLGNN DGENLSAPLQ FRSKEVSKSN VVDDMVQSNP
Leishmanim  GDD--FKSGQ TKMKSALVEF FVGAGIKPEC IASYNHLGNN DGYNLSSHKQ FCSKEITKSN VVDDMIKSNQ
C.elegans   GDD--FKSGQ TKFKSAFVDF LVSSGMKPES IVSYNHLGNN DGKNLSEARQ FRSKEISKSS VVDDMVKSNQ
Pichiapast  GDD--FKSGQ TKLKSVLAQF LVDAGIRPVS IASYNHLGNN DGYNLSAPQQ FRSKEISKAS VVDDMIESNE
Saccharomc  GDD--LKSGQ TKLKSVLAQF LVDAGIKPVS IASYNHLGNN DGYNLSAPKQ FRSKEISKSS VIDDIIASND
         318                                             # *   *   *          *   *      385

                                          #                                 #
           265                            *                                 **           327
Mycobtuber  -------KTK DVHIGPSDHV GWLDDRKWAY VRLEGRAFGD VPLNLEYKLE VWDSPNSAGV IIDAVRAAKI
Pyroc-abys  -------PH- --YIKPTGYL EPLGDKKFIA MHIEYVSFNG AHDELIITGR INDSPALAGL LVDLARLGKI
Thermomari  -------PH- --YIKPTGYL EPLGDKKFIA IHIEYVSFNG ATDELMINGR INDSPALGGL LVDLVRLGKI
Streptomyc  -------GHA VEGGVHIHHV PDLGEWKTAW DHVTFEGFLG ARMTLQFTWQ GCDSSLAAPL VLDLARFMAL
Archaeoglo  -------GYS PYSITEIQYF PSLVDNKTAF DFVHFKGFLG KLMKFYFIWD AIDAIVAAPL ILDIARFLLF
Entamoebah  IMY-KA-GEH PDHVIVITYV PYVGDSHRAM DDYTSHIFLR GHNTIALHNT CEDSLLAAPL MIDLAVLMEF
Arabidopst  ILF-EP-GEH PDHVVVIKYV PYVADSKRAM DEYTSEIFMG GKNTIVMHNT CEDSLLAAPI ILDLVLLAEL
Mesembcrys  ILY-EP-GEH PDHVVVIKYV PYVGDSKRAM DEYTSEIFMG GTNTIVMHNT CEDSLLAAPI ILDLVLLAEL
SesamumInd  ILY-EP-GEH PDHIVVIKYV PYVGDSKRAM DEYTSEIFMG GKSTIVLHNT CEDSLLAAPI ILDLVLLAEL
GlycineMax  ILY-EP-GEH PDHVVVIKYV PYVGDSKRAM DEYTSEIFMG GKSTIVLHNT CEDSLLAAPI ILDLVLLAEL
Oryzasativ  ILY-EL-GEH PDHVVVIKYV PYVGDSKRAM DEYTSEIFMG GKSTIVLHNT CEDSLLAAPI ILDLVLLAEL
Drosophila  LLY-GP-DEH PDHVVVIKYV PYVGDSKRAM DEYTSEIMMG GHNTLVIHNT CEDSLLATPL ILDLVILGEL
HomoSapien  VLY-TP-GEE PDHCVVIKYV PYVGDSKRAL DEYTSELMLG GTNTLVLHNT CEDSLLAAPI MLDLALLTEL
Leishmanim  VLF-PEGARK PDHCIVIKYI PYVGDSKRAL DEYTFSIFMG GQQTVVLHNT CEDSLLAAPL IIDLIVLTEL
C.elegans   ILF-PD-AKN PDYCVVIKYV PYVADSKRAM DEYICSIFMG GKQTFVVHNT CEDSLLASPL IYDLAILTEL
Pichiapast  ILYNEKNGNT IDHCIVIKYM KAVGDDKVAM DEYHSELMLG GHNTISIHNI CEDSLLATPL IIDLVVMAEF
Saccharomc  ILYNDKLGKK VDHCIVIKYM KPVGDSKVAM DEYYSELMLG GHNRISIHNV CEDSLLATPL IIDLLVMTEF
          386                  * *        *                                 * *          455

                                            #
           328                              *  181
Mycobtuber  AKD------- ---------R GIGGPVIPAS AYLMKSPPEQ
Pyroc-abys  AVE------- ---------K KAFGTVYEVN AFYMKNPGPR
Thermomari  ALD------- ---------R KEFGTVYPVN AFYMKNPGPA
Streptomyc  AHR------- ---------- AGVAGPVPEL GFFFKDPVGS
Archaeoglo  AKK------- ---------- KGVKGVVKEM AFFFKSPMDT
Entamoebah  MTRVTY---S IDGKE----F KNFNSVMSMI SYLLKAPVVP
Arabidopst  STRIQFK-SE GEGK-----F HSFHPVATIL SYLTKAPLVP
Mesembcrys  STRIQLK-AE EEDK-----F HSFHPVATIL SYLTKAPLVP
SesamumInd  STRIQLK-AE GEGK-----F HSFHPVATIL SYLTKAPLVP
GlycineMax  STRIEFK-AE NEGK-----F HSFHPVATIL SYLTKAPLVP
Oryzasativ  STRIQLK-AE GEEK-----F HSFHPVATIL SYLTKAPLVP
Drosophila  STRIQLRNAE KESAP----W VPFKPVLSLL SYLCKAPLVP
HomoSapien  CQRVSF--CT DMDPE----P QTFHPVLSLL SFLFKAPLVP
Leishmanim  MERVTISASD DTQTPPPASF EHMETVLSIL SYLLKAPAVP
C.elegans   ASRVSYKVDD E--------Y KPFHSVLSIL SLLLKAPVVP
Pichiapast  LSRVSYKK-- --KGDA--EY ESLHSVLSFL SYWLKAPLTR
Saccharomc  CTRVSYKKVD PVKEDA-GKF ENFYPVLTFL SYWLKAPLTR
           456                              *  494
```

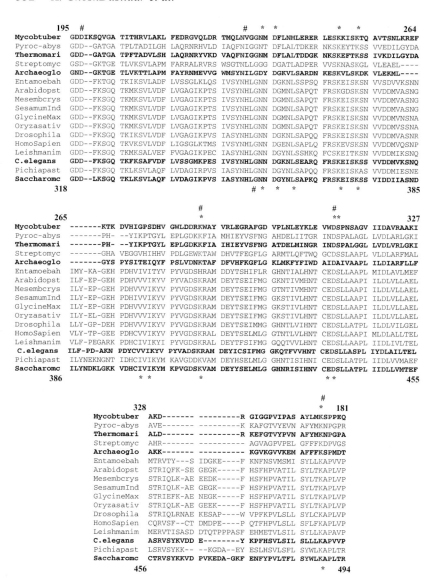

Figure 2. Multiple sequence alignment of the proposed "core catalytic region" of MIPSs from 17 representative organisms from different spectra across the life kingdom. Organisms whose MIPSs have so far been crystallized made bold. The number at the top most row and the (*) indicate the proposed or identified active site amino acid residues for *M. tuberculosis* and those at the bottom indicate the same for *S. cerevisiae*. Amino acid indicated as (#) are the proposed or identified residues at the active site of *A. fulgidus*.

representative prokaryotes, it reveals that even with respect to this highly similar "core catalytic region" in eukaryotes, prokaryotic MIPS are quite diverse. However, all the active site residues implicated in different crystal structures of MIPS till date are remarkably conserved all through and this is the major reason in favor of postulating a "core catalytic region" in MIPS which has been maintained in the time course of evolution in spite of bearing tremendous evolutionary pressure for the different MIPS sequences to be diverged in a complex manner. However, there are many other amino acid residues belonging to the "core catalytic region" which are absolutely conserved all through but have not been implicated in the catalytic mechanism. Then what are those for? Are they still somehow involved in the substrate binding or actual catalysis? This requires extensive structure-function analysis with different potential active site mutants of MIPS. Another possibility of involvement of those residues in some specialized functions such as thermal and/or salt stability can be postulated. The situation is very different in a recently identified stress tolerant MIPS from *Porteresia* (Majee *et al.*, 2004) wherein a stretch of 37 amino acid residues, physically upstream of the core catalytic region in the primary structure, has been identified as the probable structural element for high salt stability. However, the probability of having some important interaction in a three dimensional space between the salt-tolerance domain and the core catalytic region cannot be ruled out. Similarly, considering the nonredundancy of biological systems, a big question about the possibility of some other essential function(s) of the other conserved stretches in eukaryotes still remains. Variation is also observed among the eukaryotic conserved stretches mainly in case of lower eukaryotes and higher plants but in most of the cases these changes are conservative replacements. Likewise, there is a slight greater propensity for valine which replaces isoleucine from most of those highly conserved stretches in all higher plant MIPS but as this is a conservative replacement it may be of less importance. Phylogenetically this seems to be important as in higher plants greater inclination for some amino acids in those conserved stretches branches them out as a separate cluster with respect to MIPS. The situation is quite different in case of prokaryotes where different MIPS sequences are quite divergent among themselves and far more any of the known eukaryotic MIPS sequences. This divergence is also remarkable at the prokaryote – eukaryote transition. One part of MIPS evolution is definitely the incorporation and maintenance of apparently unimportant conserved stretches across the eukaryotic domain under evolutionary pressure with time. Do eukaryotic MIPS(s) exhibit any other vital cellular role? Is there any possibility of them thus being multi-functional enzymes? Indications of such "vital role" for MIPS for cellular regulations might come from studies on the phenomenon of "inositol – less death" well documented in fungal system (Majumder, 1981). At the present situation these questions are certainly puzzling although worth investigation.

7. CHLOROPLAST INOSITOL SYNTHASE(S): *SYNECHOCYSTIS* AS PROBABLE ANCESTOR?

The presence of an MIPS isoform in the photosynthetic organelles of higher plants, the chloroplasts, has been reported by several workers (Adhikari *et al.*, 1987; Johnson and Wang, 1996; Lackey *et al.*, 2003; RayChaudhuri *et al.*, 1997). Two forms of enzymatically active plastidial MIPS have been reported from *Oryza sativa*. These two forms have been identified as comprising of ~80 and ~60 kDa subunits. The ~80 kDa subunit is predominantly present in dark grown etioplasts while the light/dark grown chloroplasts accumulate the ~60 kDa subunit. Interestingly in seedlings of salt tolerant rice varieties this ~60 kDa subunit protein is phosphorylated in a Ca^{2+} dependent manner by a putative protein kinase resulting in increased activity of the enzyme in presence of light and salt (Hait *et al.*, 2002). Although MIPS from algal and higher plant chloroplasts have been reported from several sources (*Oryza, Euglena, Pisum* etc.), no evidence of the gene(s) coding for this isoform has been reported so far. In a recent work (Lackey *et al.*, 2003), Johnson's group has suggested that at least in case of beans, the organellar MIPS could be a splice variant of the cytoplasmic MIPS isoform, which is targeted to different compartments of the cell (Lackey *et al.*, 2003). They have studied MIPS gene products from chloroplasts, mitochondria, microsome, and plasma membrane of root and leaf cells of *Phaseolus vulgaris*. Comparative analysis of three *Arabidopsis* and one *Phaseolus* MIPS gene sequences have revealed that all four MIPS primary structures are type II membrane proteins with most of the primary structure associated with cytoplasm. A predicted transmembrane motif is present approximately 68 amino acids from the carboxy terminus, which is highly conserved in representatives from both plant and animal kingdom, from yeast to human. The sequences also showed striking similarities in the upstream regions of these MIPS sequences containing short ORFs which code transit peptides capable of targeting MIPS isoforms to a variety of subcellular compartments. N glyosylation, N myristoylation, and phosphorylation by protein kinase C and casein kinase II have also been identified as possible post translational modifications for these MIPS and some of its target peptides. Several new inositol utilizing pathways have been identified in organelles (Irvine and Schell, 2001; York *et al.*, 1999) which are suggested to draw inositol from the same pool generating a large requirement of readily available inositol phosphates that can be easily supplied and regulated at the level of inositol phosphate biosynthesis. The two forms of enzymatically active MIPS detected from rice plastid are mostly membrane bound, although the ~60 kDa subunit form is present in stromal fraction as well (Hait *et al.*, 2002). Such transmembrane orientation of MIPS places the enzyme in a key position, which regulates both membrane, and non-membrane bound aspects of the complex metabolic flux of inositol (Lackey *et al.*, 2003). It is proposed that a complex cellular mechanism functions to control inositol phosphate biosynthesis spatially and temporally during plant growth and development. Transcription of individual MIPS genes is regulated by development (Johnson and Wang, 1996), hormones (Smart and Fleming, 1993; Yoshida *et al.*, 2002), photoperiod (Hayama *et al.*, 2002) and salt stress (Ishitani *et al.*, 1996).

A logical expectation in the presence of MIPS in chloroplasts is the presence of the same enzyme in the evolutionary ancestors of the chloroplasts, the cyanobacteria. The first report of the presence of a MIPS protein from a cyanobacterium such as *Spirulina,* was by RayChaudhuri *et al.* (1997). Recently, a hitherto unidentified ORF, sll1722 has been identified in *Synechocystis sp* PCC6803 as the gene coding for a ~50 kDa MIPS subunit protein using biochemical, molecular, and bioinformatic tools (Chatterjee *et al.*, 2004). Functionality of the sll1722 gene product was confirmed by genetic complementation and by biochemical characterization of the overexpressed protein, which shows similar enzymatic activity as other known MIPS proteins. The gene product also contain a NAD^+/$NADP^+$ binding Rossman fold which confirms that it catalyzes an internal oxidation reduction of the substrate involving NAD^+. Comparison of sll1722 sequence with other MIPS sequences and its phylogenetic analysis revealed that this gene is quite divergent from eukaryotic MIPS, although conservation of some of the amino acids adjudged critical for the catalytic activity of the enzyme is evident (Chatterjee *et al.*, 2004).

The presence of the second active form of MIPS with a ~65 kDa subunit indicated that the inositol requirement of the organism might be provided by the interplay of two different MIPS enzymes, probably by differential expression through time and space. The results indicate that the ~65 kDa MIPS protein of *Synechocystis* might be coded by the ORF sll1981, annotated as a putative acetolactate synthase (unpublished data from this laboratory). This gene also functionally complements the inositol auxotrophic yeast strain *FY250* and *Schizosaccharomyces pombe*, a natural inositol auxotroph, and the expressed protein is immunoreactive to anti MIPS antibody.

The two different MIPS enzymes of cyanobacterium *Synechocystis sp.* PCC 6803 are not only unique among them, but also are unique in their primary sequences from any of the other known MIPS sequences till date. Neither of them shows any striking sequence similarity to the known MIPS sequences from any of the known prokaryotic or eukaryotic organisms reported. Whatever local similarity is evident, is probably enough to maintain a 3-dimensional "core structure" for the specific catalytic activity of the MIPS enzyme (Majumder *et al.*, 2003). The presence of these two MIPS enzymes point to the possibility of independent origin of MIPS enzymes in cyanobacteria from other eubacteria, or large-scale secondary changes of the original prokaryotic MIPS sequence in the cyanobacterial genome. The fact that an "acetolactate synthase" homologue can function as MIPS, gives food for thought as to whether homologous acetolactate synthase proteins from organisms which lack a regular MIPS (*e.g.*, E. coli), can also function as an MIPS enzyme under special conditions.

At present it is debatable whether all chloroplastic MIPS are splice variant of the cytosolic form (Lackey *et al.*, 2003) or they are descendants of the cyanobacterial form. It might even be possible to find more than one chloroplastic MIPS enzyme with different origins. However, until the chloroplastic gene is cloned and the recombinant protein is characterized for its enzymatic properties, the debate remains open with lively and exciting discoveries ahead.

8. FUTURE PERSPECTIVES

The current investigation on the origin and evolution of the MIPS gene/protein is largely based on the multiple approaches and tools of biochemistry, genetics, genomics, and bioinformatics. This is largely aided by the recent genome sequencing projects, particularly of the prokaryotes, coupled with functional genomic studies. While providing newer insight, the information gathered so far raise more questions for which the answers are far away.

The question of identity of the MIPS "cenancestor" is still unanswered despite an inkling of the possibility of an acetolactate synthase/MIPS bifunctional protein as a probable one. An equally interesting question is the linkage between the cyanobacterial MIPS(s) and the chloroplast MIPS which is yet to be established. Research in these areas is expected to unravel such mystery in the near future.

The MIPS sequences of the pathogenic prokaryotes offer opportunities as drug target(s). Initial ecstasy over the mycobacterial Ino1 as a possible drug target (Bacchawat and Mande, 1999) is to be reconsidered on identification of the "core catalytic domain" for MIPS(s) from all sources including *Mycobactium* (Majumder *et al.*, 2003). Interestingly however, the *Plasmodium* MIPS is unique in having an extra stretch of ~40 amino acids outside the "core catalytic domain" and undetected in any known MIPS proteins. Can this stretch be a locus for such studies? Future research may be directed toward this possibility.

Finally, an intriguing question about possible role of the MIPS protein (or the Ino 1 gene) apart from synthesis of inositol keeps coming back. Is it possible that MIPS protein performs some other vital functions pertaining to cellular physiology as well? Probing the molecular mechanism of "inositol-less death" might be one way toward answering such questions. An in-depth cell biological study correlating "inositol-less death" with cell-cycle regulation in eukaryotes may be a reasonable start. Research in this yet unexplored area may turn out to be intellectually highly rewarding.

REFERENCES

Abu-abied, M., and Holland, D., 1994, The gene cINO1 from *Citrus paradisi* is highly homologous to tur1 and Ino1 from the yeast and *Spirodela* encoding for *myo*-inositol 1-phosphate synthase. *Plant Physiol.* **106:** 1689.

Adhikari, J., Majumder, A.L., Bhaduri, T.J., Dasgupta, S., and Majumder, A.L., 1987, Chloroplast as a locale of L-*myo*-inositol 1-phosphate synthase. *Plant Physiol.* **85:** 611–614.

Bachhawat, N., and Mande, S.C., 1999, Identification of the *INO1* gene of *Mycobacterium tuberculosis* H37Rv reveals a novel class of inositol 1-phosphate synthase enzyme. *J. Mol. Biol.* **291:** 531–536.

Bachhawat, N., and Mande, S.C., 2000, Complex evolution of the inositol-1-phosphate synthase gene among archaea and Eubacteria. *Trends Genet.* **16:** 111–113.

Benaroya, R.O., Zamski, E., and Tel-Or, E., 2004, L-Myo-inositol 1-phosphate synthase in the aquatic fern *Azolla filiculoides*. *Plant Physiol. Biochem.* **42:** 97–102.

Chatterjee, A., Majee, M., Ghosh, S., and Majumder, A.L., 2004, sll1722, an unassigned ORF of *Synechocystis* PCC 6803, codes for L-myo-Inositol 1-Phosphate Synthase. *Planta* **218:** 989–998.

Chen, L., Zhou, C., Yang, H., and Roberts, M.F., 2000, Inositol 1-phosphate synthase from *Archaeoglobus fulgidus* is a class II aldolase. *Biochemistry.* **39:** 12415–12423.

Chun, J.-A., Jin, U.-H., Lee, J.-W., Yi, Y.-B., Hyung, N.-I., Kang, M.-H., Pyee, J.-H., Suh, M.C., Kang, C.-W., Seo, H.-Y., Lee, S.-W., and Chung, C.-H., 2003,. Isolation and Characterization of a myo-Inositol 1-phospahte synthase cDNA from developing sesame (*Sesamum indicum* L.) seeds: Functional and differential expression, and salt-induced transcritption during germination. *Planta* **216:** 874–880.

Corpet, F., 1988, Multiple sequence alignment with hierarchical clustering. . *Nucleic Acids Res.* **16:** 10881–10890.

DasGupta, S., Adhikari, J., and Majumder, A.L., 1984, L-*myo*-inositol 1-phosphate-synthase from lower plant groups: Partial purification and properties of the enzyme from *Euglena gracilis*. *Physiol. Plant.* **61:** 408–416.

Donahue, T.F., and Henry, S.A., 1981, *Myo*-inositol 1-phosphate synthase: Characteristics of the enzyme and identification of its structural gene in yeast. *J. Biol. Chem.* **256:** 7077–7085.

Funkhouser, E.A., and Loewus, F.A., 1975, Purification of *myo*-inositol 1-phosphate synthase from the rice cell culture by affinity chromatography. *Plant Physiol.* **56:** 786–790.

Hait, N.C., Ray Chaudhury, A., Das, A., Bhattacharyya, S., and Majumder, A.L., 2002, Processing and activation of chloroplast L-myo –inositol 1-phosphate synthase from *Oryza sativa* requires signals from both light and salt. *Plant Sc.* **162**(4)**:** 559–568.

Harris, H., 1966, Enzyme polymorphisms in man. *Proc. R. Soc. Lond.* B **164:** 298–310.

Hayakawa T., and Kurasawa F., 1976, Biochemical studies on inositol in rice seed. Part V. Some enzymatic properties of myo inositol 1 phosphate synthase in the milky stage of rice seed. *Nippon Nogel Kagaku Kaishi.* **50:** 339–343.

Hayama, R., Izawa, T., and Shimamoto, K., 2002, Isolation of rice genes possibly involved in the photoperiodic control of flowering by a fluorescent differential display method. *Plant Cell Physiol.* **43:** 494–504.

Hinchliffe, K., and Irvine, R., 1997, Inositol lipid pathways turn turtle. *Nature* **390:** 123–124.

Hirsh, A.E., and Fraser, H.B., 2001, Protein dispensability and rate of evolution. *Nature* **411:** 1046–1049.

Hirsh, A.E., and Fraser, H.B., 2003, Genomic function: Rate of evolution and gene dispensability. *Nature* **421:** 497–498.

Holsinger, K.E., 2004, The neutral theory of molecular evolution. Creative Commons Attribution. 1–6.

Hubby, J.L., and Lewontin, R.C., 1966, A molecular approach to the study of genic heterozygosity in natural populations. I. The number of alleles at different loci in *Drosophila pseudoobscura*. *Genetics* **54:** 577–594.

Imhoff, V., and Bourdu, R., 1973,. Formation, d' inositol par les chloroplasts isoles de pois. *Phytochemistry* **12:** 331–336.

Iqbal, M.J., Afzal, A.J., Yaegashi, S., Ruben, E., Triartayakorn, K., Njiti, V.N., Ahsan, R., Wood, A.J., and Lightfoot, D.A., 2002, A Pyramid of loci for partial resistance to *Fusarium solani* f. sp. *Glycines* maintains *Myo*-inositol-1-phosphate synthase expression in soybean roots. *Theor. Appl. Genet.* **105:** 1115–1123.

Irvine, R.F., and Schell, M.J., 2001, Back in the water: The return of the inositol phosphates. *Nat. Rev. Mol. Cell Biol.* **5:** 327–338.

Ishitani, M., Majumder, A.L., Bornhouser, A., Michalowski, C.B., Jensen, R.G., and Bohnert, H.J., 1996, Co-ordinate transcriptional induction of *myo*-inositol metabolism during environmental stress. *Plant J.* **9:** 537–548.

Jin, X., Foley, K.M., and Geiger, J.H., 2004, The structure of the 1L-myo-inositol-1-phosphate synthase-NAD+-2-deoxy-D-glucitol 6-(E)-vinylhomophosphonate complex demands a revision of the enzyme mechanism. *J. Biol. Chem.* **279:** 13889–13895.

Jin, X., and Geiger, J.H., 2003, Structures of NAD+- and NADH-bound 1-l-myo-inositol 1-phosphate synthase. *Acta Crystallogr. D* **59**, 1154–1164.

Johnson, M.D., 1994, The *Arabidopsis thaliana myo*-inositol 1-phosphate synthase (EC 5.5.1.4). *Plant Physiol.* **105:** 1023–1024.

Johnson, M.D., and Henry, S.A., 1989, Biosynthesis of inositol in yeast: Primary structure of *myo*-inositol 1-phosphate synthase locus and functional characterization of its structural gene, the Ino1 locus. *J. Biol. Chem.* **264:** 1274–1283.

Johnson, M.D., and Sussex, I.M., 1995, L-*myo*-inositol 1-phosphate synthase from *Arabidopsis thaliana*. *Plant physiol.* **107:** 613–619.

Johnson, M.D., and Wang, X., 1996, Differentially expressed forms of *myo*-inositol 1-phosphate synthase (EC 5.5.1.4) in *Phaseolus vulgaris*. *J. Biol. Chem.* **271:** 17215–17218.

Joint Center for Structural Genomics, (Jcsg), To be Published Crystal Structure Of Myo-Inositol-1-Phosphate Synthase- Related Protein (Tm1419) From Thermotoga Maritima At 1.70 A Resolution, (Release date Mar 23, 2004)

Joint Center for Structural Genomics, (Jcsg), To be Published Crystal Structure Of Inositol-3-Phosphate Synthase (Ce21227) From Caenorhabditis Elegans At 2.30 A Resolution, (Release date Aug 17, 2004)

Keller, R., Brearly, C.A., Tretheway, R.N., and Müller-Röber, B., 1998, Reduced inositol content and altered morphology in transgenic potato plants inhibited for 1D-*myo*-inositol 3-phosphate synthase. *Plant J.* **16**; 403–410.

Kimura, M., 1968, Evolutionary rate at the molecular level. *Nature* **217:** 624–626.

Kimura, M., and Ohta, T., 1974, Probability of gene fixation in an expanding finite population. *PNAS*, **71:** 3377–3379.

King, J.L., and Jukes, T.L., 1969, Non-darwinian evolution. *Science* **164:** 788–798.

Kleiger, G., and Eisenberg, D., 2002, GXXXG and GXXXA motifs stabilize FAD and NAD (P)-binding Rossmann folds through C (alpha)-H. O hydrogen bonds and van der waals interactions. *J. Mol. Biol.* **323:** 69–76.

Klig, L.S., and Henry, S.A., 1984, Isolation of *INO1* gene on an autonomously replicating plasmid, the gene is fully regulated. *Proc. Natl. Acad. Sci. U.S.A.* **81:** 3816–3820.

Kniewel R., Buglino JA., Shen V., Chadha T., Beckwith A., Lima CD. Structural analysis of Saccharomyces cerevisiae myo-inositol phosphate synthase. *J Struct Funct Genomics.* 2002;**2**(3): 129–34.

Koshland, D.E., 1958, Application of a theory of enzyme specificity to protein synthesis. *Proc. Natl. Acad. Sci. U.S.A.* **44:** 98–105.

Lackey, K.H., Pope, P.M., and Johnson, M.D., 2003, Expression of 1L-myoinositol-1-phosphate synthase in organelles. *Plant Physiol.* **132:** 2240–2247.

Larson, S.R., and Raboy, V., 1999, Linkage mapping of maize and barley *myo*-inositol 1-phosphate synthase DNA sequences: Correspondence with low phytic acid mutation. *Theor. Appl. Genet.* **99:** 27–36.

Majee, M., Maitra, S., Dastidar, K.G., Pattnaik, S., Chatterjee, A., Hait, N.C., Das, K.P., and Majumder, A.L., 2004, A novel salt-tolerant L-myo-inositol-1-phosphate synthase from *Porteresia coarctata* (Roxb.) Tateoka, a halophytic wild rice: Molecular cloning, bacterial overexpression, characterization, and functional introgression into tobacco conferring salt tolerance phenotype. *J. Biol. Chem.* **279:** 28539–28552.

Molecular cloning, Bacterial overexpression and characterization of L-myo-inositol 1 Phosphate Synthase from a monocotyledonous Resurrection Plant, Xerophyta viscosa Baker Manoj Majee, Barunava Patra, Sagadevan G., Mundree and Arun Lahiri Majumder. *J. Plant Biochemistry & Biotechnology*, vol-14, 71–75, July 2005.

Majee, M., Majumder, A.L., and Mundree, S., 2003, *Xerophyta viscosa myo*-inositol-1-phosphate synthase complete cds (Gen Bank accession no. AY323824).

Majumder, A.L., 1981, Coupling between fructose 1,6-bisphosphatase and myo-inositol synthase: An hypothesis for 'rescue synthesis' of myo-inositol. *FEBS Lett.* **133:** 189–193.

Majumder, A.L., and Biswas, B.B., 1973, Metabolism of inositol phosphates: Part V Biosynthesis of inositol phosphates during ripening of Mung bean (*Phaseolus aureus*) seeds. *Ind. J. Exp. Biol.* **11:** 120–123.

Majumder, A.L., Chatterjee, A., GhoshDastidar, K., and Majee, M., 2003, Diversification and evolution of L-*myo*-inositol 1-phosphate synthase. *FEBS Lett.* **553:** 3–10.

Majumder, A.L., Johnson, M.D., and Henry, S.A., 1997, 1L-myo-inositol-1-phosphate synthase. *Biochim. Biophys. Acta.* **1348:** 245–256.

Neelon, K., Wang, Y., Stec, B., and Roberts, M.F., 2005, Probing the mechanism of the *Archaeoglobus fulgidus* inositol-1-phosphate-synthase. *J. Biol. Chem.* **280:** 11475–11482.

Norman, R.A., McAlister, M.S., Murray-Rust, J., Movahedzadeh, F., Stoker, N.G., and McDonald, N.Q., 2002, Crystal structure of inositol 1-phosphate synthase from *Mycobacterium tuberculosis*, a key enzyme in phosphatidylinositol synthesis. *Structure* **10:** 393–402.

Pal, C., Papp, B., and Hurst, L.D., 2003, Genomic function: Rate of evolution and gene dispensability. *Nature* **421:** 496–497.

Park, S.H., and Kim, J.I., 2004, Characterization of recombinant *Drosophila melanogaster* myo-inositol-1-phosphate synthase expressed in *Escherichia coli*. *J. Microbiol.* **42:** 20–24.

Pittner, F., Tovorova, J.J., Karnitskaya, E.Y., Khoklov, A.S., and Hoffmann-Ostenhof, O. 1979, Myo-inositol 1-phosphate synthase from *Streptomyces griseus*. Studies on the biosynthesis of cyclitol XXXVIII. *Mol. Cell. Biochem.* **25:** 43.

RayChaudhuri, A., Hait, N.C., DasGupta, S., Bhaduri, T.J., Deb, R., and Majumder, A.L., 1997, L-*myo*-Inositol 1-phosphate synthase from plant sources: Characteristics of the chloroplastic and cytosolic enzymes. *Plant Physiol.* **115:** 727–736.

Rocha, E.P., and Danchin, A., 2004, An analysis of determinants of amino acids substitution rates in bacterial proteins. *Mol. Biol. Evol.* **21:** 108–116.

Sarich, V.M., and Wilson, A.C., 1967, Immunological time scale for hominid evolution. *Science* **158:** 1200–1203.

Smart, C.C., and Fleming, A.J., 1993, A plant gene with homology to D-myo-inositol-phosphate synthase is rapidly and specially upregulated during an abscsic acid induced response in *Spirodela polyrrihiza*. *Plant J.* **4:** 279–293.

Stein, A.J., and Geiger, J.H., 2002, The crystal structure and mechanism of 1-L-myo-inositol-1-phosphate synthase. *J. Biol. Chem.* **277:** 9484–9491.

Stieglitz, K.A., Yang, H., Roberts, M.F., and Stec, B., Reaching for mechanistic consensus across life kingdoms: Structure and insights into catalysis of the inositol-1-phosphate synthase (mIPS) from *Archaeoglobus fulgidus*. *Biochemistry,* (2005), 44, pp. 213–224.

Vieille, C., and Zeikus, G.J., 2001, Hyperthermophilic enzymes: Sources, uses, and molecular mechanisms for thermostability. *Microbiol. Mol. Biol.* Rev. **65:** 1–43.

Wilson, A.C., Carlson, S.S. and White, T.J., 1977, Biochemical evolution. *Ann. Rev. Biochem.* **46:** 573–639.

Yang, J., Gu, Z., and Li, W.H., 2003, Rate of protein evolution versus fitness effect of gene deletion. *Mol. Biol. Evol.* **20:** 772–774.

York, J.D., Odom, A.R., Murphy, R., Ives, E.B., and Wente, S.R., 1999, A phospholipase C dependent inositol polyphosphate kinase pathway required for efficient messenger RNA export. *Science* **285:** 96–100.

Yoshida KT., Wada T., Koyama H., Mizobuchi-Fukuoka R., Naito S. Temporal and spatial patterns of accumulation of the transcript of Myo-inositol-1-phosphate synthase and phytin-containing particles during seed development in rice. *Plant Physiol.* 1999 Jan;**119**(1)**:** 65–72.

Yoshida, K.T., Fujiwara, T., and Naito, S., 2002, The synergistic effects of sugar and abscisic acid on myo-inositol-1-phosphate synthase expression. *Physiol. Plant.* **114:** 581–587.

Zhang, J., and He, X., 2005, Significant impact of protein dispensability on the instantaneous rate of protein evolution. *Mol. Biol. Evol.* **22:** 1147–1155.

Zuckerkandl, E., and Pauling, L., 1965, Evolutionary divergence and convergence in proteins. In: Bryson V. and Vogel H. J. (eds.), Evolving Genes and Proteins, Academic Press, New York, NY, pp. 97–166.